G000022552

Molecular Mechanisms for Sensory Signals

Molecular Mechanisms for Sensory Signals

Recognition and Transformation

EDWARD M. KOSOWER

Princeton University Press
Princeton, New Jersey

Copyright © 1991 by Princeton University Press

Published by Princeton University Press, 41 William Street, Princeton, New Jersey 08540
In the United Kingdom: Princeton University Press, Oxford

All Rights Reserved

Library of Congress Cataloging-in-Publication Data

Kosower, Edward M.
Molecular mechanisms for sensory signals : recognition and transformation / Edward M. Kosower.
p. cm. Includes bibliographical references.
ISBN 0-691-08553-6 (alk. paper)
1. Molecular neurobiology. 2. Cellular signal transduction.
I. Title. QP356.2.K67 1990 591.1′88—dc20 89-70213

This book has been composed in Linotron Times Roman

Princeton University Press books are printed on acid-free paper, and meet the guidelines for permanence and durability of the Committee on Production Guidelines for Book Longevity of the Council on Library Resources

Printed in the United States of America by Princeton University Press, Princeton, New Jersey

10 9 8 7 6 5 4 3 2 1

Designed by Laury A. Egan

Contents

List of Illustrations viii

List of Tables xiii

Preface xiv

1. Hierarchies in Natural Systems 3

Levels 4

The Nervous System 7

Neurons 11

Neuronal Communication 16

Conclusions 18

2. Chemotaxis 19

Behavior 19

Chemotaxis 23

Chemotactic Response 27

Chemotaxis: Adaptation via Methylation 32

Enzyme II Receptors 35

Oxygen Chemotaxis (Aerotaxis) and Other Taxes 36

General Schemes for Chemotactic Signal Generation 39

Flagellar Motor Structure 47

Membrane Potential and Chemotaxis 53

Structure and Function of MCP Chemoreceptors 54

Overall Scheme for Bacterial Chemotaxis 56

Conclusions 57

3. Sensory Transduction: Pheromones and Taste 58

General Classification of Stimuli for Biological Organisms 58

Chemoreception in Insects: Pheromones 60

Chemotactic Responses in Higher Vertebrates 72

Classification of Taste Stimuli 77

Peptides and Proteins 85

Bitter and Other Tastes 93

Mechanism of Taste 95

4. Olfactory System 98
Sensory Transduction: Olfaction (Smell) 98
Classification of Olfactory Stimuli 103
Olfactory Neural Signals 114
Origin of Receptor Specificity 122

5. Visual System 129
Biochemical System: Rhodopsin-Transducin System 138
Polymolecule: Rhodopsin 143
A Structural and Dynamic Model for Rhodopsins and Lodopsins 148
Genetics of Color Blindness 161
Molecule: 11-*cis*-Retinylidene-ϵ-lysine 161
Further Biochemical Reactions: Retinal Formation 171

6. The Nicotinic Acetylcholine Receptor 172
Introduction to Neurotransmitter Receptors 172
Acetylcholine Receptor: Structure 175
Single Group Rotation (SGR) Theory 183
Nicotinic Acetylcholine Receptor: Summary of Model 192
AChR Models 195
Dynamics of the AChR According to the Model 215
Binding Site for Noncompetitive Antagonists (NCA) 216
Ion Pathway through Exobilayer Cup 224
Tests of Model 225
Desensitization and Resensitization 231
Antibodies 233
Snake Neurotoxins 234
Potential Artifacts in Structural Analysis 234
Conclusions 235

7. Molecular Models for Sodium Channels 237
Introduction to Channels 237
Constructing the Model: Folding Proteins 240
Channel Dynamics 251
Toxin Effects and Toxin Classes 255
Antibody Effects: Antipeptide Antibodies 259
Anesthetics 259
Specific Groups 263
Shape and Biological Stability 264

Other Models 265
Conclusions 266
A Molecular Graphical Model of the Eel Sodium Channel 266

8. Receptor and Channel Superfamilies 270
Receptor Superfamilies I 270
Receptor Superfamilies II: Ligand-activated G-protein Receptors 281
Ligand-activated Phospholipase C-related Receptors 292
Neuropeptides 295
Channel Superfamilies 295
Conclusions 298

9. Learning and Memory 299
Model Systems for Learning 299
A Generalized View of Learning 311
Higher-order Representations 314
Representations in Culture 318

References 321
Author Index 402
Subject Index 429

List of Illustrations

Figures

Fig. 1.1. The principal sections of the human brain 12
Fig. 1.2. Several typical neurons of the cerebrum 12
Fig. 1.3. Cell arrangements in the outer layers of the cerebellum 13
Fig. 1.4. Detailed neuron structure 14
Fig. 1.5. A schematic diagram for a neuron-neuron synapse 14
Fig. 2.1. A stimulus-response scheme for a multicellular organism 20
Fig. 2.2. A stimulus-response scheme for a single cell 23
Fig. 2.3. Phylogenetic relations among prokaryotes 24
Fig. 2.4. Formulas for chemotactic stimulants 28
Fig. 2.5. Rapid mixing and detection apparatus for bacteria 30
Fig. 2.6. Swimming behavior of *E.coli* after concentration changes 31
Fig. 2.7. Overall general scheme for MCP chemotaxis 40
Fig. 2.8. Chemotactic response to pH change 41
Fig. 2.9. The origin of signals from ''complex receptors'' in chemotaxis 42
Fig. 2.10. Methyl-accepting chemotaxis proteins (MCP) as chemoreceptors 44
Fig. 2.11. Signal protein transfer between MCP receptors and motor 46
Fig. 2.12. The motor machinery at the flagellar attachment site for *E.coli* 47
Fig. 2.13. Structure of the basal region of a bacterial motor 48
Fig. 2.14. Relation of rotor and stator in bacterial motor 50
Fig. 2.15. Molecular mechanism for operation of bacterial motor 51
Fig. 2.16. Molecular mechanism for switching of bacterial motor 52
Fig. 3.1. Attractant search patterns generated by nematodes 61
Fig. 3.2. Silkworm moth, *Bombyx mori* L. 66
Fig. 3.3. Fine-structure of a sensory hair of the antenna of a silkworm moth 67
Fig. 3.4. The location of taste buds on a circumvallate papilla of the tongue 74
Fig. 3.5. Longitudinal section through a taste bud on a circumvallate papilla 75
Fig. 3.6. X-ray crystallographic structure of thaumatin I 88
Fig. 3.7. Backbone structures of monellin and thaumatin 89
Fig. 4.1. Elements of the olfactory system 100
Fig. 4.2. The olfactory neuron 101

Fig. 4.3. Estimated dimensions for odorant receptor sites 107
Fig. 4.4. Formulas of enantiomeric odorants 109
Fig. 4.5. Signals from various regions of the olfactory system after stimulation by odorants 115
Fig. 4.6. Overall molecular scheme for olfactory response 119
Fig. 4.7. An olfactory receptor model 125
Fig. 4.8. A side view of the olfactory receptor model 127
Fig. 5.1. The human visual system 131
Fig. 5.2. The human eye 132
Fig. 5.3. Structure of the human retina 133
Fig. 5.4. Rod and cone cells 135
Fig. 5.5. Light sensitivity of the retinal cells after dark adaptation 135
Fig. 5.6. The wavelength sensitivity of rod and cone cells 136
Fig. 5.7. Biochemical cycles related to light amplification 141
Fig. 5.8. Parallelism between hormone and light responses 142
Fig. 5.9. Dichroic absorption of rhodopsin in retinal rod cells 144
Fig. 5.10. Rotational relaxation of rhodopsin in rod disk membranes 145
Fig. 5.11. Intermediates produced by light excitation of rhodopsin 147
Fig. 5.12. Glycosyl sequences near N-terminal end of bovine rhodopsin 151
Fig. 5.13. Amino acid sequences and structures of human iodopsins (cone pigments) and human and bovine rhodopsins (rod pigments) 152
Fig. 5.14. Arrangement of critical groups in the central bilayer region of iodopsins and rhodopsins 155
Fig. 5.15. Geometric change accompanying photoisomerization of rhodopsin to photorhodopsin 157
Fig. 5.16. A mechanism for the catalysis by metarhodopsin II (RX) of the exchange of GTP for GDP bound to T_α 160
Fig. 5.17. A protonated 6-s-*cis*-11-*cis*-12-s-*trans*-retinylidene-ϵ-lysine group, the chromophore of rhodopsin 161
Fig. 5.18. Solvent polarity parameters, **Z** and $\mathbf{E_T}$(30) 167
Fig.5.19. Adjacent-charge effects on absorption maxima in α,β-unsaturated iminium ions 169
Fig. 6.1. Electron micrograph of nicotinic acetylcholine receptor (nAChR) from *Torpedo californica* 177
Fig. 6.2. A topographic model and electron micrographic image analyses of the nicotinic acetylcholine receptor 178
Fig. 6.3. A general helix representation along with its projection on a ''surface net'' or cylindrical plot 181
Fig. 6.4. Helical arrangements of polypeptide chains and arrangements for parallel and antiparallel strands 184
Fig. 6.5. Single group rotations (SGRs) for polar amino acid side chains of proteins 187

Fig. 6.6. Four conformations of the neurotransmitter molecule acetylcholine 188
Fig. 6.7. Model combining site for acetylcholine 189
Fig. 6.8. Model AChR binding site ''combined'' with *d*-tubocurarine 190
Fig. 6.9. Molecular model of a plausible ion channel based on polypeptides 191
Fig. 6.10. Sequence homology between nicotinic AChR α-, β-, γ-, and δ-subunits 196
Fig. 6.11. The ion channel elements of the nicotinic acetylcholine receptor 200
Fig. 6.12. Sequences assigned as bilayer helices of the α-subunit of the nicotinic acetylcholine receptor 201
Fig. 6.13. Sequences assigned as bilayer helices of the β-subunit of the nicotinic acetylcholine receptor 202
Fig. 6.14. Sequences assigned as bilayer helices of the γ-subunit of the nicotinic acetylcholine receptor 203
Fig. 6.15. Sequences assigned as bilayer helices of the δ-subunit of the nicotinic acetylcholine receptor 204
Fig. 6.16. The exobilayer arrangement of the nAChR α-subunit 206
Fig. 6.17. The exobilayer strand arrangement of the nAChR α-subunit 207
Fig. 6.18. The exobilayer arrangement of the nAChR β-subunit 208
Fig. 6.19. The exobilayer arrangement of the nAChR γ-subunit 209
Fig. 6.20. The exobilayer arrangement of the nAChR δ-subunit 210
Fig. 6.21. Cytoplasmic section of the nicotinic acetylcholine receptor α-subunit 211
Fig. 6.22. Cytoplasmic section of the nicotinic acetylcholine receptor β-subunit 212
Fig. 6.23. Cytoplasmic section of the nicotinic acetylcholine receptor γ-subunit 213
Fig. 6.24. Cytoplasmic section of the nicotinic acetylcholine receptor δ-subunit 214
Fig. 6.25. Schematic drawings of the resting (closed) and active (open) forms of the nicotinic acetylcholine receptor 217
Fig. 6.26. The exobilayer region of the nicotinic acetylcholine receptor as viewed from the outside of the cell 218
Fig. 6.27. Model for the nicotinic acetylcholine receptor ion channel 219
Fig. 6.28. Bilayer helices in nicotinic acetylcholine receptors 222
Fig. 6.29. Intrabilayer polar interactions in nicotinic acetylcholine receptors 223
Fig. 6.30. Relationship of exobilayer strands to the bilayer helices and the cytoplasmic segments in an α-subunit 225
Fig. 6.31. Schematic summary of nicotinic acetylcholine receptor states 232

Fig. 6.32. Three-dimensional model (''flower'') representing the operation of the nicotinic acetylcholine receptor 233
Fig. 7.1. Eel sodium ion channel elements 242
Fig. 7.2. Hydrophobic helices H1–H10 and H11–H20 selected from the eel sodium ion channel sequence 244
Fig. 7.3. Ion channel active amino acids and charge distributions 246
Fig. 7.4. The level-1 arrangement of the sodium channel 248
Fig. 7.5. Arrangement of eel ion channel elements and hydrophobic helices 249
Fig. 7.6. Compact representation of sodium channel organization 250
Fig. 7.7. The sequences of the outer and inner control elements of the sodium channel 252
Fig. 7.8. Dynamics of sodium channel operation 253
Fig. 7.9. ''Oxygen triads'' of various toxins that prolong sodium channel opening 256
Fig.7.10. Equilibria involving local anesthetics and the sodium channel at various stages of sodium channel operation 262
Fig. 7.11. A ''stick'' model of the sodium channel 267
Fig. 7.12. Side view of a ''stick'' model of the sodium channel 268
Fig. 7.13. A color view of a ''stick'' model of the eel sodium channel blocked by cocaine from the cytoplasmic side 268A
Fig. 7.14. Color stereo side views of a space-filling model of the eel sodium channel, with a cocaine molecule bound to the cytoplasmic side of the channel 268A
Fig. 7.15. Color stereo views of a space-filling model of the eel sodium channel from the cytoplasmic side, with a cocaine molecule blocking the channel 268A
Fig. 7.16. Side view of a ''stick'' model of the eel sodium channel with a molecule of lidocaine, a local anesthetic 269
Fig. 8.1. Exobilayer strand arrangements for the α- and β-subunits of the γ-aminobutyric acid ($GABA_A$) receptor 274
Fig. 8.2. Activation and desensitization mechanism for the $GABA_A$ receptor 275
Fig. 8.3. Mechanism of activation of the $GABA_A$ receptor by GABA 276
Fig. 8.4. Homologies between membrane anion exchange proteins (AEP) and $GABA_A$ receptor 279
Fig. 8.5. Comparison of sequences for substance K receptor, α_1-, α_2-, β_1-, and β_2-adrenergic receptors, muscarinic M1 and M2 receptors, and rodopsin, all G-protein receptors 282
Fig. 8.6. A general model for a G-protein receptor 285
Fig. 8.7. Seven transmembrane helices of a G-protein receptor 286
Fig. 8.8. Hydrogen-bonding arrangements for a catecholamine derivative 286

Fig. 8.9. Relationships between phosphatidylinositol diphosphate ($PtdIP_2$), diacylglycerol (DAG), and inositol phosphates (IP_3, IP_4, IP_5, and IP_6) 293
Fig. 9.1. The response of the sea hare *Aplysia californica* to a tactile stimulus 300
Fig. 9.2. Partial neural circuit for the transformation of a tactile stimulus into gill withdrawal in *Aplysia californica* 301
Fig. 9.3. Molecular processes in habituation 303
Fig. 9.4. The neural circuit that sensitizes the siphon reflex pathway in *Aplysia californica* 303
Fig. 9.5. Molecular processes in sensitization 304
Fig. 9.6. The neural system of the marine snail *Hermissenda*, which couples light stimulation to the detection of motion 306
Fig. 9.7. Molecular processes in *Hermissenda* B-cell excitability 307
Fig. 9.8. Catecholamine biosynthesis 309
Fig. 9.9. A simplified diagram of the neural circuits in a part of the visual system 310
Fig. 9.10. Molecular stimulus-response scheme for a multicellular system 312
Fig. 9.11. Spiral Belousov-Zhabotinsky (B-Z) wave 315
Fig. 9.12. A molecular scheme for learning and memory 319

Schemes

Scheme 2.1. Reaction schemes for phosphotransferase (PTS) systems 37
Scheme 2.2. General scheme for chemotaxis 56
Scheme 4.1. Kinetic factors in olfactory receptor operation 120
Scheme 6.1. Purification of acetylcholine receptor 175
Scheme 6.2. Sequence determination of nAChR α-subunit 180
Scheme 6.3. Summary of holistic approach to modeling nAChR 199
Scheme 6.4. Geometric relationships between bilayer helices 221
Scheme 7.1. Summary of sodium channel states 254
Scheme 7.2. Summary of possible epitopes present in the eel sodium channel 260
Scheme 8.1. Summary of the states for the γ-aminobutyric acid receptor (GR): γ-aminobutyric acid (GABA) combination 275
Scheme 8.2. Activation and desensitization through phosphorylation of adrenergic receptors 285
Scheme 8.3. Combination of acetylcholine (ACh) with the muscarinic acetylcholine receptor (mAChR) 289
Scheme 9.1. Short- and long-term changes arising from stimuli to the nervous system 314
Scheme 9.2. Short- and long-term changes leading to complex responses from the nervous system 317

List of Tables

Table 1.1 Hierarchy in the Taxonomy of Dogs 5
Table 1.2 Hierarchy in the Physical/Inorganic World 6
Table 1.3 Levels of the Biological World 8
Table 1.4 Levels of the Nervous System 9
Table 1.5 Organizational Levels of the Neuron 15
Table 2.1 Chemosensors for *E.coli* 26
Table 2.2 Complex Receptors in Chemotaxis 42
Table 3.1 Sex Pheromones 64
Table 3.2 Alarm Pheromones 65
Table 3.3 Recruiting Pheromones 65
Table 3.4 Aggregating Pheromones 66
Table 3.5 Classes of Sweet-tasting Compounds 80
Table 3.6 Taste Intensity of Sweet Substances 80
Table 3.7 Common Sequences for Sweet-tasting Proteins, Thaumatin and Monellin 87
Table 4.1 Odorant Classes 105
Table 4.2 Olfactory Properties of Enantiomeric Odorants 108
Table 4.3 Threshold Values for Detection of Odors 110
Table 4.4 Candidate Receptor Groups 124
Table 4.5 Primary Olfactory Receptor Groups 124
Table 5.1 Level Analysis of the Visual System 130
Table 5.2 Proteins Associated with Disks 137
Table 5.3 Absorption Maxima and Intensities for Polyenals in Dioxane 163
Table 5.4 Absorption Maxima of Retinals and Retinylidene N-Butylimines 163
Table 5.5 Visible Absorption Maxima of All-*trans*-Retinylidene Pyrrolidinium Ion Perchlorate in Various Solvents 166
Table 5.6 Absorption Maxima of Retinylidene Iminium Ions in Dichloromethane Containing Trifluoracetic Acid (TFA) 169
Table 5.7 Absorption Maxima for Visual Pigments 171
Table 6.1 The Genetic Code 181
Table 6.2 Polypeptide Linear Groups/Helices 182
Table 6.3 Homologies among Exobilayer ''Active'' Amino Acids 227
Table 6.4 Homologies among Ion Channel ''Active'' Amino Acids 228
Table 8.1 Neurotransmitters 271
Table 8.2 Mammalian Tachykinins 291

Preface

Dawn to Dusk
Fresh green on rose clouds
Sweet kisses shower symbols
Stone dust on still hope.
—1988:24 September

Man fears the unknown and also fears knowledge. Not so long ago, the physical world seemed unbearably complex, full of unicorns and epicycles, to be described in terms specified by the powerful. Within a few centuries, discoverers, analyzers, and thinkers clarified our understanding in a multitude of disciplines. Today, much of the physical world is within our ken and we have begun to penetrate the more ephemeral world of the mind. The present book introduces a novel and rich approach to connecting molecular properties with the biological properties that make it possible for man to write and read, to create culture and ethics, and to think. What will emerge from our knowledge of the molecular basis of learning and memory is difficult to imagine. After all, the cultural evolution of the human species has lasted but 15,000 years; its issue, the scientific and industrial age, only 400 years. For the future, there are a plenitude of projects: searching for others in the galaxy, creating multisensual art, probing the attraction of the arts, finding answers, no matter how transient, to man's physical, biological, psychological, sociological, political, and cultural problems. The challenge is to understand and to improve on what biological and cultural evolution have produced. To do otherwise would be to accept blindly the statistical accidents of natural evolution and the myriad of dogmatic opinions that masquerade as natural truth in the name of countless religions, however successful the latter are in directing and guiding social energies.

The mind and its inner workings is a subject that has fascinated many people; to provide a reason for this pursuit, one needs only to extend the meaning of the adjuration, ''Know thyself.'' Some years ago, I wrote a book, *Molecular Biochemistry* (New York: McGraw-Hill, 1962), in which the ideas of physical organic chemistry (some found in another book of mine, *Physical Organic Chemistry*, [New York: Wiley, 1969]) were applied to biochemical problems. From time to time, I taught a course in biophysical organic chemistry and introduced more and more material related to molecular neurobiology. A theoretical analysis of hierarchies in our perception of nature made the organization and choice of material more facile. At the end of 1979, in thinking about the forthcoming semester's lectures, I asked myself, "Why are amino acids and acetylcholine neurotransmitters?" I drew the formulas and, after some time, saw that these molecules could undergo small conformational changes of sufficient magnitude to induce a meaningful change in the receptor. I called this change single group rotation (SGR) and realized that this simple idea had broad implications for many problems in neurobiology and biochemistry. The development of the ideas about SGR led to models for the nicotinic acetylcholine receptor and the sodium channel. A book was the appropriate platform to present these ideas together with an analysis of the mechanisms of chemotaxis, taste, smell, and vision. The strength of the organizational scheme has been confirmed by my finding that it was easy to incorporate the surge of new information that emerged during the last stages of writing. A new molecular scheme for memory was also created at the final stage. The present volume thus reflects the intense activity in the field of molecular neurobiology. Of necessity, the book has features with which not all might agree, but I hope that the approach helps to achieve a truly comprehensive understanding of mental processes at the molecular level.

The book begins (Chap. 1) with the construction of hierarchies in classifying various physical and biological objects and some general information about the nervous system. An analysis of single-cell behavior is focused on chemotaxis, the response of a single cell to an extracellular signal (Chap. 2). Much behavior is mediated by pheromones, the molecular signals used by organisms from single cells to human beings. The molecules that stimulate taste receptors (Chap. 3) and olfactory receptors (Chap. 4) are considered. A hierarchy of levels for the visual system helps to relate the various complex subsystems to one another without losing sight of the need to explain the whole system. A detailed analysis of vision (Chap. 5) reveals the important result that the light-responsive protein, rhodopsin, is a G-protein receptor. Dramatic breakthroughs in molecular neurobiology came with the isolation and sequence determinations of the nicotinic acetylcholine receptor (Chap. 6) and the sodium channel (Chap. 7). These complex molecules must be assigned structures, using both experiment and model-building to see the system response that might be generated. A revolution in drug design is under way on the basis of the new structural and functional knowledge about the targets for drugs. Structural analysis will be made somewhat easier by the family relationships of many of the important proteins (Chap. 8). In the end, a collection of interlocking systems is suggested as the basis for memory and its expression (Chap. 9). The book yields a coherent view of the molecular basis of learning.

My efforts were aided by many people, far and near. Greg Petsko (Massachusetts Institute of Technology) and Richard J. Feldmann (National Institutes of Health, Bethesda) were my mentors and colleagues in the molecular graphics studies. Hamutal Meiri (Rappaport Institute, Technion) initiated work on the anti-Cl^+ peptide antibody for the sodium channel before the sequence had been characterized as a common theme in membrane ion channels. Bernhard Witkop (National Institutes of Health, Bethesda) gave much early encouragement to my approach to receptor and ion channel models. Rachel Magen and Nancy Gunkelman provided much of the artwork. Various kinds of help came from many others, among whom are Carl and Agnes Pearlman, Max and Frances Morris, Richard and Florence Weiner. My wife, Nechama S. Kosower, offered useful comments when she could escape from her position as chairman of the Human Genetics Department, Sackler Faculty of Medicine, Tel-Aviv University. M. Eisenbach (Chap. 2) and M. Cheves (Chap. 5), both of the Weizmann Institute, commented in detail on two chapters. My students and postdoctoral fellows deserve credit for continuing our research when the exigencies of various deadlines connected with the book prevented me from giving them my undivided attention. The hospitality that I enjoyed from time to time at State University of New York, Stony Brook, at which I am Adjunct Professor of Chemistry, helped in many ways, as did sabbatical stays at the University of California, San Diego, the University of California, Berkeley, Massachusetts Institute of Technology, and the Marine Biological Laboratory, Woods Hole. A J. S. Guggenheim fellowship supported some of my early work on the book.

Tel-Aviv, Israel, 9 May 1989

Molecular Mechanisms for Sensory Signals

1 Hierarchies in Natural Systems

The stars of early evening appear behind a spiraling cloud of dark gray smoke rising from a campfire, a miniature of the black funnel of a tornado in the near distance. In the darkening blue sky, the giant spiral arms of distant galaxies reflect far suns, and also move, but on a time scale too long to be perceived directly. The geometry of the curving shapes and their real or imagined motion evoke in the pensive human observer a joy of discernment and a fear of the maelstrom.

How does one recognize stars, smoke, fire, color? How do creatures, raised through evolution from cosmic molecules on the condensed debris of the "big bang," know anything? How does one learn, and in what way is memory made?

In the quest for answers, one must find the best way to ask the right questions. In any complex system, there are levels of organization. The inquiries should be appropriate to the level being studied. Grinding can reduce the most highly integrated circuit chip to an impure heap of useless dust. And yet, if one had to discover how the chip were made, it would be dissected and decomposed by all possible means to discover its secret. One might not find the very tiny amounts of those elements on which the conductivity of the chip depends unless one had a good idea of what to look for. Reconstructing a chip is simple compared to reassembling even a simple nervous system, to say nothing of understanding how the system senses the outside world and responds to that input.

Learning and Memory

An external or internal input to any complex system will produce a *trace*. The trace is equivalent to a *representation* of the input for the system. In a nervous system, that representation will be multicellular, consisting of all of the changes occurring in the single cells comprising the multicellular set. The initial representation and any subsequent transformations are *learning* by the system. The initial representation will normally consist of *transient* changes in local ionic or molecular concentrations. The local changes can disappear without further effect, or the representation can be retained by being recycled or reinforced, presumably by an internal input that activates the same representation. At some point, some molecules or molecular systems are permanently transformed into a memory. *Memory* is the *record* of the representation. There may be several stages in the molecular storage or memory formation, one *short term* and one *long term*. For Changeux, the representation is a ''mental object'' that can be manipulated by the mind (1), i.e., in our terms, be transformed by an internal input into another representation.

In 1978, Joel S. Hildebrand, then a 98-year-old professor emeritus of chemistry at the University of California, Berkeley, responded to a remark about William Jennings Bryan (American politician and presidential candidate, 1860–1925), by saying that he had heard him speak in 1916. He was then able to recount in detail what Bryan had said. Apart from our admiration for the acuteness with which Hildebrand was able to recount the contents of the lecture, we have to be struck by the longevity and fidelity of the molecular basis of memory.

Identifying the memory molecules and molecular systems and the learning processes that produce the memories is one of the ultimate goals of modern neurobiology. The present volume is concerned mostly with processes within a single cell. Understanding what happens will yield a molecular basis for the representation in the single cell in transient, short-term, and long-term forms. The circuits underlying the total (multicellular) representation may be definable (2) without knowing the molecular basis for the interconnections, but a complete theory for learning and memory must describe both the molecular transformations and the intercellular connectivity.

Levels

The first problem to be addressed is that of the identification and properties of levels. The choice of elemental units for levels is often, but not always, made by size. First, there must be a sufficiently detailed catalogue of the units so that the relationships between them can be identified and tested. Second, levels that are properly related to one another must be found. It is true, but not especially useful, to state that a block of wood is made of atoms.

Hierarchies in Large Systems

In any system composed of a sufficiently large number of units in communication with one another, special properties characteristic only of the system may be identified. Units that are in proximity may interact (''communicate'') and, under certain conditions, act as a system. System formation can occur without physical contact. The individual systems created in this way may themselves be components of a higher-order system. In the *hierarchy* of systems thus created, the behavior of the metasystems is determined by the nature and structure of the unit systems of which they are composed.

What does ''sufficiently large'' mean? A proton and an electron, a system of two units, constitute a hydrogen atom. In this case, two units of the appropriate type are sufficient to create a system that behaves differently from its component parts, i.e., has special properties identified only with the system.

A large molecule (a molecular system) is composed of many more units. The units may be identical, as in the case of a single crystal, or different, as would be true for an organism.

Constructing an appropriate hierarchy of systems constitutes one of the most effective ways of analyzing relationships in the real world. In this approach, choosing a *level* means formulating a description of the units for which a set of distinctive rules of behavior can be stated. The level can be related in a definite way to higher and lower levels within the hierarchy. The taxonomic classification of dogs (3) illustrates the construction of a hierarchy (Table 1.1). The canine hierarchy is one solution to the problem of relating individual organisms to one another and, in effect, to all other organisms.

Table 1.1. Hierarchy in the Taxonomy of Dogs

Level Number	Level Classification	Level Description
25	Biological World	
24	Kingdom	ANIMALIA
23	Subkingdom	Metazoa
22	Phylum	Chordata
21	Subphylum	VERTEBRATA
20	Superclass	Tetrapoda
19	Class	Mammalia
18	Subclass	THERIA
17	Infraclass	Eutheria
16	Cohort	Ferungulata
15	Superorder	FERAE
14	Order	Carnivora
13	Suborder	Fissipeda
12	Superfamily	CANOIDEA
11	Family	Canidae
10	Subfamily	Caninae
9	Genus	CANIS
8	Subgenus	———
7	Species	Canis familiaris
6	Subspecies	Canis Familiaris Metris-Optimae
5	Race	SHEEPDOG
4	Breed	German Shepherd dog
3	Population	Israel
2	Subpopulation	Tel-Aviv
1	Individual	Your neighbor's watchdog

Source: Reference 3.

The universe and its component parts may be analyzed as a hierarchy of systems, in which the identification of certain levels is still open (4–6). Two lines of hierarchical organization for the physical/inorganic world are shown in Table 1.2.

Relationships between Levels

The biological world may be formulated as a hierarchy of systems (Table 1.3). The rules governing the relationships between the units that comprise a given level is an *outer program*; the rules for the relationships within a given unit can be called an *inner program*. The inner program for the units at a given level will then be the outer program for the units on the level below. Many of the programs identified through our level analysis are well known under other names. Some of the programs do not have specific names, either because a need for them has not yet arisen or because the same name is used to describe

Table 1.2. Hierarchy in the Physical/Inorganic World

Level	Unit	Alternative Unit
20	Universe[a]	
19	Large-scale metagalactic group	
18	Local metagalactic group	
17	Metagalaxies[b]	
16	Large-scale galactic groups	
15	Local galactic group	Solar system
14	Galaxy[c]	Solar company
13	Large-scale cluster	Planetary system
12	Local cluster	Inner planets
11	Sun	Earth
10	Convection zone	Lithosphere
9	Giant plasma cells	Metamagmatic flows
8	Super-granular plasma cells	Magma
7	Small plasma cells	Liquid cells
6	Large-scale plasma masses	Fluid rock
5	Small-scale plasma masses	Polymolecules
4	Plasma	Molecules
3	Multi-ionized atoms, electrons	Atoms
2	Nucleons	Nucleons, electrons
1	Quarks	Quarks

[a] Reference 4.
[b] Reference 5.
[c] Reference 6.

the rules at a different level in the hierarchy. The names for the programs that control the relationships between the units are given in Table 1.3.

It is possible to carry the formal analysis still further by defining the energy states of the levels I have identified (for example, electronically excited molecules, emotionally excited individuals), and considering the time dependence of the state distribution of the levels (excited molecule emits light, excited individual calms down). A hierarchical analysis supported by statistical mechanical arguments has been given by Iberall and Soodak (7). A ''general'' hierarchical theory by Miller (8) suffers from the use of anthropomorphic (and macroscopic) categories. Explanations for ordering within states by Haken (9) and catastrophe theory (10–12) might well be extended through level analysis. Koestler identified a level unit as a ''holon,'' and titled a book after the two-faced Greek god, Janus, who faced simultaneously in opposite directions (to upper and lower levels) (13). Previous analyses of hierarchical concepts and organization are summarized by Wilson (14).

The Nervous System

A background sketch of some of the structural elements of the nervous system will provide some reference points for our molecular-level examination. A brief look at the past helps bring into perspective how recently our current views have developed.

Historical Note

The !Kung San of the Kalahari desert in Botswana (Africa), who are among the most primitive human beings known to us today, refer to themselves as Zhu/twasi (''real people''), a term that reflects a well-defined self-awareness (15). Human beings have been aware of themselves as individuals for longer than 5000 years, to judge from the descriptions given in recorded history and in folktales that may be relics from a time antedating the creation of any written language. The Greek, physician Alcmaeon (ca. 500 B.C.) was the first to recognize the brain as a center of perception. Explicit recognition of the nervous system as an entity was expressed by Rufus of Ephesus (ca. A.D. 100). Opinion among the ancients was divided about whether the brain or the heart was the seat of consciousness. Aristotle (350 B.C.) imagined that the brain was for cooling the body, possibly the origin of the expression that ''cooler heads prevail.'' Galen (A.D. 150) favored the brain as the center of consciousness, a dogma changed back to the heart by the Persian Moslem Ibn Sīnā (Avicenna) about A.D. 1000. The modern belief that the brain is the seat of consciousness (and the ''soul'') dates from about A.D. 1700, and has been formalized legally in the specific recognition that death of a person means brain death (16).

Table 1.3. Levels of the Biological World

Level	Description	Program
22	World culture	
		Multifaceted interactions
21	Civilizations	
		Supranational system
20	Nations	
		Cultural networks
19	Mininations	
		Superhistorical system
18	Regions	
		Historical system
17	Villages (cities)	
		Economic system
16	Communities	
		Belief system
15	Clans	
		Residual language
14	Family groups	
		Language
13	Organisms	
		Organism organization
12	Organ systems	
		Organ system organization
11	Organs	
		Organ structure
10	Tissues	
		Tissue organization
9	Cells	
		Cell structure
8	Organelles	
		Organelle assembly
7	Biochemical systems	
		System organization
6	Polymolecules	
		Polymolecular structure
5	Molecules	
		Molecular structure
4	Atoms	
		Electronic structure
3	Nuclei	
		Internucleon forces
2	Nucleons, electrons	
		Quantum chromodynamics
1	Quarks	

The Nervous System

Learning and memory are associated with the nervous system. The nervous system is extremely complex, and is described in many books (17–25). A level scheme (Table 1.4) and a few sketches are helpful in sorting out some of the complexity of the nervous system in higher organisms. The nervous system can be divided into two subsystems, the central nervous system (CNS) and the peripheral nervous system (PNS). An organ is a part of an organism adapted for a function, and thus may be localized, like the liver, or distributed, like the PNS. The organs of the PNS include the nerves that mediate sensory and

Table 1.4. Levels of the Nervous System

Level	Components[a]
Organ system	Nervous system
Organ subsystem	Central nervous system (CNS) Peripheral nervous system (PNS)
Organ (CNS)	Brain Spinal cord
Organ regions	Forebrain (prosencephalon, PR) Midbrain (mesencephalon, ME) Hindbrain (rhombencephalon, RH)
Tissue regions	
(PR)	Diencephalon (Di) Telencephalon (Te)
(RH)	Metencephalon (Mt) Myelencephalon (My)
Tissues	
Di	Thalamus, lateral geniculate body (LGS) Hypothalamus, epithalamus (pineal body)
Te	Hippocampus, olfactory bulb, amygdala Cerebrum
Mt	Cerebellum, pons
My	Medulla
ME	Inferior colliculus, superior colliculus Red nucleus, *Substantia nigra*
Tissue lobes (cerebrum)	Frontal, parietal, temporal, occipital
Lobal regions	Precentral, prefrontal
Gyri (precentral)	Superior, middle, inferior
Cells	Neurons, neuroglia
Neurons (cerebrum)	Pyramidal, fusiform, stellate, bipolar
(cerebellum)	Purkinje, granule, Golgi, basket, stellate

[a] Reference 18.

motor activities, as well as the autonomic (sympathetic and parasympathetic) nervous system. The organs of the CNS are the brain and the spinal cord. The brain is divided into some fairly well defined organ regions—the forebrain (prosencephalon), the midbrain (mesencephalon), and the hindbrain (rhombencephalon)—which are further subdivided into tissue regions. The subdivisions of the tissue regions are tissues that include familiar names like the thalamus, olfactory bulb, and hippocampus. The names for the tissues clearly represent a classification based on anatomical distinctions, but reflecting a high degree of functional separation which has survived a long evolutionary period of variation and selection. The development and many of the structural and functional subdivisions of the brain are similar over the range of vertebrates from birds to primates, except for the most advanced part of the brain, the telencephalon (26). It is interesting that the brain of an advanced polychaete worm, for example, shows a separation into a forebrain, a midbrain, and a hindbrain. Neuroscientists have made use of the parallelisms between all sorts of nervous systems and brains to choose systems for detailed study, the phylogenetic comparisons yielding insights into the evolution of nervous systems (22). The similarities extend to the cellular and molecular level, as shall be seen later in the cases of the nicotinic acetylcholine receptor, rhodopsin, and the sodium channel. Our discussion will focus on cells and the molecular basis of cellular processes.

Neural Cells

At the lowest level of the organization table for the nervous system (Table 1.4) are neural cells, which are themselves highly complex systems. Since the cells are the elementary units of the nervous system, it is worth knowing how many there are. There are various estimates for the number of cells in the brain, ranging from 10^{11} and up. The upper limit is easy to set, since the average human brain is about 1500 cm^3 in volume and a small nerve cell about 10×10^{-4} cm (10 μ) in diameter. If the brain were packed solid with small nerve cells, the number of cells would be about 3×10^{12}. Some nerve cells are quite large (Purkinje cell, 60 μ) but probably most are 10–15 μ in diameter, leading to a maximum cell count of 10^{12} cells. A reasonable estimate is that there are 5×10^{11} cells in the brain.

The two major types of cells in the brain are neurons and neuroglia (''neural glue'') in the ratio 1:2. Glia can be classified into three groups, astrocytes, oligodendrocytes, and microglia. The first two types both develop from neuroendothelial cells and are thus closely related to neurons in origin. The microglia, however, appear to be related to phagocytes, with a function that reflects their origin. The astrocytes, which have an internal fibrillar ''skeleton,'' serve structural, insulating, and transfer functions. The astrocytes produce a complete membrane over brain and blood vessels into the central nervous system, serve to separate neurons from one another except at synapses, and provide a pathway for intermediates, electrolytes, and gases between the neurons and the blood. There are many ribosomes present in oligodendrocytes, a sign that protein synthesis is an important function. Since transfer of proteins between nerve cells is a known process, the oligodendrocytes may be a source of proteins for other cells. In addition, these cells contribute to the formation of the myelin sheath of nerve cell connections. Microglia appear to act as scavenger cells within the brain. Glial cells retain their ability to divide, unlike neurons, a circumstance that may underlie the fact that most brain tumors are gliomas, that is, they arise from the uncontrolled division of glial cells.

Neurons

Biological cells could readily be seen with the achromatic light microscope, perfected around 1820. The results led to the "cell theory" of Schwann in 1839. The cell staining methods then available led to the view that nerves formed a continuous network ("reticularist" theory). Golgi found that silver nitrate, under certain conditions, marks only particular classes of neurons. By making critical improvements in the Golgi methods, and by using the most elementary technical means, the remarkable Spanish anatomist Ramon y Cajal established that nerve cells carrying dendrites and axons were independent units (27). The name for the cells, neuron, was first used by Waldeyer in 1891. Although the absolute number of neurons in the brain is very large, the variety in types is limited.

Components of the Brain

A glimpse at the major components and divisions of the brain shown in Figure 1.1 indicates how much larger the cerebrum is than other parts of the brain. The cerebrum has developed both in size and in internal structure during the course of evolution. Apart from the obvious increase in size, there is a spectacular decrease in neuron density, accompanied by a tremendous increase in dendritic arborization (branching). Several of the principal cerebral neuron types are illustrated (Fig.1.2).

The cerebellum is divided into three parts on a phylogenetic basis—the archicerebellum, the paleocerebellum, and the neocerebellum—which is a mammalian development connected to the control of precise movements (22). The structure has been analyzed in great detail. Two features dominate the structure of the cerebellum. First, the very large and spectacularly arborized Purkinje cells are among the largest neurons in the brain. The branches on the Purkinje cells belong to the dendritic portion of the cell, and are further subdivided into small branches or "spines." Second, there is an enormous number ($>10^{10}$) of granule cells from which parallel fibers emanate. Each spine on a Purkinje cell makes one or two contacts (synapses) with a parallel fiber. The

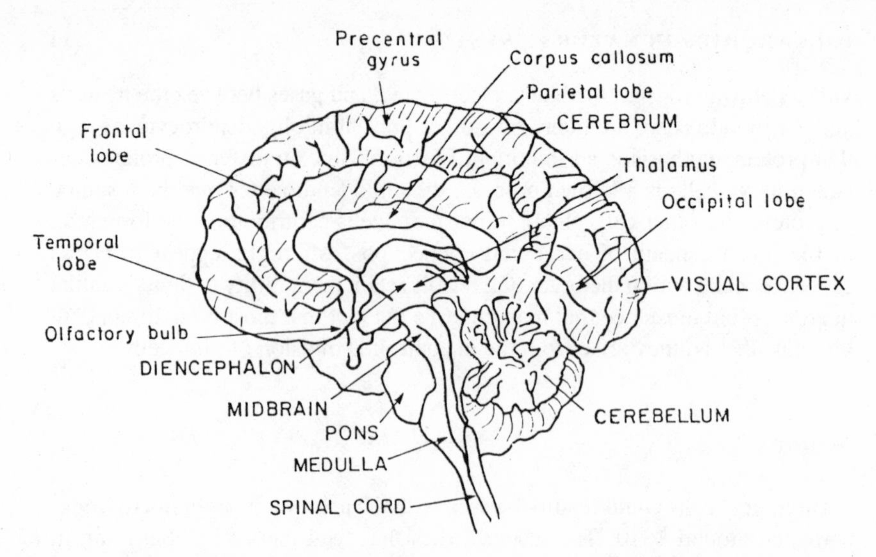

Fig. 1.1. The principal sections of the human brain. The locations of the temporal, frontal, parietal, and occipital lobes are indicated. The highly ramified cerebrum lies above and around the diencephalon and midbrain. The cerebellum is at the back of the brain, behind the pons, medulla, and the top of the spinal cord. The olfactory bulb is directly connected to the olfactory neurons in the nasal cavity. The visual cortex is at the back of the cerebrum, on the side of the brain opposite that of the eyes.

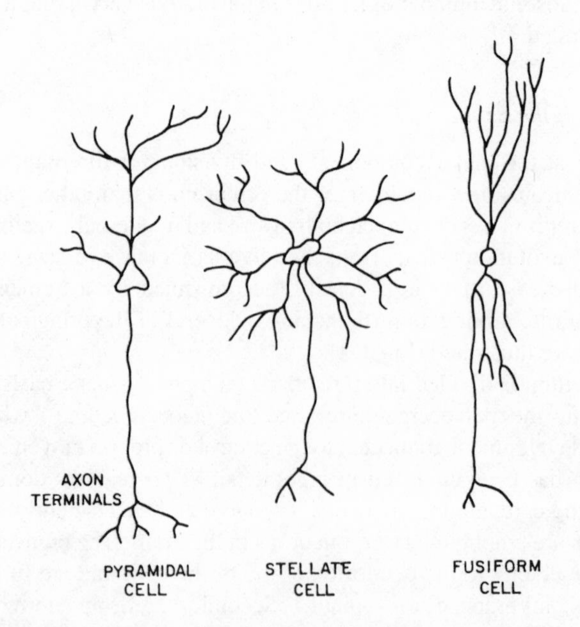

Fig. 1.2. Several typical neurons of the cerebrum. The shapes and arborization of pyramidal cells, stellate cells, and fusiform cells illustrate the varied forms of neurons.

number of parallel fibers in contact with each Purkinje cell is extremely large, and increases from 4000 in the frog to over 100,000 in the cat. Some cerebellar neuron types are shown in Figure 1.3.

Neuron Organelles

The features of a typical neuron are illustrated in Figure 1.4. The cell is divided into four distinct regions; dendrites, cell body (or soma), axon, and axon terminals. Dendrites (usually 700μ or less in length, and from two to many in number) enter the cell body from many directions. Only one axon leaves the cell body, beginning at the axon hillock (trigger region), often branching and extending to sometimes numerous axon terminals (200 μ to 1 meter distant).

The functional connections between one neuron and another were named synapses by Sherrington in 1906 (Fig.1.5). Synapses can be electrical or chemical in character, and are complex in structure. Since synapses are found between two cells, each cell contributes only a portion of the synaptic structure. I have chosen *postsynapton* to name the organelle complex that receives a signal from another neuron and *presynapton* for the organelle complex that produces a signal for another cell.

The cell body contains the nucleus, some highly visible and active protein synthesizing systems (''Nissl bodies'' or rough endoplasmic reticulum, ribosomes), and many postsynaptons. The axon contains many neurotubules and neurofilaments, and is bounded by the axon membrane, which contains the ion channels responsible for the motion of the nerve signal. The axon is often surrounded by a myelin sheath, produced by another cell wrapping itself around the axon so as to produce a multilayer structure. The sheath insulates

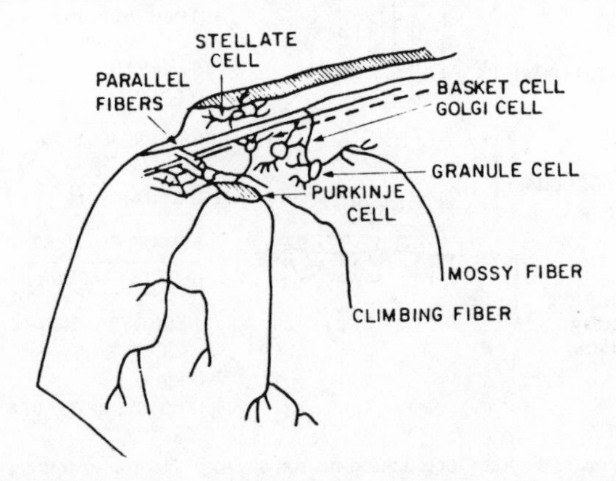

Fig. 1.3. Cell arrangements in the outer layers of the cerebellum, showing Purkinje cells, granule cells and their associated parallel fibers, Golgi cells, basket cells, and stellate interneurons. Climbing fibers make many contacts with an individual Purkinje cell; mossy fibers contact granule cells and Golgi cells.

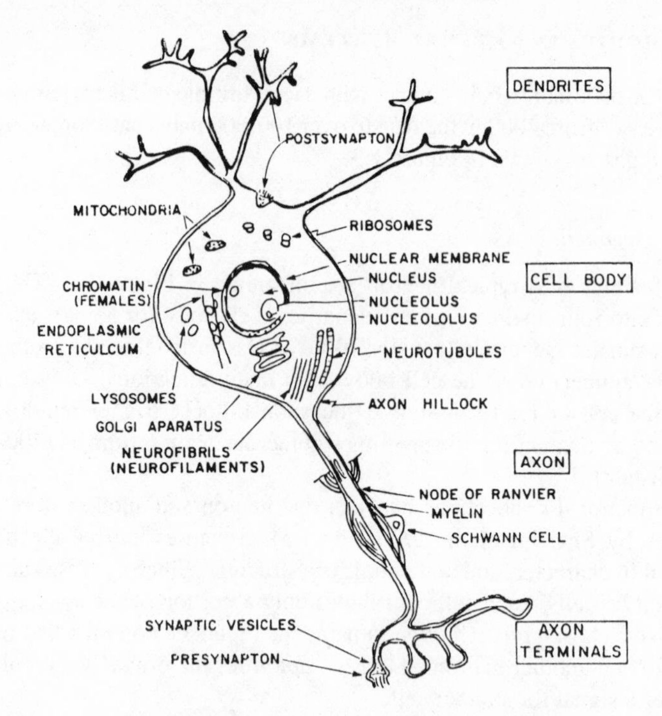

Fig. 1.4. Detailed neuron structure. Neuronal regions are dendrites, cell body (soma), axon, and axon terminals. Many neuronal organelles are shown. The presynapton is a complex of proteins on the presynaptic side of a synapse; the postsynapton is a protein complex on the postsynaptic side of a synapse (see Fig. 1.5).

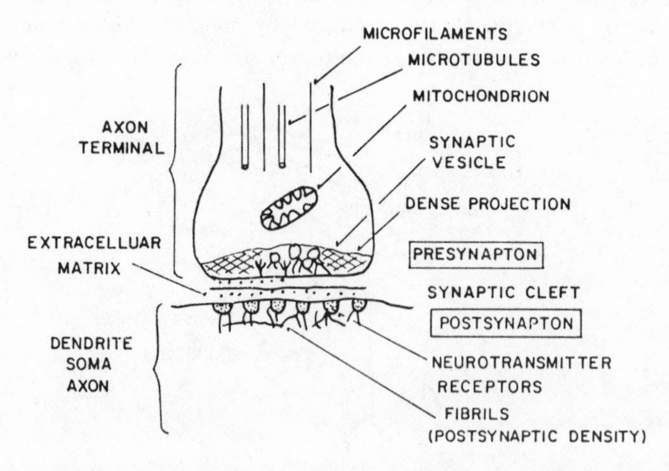

Fig. 1.5. A schematic diagram for a neuron-neuron synapse. Neurotransmitters are the active agents of signal transfer from the presynapton to the postsynapton through the synaptic cleft. Synapses based on the transfer of chemical signals are often called ''chemical synapses.'' The representation of the presynapton is based on electron micrographs of unfixed frozen preparations (Reese, T. S., in reference 55).

the axon to the motion and transport of ions, and augments conduction by decreasing the electrical capacitance of the axon membrane.

The axon terminals contain synaptic vesicles (complex structures that are loaded with neurotransmitter and ATP), mitochondria, and presynaptons. The synaptic vesicles contain the neurotransmitter and are released upon depolarization of the axon terminal. Mitochondria in the axon terminal provide energy to drive processes like recovery of synaptic vesicle membrane (via coating with clathrin, a protein), reuptake of neurotransmitter or precursors, and synthesis of neurotransmitter. Fast transport to and from axon terminals occurs through an active process promoted by motility proteins, kinesin and retrograde kinesin, which propel materials to and from the cell body along the microtubules (28–30). The videotechnique central to the observation of organelle motion has been reviewed (31). The presynapton is an organelle region that seems to be bounded by a meshwork of filaments derived from the protein synapsin I. The synapsin I becomes attached to the synaptic vesicle and aids in the process of bringing the vesicle to the release site at the membrane. Structures near the release site (drawn in Figure 1.5 as a triplet near the synapse membrane) may include a calcium storage organelle.

Certain labeled portions of Figure 1.5, illustrating the components of neurons, are included in Table 1.5. Schwann cells are a vital component of the system to which the neuron belongs at the next higher level of organization. The Schwann cell wraps itself around the axon and thus produces the myelin sheath which electrically insulates the axon. The exposed regions between Schwann cell covered regions are called "nodes of Ranvier," and should be

Table 1.5. Organizational Levels of the Neuron

Level	Component
Cell	Neuron
Cell region	Dendrites, cell body (soma), axon, axon terminals
Organelles	Postsynapton, nucleus, mitochondria, ribsomes, lysosomes, smooth endoplasmic reticulum, rough endoplasmic reticulum (Golgi apparatus), axon hillock, neurotubules, neurofilaments, presynapton, synaptic vesicles, dendritic spikes, dendritic branch points
Suborganelle systems	Nuclear membranes, nucleolus, nucleololus, chromatin (females), neurotransmitter receptors, ion channels
Biochemical systems	Protein synthesis Neurotransmitter: uptake, synthesis, vesicle loading, release Calcium uptake, release Ion transport

classified at a level of organization that involves at least two cells. Schwann cells do not occur in the brain. In the brain, all large axons are myelinated through laminar wrapping by the glial cells, the oligodendrocytes.

Neuronal Communication

In order to produce the *trace* of an input to a complex system, the input must be received and transferred throughout the relevant parts of the system. For the human, the sensory inputs are transformed by complex molecular systems into a transfer signal. In subsequent chapters, I will consider many of the important types of sensory receptors and an important component of the transfer mechanism within a neuron. The overall transfer signal from the receptor cell to a following cell is effected mostly by chemical means, but also by purely electrical signals.

Understanding the molecular basis of learning requires knowledge of how input transformation and signal transfer occur on the molecular level.

Input Transformation and Transfer

A suitable signal (a molecule or a photon) acts on the receptor molecule to promote the entry of ions into or out of cells. The change in the membrane potential thus produced decreases away from the location at which the signal acts. The gradient in membrane potential varies according to the magnitude of the signal. In a neuron, the gradient of change in the membrane potential will decrease from the region of the dendrites to the ''axon hillock'' or ''trigger region'' at the junction of the cell body and the axon. At this point, the signal-induced change in membrane potential is replaced by a constant signal in the axon due to a voltage-activated channel which transfers the signal to the axon terminals. Depolarization of the terminal leads to release of neurotransmitter, providing input signals for succeeding cells in the system. These processes are the *transient representation* of the input.

Long-term Processes

Once the transient representation of the input has been produced, reactivation of the pathway must occur to yield a short-term memory of the input. It is likely that part of the output of each neuron in the pathway reaches the hippocampus or some other part of the limbic system (that portion of the brain concerned with ''emotional'' responses around the diencephalon and telencephalon). The hippocampus might, in response, decrease the signal to inhibitory neurons on the pathway, and thus facilitate the reactivation of a pathway already primed to respond by the partial depolarization of the initial trace. Reiteration of the representation may ultimately affect dendrite growth and synapse formation. The representation would thus be (a) strengthened, (b) more readily evoked, and (c) more effective at activating connected (associated) pathways. The hippocampus, or equivalent regions active in motivation, could also be further activated. In the limit, a substantial portion of the brain might be activated, or become *conscious* of the representation. Failure of long-term memory formation was the result in a famous case in which the hippocampus and medial temporal lobe were resectioned to alleviate severe epileptic attacks. Thus, the intervention of the hippocampus in long-term memory formation seemed likely. The intelligent individual survived many years but had only short-term memory in addition to already established memories. He was completely unable to be aware of anything new, although he was able to master several new skills without being aware of his mastery (26, 32). Another individual who exhibited a failure of conversion of short-term to long-term memory turned out—on pathological examination after death—to have ischemic (stroke) damage to the hippocampus (33). Amnesia in macaque monkeys (*Macaca fascicularis*) is induced by lesion of the amygdala-hippocampal section of the brain (34). The hippocampus has been described as a temporary memory store (35). A better description might be that the hippocampus is the locus for reiteration of representations. Hippocampal reiteration could be a specific element in computing in neural networks (36–38). Long-term potentiation (LTP) in rat hippocampal slices is inhibited by a monoclonal antibody selected from those generated against dentate gyri of 5-day-old rat brain (39) without effect on normal synaptic transmission. LTP in the hippocampus is promoted by other inputs to that region (40, 41). Associative interactions between synapses during LTP induction may occur in local dendritic domains (42).

Classic experiments in sensory deprivation using rats in "enriched environments" versus those in "dull" and restricted surroundings have demonstrated that dendritic growth and synapse number depend on inputs to the system (43). Synapse turnover in the central nervous system can be fairly rapid, with substantial changes taking place in target neurons of the hippocampus within two days after lesions in the entorhinal region, which produce the input (44). Regeneration of function through neuron growth ("sprouting") and synapse formation can occur without the regeneration of all connections, so that some of the deficits in Alzheimer's cases might be partially replaced by the creation of other, less effective pathways (44, 45).

The representations of the input can be extremely complex based on the neurons activated by particular signals such as moving lines, edges, etc., an example being sets for edge, black bar, and edge denoting an "I" (or perhaps the number "1") (46). The representations are related to an input, external or internal. Assemblies, as defined by Dudai and colleagues (47), are coactive neurons which may or may not be equivalent to a representation. The representations I have described are multiple, the distinguishable sets formed according to connectivity and other existing states of the system. Detailed models are being constructed (48), but should be regarded with a certain skepticism until tempered by biological and biochemical information (49).

The long-term stability of memories is then somewhat mysterious in the face of the dynamic character of the physical and electrical characteristics of the cells in the nervous system.

Conclusions

An input to the nervous system produces a trace, which is a representation of the input. The trace activates a system that causes the trace to reappear. With sufficient activation or collateral inputs, the trace is repeated, strengthened, expanded, and eventually converted into some permanent change or changes in the cells of the nervous system (50, 51). Strengthening the synapse through addition of neurotransmitter receptors (part of the process of "consolidation") may be stimulated by electrical activity and calcitonin gene-related peptide (CGRP). Both activity and CGRP lead to an increase of the messenger RNA for the α-subunit of the acetylcholine receptor at the neuromuscular junction (52–57). Agrin, a peptide complex released by nerve endings, stimulates the organization of synaptic components at the neuromuscular junction (58, 59).

A detailed molecular understanding of what happens in a single cell must underlie any attempt to understand what goes on in sets of cells. There is a very long way to go, both experimentally and theoretically (32), but there is hope. Amino acid sequences are known for some of the most important molecular components of the system: the nicotinic acetylcholine receptor, the sodium channel, the visual receptor, the γ-aminobutyric acid ($GABA_A$) receptor, the muscarinic acetylcholine receptor, and others. Understanding the structure and dynamic behavior of these large protein molecules constitutes an important part of a comprehensive theory for the molecular basis of learning and memory.

2 Chemotaxis

Behavior

In psychology, memory is classified according to the way in which it is manifested by some form of behavior. Memory can be defined as the faculty by which things are remembered; memory is also the remembrance itself (60). In the face of a variety of meanings for the central term, memory, I shall restrict my usage to the physical or logical record of the trace produced by an input. For comparison, the physical memory record in a computer diskette is the magnetization of a cluster of iron oxide particles and the logical record is the string of 0's and 1's which the magnetized clusters represent.

This chapter will briefly touch upon the appearance of neurons in evolution and amplify to some extent definitions for some of the elements involved in complex behavior. The main focus is on the behavior of certain single-celled organisms, the goal being to understand the behavior in terms of elementary molecular processes.

Evolution

Microfossils of filamentous and colonial microorganisms have been found in 3- to 3.5-billion-year-old Pre-Cambrian rocks (61). Such cells appear to be the earliest known forms of life on this planet, but bear no resemblance to neurons. At what point did neurons appear in the course of evolution? Sponges (*Porifera*), the most primitive multicellular organism, are aggregations or colonies of independent cells. Although there is a limited degree of differentiation among the cells, there are no neurons. The next higher groups of animals are the *Coelenterates* (hydra, jellyfish) and *Ctenophores* (sea gooseberries, Venus' girdle), which do have neuron-like cells and neural networks. The neuron thus appeared at an extremely early stage in evolution.

The brain is harder to recognize than a neuron, but flatworms (*Platyhelminthes*) have a cerebral ganglion, a simple brain. The advantages conferred by a brain on the success of an individual organism were apparent early in evolution, and have become overriding at the highest level, for human beings. "One-neuron" cells (independent, responsive cells) evolved very quickly through two-neuron systems (sensory cell, motor response) to three-neuron systems (input cells, intermediate cells, and motor cells). The brain continued to evolve after the appearance of the simplest neural networks, becoming more and more complex. Primate brains are the most elaborate with respect to the number of connections. On the biochemical level, neurons of organisms at opposite ends of the evolutionary scale are similar but different in the complexity of their interconnections.

Elements Responsible for Complex Behavior

An organism responds to stimuli in the environment through the operation of a sequence of internal processes. A stimulus-response scheme illustrating the minimal number of operations and components for the behavior of the organism is shown in Figure 2.1. One can expand the fundamental definitions of trace, representation, learning, and memory (Chap. 1) to characterize the elements contributing to observed behavior.

The *receptor cell* contains a molecular system that is affected by an input signal. Retinal rod cells absorb visible light; olfactory and taste cells complex with certain molecules. The changes in these cells in response to the inputs constitute *data*, which is transformed into outputs that are *transferred* through other cells to the *central processor*. A trace of the input appears in the cells that are activated by the act of data transfer. The trace is a *representation* of the input for the system. The central process may operate with most or all of the cells involved in the representation (Chap. 1).

On the molecular level, *memory* is the physical record of the representation. For the system, *memory* is the logical record of the inputs. The physical record can have different features for short-term and long-term memories. The logical record presumably is similar in the short and long term, although changes over long times (processes such as consolidation, or fixation of memory) suggest that internal inputs can augment both the physical and logical records.

A separate central processor is at the heart of a von Neumann computer and is a device distinct from that in which the data are stored. The central processing element of higher nervous systems, the brain, operates differently. The manipulation of representations is carried out via overlapping circuits in which

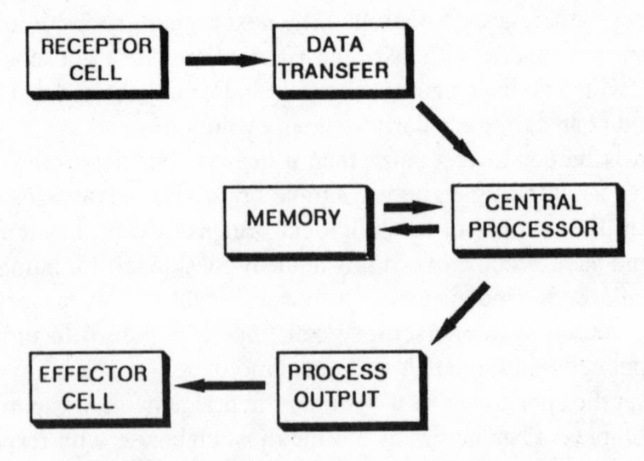

Fig. 2.1. A stimulus-response scheme for a multicellular organism.

the storage directly affects the outcome. Recognition of the problem being addressed can evoke the answer. The set of representations that can solve the problem are activated by an overlap between the representation of the problem and those of the solution. Thinking about a problem is often an exercise in distorting the nature of the original representation until it activates a previously resident set of representations. Ultimately, output representations—traces (*process output*) that effect the speaking, writing, or movement—are produced. Intelligence may well be a matter of how efficiently the output representations for certain classes of problems are found.

The process output is transferred to an effector cell, which then produces an expression of the operation of the central processor on data acquired by the receptor cell.

Single-Cell Behavior

The carnivorous flagellate *Peranema* exhibits a startle response. The tip of its flagellum beats with a motion that propels the cell in the direction toward which the organelle is pointed. If the tip encounters an obstacle, it bends. The cell moves so as to straighten out the flagellum, and the cell then moves off in the new direction. This avoidance behavior is accomplished with the biological mechanisms available within a single cell.

The predatory protozoan *Didinium* responds in different ways to different external stimuli, using its cilia to move in a manner determined by whether the stimuli it encounters are judged to be prey, indifferent, or noxious.

Vorticella withdraw abruptly in response to a stimulation of the head, and responses are graded according to the magnitude of the stimulation. The withdrawal involves the contraction of a single fiber, which is apparently induced by a change in the intracellular level of calcium ion.

If one disturbs the algal dinoflagellate *Noctiluca* by splashing around in the ocean late at night, these organisms produce bright flashes.

Some single-celled organisms possess light-sensitive structures, an "eye-spot." Those for flagellates like *Euglena*, *Volvox*, and *Ponchetia* are located in a cup immediately beneath the cell membrane and have a thickened lens-like cover. The flagellae respond to a stimulus of the eye-spot. The mechanism for the conversion of the light stimulus into motion is unknown. The single-celled alga *Chlamydomonas* utilizes rhodopsin as the light-sensing material in its photoreceptor (62).

Bacteria respond to external stimuli and bias the direction of random motion by changing the way in which their flagella turn. Chemotaxis is an example of this response.

The examples I have cited are sufficient to demonstrate that single-celled organisms can exhibit complex behavior (63). The general scheme set forth for higher organisms must be translated into an analog involving molecules and molecular systems within a single cell.

Stimulus-Response Scheme for Single Cells

Single cells detect certain physical or chemical changes in their surroundings, evaluate the nature of the information inherent in the changes, and respond to the changes. The physical "device" that receives the signals is a receptor or sensor. In response to the external stimulus, the receptor undergoes a physical or chemical change, producing a message (or data) in the form of a molecular change. The molecular change is detected or transferred to a data processing unit, a system that can respond to the molecular change. Under certain conditions, the data, either in the form of the original molecular changes or in the form of some other types of molecular changes, can be stored. Data processing produces another set of molecular changes, which can be transferrable to an effector, which then produces the response of the cell. The events described as molecular changes can be amplified, can occur as a cascade of molecular changes, and can produce data processing on many different levels and time scales simultaneously. Amplification may result from the activation of an enzyme, cascades might be the series of molecular conformational changes occurring after absorption of a photon, and different time scales can be generated by pH changes, or changes in the thiol-disulfide status of the cell or cell membrane after exposure to acids or oxidizing agents, respectively.

A stimulus-response scheme summarizing these generalizations is shown in Figure 2.2. There are obvious similarities between the processes and intermediates proposed for the single cell and those set forth for the multicellular organism (Fig. 2.1). If the molecular basis of the stimulus-response scheme for a single cell were understood, one might then be in a position to better understand the problems of stimulus and response in a multicellular organism.

The behavior of single cells has been studied intensively over the last decade with a view to revealing the molecular basis for their responses to environmental stimuli, especially molecular stimuli. The study of chemotaxis has brought new insight on the molecular level into behavioral responses, as shall be seen in this chapter. A thorough review of the molecular components of bacterial chemotactic systems extends many of the points covered here (64).

Prokaryotic Phylogeny

Some perspective on the relationship of known chemotactic organisms to other unicellular species can be helpful in choosing further subjects for study. Analysis of the linkages between ribosomal RNAs of organisms leads to a general picture of the phylogeny of prokaryotes (65, 66), which can be extended to higher organisms (67). Bacterial phylogeny is illustrated in Figure 2.3.

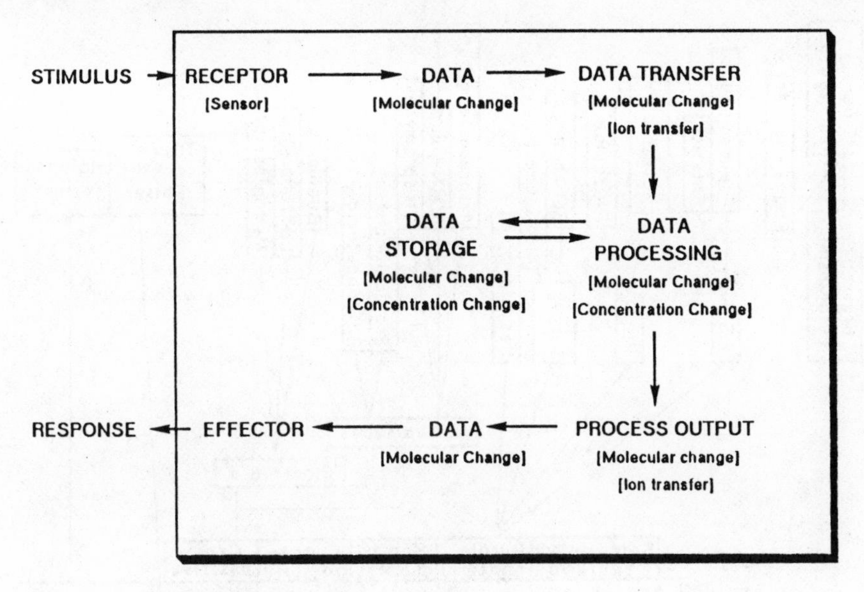

Fig. 2.2. A stimulus-response scheme for a single cell.

Chemotaxis

Freely motile organisms frequently respond to external stimuli by a change in the way in which they are arranged or distributed: a *taxis* (Greek, *tattein*, to arrange). Motion in response to light, for example, is phototaxis. Motion in response to chemicals is *chemotaxis* (68–76). Leeuwenhoek, who made the first microscope, noted before 1700 the attraction of food particles for bacteria. Systematic studies by Engelmann and Pfeffer in 1881–84 showed that bacteria were attracted to light, and either attracted or repelled by particular chemicals, i.e., asparagine or potassium nitrate, respectively.

Chemotaxis is common among single cells as diverse as bacteria, protozoa, slime molds, and polymorphonuclear leukocytes. The responses of higher organisms like insects, fish, dogs, and humans to chemicals follow detection by receptor cells. Various activities of the cell (or higher organisms) may be mediated by chemotaxis, including the search for nutrients, avoidance of danger, aggregation to complex structures (slime mold [77], myxobacteria [78, 79], cyanobacteria [''blue-green algae'']–bacteria complex [80]), and accumulation at sites of injury (phagocytes).

Detection of Chemotaxis

A capillary containing a test solution is inserted into a suspension of bacteria; microscopic examination reveals whether or not the gradient of test mate-

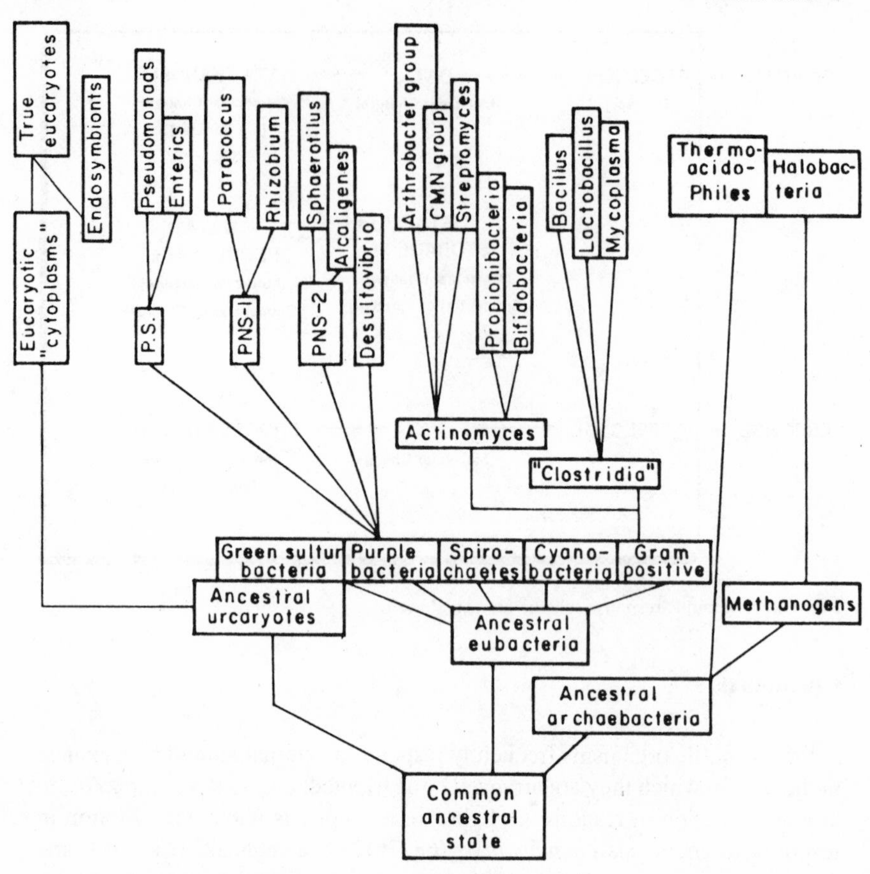

Fig. 2.3. Phylogenetic relations among prokaryotes.

rial established by normal diffusion at the exit of the capillary has created an accumulation (attraction) or a deficiency (repulsion) of bacteria near the exit. Nonmetabolizable materials can be tested in this way just as well as potential nutrients (81).

Another technique for detecting chemotaxis involves the use of a soft agar plate containing a metabolizable attractant which is inoculated with bacteria. Consumption of the attractant leads to a gradient, and the bacteria form a ring around the original inoculum. Alternatively, a hard agar cylinder containing repellent may be placed on a soft agar plate containing a sufficient number of bacteria to be visibly turbid; the region around the cylinder becomes more or less clear in accordance with the strength of the repellent. The use of a plate in the evaluation of chemotactic behavior makes the search for mutant bacteria that are not attracted (or repelled) fairly simple, such mutants being found in regions in which there are few ''normal'' bacteria (82).

Classes of Chemosensors

A large number of substances have been studied for their ability to attract or repel bacteria, especially *E.coli* and *S.typhimurium*. The number of different chemosensors is far less than the number of different substances that attract or repel the bacteria, as shown by competition, genetic, and inducibility criteria (83–85). Specifically nonchemotactic mutants are not attracted (or repelled) by all members of a particular chemosensory class. Other stimulants attract or repel these mutants in the usual way.

There are at least a dozen different chemosensors for attractants (compounds that produce positive chemotaxis) and at least four different chemosensors for repellents (compounds that produce negative chemotaxis). These are listed in Table 2.1, together with pT, the negative logarithm of the threshold concentration for the observation of activity. The formulas for the chemotactic stimuli are shown in Figure 2.4. The compounds are grouped according to the ultimate receptor for transduction, that is, the membrane protein that creates the intracellular signal. Some compounds, like phenol (86), are attractant for *E.coli* and repellent for another species. Glycerol is repellent like alcohols and diols (87). The groups are MCP (subgroups: I, II and III, methyl-accepting chemotaxis proteins), EII (subgroups according to the sugar substrate for the particular enzyme II), and "other."

Removable Receptors

Removable receptors are binding proteins that can be released from the cell periplasm by a sudden change in osmolarity ("osmotic shock"). These removable receptors are involved in certain types of chemotaxis. A mutant, not attracted to ribose, does not bind antibody to purified ribose receptor protein of *S.typhimurium*. Revertants that exhibit chemotaxis toward ribose bind the antibody. Chemotaxis toward other compounds is unaffected by absence of the ribose receptor. The binding proteins also serve as the receptors for sugar-specific membrane transport systems, but transport is essentially independent of chemotaxis beyond the stage of binding for the MCP group of stimulants (88).

A typical periplasmic receptor is that for galactose. *E.coli* mutants that lack the galactose binding protein neither exhibit chemotaxis toward galactose nor transport galactose. A permease-negative mutant shows a chemotactic response to galactose even though the sugar is not transported. Another mutant that cannot metabolize galactose is nevertheless chemotactic toward and transports the sugar (88).

Chemoreceptor Density and Kinetic Properties

The number of receptors present in the membrane of a single cell ranges from 10^3 (aspartate, serine) (89) or 10^4 (ribose) (90) up to 5–8 $\times$ 10^4 (galac-

Table 2.1. Chemosensors for *E.coli*

Attractant	pT[a]	Repellant	pT[a]
MCP I[b,c]		**MCP I[b]**	
L-Serine	6.5	Hydrophobic amino acids	
L-Cysteine	5.4	L-leucine	4.0
L-Alanine	4.2	L-Isoleucine	3.8
Glycine	4.5	L-Valine	3.6
		L-Tryptophan	3.0
MCP II[b]		L-Phenylalanine	2.5
		L-Glutamine	2.5
L-Aspartate	7.2	L-Histidine	2.3
L-Glutamate	5.3		
Maltose	5.5	Indole	6.0
		Skatole	6.0
MCP III[b]			
		MCP II[b]	
D-Galactose	6.0		
D-Glucose	6.0	Metal cation	
D-Fucose	4.7	Co^{++} ($SO_4^{=}$)	3.7
D-Ribose	5.2	Ni^{++} ($SO_4^{=}$)	4.7
N-Acetyl-D-glucosamine	5.0	**pH shift[d]**	
D-Fructose	5.0	H^+ (pH)	6.5
		OH^- (pH)	7.5
EII[b]			
		Fatty acids[e]	
D-Glucose	5.5		
Methyl alpha-D-glucoside	5.5	Acetate	3.5
Methyl beta-D-glucoside	5.0	Propionate	3.7
		n-Butyrate, isobutyrate	4.0
D-Mannose	5.5	*n*-Valerate, isovalerate	4.0
D-Glucose	5.5	*n*-Caproate	4.0
2-Deoxy-D-glucose	5.0	*n*-Heptanoate	2.2
D-glucosamine	4.5	*n*-Caprylate (CB)	1.5
D-Mannitol	5.2		
D-Glucitol (Sorbitol)	5.0	**Aromatic acids[e]**	
Trehalose	5.2	Benzoate	4.0
		Salicylate	4.0
Other			
		Other	
Oxygen			
Phenol[f]		Sulfide (SH^-)	2.5
		1-Propanethiol	2.5
		Alcohols	
		Methanol	1.0
		Ethanol	3.0
		1-Propanol	2.4
		2-Propanol	3.2
		Isobutanol	3.0
		Isoamyl alcohol	2.2
		1,2-Ethanediol, glycerol[g]	

Table 2.1 cont.

Note: Sensor class in boldface (for EII sensors, strain K12 was used).
[a] Negative logarithm of the threshold concentration for activity.
[b] Chemosensor group according to protein responsible for signal transduction across the cell membrane: MCP, methyl-accepting chemotaxis proteins I, II, and III (genes: tsr, tar, trg); EII, enzyme II.
[c] Also α-aminoisobutyrate.
[d] MCP I is involved in transduction of pH changes.
[e] Acid anions (acid dissociation constants, p*K* 3.0–5.0).
[f] Phenol is a repellant for *S.typhimurium* (reference 86).
[g] Reference 87.

tose) (91). The ribose receptors as well as the other periplasmic receptors are thought to be evenly distributed over the cell surface. The shock-releasable receptors for galactose, ribose, and maltose have been isolated and purified. The rate constant for galactose binding to galactose binding protein is 3–4 × 10^7 $M^{-1}sec^{-1}$, while the rate constant for galactose dissociation from the complex is 4.5 sec^{-1}, these values being consistent with roles for the binding protein in both chemotaxis and transport (92). The dissociation constants for both the galactose and ribose receptors are similar, ca. 10^{-7} M.

Galactose chemotaxis has been reconstituted from galactose binding protein and shocked wild-type cells, but not from mutants lacking the binding protein (88). Incubation of purified maltose binding protein with mutants lacking binding protein under conditions that permeabilize the outer membrane (250 mM $CaCl_2$ at 0°C) resulted in reconstitution of maltose chemotaxis (93).

Chemotactic Response

What is the physical basis for the migration of bacteria in response to a stimulant? Tracking *E.coli* in three dimensions under the microscope reveals that the bacteria move in two different ways: straight runs (smooth swimming) and tumbles (vigorous random, chaotic motion) (94, 95). Within a solution in which a positive gradient of an attractant is present, bacteria move in straight runs more frequently than in a uniform (isotropic) medium, resulting in net motion toward the higher concentration of attractant. Bacteria moving in a positive gradient of repellent tumble more frequently than in an isotropic medium, and straight runs predominate in the direction away from the repellent. The bacteria move in the direction in which straight runs predominate over tumbles.

By rapidly changing the concentration of stimulant using rapid mixing and stopped flow, a new feature of behavior is easily seen (96) (Fig. 2.5). A sudden increase in attractant concentration causes the bacteria to swim at first only in straight runs, with almost no tumbling. The unusual frequency of straight

Attractants

MCP1

$HOCH2CH(NH_3^+)COO-$

Serine

$HSCH_2CH(NH_3^+)COO^-$

Cysteine

$CH_3CH(NH_3^+)COO^-$

Alanine

$CH_2(NH_3^+)COO^-$

Glycine

MCP2

$^-OOCCH_2CH(NH_3^+)COO^-$

Aspartate

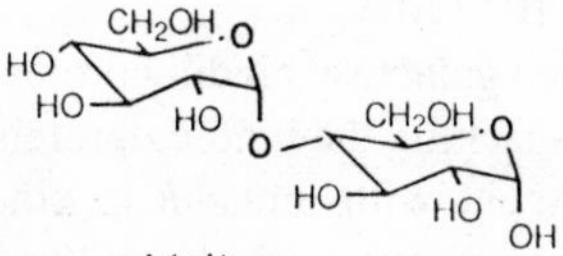

CH_2OH, O, HO, HO, HO, O, CH_2OH, O, HO, HO, OH

α-Maltose

MCP3

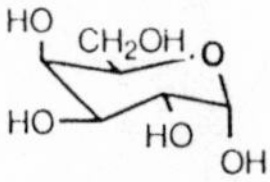

HO, CH_2OH, O, HO, HO, OH

D-Galactose
α-anomer

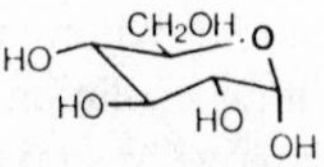

CH_2OH, O, HO, HO, HO, OH

D-Glucose
α-anomer

Repellants

MCP1

$(CH_3)_2CHCH_2CH(NH_3^+)COO^-$

Leucine

$CH_3CH(CH_3)CH_2CH(NH_3^+)COO^-$

Isoleucine

$(CH_3)_2CHCH(NH_3^+)COO^-$

Valine

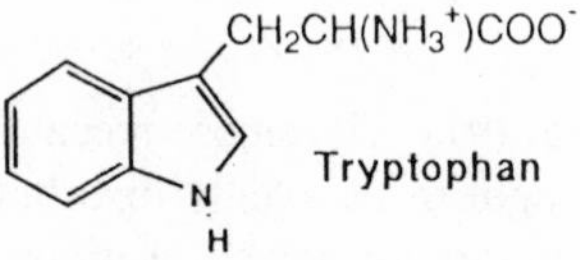

$CH_2CH(NH_3^+)COO^-$, N, H

Tryptophan

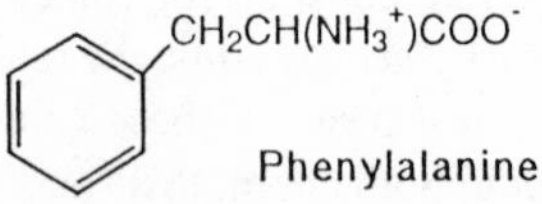

$CH_2CH(NH_3^+)COO^-$

Phenylalanine

$NH_2COCH_2CH_2CH(NH_3^+)COO^-$

Glutamine

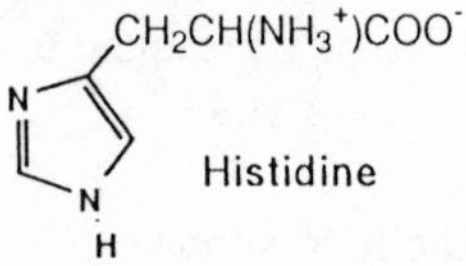

$CH_2CH(NH_3^+)COO^-$, N, N, H

Histidine

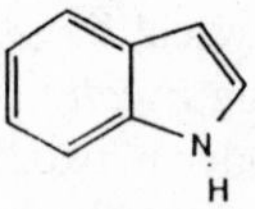

N, H

Indole

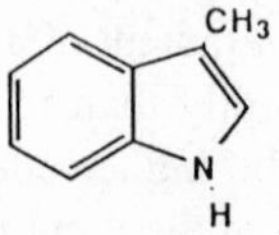

CH_3, N, H

Skatole

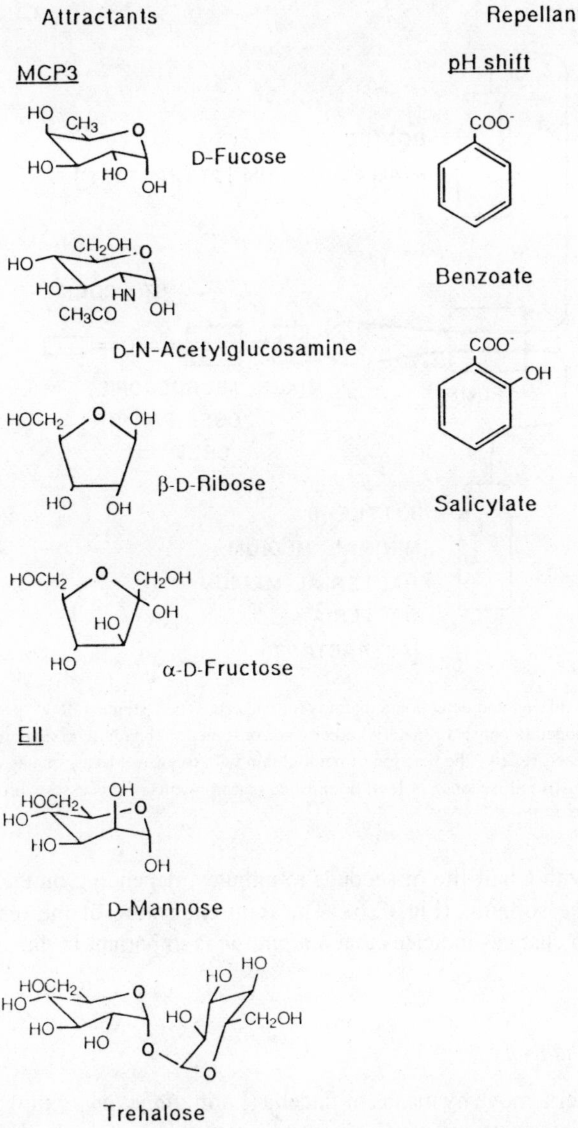

Attractants

MCP3

D-Fucose

D-N-Acetylglucosamine

β-D-Ribose

α-D-Fructose

EII

D-Mannose

Trehalose

Repellants

pH shift

Benzoate

Salicylate

Fig. 2.4. Formulas for chemotactic stimulants, grouped according to the nature of the receptor. Trehalose = O-α-D-glucopyranosyl-(1→1)-α-D-glucopyranoside.

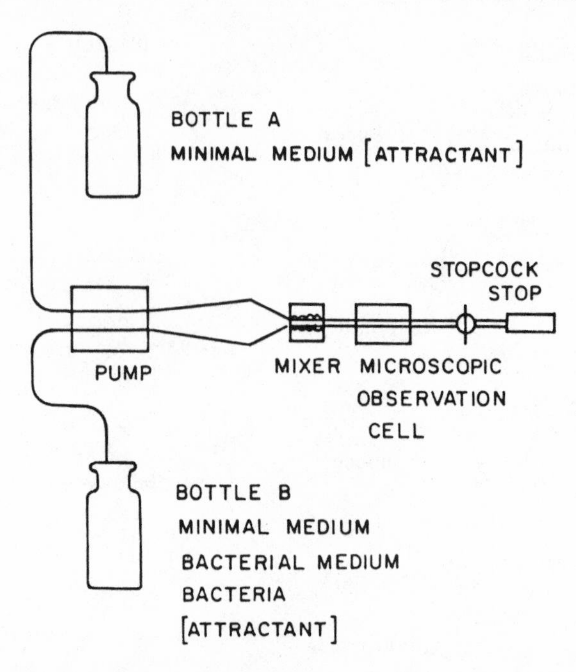

Fig. 2.5. Rapid mixing and detection apparatus for bacteria. The "stopped flow" principle was originally developed to conserve material otherwise consumed in the rapid mixing apparatus for fast chemical processes like the reaction of hemoglobin with oxygen. Mixing can be accomplished within 5–10 milliseconds or less, permitting measurements of processes with half-lives of 10 msec or more.

runs is lost with a half-life of seconds to minutes, depending on the size and identity of the stimulus (Fig. 2.6). The transient nature of the response to concentration changes indicates that *adaptation* is important in the chemotactic response.

Motility Mechanism

Many bacteria move by means of flagella (Latin, *flagellum*, whip), long (3–10 μ), filamentous structures which may be at one or both ends of the bacterium ("polar") or distributed at random ("peritrichous," from the Greek, *trichos*, hair). The flagella of *E.coli* and *S.typhimurium* are peritrichous; the bacteria possess 5 or more flagella.

In straight runs ("smooth swimming"), the flagella of these two types of bacteria function as single helical bundles projecting in the direction of the long axis of the cell body. Each flagellum is attached to a structure, which resembles and is often called a motor. The helical bundle which propels the bacterium rotates in the counterclockwise (CCW) direction, as viewed from the end of the bundle looking toward the cell (97). If the flagella rotate in the

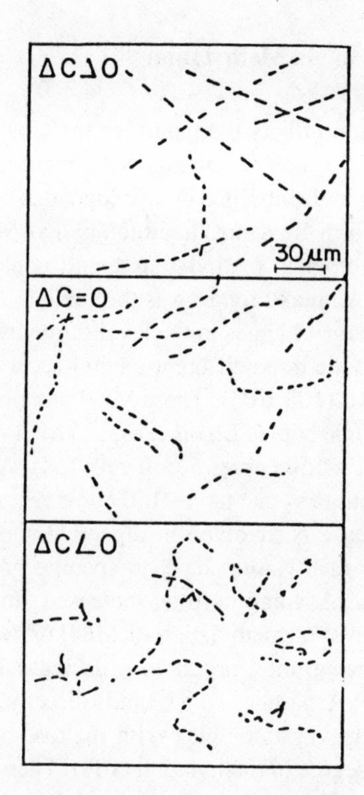

Fig. 2.6. Swimming behavior of *E.coli* after exposure to sudden changes in concentration. Top: $\Delta C > 0$, smooth swimming; middle: $\Delta C = 0$, normal mixture of smooth and random swimming; bottom: $\Delta C < 0$, random swimming.

opposite direction, the bundle becomes uncoordinated, and the bacterium moves randomly or "tumbles."

The flagella can exist in at least two forms, the normal, a left-handed helix (pitch 2.1 μ) (98) and a "curly," right-handed helix (pitch ca. 1.1 μ) (99). With CCW rotation, the left-handed helical flagella can bundle and generate smooth swimming. On reversal of the direction of rotation, the flagella become splayed and motion is random. In a mutant, however, after a period of CW rotation, the flagella are converted into the "curly" form, and can bundle and produce smooth swimming. Reversal of rotation leads to random motion (100).

Monoflagellar bacterial cells may be grown in special, high-glucose media, or may be produced by mechanical means. These cells may then be "*tethered*" to the surface of a slide with antibodies to the flagellar protein (101). (Antibody-coated flagellum sticks to the glass microscope slide.) Stimulation of the cell now results in rotation of the cell body in a direction opposite to that taken by the flagellum in the free state.

Chemotaxis: Adaptation via Methylation

Methionine deprivation affects the nature of the chemotactic response. A methionine auxotroph of *E.coli*, a mutant that cannot synthesize the amino acid, is nonchemotactic in the absence of an exogenous supply. Methionine is utilized in the cell to form S-adenosylmethionine (SAM); the synthesis and turnover of SAM, a biological methylating agent, could be correlated with chemotactic behavior. A breakthrough was the finding that there were membrane proteins that undergo changes in the level of methylation in response to chemical stimuli. There are three different methyl-accepting chemotaxis proteins (MCP): I, II, and III (102, 103). These MCP proteins are associated with the genes tsr (MCP-I, 1600 copies/*E.coli* cell), tar (MCP-II, 900 copies/*E.coli* cell), and trg (MCP-III, $\approx$150 copies/*E.coli* cell) (64). A fourth MCP-related protein is associated with the gene tap ($\approx$150 copies/*E.coli* cell) and together with a dipeptide permease is involved in dipeptide chemotaxis (104, 105). Each MCP changes in methylation level to specific groups of chemotactic stimulants or repellents. Adaptation to an increased attractant concentration parallels a rise in the level of methylation of MCP proteins; adaptation to an increased repellent concentration produces a decrease in the level of MCP methylation (106, 107). Aspartate, which binds directly to MCP II, and maltose, which acts directly via a complex with maltose-binding protein, have additive effects on the degree of methylation (108). There are 4–5 methylation sites per MCP. The adaptive responses are reversed in gram-positive *Bacillus subtilis*, with methylation decreasing in response to attractants (109–111). The site of methylation has been identified as a γ-glutamyl carboxylic acid group (Eq. [2-1]); demethylation is accomplished through hydrolysis of the ester (112, 113).

(2-1)

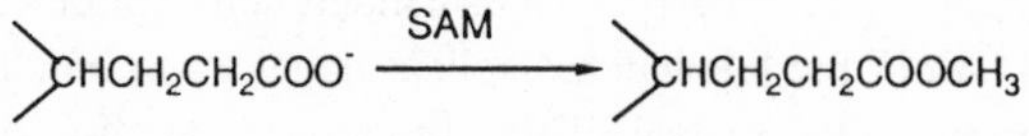

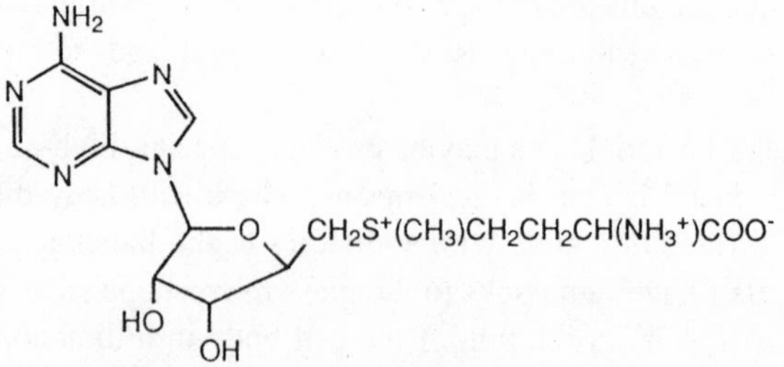

$$>CHCH_2CH_2COO^- \xrightarrow{SAM} >CHCH_2CH_2COOCH_3$$

NH_2 ... $CH_2S^+(CH_3)CH_2CH_2CH(NH_3^+)COO^-$ ... HO ... OH

S-adenosylmethionine (SAM)

The degree of methylation for each MCP depends upon the level of attractant or repellent to which the bacteria have become adapted (114–121). Up to sixteen different methylated forms of MCP I have been detected, each carrying a different combination of COOH, $COOCH_3$, and $CONH_2$ groups. The number of combinations in MCP II is lower but not yet defined (118, 122–128).

The relationship of function and gene product has been established in many mutants of *E.coli* and *S.typhimurium* (129, 130). Mutants that fail to respond to specific attractants or repellents lack viable receptors for those stimuli. Mutants that respond to only one stimulus lack viable copies of one or more of the other MCPs (85). The stimuli that have been associated with a particular MCP are grouped accordingly in Table 2.1.

Mutants that are generally nonchemotactic could have a defect in proteins other than the MCPs. Methylation of the MCPs did not occur in a particular mutant of *S.typhimurium*, suggesting that a methyltransferase enzyme was lacking. The methyltransferase was isolated from the wild type, shown to have a molecular weight of ca. 40,000 and to catalyze the methylation of the MCPs from both *S.typhimurium* and *E.coli* (131). The methyl esterase that catalyzes hydrolysis of the methyl esters was identified in cytoplasmic extracts of both bacterial species (132). The methylesterase of *B.subtilis* hydrolyzes the methyl esters of *E.coli* MCPs (111).

A cleavable cross-linking agent, dithio-bis(succinimidyl propionate), has been used to show that MCP I and MCP II are present as functional tetramers in the membrane of *E.coli*. The chemical basis for the experiment is shown in Eq. 2-2 (133). A different, novel cross-linking procedure, involving cysteines introduced by mutagenesis, leads to the conclusion that aspartate receptors (tar, MCP II) are present as functional dimers in the presence or absence of ligand (134, 135).

(2-2)

$$ProtNH_2 + SucN\text{-}O\text{-}COCH_2CH_2\text{-}S\text{-}S\text{-}CH_2CH_2COO\text{-}NSuc \longrightarrow$$

$$ProtNHCOCH_2CH_2\text{-}S\text{-}S\text{-}CH_2CH_2CONHProt \xrightarrow{DTT} ProtNHCOCH_2CH_2SH$$

Prot = protein

Suc - (cyclic succinyl group: four-membered chain with two C=O)

Gel electrophoretic analyses of MCPs reveal sets of proteins having apparent small molecular weight differences produced by labeling with between 0 to 4 or 5 methyl groups. An increase in methylation correlates with adaptation to increased attractant (107, 114–121). The methyl groups added in response to attractants become part of a kinetically equivalent pool of methyl esters (136).

Sensory adaptation in bacteria apparently requires the presence of a methyltransferase, to promote the formation of methyl esters, and a methyl esterase, to catalyze the hydrolysis of methyl esters. Some types of bacterial behavior can be rationalized in terms of the activity of the two enzymes. Mutants defective in the methyltransferase usually swim smoothly, but in the presence of repellents the cells tumble. Since the transition from smooth swimming to a mixture of smooth swimming and tumbling is the sign of adaptation, a mutant that does not change its behavior naturally makes adaptation to attractants difficult to demonstrate directly. Benzoate (.03 M), for example, produced tumbling for as long as 36 hours. Addition of attractant to repellent-activated cells led to smooth swimming with no sign of adaptation (137). The methyl esterase can, in the presence of repellents, lead to covalent modifications of MCPI and MCPII. This is due to the hydrolysis of glutamine, since glutaminases also have methyl esterase activity (138, 139). The same modification may have been observed without specific addition of repellents (140). The integration of opposing stimuli to different MCPs appears to control demethylation (124).

Adaptation in Other Bacteria

The monotrichous flagellate *Pseudomonas aeruginosa* is biochemically distinct from the enteric bacteria like *E.coli* and *S.typhimurium*, and exhibits chemotaxis toward glucose, organic acids, and amino acids (141). The periplasmic glucose-binding protein is the glucose chemotaxis receptor (142). Adaptation to L-serine, L-arginine, and α-aminoisobutyric acid increases the level of SAM-dependent methylation of a 73,000 MW membrane protein, and removal of the attractant leads to a decrease in methylation of that protein (142). However, in *Bacillus subtilis*, attractants lead to demethylation of the γ-methyl glutamate ester groups (109–111, 144). Methyl group transfers occur between MCPs in *Bacillus subtilis*, a reaction involving the methyltransferase (145). Methylation in response to chemical stimuli has also been detected in *Halobacterium halobium* (146, 147), *Caulobacter crescentus* (148), and *Spirochaeta aurantia* (149). The methyl-accepting portions of many flagellated chemotactic bacteria have antigenic determinants that interact with anti-MCP III antibody, indicating a common primordial origin for this portion of the MCPs (149).

Spirochaetes are motile bacteria without flagella, but with axial periplasmic fibrils which may function in an analogous way. *Spirochaeta aurantia* exhibits a temperature-dependent, L-cysteine-stimulated chemotaxis toward D-glucose, accompanied by SAM-derived D-glucose-stimulated methylation of two polypeptides (149, 151).

Protein Methylation in Biology

Methylation of γ-glutamyl carboxylic acid groups is apparently limited to proteins involved in the chemotactic response of prokaryotes. A different type of protein carboxymethyl transferase, which produces a nonstoichiometric (β)-methylation of D-aspartic (or L-isoaspartic) residues, has been detected in red-blood-cell membranes. The most significantly methylated proteins are "band 3" anion exchange protein, and the cytoskeletal proteins 2.1 and 4.1 (152–154). The conversion of L-isoaspartyl peptides into L-aspartyl peptides is facilitated by methylation via the formation of a succinimide intermediate (155) (Eq. [2-3]).

(2-3)

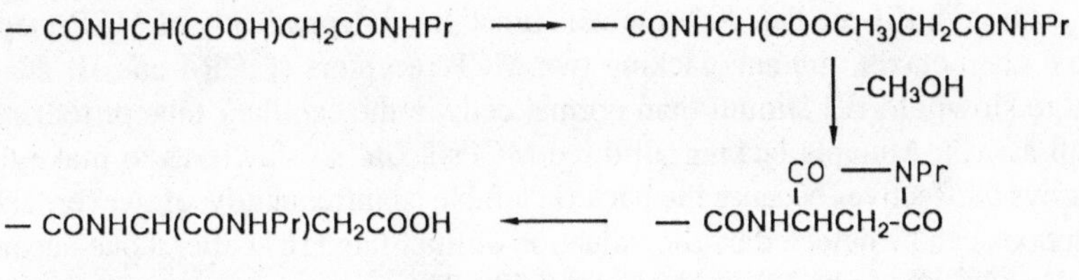

Pr = Protein

Other amino acids that are methylated in posttranslational modification of proteins are lysine, arginine, and histidine (156). The methylation and demethylation of the cell nuclear envelope protein, lamin B, during mitosis in dividing cells may be related to removal and reformation of the envelope, since methyl ester group turnover does not occur in nondividing brain cells (157). Methylation of carboxyl groups in proteins of archaebacteria has been found, suggesting a general biological importance for this type of reaction (158).

Enzyme II Receptors

Receptors that adapt to stimuli via methylation or demethylation constitute a general class for which the transmembrane transfer of information ("transduction") occurs via methyl-accepting chemotaxis proteins (MCPs).

A major sugar transport pathway leads to the appearance of sugar phosphates within the cell (159, 160). Sugar transport across the membrane is coupled to phosphorylation via phosphotransferase (PTS) systems, each sugar having a specific receptor. Mutants lacking the receptor portion of the system for glucose are not chemotactic toward glucose, but are chemotactic toward another sugar subject to PTS transport like mannose (161–165). The PTS systems constitute a second general class of chemosensors (166). Chemotaxis promoted by the PTS system does not involve methylation, since mutants lacking MCPs or methyltransferase or demethylase still show chemotaxis toward sugars (167). A PTS system consists of a specific membrane receptor enzyme (II), a heat-stable phosphoryl carrier protein (HPr) and an enzyme I, and, for many sugars, a soluble, somewhat hydrophobic enzyme III. Enzyme I is a cytoplasmic enzyme that catalyzes the phosphorylation of HPr with phosphoenolpyruvate. In the normal *E.coli* K12, phosphorylation of glucose enzyme II (EII^{glc}) requires another protein, glucose enzyme III ($EIII^{glc}$) (168). Glucose enzyme II is quite easily inhibited by one-electron oxidizing agents, apparently through conversion of dithiols to disulfides (169). The PTS system is shown in Scheme 2.1. Separate PTS systems have been found for glucose, mannose, and up to ten additional carbohydrates, including D-fructose, mannitol, D-glucitol, and N-acetyl-D-glucosamine. These are grouped under EII in Table 2.1.

Although EII-mediated chemotaxis functions independently of MCP-mediated chemotaxis, mutants lacking two MCP receptors (MCP I and II) adapt more slowly to EII stimuli than normal cells in the capillary tube or tethered cell assays. Mutants lacking all three MCPs adapt so slowly as to make the assays ineffective, because the bacteria tumble so infrequently. However, chemotaxis can be detected by the induction of tumbling (167) after a one-second pulse of ''blue light'' ($\lambda = 450$ nm) (170). The presence of normally operating MCPs is thus essential for the cells to exhibit normal behavior. Mutants that lack cyclic AMP exhibit normal PTS chemotaxis (171), but a role for cyclic guanosine monophosphate remains to be tested. A diffusible compound for which the synthesis is ATP-dependent has been implicated (160). Guanosine diphosphate and guanosine triphosphate may then be directly or indirectly involved in the operation of the PTS and other chemotactic responses.

Oxygen Chemotaxis (Aerotaxis) and Other Taxes

The taxis of bacteria toward oxygen has been known for a long time (68). Does oxygen taxis involve MCPs (methyl-accepting chemotaxis proteins)? Since MCP-deficient mutants swim smoothly, it was necessary to use the ''blue light assay,'' in which tumbling is the response in the absence of an attractant. Oxygen taxis occurred whether or not MCPs were present and adaptation to oxygen was therefore not directly connected to methylation or demethylation of MCPs. It has been proposed that oxygen interacts with the terminal oxidase of the respiratory chain (cytochrome c oxidase), stimulating electron transport and the transport of protons across the cell membrane out of the cell (170, 172).

Higher pH in the cytoplasm activates a CCW motor response and the positive ΔpH induced by oxygen on a short time scale could produce a CCW motor response and smooth swimming. Adaptation would involve return to the original equilibrium pH within the cell. Transient pH changes may affect the state

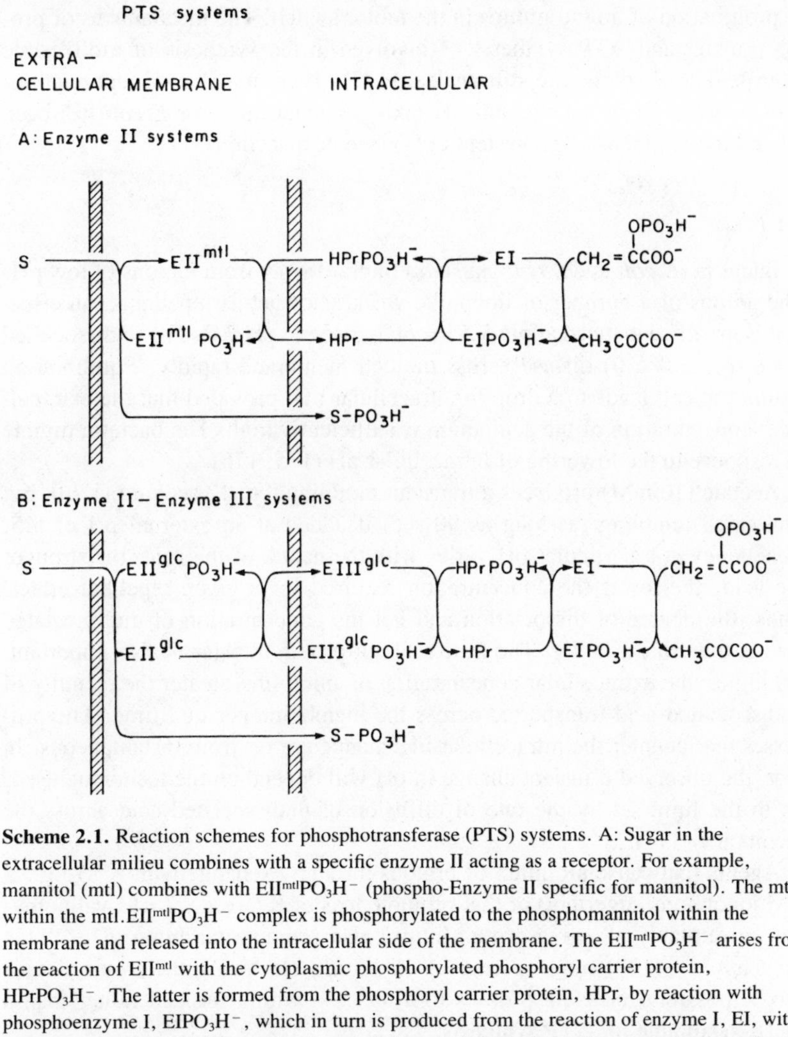

Scheme 2.1. Reaction schemes for phosphotransferase (PTS) systems. A: Sugar in the extracellular milieu combines with a specific enzyme II acting as a receptor. For example, mannitol (mtl) combines with $EII^{mtl}PO_3H^-$ (phospho-Enzyme II specific for mannitol). The mtl within the mtl.$EII^{mtl}PO_3H^-$ complex is phosphorylated to the phosphomannitol within the membrane and released into the intracellular side of the membrane. The $EII^{mtl}PO_3H^-$ arises from the reaction of EII^{mtl} with the cytoplasmic phosphorylated phosphoryl carrier protein, $HPrPO_3H^-$. The latter is formed from the phosphoryl carrier protein, HPr, by reaction with phosphoenzyme I, $EIPO_3H^-$, which in turn is produced from the reaction of enzyme I, EI, with phosphoenolpyruvate (PEP), a tricarboxylic acid cycle intermediate. The sugars known to be phosphorylated by the EII PTS system are nag (N-acetylglucosamine), mtl (mannitol), gat (galactitol). B: Many sugars are recognized, phosphorylated, and transported from the extracellular milieu into the cell by a system like the EII system, but which has one additional enzyme (enzyme III) specific for particular sugars. The sugars that are accepted by the Enzyme II–Enzyme III PTS system are glucose (glc), mannose (man), β-glucoside (bgl), glucitol (gut), fructose (fru) ($EIII^{fru}$ may require the factor FPr instead of HPr) (160).

of protonation of amino groups in the motor switch. The mechanisms of proton transfer and ATP synthesis, if involved in the synthesis of a diffusible intermediate for taxis, are still unclear (173). There may be a direct relationship between oxygen taxis and pH taxis. Oxygen taxis for *E.coli* has been related to the glucose PTS system (174) (see next section).

pH Taxis

Bacteria (*E.coli* and *S.typhimurium*) migrate away from regions of low pH. The anions of a number of lipophilic weak acids act as repellents, an effect that is greatly enhanced at pH 5.5 in comparison to pH 7.0. The undissociated acids (pK 3.0–5.0) diffuse across the cell membrane rapidly. Equilibration within the cell leads to a drop in intracellular pH, provided that the extracellular concentration of the acid anion is sufficiently high. The bacteria tumble in response to the lowering of intracellular pH (175, 176).

Acetate (10 mM) produces a transient tumbling (ca. 12 sec) at pH 7.0, but prolonged tumbling (as long as 90 sec) if added at an external pH of 5.5. Repellency at a particular pH varies with the nature of the acid: the stronger the acid, the lower the concentration required for a given repellent effect. Thus, the degree of dissociation and not the concentration of undissociated acid is the critical factor. The time scale of proton transfer is also important: the higher the extracellular concentration of anion, the greater the quantity of undissociated acid transported across the membrane per unit time. The processes that counter the intracellular pH change are far from instantaneous. In fact, the observed transient change in pH will depend on the technique used, up to the limit set by the rate of diffusion of undissociated acid across the membrane (177).

Agents that cause an influx of protons through exchange with K^+ (e.g., 5 μM ionophore, nigericin) or that promote loss of K^+ (e.g., 2 μM valinomycin, presumably allowing proton influx) also promote tumbling. The "blue light assay" (178) may operate by inhibiting electron transfer via flavins (167), possibly releasing protons to a location within the cell rather than a location suitable for ATP synthesis.

Increasing the pH within the cytoplasm by raising the external hydroxide ion concentration, or that of an amine salt such as ethanolamine hydrochloride, leads to smooth swimming. MCPI-mutants (high unstimulated tumbling frequency) show responses opposite to those seen for normal cells, i.e., swimming smoothly in the presence of repellents and increasing their tumbling frequency with attractants, possibly by a mechanism like that responsible for the "blue light" effect (175). After lowering the extracellular pH to 5.5, the initial inverted response is apparently overcome by the net change in cytoplasmic pH, with tumbling as the final result (176).

In some experiments on the effects of pH, the membrane potential and pH gradients were measured using flow dialysis to equilibrate the measuring agents (triphenyl-[^{3}H]-methylphosphonium bromide plus sodium tetraphenylboride for potential, [^{14}C]-benzoate and [^{3}H]-ethanolamine for pH) which were loaded into the cells, then analyzed in the dialysate (175).

Proline Taxis

L-Proline chemotaxis (179) can be observed after the induction of proline transport and proline dehydrogenase, a cytoplasmic enzyme, even for mutants lacking MCPs. Oxidation of proline produces NADPH; proline taxis may thus function by the same mechanism as that obtaining in oxygen taxis.

Other Effects on Taxis

Decreasing the internal Mg^{++} or Ca^{++} concentration (180, 181) with A23187 and EDTA suppresses tumbling, and even produces smooth swimming for nigericin-treated tumbling cells.

Changing membrane properties through a change in chemical composition or temperature may influence chemotaxis. *S.typhimurium* and *E.coli* grown in the presence of agents for raising the phase transition temperature of the membrane phospholipid (e.g., elaidic acid, *trans*-9-octadecenoic acid) exhibit tumbling frequencies like those of normal cells, although the *E.coli* fatty-acid auxotroph strain used in the research had a very low motility (182). Tumbling was maximal at the growth temperature, reflecting the control that the cell exerts on the chemotactic response mechanism. To a first approximation, membrane fluidity has little influence on chemotaxis. In methylase (cheR) deficient mutants of *E.coli*, smooth swimming frequency rose from 2% to 100% as the temperature was increased from 20 °C to 35 °C in the presence of the MCP-I (tsr) repellent, L-leucine. The change in smooth swimming frequency could be reversed by lowering the temperature. Thus, unmethylated MCPs induced a signal (183) that varied with temperature (184). High methylation levels induced by the MCP-I (tsr) attractant, serine, remove the temperature response. Aspartate inverts the temperature response for MCP-II (tar) receptors (185).

Phototaxis has been related to changes in membrane potential and the K^+/Na^+ gradient in *Halobacterium halobium* (186, 187) but actually depends on a ''slow'' rhodopsin (188, 189).

General Schemes for Chemotactic Signal Generation

I have implied that there might be a common thread running through a number of the chemotactic responses. Three general schemes for signal generation appear to account for the action of most stimuli.

A scheme for MCP-mediated chemotactic responses is shown in Figure 2.7.

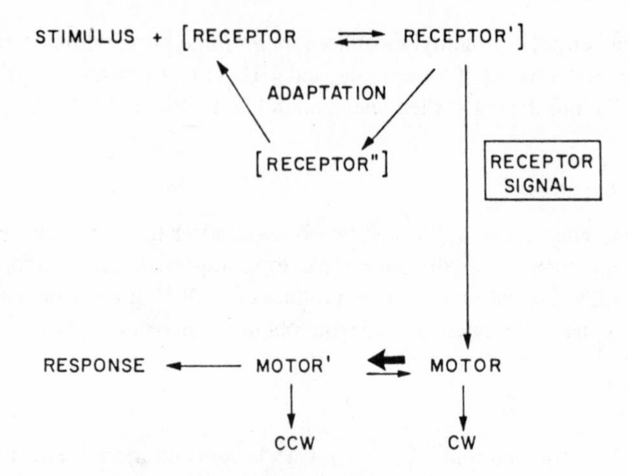

Fig. 2.7. Overall general scheme for MCP chemotaxis. The receptor is an MCP protein. In the case of certain sugars and amino acids, the binding site might be another protein. The binding protein either diffuses to or is attached loosely to the MCP within the membrane. The receptor.stimulus complex exists in two forms; one with, and one without, catalytic ability for the conversion of active signal protein to inactive signal protein. The active signal protein ''interferes'' with the natural propensity of the bacterial motor to maintain smooth swimming and causes switching and, therefore, tumbling.

Combination of the stimulus with the receptor changes the state of the receptor, and affects the magnitude of the signal induced by the receptor. (If the receptor is not part of the MCP, association of the receptor precedes the change in state.) The signal is transmitted to the motor, there affecting the switching between counterclockwise rotation (smooth swimming) and clockwise rotation (tumbling). The altered receptor is methylated or demethylated, the choice depending on the nature of the stimulus, and thus returns to its original level of signal induction, the overall result being adaptation. The signal from the receptor changes the preference of the motor for one of the two directions. For attractants, the preferred direction is CCW; for repellents, the preferred direction is CW. The cooperative rotation of the flagella of the bacteria as a helical bundle occurs only for the CCW rotation, and smooth swimming (straight runs) is the result (97, 190). The uncoordinated rotation of the flagella turning in the CW direction results in tumbling. A more detailed scheme for MCP-mediated signal generation is given in the next section.

The generation of a signal through pH change is shown in Figure 2.8. Weak acids in the extracellular milieu lead to a rapid change in the intracellular pH, leading to a behavioral response like that for a repellent. The variety of agents responsible for positive chemotaxis, together with the thought that pH might play a role via proton loss, led to the idea for a common scheme. The formulation shown in Figure 2.9 indicates that loss of H^+ (formation of OH^-) serves

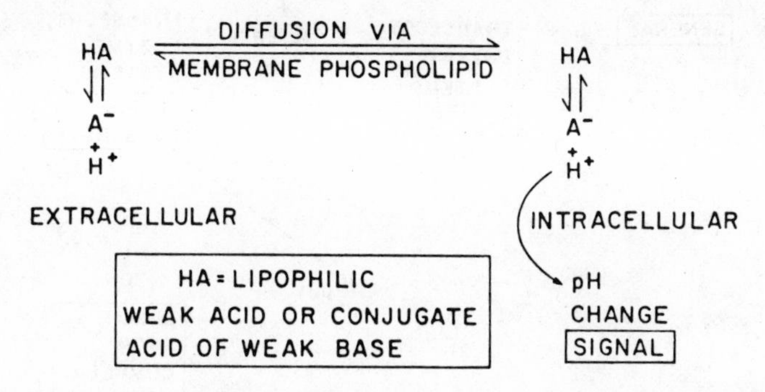

Fig. 2.8. Chemotactic response to pH change. Stimulus molecules that exist either as neutral lipophiles or as dissociated species, including protons, produce tactic responses. The times of diffusion of small lipophilic molecules across the cell membrane bilayer are much shorter than the times involved in tactic responses. The conjugate acids of weak bases (which cross the membrane as neutral bases) produce a rise in intracellular pH and smooth swimming in normal *E.coli* or *S.typhimurium* cells. Weak acids produce a fall in intracellular pH and tumbling. The change in pH is presumed to affect the signal molecule, either directly or by hydrolysis of the ST.T complex (see Fig. 2.11).

as the signal emanating from a "receptor complex." The specific stimuli and the components of each "receptor complex" are listed in Table 2.2. In each case, the perturbation of local metabolic equilibrium leads to a signal; adaptation involves the recovery of the equilibrium (local = the place in which the signal to the motor is produced).

Signal Generation by MCP species

The signal generated by the MCP appears to be a protein, probably the cheY product (11 kDa), as judged by the decay of the signal within a few μm from L-aspartate applied iontophoretically to the flagellar antibody-labeled filamentous form of *E.coli* (191). The bare mechanism outlined in Figure 2.7 requires a more detailed model (Figure 2.10) to explain the phenomenology of MCP-mediated chemotaxis. An equilibrium between catalytically active "open" forms (active for signal protein activation and as substrate for methylase) and catalytically inactive "closed" forms (ineffective in signal protein activation and active as substrate for esterase) is proposed. The "open" form catalyzes the transformation of the signal molecule from a motor-active form to a motor-inactive form. Catalysis involves an autophosphorylating protein kinase, in which his 48 is the phosphoryl group carrier, and also acetyladenylate (191). The natural, uncontrolled direction of rotation for flagella is CCW, and smooth swimming should be the consequence of such rotation (192, 193). The activated receptor must then induce a "positive" signal which interferes less with the operation of the motor.

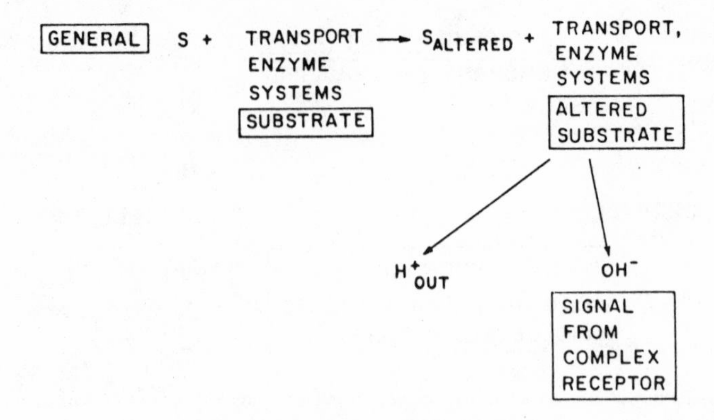

Fig. 2.9. The origin of signals from "complex receptors" in chemotaxis. The stimuli that produce a transfer of protons out of the cell are listed in Table 2.2. The signal is probably a pH change which then affects the signal molecule (see Fig. 2.8).

Table 2.2. Complex Receptors in Chemotaxis

Stimulus	Complex Receptor (Substrate)	Altered Stimulus	Altered Substrate
Oxygen (O_2)	(Reduced cytochrome c oxidase)	H_2O	Oxidized cytochrome c oxidase
Proline	Proline transport Proline dehydrogenase (NADP)	Dehydroproline	NADPH
Sugar	PTS System (PEP)	Sugar phosphate	$CH_3COCOOH$
	Glucose-6-phosphate dehydrogenase (NADP)	D-6-phospho-δ-gluconolactone	NADPH

The equilibrium state of a wild-type MCP-receptor should exist in an equilibrium between a half-methylated open form and a half-methylated closed form. Combination of the receptor with an attractant molecule (S) leads to a conformational change which shifts the equilibrium to the open form. The rate of induction of positive signal molecules is increased, the motor direction is switched less often, and more smooth swimming results. The open form of the MCP-receptor is subject to methylation. The methylated, attractant-occupied MCP-receptor has the "natural" equilibrium between the open and closed forms, and the system has thus adapted. Dissociation of the attractant shifts the equilibrium to favor the closed form. The closed form is subject to demethylation. The esterase (demethylase) removes a methyl group, returning the MCP-receptor to its starting state of activity and methylation.

Occupancy by a repellent (R) favors a shift in the open-closed equilibrium to the closed form, followed by demethylation, dissociation of the repellent, a shift in the conformational equilibrium of the MCP-receptor, and methylation to the starting MCP-receptor state. The adaptation via change in methylation status is needed for substantial changes in stimulus level. Mutants that lack both methylating and demethylating enzymes respond chemotactically in a normal fashion to modest changes in stimulus level (126, 194, 195). In our model, cheD mutations alter the conformation of MCP I so as to favor the open form, even after methylation, thus overproducing (with respect to a wild-type cell) the positive signal (ST.T). This leads to a bias of the motor toward the natural rotation, CCW (123, 196).

Signal Transfer

The signal transfer protein (ST-protein) is probably the product of the cheY gene (*E.coli*), crystals of which have yielded some X-ray data (191, 197, 198, 199). An interesting connection between the ST-protein and the methylesterase has been discovered. The 129-amino-acid protein sequence (200), Mr 14,000, of the *Salmonella typhimurium* ST-protein is completely homologous to the N-terminal sequence (the regulatory domain) of the cheB product, the methylesterase (201). Without the regulatory domain, the 21 kDa C-terminal fragment of the methylesterase is 15-fold more active in demethylation than the whole 37 kDa protein (202), but is completely inhibited by attractants. A proteolytic cleavage from the 37 kDa form to the 21 kDa form may be stimulated by attractants (203). A suggested relationship between chemotaxis and sporulation (204) is confirmed by the sequence similarities between the ST-protein and a sporulation regulatory factor (201).

There is a possibility that a phosphorylated nucleotide may be involved in signal transfer (205). The biochemical system operating in the rod outer segment (Chap. 5) converts a photon stimulus into a neuronal impulse via a receptor-catalyzed replacement of GDP with GTP in the protein transducin. Eventually, a transient negative change in the current carried by the rod cell is the result of the hydrolysis of cyclic GMP. By analogy with the first step of the transducin system, I propose that the motor-active form of the ST-protein (ST.D = ST.GDP) is converted into the motor-inactive form (ST.T = ST.GTP) by the open form of the MCP-receptor. The cheZ product catalyzes the conversion of ST.T into ST.D (206) (Fig. 2.11). A protein, cheW, with Mr of 35,000 as a homodimer is essential for chemotaxis and may be involved in nucleotide binding (207). According to the scheme shown in Figure 2.11, the ST-protein and the methyl esterase must recognize similar regions on the MCP-receptor. It can be no coincidence that the entire sequence of the cheY product (the ST-protein) is identical to the N-terminal regulatory sequence of

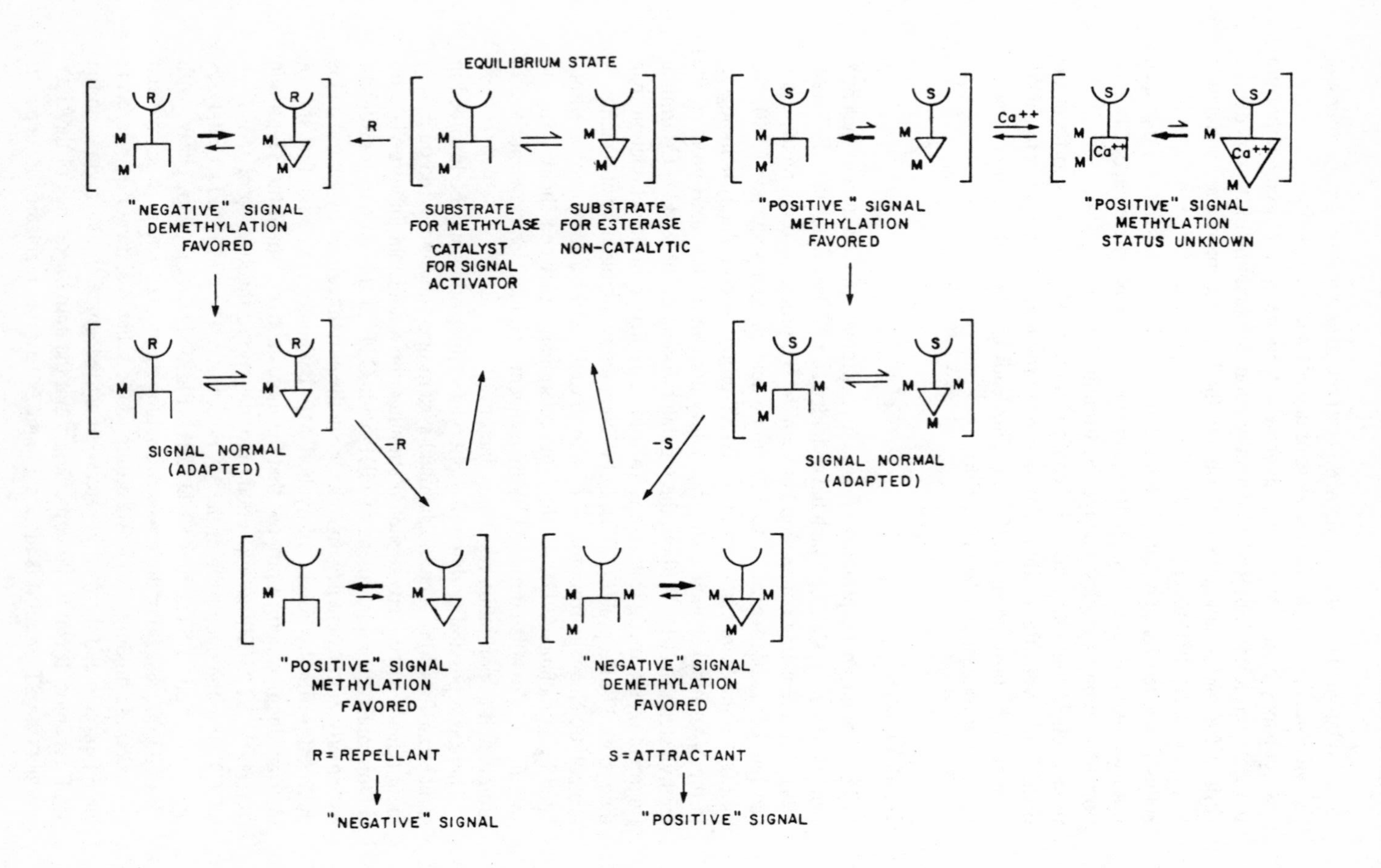
EQUILIBRIUM STATE
R
Ca++
"NEGATIVE" SIGNAL
DEMETHYLATION
FAVORED
SUBSTRATE
FOR METHYLASE
CATALYST
FOR SIGNAL
ACTIVATOR
SUBSTRATE
FOR ESTERASE
NON-CATALYTIC
"POSITIVE" SIGNAL
METHYLATION
FAVORED
"POSITIVE" SIGNAL
METHYLATION
STATUS UNKNOWN
SIGNAL NORMAL
(ADAPTED)
-R
-S
SIGNAL NORMAL
(ADAPTED)
"POSITIVE" SIGNAL
METHYLATION
FAVORED
"NEGATIVE" SIGNAL
DEMETHYLATION
FAVORED
R = REPELLANT
"NEGATIVE" SIGNAL
S = ATTRACTANT
"POSITIVE" SIGNAL

Fig. 2.10. Detailed scheme for operation of the MCP-receptors. The usual wild-type MCP-receptor appears to exist as an equilibrium mixture of half-methylated open and half-methylated closed forms. Combination of the receptor with an attractant molecule (S) leads to a conformational change which shifts the equilibrium to the open form. The rate of induction of positive signal molecules is increased, the motor direction is switched less often, and more smooth swimming results. The open form of the MCP-receptor is subject to methylation. The methylated, attractant-occupied MCP-receptor has the ''natural'' equilibrium between the open and closed forms, and the system has thus adapted. Dissociation of the attractant shifts the equilibrium to favor the closed form. The closed form is subject to demethylation. The esterase (demethylase) removes a methyl group, returning the MCP-receptor to its starting state of activity and methylation. It is significant that the probable signal protein (the cheY gene product) has exactly the same 129-amino-acid sequence as the regulatory portion of the cheB methylesterase, presumed in our model to interact at the same set of sites as the signal protein. Occupancy by a repellant (R) favors a shift in the open-closed equilibrium to the closed form, followed by demethylation, dissociation of the repellant, a shift in the conformational equilibrium of the MCP-receptor, and methylation to the starting MCP-receptor state.

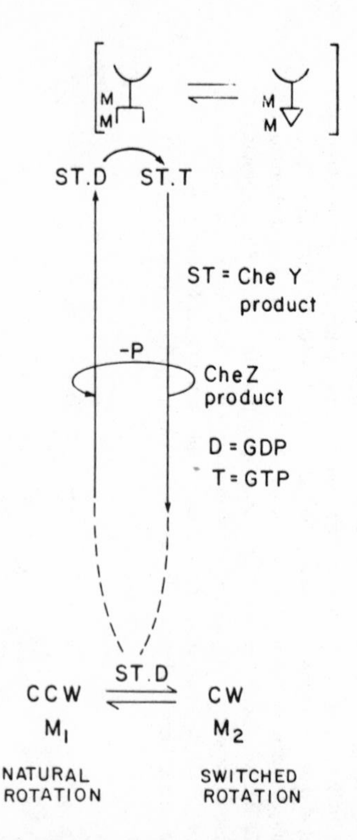

Fig. 2.11. Signal protein transfer between MCP receptor and motor. The motor-active form of the ST-protein (ST.D = ST.GDP) is converted into the motor-inactive form (ST.T = ST.GTP) by the open form of the MCP-receptor. The cheZ product catalyzes the conversion of ST.T into ST.D. The ST protein may function alone in the absence of D or T cytoplasmic factors to convert counterclockwise motion to clockwise motion (see reference 197).

the cheB product, the methyl esterase. The complementary roles of the cheY and cheZ products have been evaluated by overproducing them in *E.coli* cells (208). The integrated signal controls demethylation (124). Cells with mutant MCP-III (trg-21) show defective taxis to galactose and ribose but normal taxis to other attractants. Overproduction of the mutant MCP-III favored CCW swimming and nonchemotactic behavior. The mutation involves the substitution of a threonine for ala-419, a change that decreases deamidation of glutamine to glutamate in the cytoplasmic region of the protein and thus the number of methyl-accepting sites (209). The glutamines that are normally deamidated are those at 311 and 318 in MCP-III (210). The identity of the methyl acceptor in demethylation during chemotaxis of *B.subtilis* would be of interest (211).

The signal transfer system may respond to pH change by promoting the hydrolysis of GTP to GDP and to the PTS system through synthesis of the phosphorylating agent for conversion of GDP to GTP. This ''GUT'' (grand unified theory) view of chemotaxis lacks many molecular details, but does provide a complete overall picture, consistent with careful measurements of temporal responses to chemotactic stimuli (212).

Difference between Open and Closed Forms of MCP-receptor

Flooding a chemotactic system with Ca^{++} (0.5 mM ion + ionophore A23187) resulted in smooth swimming (180, 181), an effect interpreted as due to increased formation of the motor-inactive form of the signal transfer protein (ST.T). Complexation of pairs of carboxylate or carboxyethyl groups with Ca^{++} might be the mechanism for maintaining the conformation of the MCP-receptor in the catalytically active open form (see Fig. 2.10).

Flagellar Motor Structure

The bacterial apparatus responsible for flagellar rotation has a structure in electron micrographs very much like that of an electrical motor (213, 214) (Fig. 2.12). The flagellum is connected through a hook to a rod that occupies the central hole in two rings (gram-positive bacteria) or four rings (gram-negative bacteria, e.g., *E.coli*). The outer rings (the L- and P-rings) are located in the peptidoglycan layer (the cell wall). The inner rings (the S- and M-rings) are in the cytoplasmic membrane. A ring of 14–16 "studs" surrounds the M-ring. Approximately 16 force-generating units may contribute to the rotation of bacterial motors (215). The M-ring itself is located immediately below the membrane in *Aquaspirillum serpens*. Below the M-ring is a set of fibrils, lo-

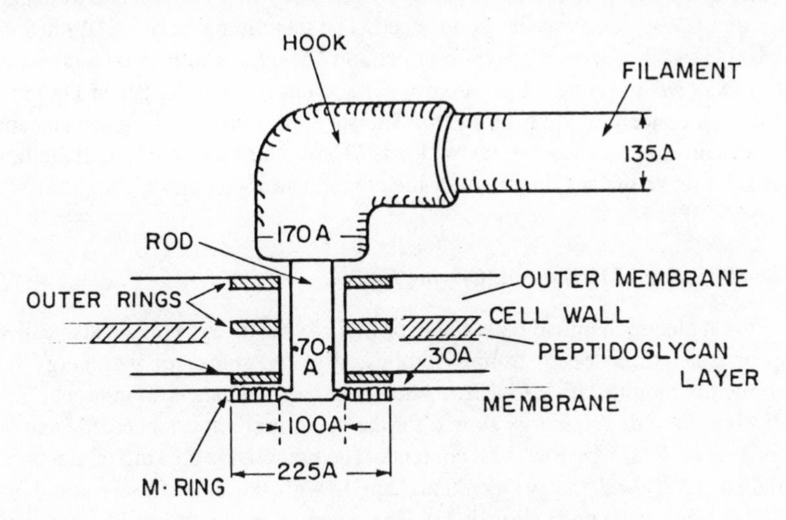

Fig. 2.12. The motor machinery at the flagellar attachment site for *E.coli*.

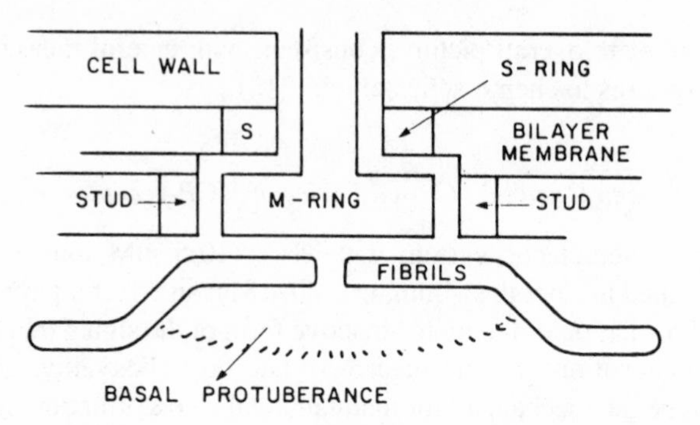

Fig. 2.13. Detailed structure of the basal region of a bacterial motor based on electron micrographs of the motor apparatus of *Aquaspirillum serpens*.

cated on a protuberance. These special features of the motor apparatus (Fig. 2.13) may be present in many bacteria (216). The genetics of motor formation have been partially worked out, and some sequences of the motor structure are known (217, 218). The "normal" form of hook components is a right-handed helix whereas the flagellum is left-handed, suggesting that the conformational changes of the bacterial propulsive system may be quite complex (219, 220).

Flagellum: Nature

Bacterial flagellae only rotate, rather than wave or beat, and thus have no intrinsic motile power, deriving their driving force from motors (221–223). Tethered cells (flagella fixed by antibody:cell-body motion opposite to flagellar) spin CW or CCW at the same speed. The flagellum can rotate at rates up to 100 Hz with abrupt changes in direction (224). Lags in motor response of *Streptococcus* in dilute buffer are ca. 0.1 sec but become longer in D_2O and shorter in concentrated buffer (225). The speed is constant at light loads and shows constant torque with heavy loads (226). The motion of the flagellum has been related to the dynamics of the component molecules (227).

Flagellar Motor Mechanism: Driving Force

Proton electrochemical potential ("proton motive force") (222), also called Δp or $\Delta\mu H^+$, has been strongly implicated as the source of the energy for turning the motor (226, 230–236). Adenosine triphosphate is an indirect (237, 238) but not a direct energy source for the bacterial motor, since cells can be motile even with very low ATP contents. The potential is the sum of electrical gradient ($\Delta\Psi$) and the pH gradient (ΔpH), with the constants required for voltage (V or mV) units (Eq. [2-4]). The potential is that utilized by Mitchell in the "chemiosmotic hypothesis" (174).

$$(2\text{-}4) \qquad \Delta p = \Delta\Psi - 2.3\,(RT/F)\,(\Delta pH)$$

The electrical potential is derived from the equilibrium distribution of $[^3H]$-CH_3-$TPMP^+$ (triphenylmethylphosphonium ion) across the bacterial membrane (186), equilibration being accelerated by tetraphenylboride ion (TPB^-) especially in lipopolysaccharide A phosphateless mutants of *E.coli* (239) (Eq. [2-5a]). The pH gradient is reflected in the distribution of a radioactive weak acid (or weak base) across the membrane (Eq. [2-5b]).

$$(2\text{-}5a) \qquad \Delta\Psi = 2.3(RT/F)\log\{[TPMP^+]_{out}/[TPMP^+]_{in}\}$$

$$(2\text{-}5b) \qquad \text{pH gradient (in V)} = -2.3(RT/F)(\Delta pH)$$

Bacteria like *B.subtilis* apparently require a Δp threshold of 30 mV for counterclockwise (CCW) motion and motility, and at least 45 mV for clockwise (CW) motion (231, 233). The difference is consistent with the notion that CCW motion is "natural." Maximum motor speed is reached with a Δp of 60 mV. Other bacteria (e.g., *Streptococcus* strain V4051 [226, 230]) do not have a Δp threshold for motor operation. The torque estimated for each generating unit is $2\text{–}3 \times 10^{-12}$ dyne-cm and 60–90 protons pass through a unit for each revolution (232, 240).

The proton is not the intracellular chemotactic signal, since the motor operates in chemotactically defective mutants. The clockwise (CW)/counterclockwise (CCW) ratio is regulated by protons. The faster the motor rotates, the greater the proportion of time spent in the CW (tumbling) mode.

Flagellar Motor Mechanism: Molecular Model

A schematic model for the motor (upper part of Figure 2.14) shows how torque may be exerted on a flagellum (241). A modification seems to us better adapted for both operation and switching. The M-ring (like the armature of a motor) (lower part of Figure 2.14) has proton-conducting groups on chains that are "stiffly" linked at an angle to the axes of the center hole, the rod, and the studs (Fig. 2.15). The driving force is proposed to arise from a combination of the repulsion between like charges and attraction between hydrogen-bonding groups attached to the M-ring and the studs, respectively. For simplicity, the model utilizes amine and carboxylic acid groups as like pairs. Only a fraction of the groups interact at any one time. For a close approach, without an intervening water molecule, the interaction energy can be estimated as 20 kcal/mole from Eq. (2-6). An increment of rotation would diminish the interaction drastically. The space between the groups would be filled by a water molecule, greatly increasing the dielectric constant. The combined effects of the change in dielectric constant and increase in charge separation would decrease the interaction to ca. 1 kcal/mole. The proton flux (from the outside)

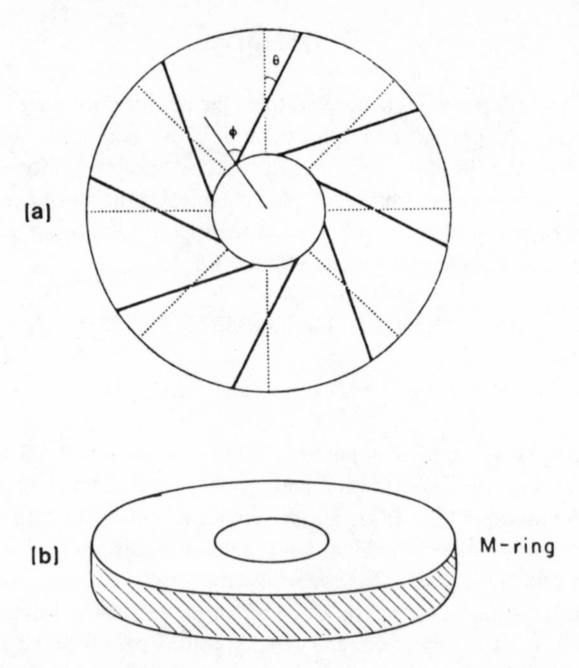

Fig. 2.14. (a) Schematic relation of rotor and stator in bacterial motor. Hypothetical arrangement of ligand groups on the S- (dashed lines) and M- (solid lines) rings. The intersection site moves radially with rotation. (b) M-ring: angular lines represent protein chains carrying ligand groups for protons (see text).

favors protonation of the upper portion of the motor and deprotonation of the lower portion. A driving force should thus be created with a direction related to the angle of the M-ring chain. The model is similar to that proposed and analyzed by Läuger (242) and seems consistent with parallelism between the stall and running torques of the Streptococcus motor (243).

(2-6) E (kcal/mole) $= 330\, q_1 q_2/Dr$ (in A),

where q_i = charge; D = dielectric constant; r = distance between charged centers (in A).

The molecular model leads naturally to a molecular model for switching. The motors on a single bacterium switch asynchronously, under the control of a local signal (the cheY product, "ST," Fig. 2.10) (244), with the bias (the fraction of time in CCW spinning) under global control (245). The time (ca. 200 msec) required for the motor to change direction after a chemotactic signal, created by electrophoretically delivering a defined pulse of attractant (α-methylaspartate), is the "latency" of the chemotactic response. Activating the MCP is not rate-limiting since there is no concentration effect on latency and

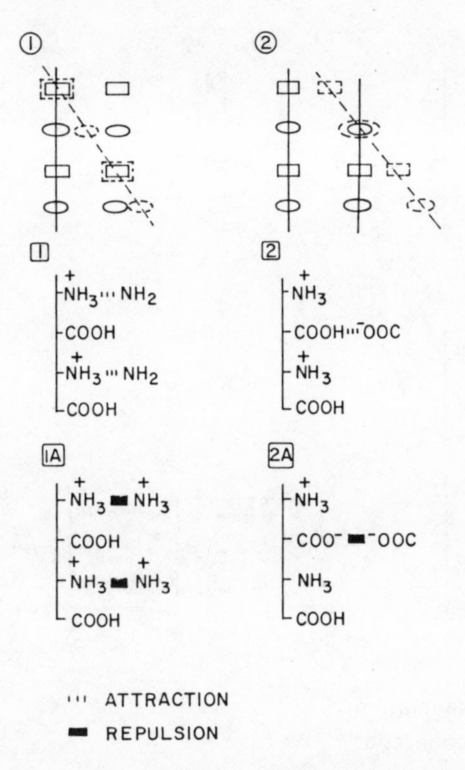

Fig. 2.15. Molecular mechanism for operation of bacterial motor. Vertical lines: protein chains from the vertical studs surrounding the M-ring. The polar groups are symbolized by rectangles (= NH_3^+/NH_2) and ellipses (= $COOH/COO^-$). Angular dashed-line: M-ring protein chain bearing groups like those on the studs (dashed-line symbols). Attractive and repulsive driving forces responsible for the motion operate between charged and neutral groups of the same class; addition or loss of a single proton converts an attractive to a repulsive interaction.

a response occurs at receptor occupancy as small as 2%. Neither is intracellular diffusion of a signal molecule likely to delay motor response for more than a few msec. Either production of the signal or the initiation of the motor response is rate-limiting (246). Difficulties in integrating the signal would make the former less probable. Combination of ST.D with the COO^- groups at the lower end would "freeze" the motion of that end of the chain while the top would continue and would undergo a conformational change to the alternative form. The favored conformation is CCW, for structural reasons which cannot be now defined. Dissociation of the ST.D would allow the motor to move (Fig. 2.16). A "frozen" ("pause") state of the motor has also been recognized, but is not shown (247).

Higher Ca^{++} produces incessant tumbling in *B.subtilis* (248), as would be expected for frequent switching in our model. Since attractants lead to demeth-

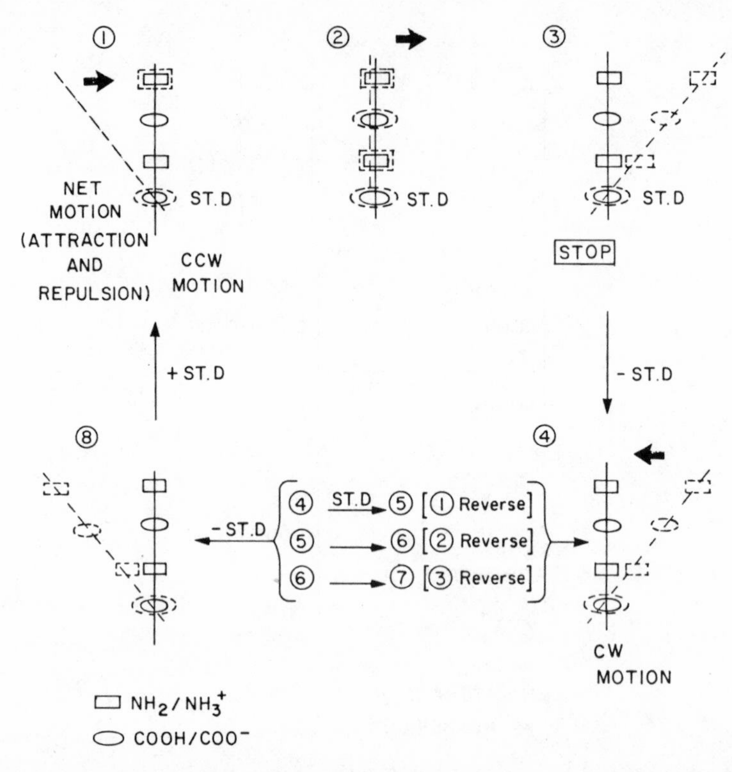

Fig. 2.16. Molecular mechanism for switching of bacterial motor. Vertical GDP-loaded signal-protein-controlled switching cycle for bacterial motor direction. The momentum of the moving chain carries it into the reverse orientation after the motor is stopped.

ylation of the MCP-receptor in this organism (144), one can infer that the closed form of the receptor is catalytic for the conversion of ST.D to ST.T.

Lower intracellular pH may have several effects on the signaling system, either decreasing the rate of change of ST.D to ST.T or increasing the rate of change of ST.T to ST.D. These factors may explain the frequent tumbling (repellent effect).

The model resembles one proposed by Berg for tethered cells (249). Some features of motor operation are clarified, including (a) constant stall torque for the motor over a substantial temperature range (4°–38°C) for a particular protonmotive force, and (b) some decrease in zero-load speed in D_2O versus H_2O (240). The momentum of the rotor and the electrostatic interactions of the charged groups would be quite similar in D_2O and H_2O, but the rate of separating and forming D-bonds would be lower than for H-bonds. Earlier measurements had indicated no difference between motor properties in D_2O and H_2O (249). Our model utilizes the ''studs'' as stator (rather than the S-ring), explains the CCW bias and why the motor might be driven in the opposite direction for an outward proton flux, and has a realistic conformational change for the change in motor direction. Another model that depends upon offset sites also uses the S-ring as stator (250).

Membrane Potential and Chemotaxis

A report that changes in membrane potential accompanied chemotaxis (251) was questioned (252, 253), but later confirmed, the difficulty being that metabolism could mask small potential changes (254). Hyperpolarization (ca. -10 mV) by the attractant (galactose, for *E.coli*) is correlated with chemotaxis (255); the transient change is not due to changes in surface charge (256). However, fixing the potential at either high or low values with appropriate external K^+ concentrations and valinomycin addition did not affect chemotaxis (257). It appears that changes in membrane potential are not critical for the chemotactic responses of small bacteria (181).

Membrane potential changes are observed for D-xylose chemotaxis by *Spirochaeta aurantia*, and the process is inhibited by various neurotoxins which either block sodium channels like tetrodotoxin, or activate sodium channels like sea anemone toxin, scorpion toxin, and aconitine (258–260). The mechanism of the inhibition on the molecular level would be of considerable interest in connection with the evolution of sodium channels (Chap. 7).

The change in membrane potential corresponds to a loss of positive ions from the cell. Biological membranes typically have a capacitance of 1 $\mu F/cm^2$ (261). The change in membrane potential can be accounted for (Eq. [2-7]) by the movement of 7000 K^+ ions out of the cell. This is a negligible fraction of the total K^+ (21).

(2-7a) $Q = C \times V$
Coulombs $=$ [Farads/cm^2] $\times$ Volts

(2-7b) No. of ions $= (Q/F) \times A \times N$
1 Faraday $=$ 96500 Coulombs/mole; Area of 1-μm cell $= 10^{-7}cm^2$
$N = 6 \times 10^{23}$ molecules/mole; volume of cell $= 4 \times 10^{-12}cm^3$

Permeabilized cells appear to behave normally in chemotaxis experiments. The Ca^{++} concentration can be diminished from 33 μM to 1.9 μM in 15 minutes by incubation with external EGTA, a Ca^{++} complexing agent (253). About 5000 Ca^{++} ions/minute pass out of the cell, possibly by a natural channel. The transient changes in potential may involve the exit of some Ca^{++} from the MCP receptors (there are probably fewer than 10^5 MCP receptors per *E.coli* cell), possibly by a channel formed by an MCP tetramer or dimer (see below).

Structure and Function of MCP Chemoreceptors

The three MCPs respond to a variety of molecules and act as receptors for some of them. Other membrane binding proteins (BP) (e.g., ribose, galactose, maltose receptors [262, 263]) combine with particular molecules (M) and then the M.BP combination associates with the appropriate MCP.

Activation of the MCP shifts the equilibrium between the catalytic (open) and noncatalytic (closed) forms of the cytoplasmic region of the MCP (Fig. 2.10). Catalysis involves the conversion of the GDP form of the ST protein (ST.D) to the GTP form (ST.T) (Fig. 2.11). The mechanism by which a conformational change in the extracellular portion of a membrane protein is transmitted to the cytoplasmic portion is not known, although of central importance in G-protein receptors like those for taste (Chap. 3), olfaction (Chap. 4), rhodopsin (Chap. 5), and the muscarinic acetylcholine receptor.

Two membrane-spanning hydrophobic amino-acid sequences are readily recognized in each of the MCPs. Such sequences lie within the membrane in an α-helical form to minimize the contact of polar groups (peptides) with the hydrophobic interior of the bilayer. How such α-helices could transfer a conformational change is not obvious, but might do so through subtle changes in helix length.

If a minor alteration in the position of an α-helix is insufficient to account for the information transfer, what other possibilities exist? Two other mechanisms, (a) aggregation of MCP monomers and (b) Ca^{++} export, can be put forward. These might operate simultaneously. (a) A conformational change in the extracellular receptor region could lead to aggregation (dimerization, trimerization, or tetramerization) of the MCPs, perforce producing some conformational changes in the cytoplasmic portions of the molecule through the association thus induced. (b) The multimeric MCPs might operate as a "slow" channel, a small number of Ca^{++} ions being exported in response to an attractant. The experiments that show the lack of involvement of Ca^{++} are flawed, since the incubation solutions used after EGTA (253) or EDTA (231) treatment contain 6 μM Ca^{++}.

Methyl-accepting Chemotaxis Proteins: Signals via Aggregation

Given that the receptors are randomly distributed over the surface and that there is an equilibrium between monomers, dimers, and tetramers, one could ask how long it might take to respond to an attractant. The aggregation of the monomers would be determined by the lateral diffusion rates ($3–9 \times 10^{-10}$ cm^2/sec [264]). A typical value would lead to ca. 200 msec as the time for two monomers to diffuse together, a time that is close to the measured latency. Domains with higher MCP concentrations could decrease the MCP contribution to the latency. However, cysteine cross-linking indicates that ligand does not change the degree of receptor aggregration (108).

Structure of the Galactose Binding Protein

The crystal structure for the galactose binding protein from *Salmonella typhimurium* (S-GBP) at 3.0 Å resolution allows the construction of a polypeptide path structure for the 292 amino acids in the chain. Two similar domains are present, separated by an 8-Å cleft. The molecule is a dimer in the crystal. In solution, galactose apparently causes a shift from dimer to monomer, suggesting that the MCP-III binding sites are revealed by combination with galactose (265).

The 309-amino-acid sequence for the galactose binding protein from *E.coli* is known (266). A binding site for galactose has been proposed on the basis of the homology in the amino-acid sequences in one region of the ribose, arabinose, and galactose binding proteins (267). The galactose binding site is ca. 30 Å from a seven-liganded Ca^{++} ion. A site for interaction with the trg gene product, MCP-III, is located 45 Å from the Ca^{++} ion. Ellipsoidal binding proteins (Mr 33,500–40,500) for arabinose, leucine, isoleucine, D-maltose, sulfate, and phosphate are similar to the galactose binding protein with a binding cleft between two globular domains (268).

MCP Chemoreceptors: Isolation and Amino Acid Sequences

Aspartate chemoreceptor from *Salmonella typhimurium* and the serine chemoreceptor from *Escherichia coli* have been isolated and shown to be MCP proteins II and I, respectively (269). After selected amino-acid sequences had been determined, the mRNAs for the chemoreceptors were cloned and sequenced. The DNA and amino-acid sequences have been reported for the aspartate receptor (562 amino acids, Mr 59,416 Da) (270), the serine receptor (536 amino acids, Mr 57,483 Da) (271), and the galactose-binding protein receptor (MCP-III, 535 amino acids, Mr 57,965 Da) (272). Mutant studies have indicated that the interaction sites for galactose (G) and ribose (R) binding proteins (BP) in MCP-III are different, GBP affinity being affected by a gly 151 to asp mutation, and both GBP and RBP affinities lowered by an arg 85 to his mutation (273). Only 200 copies of the purified S-adenosylmethionine:glutamyl transferase which methylates the MCPs are present in *Salmonella typhimurium* (274).

Methyl-accepting Chemotaxis Proteins as Membrane Channels

Hydrophobic amino-acid sequences suitable as transmembrane elements are easily identified in all three MCPs. If the MCP proteins can form ion channels, plausible ion-channel element sequences should be present. Ion-channel elements might be amphiphilic sequences of about 24 amino acids. Two possible transmembrane ion-channel elements can be selected in MCP-I (SerR, *E.coli*),

MCP-II (AspR, *S.typhimurium*) (275), and with less certainty, in the MCP-III (GBPR, *E.coli*).

There must be an even number of bilayer-crossing (two) channel elements per MCP monomer, in view of the intracellular location of the methylation sites. The active ion channel might be a dimer, trimer, or a tetramer, suggested by cross-linking (108, 135). Minor export of Ca^{++} may be promoted by the MCP, perhaps limited to those ions intimately associated with the MCP at the time of activation by a molecular stimulus.

Adaptation through methylation or demethylation would follow channel operation. A fall in Ca^{++} promotes methylation. Demethylation of chemotactically labeled membrane proteins in *Halobacterium halobium* is promoted by Ca^{++} (276).

Overall Scheme for Bacterial Chemotaxis

Scheme 2.2 summarizes the processes occurring in chemotaxis and may apply to many prokaryotic organisms.

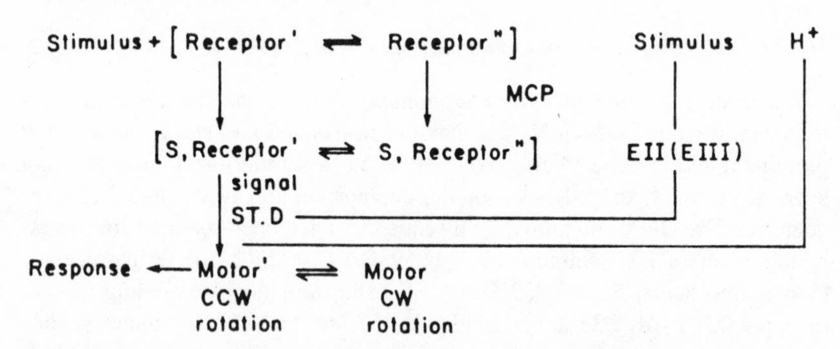

Scheme 2.2. General scheme for chemotaxis. Three general pathways for changing the behavior of bacteria are recognized: (1) Stimulus combines with methyl-accepting chemotaxis protein (MCP) and alters the equilibrium between a signal-producing and a signal-inactive form. (2) The stimulus (sugar) combines with the EII and eventually perturbs the phosphate status of the cell, altering the signal level. (3) Protons are accumulated (or lost) from the cell and the altered pH affects the signal level. (Phosphate status is the distribution among the various forms of phosphorylated compounds in the cell, including GTP, GDP, GMP, cGMP, ATP, ADP, AMP, cAMP, etc.) Symbols: ST, signal transfer protein; D, guanosine diphosphate; CCW, counterclockwise; CW, clockwise. The "pause" state of the motor is not shown.

Other Chemotactic Responses

Aromatic acids are repellents for *E.coli* (cf. Table 2.1), but attractants for *Pseudomonas putida* (277), *Bradyrhizobium japonicum*, and *Rhizobium trifolii* (278). Thaumatin, the sweetest-tasting compound known for humans (see Chap. 3), is neither an attractant nor a repellent for *E.coli*, but blocks at 0.1%

concentration the motility of the bacteria (279). The block is reversed by raising the concentration of phosphate in the buffer from 0.01 M to 0.025 M, a change that also blocks the sweet taste in humans. Acetylated thaumatins are not sweet and do not block motility. A direct effect on flagellar motion is suggested; the curious correlation of interference with bacterial motion and the human taste response may be worthy of further study.

Conclusions

What has been learned from the study of chemotaxis, a process best understood in prokaryotes? The scheme shown in Figure 2.2 is not far off the mark, albeit without the important molecular detail. Specific molecular stimuli in the extracellular milieu combine with a receptor, and data pertaining to a stimulus are transferred to the cytoplasmic region of the cell. Data for the MCP pathway are in the form of an altered rate of removal of the active motor-switching signal (ST.D). Several other stimuli may influence the quantity of switching signal by different pathways. System adaptation to molecular stimuli normally occurs via a return to normal rates of ST.D removal. The response of the bacterium is expressed through motor action. The natural direction of rotation for the bacterial motor is CCW; the active signal causes the motor to switch between CCW and CW. The CCW rotation produces smooth swimming, and progress toward the source of the stimulant gradient. CW rotation produces tumbling or random motion. A third state of the motor, ''pause,'' does nothing (247).

In large prokaryotic organisms, stimulation of the cell seems to produce a change in membrane potential (depolarization), yielding an electrical signal for the propulsion mechanism. In multicellular organisms, an electrical signal can be transferred directly (like the escape response in fish controlled by Mautner cells) or via release of a burst of neurotransmitter molecules to successor cells.

The single cell is perhaps not the best model for more complex systems, since one signal transfer system has to carry the burden of transferring the responses of all detection systems. Signal transfer by molecules to specific targets is effective over short distances. The signal molecule complex ST.D is more like a neurotransmitter than a hormone. Storage of information in the form of transient changes in ST.D levels or in the levels of those molecules directly connected to this system is only suitable for the short term. Can short-term memory in the human brain represent the result of forming a metastable molecule?

3 Sensory Transduction: Pheromones and Taste

General Classification of Stimuli for Biological Organisms

Stimuli for biological organisms can be divided into classes that reflect the manner of their interaction with receptor molecules or receptor molecular complexes: electromagnetic waves; specific molecules; nonspecific molecules.

A stimulus has an amplitude (intensity or concentration) and a topology (frequency or shape), which influence the nature of the signal produced at the site of reception. The amplitude of the stimulus determines how many receptors are activated and for how long, depending on dispersion of the stimulus and the response characteristics of the receptor system. The topology of the signal controls which receptors are activated.

Electromagnetic radiation includes near ultraviolet, visible, and infrared wavelengths. High-energy radiation (far uv, X rays) cannot produce a localized effect, or deposits only a small proportion of the energy at a given receptor site. Low-energy radiation (far infrared, radio waves) cannot be detected efficiently or is too susceptible to interference by ''noise'' in the environment.

Specific molecule receptors are used to detect neurotransmitters, molecules used to transfer information between neurons and perhaps also between secondary sensory cells and neurons. Specific molecule receptors may be involved in the detection of some ''tastes'' and ''smells.'' Nonspecific molecular receptors are those which respond to the momentum of the molecules, without regard for structure. Air molecules acting together constitute the pressure waves of sound; such changes acting on hairs in the auditory system are transduced into neural signals by a molecular structure. Electric field changes, used for probing the environment by electric fishes, are detected by translation into pressure changes in special organs.

Chemical Stimuli

An enormous number of different molecules are present in the environment of biological organisms. Some emanate from food targets, some represent means of communication between different individuals of a species, and some are warnings to competitors. There are scientific and economic reasons for trying to understand how these molecules are recognized, and how the signals arising from the presence of the molecules are produced. Neurotransmitters represent chemical stimuli that may be very similar in their mode of action to environmental chemical stimuli.

The evolution of biological systems must include steps in which both the production and reception of signal molecules between the two sexes of a particular species must have changed from that of predecessor species. I will consider some aspects of the problem, with its implications for evolution of signaling systems.

Chemotaxis in Higher Organisms

The repertoire of responses possessed by higher organisms to chemical stimuli is quite large. In spite of the wide variety of specific molecular stimuli open to study, the mechanism of the stimulus-receptor interactions is still obscure in most cases. An exception is that of one neurotransmitter-receptor combination (acetylcholine stimulus, nicotinic acetylcholine receptor, Chap. 6), for which a molecular-level structure has been proposed and a dynamic mechanism outlined. The photon-stimulus–rhodopsin-receptor combination of the visual system has been modeled (Chap. 5). Until many more individual systems have been examined in depth (molecular-level structures for the receptor, molecular interpretations of the stimulus-receptor interaction), and the appropriate rules and conclusions developed, it will continue to be necessary to seek empirical generalizations in the effects of particular molecules on the behavior of higher organisms. Now that amino-acid sequences are known for the dihydropyridine-binding protein (a putative calcium channel), the Shaker gene product (a putative potassium channel), the γ-aminobutyric acid (GABA) receptor, the muscarinic acetylcholine receptor, and the β-adrenergic receptor, one can hope that three-dimensional structures can be formulated. Discussion in later chapters will show that folding sequences into working 3D structures is not simple.

The name *semiochemicals* (Greek, *semeion*, mark or signal) has been proposed for a special class of chemical stimuli that convey messages from one organism to another (280). Pheromones are intraspecific semiochemicals, permitting communication between members of the same species. Interspecific semiochemicals that favor the producer are allomones, while those that favor the receiver are kairomones.

Chemotaxis in Nematodes

Nematodes are organisms that belong to the phylum of eel worms or nemas. *Caenorhabditis elegans* was selected because it was complex enough to yield information about the neural systems of higher organisms, yet simple enough to be cultivated and manipulated by the genetic techniques so successful with bacteria. A map of the 302 adult neurons and the 8000 synapses has been prepared; mutants of the six touch-sensitive neurons have been examined in a preliminary way. Other genetic and developmental studies suggest that local molecular rules for assembly rather than an overall developmental program determine the connections within the neural system. Knowledge of cell-cell interactions at the molecular level is then a prerequisite to understanding how the nervous system is constructed (281). The organization of the nervous system determines much of the higher-order processing.

Chemotaxis in bacteria is now understood fairly well (Chap. 2). Chemotaxis to 3′,5′-cyclic adenosine monophosphate (cAMP) is very effective (attractant pT 3.7). *E.coli* deficient in adenyl cyclase (the enzyme that catalyses the formation of cAMP) were still attractive to nematodes, implying that other molecules served as attractants. A temperature-dependent, fatty-acid pheromone-induced developmental switch forms a developmentally arrested dispersal stage in *C.elegans*, its operation inhibited by "food" (282).

Chemotaxis of nematodes can be followed by the orientation of the worms in a gradient on an agar plate as seen by the tracks left in the wake of the moving worm. The worms travel in more or less straight lines up the attractant gradient. The receptors are located near the head as shown by the curved tracks left by bent-headed individuals (Fig. 3.1). "Side to side" comparison of concentrations, presumably over time, is the simplest way to understand the motion along a curved path. Arrival at the region of highest concentration is accompanied by an increasing frequency of stops and turns which eventually bring the worm back to the lowest concentration region, to begin the whole cycle of movement over again.

Chemoreception in Insects: Pheromones

There are more species of insects (>1 million) than all other species combined. Chemical communication is a major element in the reproductive and social behavior of most insects, and a detailed knowledge of the structure and function of semiochemicals is of both scientific and economic importance (283).

The search for a signal chemical, begun in 1939, was motivated by an interest in sex hormones, insect attractants, and olfactory stimulation (284). After many years of hard work, the sex pheromone of the silkworm moth, *Bombyx mori* L., was isolated and characterized as bombykol, (10E,12Z) (*trans*-10-*cis*-12-hexadecadienol-1). The research involved the collection of more than a million silk cocoons, the excision of 500,000 female scent glands, the extraction of those glands, and the separation of the components of the mixture, all this being followed by a tedious biological assay, the "flutter-dance" of the male moth (284). The class name for intraspecific signal molecules, "pheromone," was coined in 1959 (285).

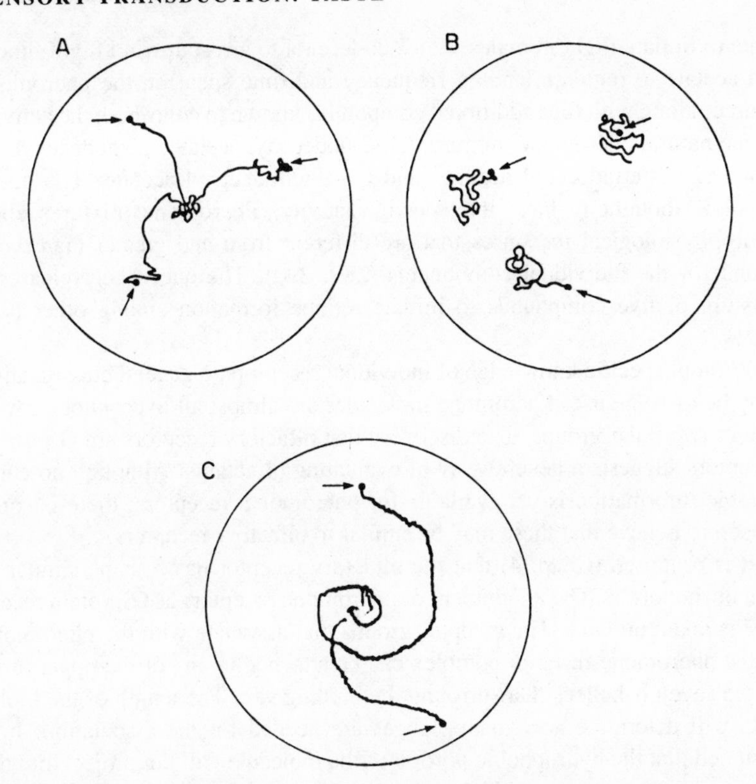

Fig. 3.1. Attractant search patterns ("tracks") generated by nematodes of the species *C.elegans* on agar plates. (A) Gradient of $NH_4^+Cl^-$ is present. (B) No attractant present. (C) Tracks produced by two bent-headed worms of strain E611 on plate with gradient of $NH_4^+Cl^-$. The less curved track belongs to the worm with the less bent head.

The techniques for isolation of chemical compounds have been enormously improved and means for identification have improved a million-fold in sensitivity since the bombykol study (286). Many hundreds of pheromones have been isolated and identified. More than 150 sex attractants are listed for the 5000 species of *Tortricidae* moths (287). Since each insect species presumably has a more or less unique set of pheromones, an extraordinary assemblage of organic molecules must be floating around in nature. However, the conclusion that sex pheromones may interact with their receptors by a mechanism different from that of human olfactants (284) may not be justified, as will be pointed out below. The species-specific character of pheromones is emphasized by the lack of development of resistance to artificial pheromone exposure in the pink bollworm moth over a period of 3–5 years (288). Pheromones have various functions. The cabbage looper moth female releases (Z)-7-dodecenyl acetate to initiate flight in males, (Z)-7-dodecenol to arrest upwind flight, dodecyl acetate to regulate landing frequency and time spent on the pheromone source, along with four additional compounds known to contribute the activity of the natural pheromone mixture, (Z)-5-dodecenyl acetate, 11-dodecenyl acetate, (Z)-7-tetradecenyl acetate, and (Z)-9-tetradecenyl acetate. The dodecenol is thought to have interspecific activity. Pheromone mixtures elicit neurophysiological responses that are different from and greater than those found for the individual components (289, 290). The queen bee releases a mixture of five compounds to initiate retinue formation among other bees (291).

Without specific knowledge of individual receptors, a general classification may be of some use. Pheromone molecules are almost all hydrophobic, with one or two polar groups. The discovery that olfactory receptors are G-protein receptors suggests a general way of evaluating olfactants. Although no comparable information is yet available for pheromonal receptors, there is some reason to believe that these may be similar to olfactory receptors.

It is proposed (Chap. 4) that the olfactory receptor has a shape similar to that of rhodopsin. The assignment of pheromone receptors as G-protein receptors is taken up later. The receptor groups that associate with the pheromone in the pheromone-receptor complex can be attached to any of the upper turns of the seven α-helices that surround the binding site. The length of the molecule will determine how many helices are needed for the association. It is assumed that the hydrophobic portion of the molecule will not project into the aqueous environment outside the receptor. I have estimated that a four- to six-carbon fragment (a "**U**" length) will bridge two helices, 8–12 carbon fragments ("2**U**") will bridge three helices, and 13–18 carbon fragments ("3**U**") will bridge four helices, presumably the normal limit for this type of binding site. The probable location for adrenergic and muscarinic binding sites is within the bilayer (Chap. 8), and the retinylidene chromophore of rhodopsin is certainly in the middle of the bilayer region of the protein (Chap. 5). An attractive alternative mode of binding for pheromones and olfactants would be within the bilayer region of the protein. Expansion of the central region through occupancy by an appropriate molecule would change the conformation of the cytoplasmic portion of the receptor so that it could catalyze a change in an associated G-protein.

Pheromones are mostly hydrophobic molecules carrying one hydrophilic or polar group. Sex, alarm, recruiting, and aggregating pheromones are listed in Tables 3.1–3.4, in order to illustrate the wide variety of molecules and the variation in behavioral responses. The tables have been subdivided according to the lengths of the molecules, with **U** = 4–6 carbons, 2**U** = 8–12 carbons, 3**U** = 13–18 carbons, and 4**U** = 19–24 carbons. The variety of pheromonal structures is reflected in Formulas **1–5**, shown on the facing page.

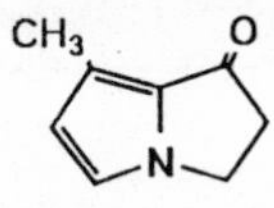

1

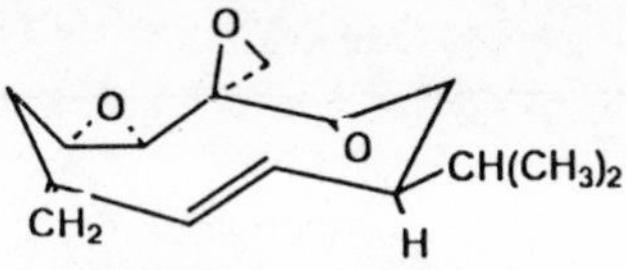

periplanone B (2)

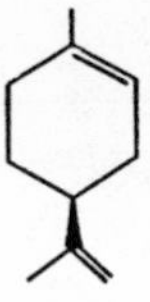

3

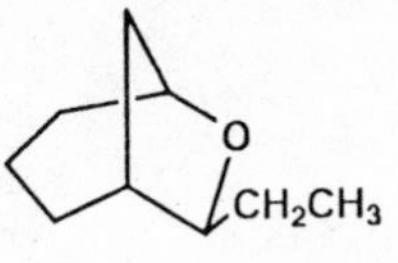

4

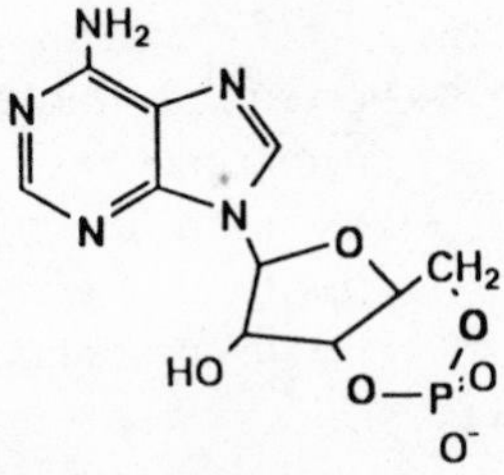

Cyclic AMP (5)

Pheromone Detector Structure

The male silkworm moth has a pair of beautiful feathery antennae that sprout from the upper part of the head, one on each side. Each antenna has many branches, and on each branch there are about 200 sensory hairs. Within each hair are the dendrites of two neurons. Each moth (two antennae) is estimated to have a total of 16,000 hairs (32,000 receptor cells) (292). The antennal arrangement is illustrated in Figure 3.2 along with a diagram of a typical antenna.

Each hair on the antenna has about 3000 pores, openings that connect the area of the dendrites to the outside through 5–10 fine tubules. The precise structure of the region between the tubules and the dendrites has not been established, but the tubules end in the ''sensillum liquor'' (293, 294). During development, the pore-tubule structures are formed before the dendrite reaches the end of the hair, showing that the tubules are part of the hair rather than the nerve. The dendrites are bathed in ''sensillum liquor.'' The tubules do not have direct contact, shown by the fact that silver-particle stains introduced into the tubules are not found in the liquid; few or no direct tubule-dendrite contacts are found in electron micrographs of the hair (295). Hair structure is shown in Figure 3.3.

Pheromonal Detector Sensitivity

Neuronal and behavioral responses to the attractant bombykol can be distinguished. Using fixed moths, the rate of delivery of tritium-labeled attractant through an airstream to antennas was estimated as 1.8 molecules/second on

Table 3.1. Sex Pheromones

Compound	Structure	Organism[a]
Class U (4–6 C)		
Valeric acid	$CH_3(CH_2)_3COOH$	Worm 1f
Phenol	C_6H_5OH	Beetle 2f
Benzaldehyde	C_6H_5CHO	Moth 3m
Class 2U (8–12 C)		
trans-Dec-9-on-2-enoic acid	$HCO(CH_2)_5HC=CHCOOH$	Bee 4f
n-Undecanal	$CH_3(CH_2)_9CHO$	Moth 5m
cis-7-Dodecenyl acetate	$CH_3(CH_2)_3CH=CH(CH_2)_5CH_2OOCCH_3$	6f
trans-7-Dodecenyl acetate	$CH_3(CH_2)_3CH=CH(CH_2)_5CH_2OOCCH_3$	Moth 7f
cis-8-Dodecenyl acetate	$CH_3(CH_2)_3CH=CH(CH_2)_5CH_2OOCCH_3$	Moth 8f
2,3-Dihydro-7-methyl-1H-pyrrolizidin-1-one **(1)**		Butterflies 9m, 10m
Class 3U (14–18 C)		
cis-9-Tetradecen-1-ol	$CH_3(CH_2)_3CH=CH(CH_2)_7CH_2OH$	Worm 11f
cis-9-Tetradecenyl acetate	$CH_3(CH_2)_3CH=CH(CH_2)_7CH_2OOCCH_3$	Moth 12f
trans-9-Tetradecenyl acetate	$CH_3(CH_2)_3CH=CH(CH_2)_7CH_2OOCCH_3$	Moth 12′f
cis-11-Tetradecenyl acetate	$CH_3CH_2CH=CH(CH_2)_9CH_2OOCCH_3$	Moth 13f, 14f
trans-3-*cis*-5-Tetradecadienoic acid	$CH_3(CH_2)_7CH=CH-HC=CHCOOH$	Beetle 15f
trans-10-*cis*-12-Hexadecadien-1-ol	$CH_3(CH_2)_2CH=CH-HC=CH(CH_2)_8CH_2OH$	Moth 16f
(−)-14-Methyl-*cis*-8-hexadecen-1-ol	$CH_3CH_2CH(CH_3)(CH_2)_4CH=CH(CH_2)_6CH_2OH$	Beetle 17f
Periplanone B **(2)**		Cockroach 18f

Source: Adapted from reference 280.

Note: In the structures, CH=CH signifies a *cis* double bond and HC=CH a *trans* double bond.

[a] Organisms from which the sex attractants were isolated are as follows (m = male, f = female): 1, sugar beet wireworm (*Limonius californicus*); 2, grass grub beetle (*Costalytra zalandica*); 3, moth (*Leucania impura*); 4, honeybee (*Apis mellifera*); 5, greater wax moth (*Galleria mellonella*); 6, cabbage looper (*Trichoplusia ni*); 7, false codling moth (*Argyroplace leucotreta*); 8, oriental fruit moth (*Grapholitha molesta*); 9, butterfly (*Danaus gillippus berenice*); 10, butterfly (*Lycorea ceres ceres*); 11, fall army worm (*Laphygma frugiperda*); 12, gale-child moth (*Bryotopha similis* [B1]); 12′, gale-child moth (*Bryotopha similis* [B2]); 13, leaf roller moth (*Argyrotaenia velutinana*); 14, european corn borer (*Oatrinia nubilalis*); 15, black carpet beetle (*Attagenus megatoma*); 16, silkworm moth (*Bombyx mori*); 17, grain beetle (*Trogoderma inclusum*); 18f, American cockroach (*Periplaneta americana*) (Nakanishi, K., Still, W.C., and Persoons, C. J., *J. Am. Chem. Soc.* 101; 2493, 2495 [1979]).

Table 3.2. Alarm Pheromones

Compound	Structure	Insect Family[a]
Class U (4–7C)		
Isoamyl acetate	$CH_3CH(CH_3)CH_2CH_2OOCCH_3$	H
2-*trans*-Hexen-1-al	$CH_3(CH_2)_2HC=CHCHO$	M
Heptan-2-one	$CH_3(CH_2)_4COCH_3$	H, D
Class 2U (8–12 C)		
Nonan-2-one	$CH_3(CH_2)_6COCH_3$	H
Citronellal	$(CH_3)_2C=CH(CH_2)_2CH(CH_3)CH_2CHO$	F
Citral	$(CH_3)_2C=CH(CH_2)_2C(CH_3)=CHCHO$	F, H
Limonene **(3)**		I, M
Class 3U (13–18 C)		
Tridecane	$CH_3(CH_2)_{11}CH_3$	F
Tridecan-2-one	$CH_3(CH_2)_9COCH_3$	F
Pentadecane	$CH_3(CH_2)_{13}CH_3$	F

Source: Adapted from reference 280.
Note: In the structures, CH=CH signifies a *cis* double bond and HC=CH a *trans* double bond.
[a] The alarm pheromones have been detected in the following insect families: F, *Formicinae*; D, *Dolichoderinae*; H, *Hymenoptera*; I, *isoptera*; M, *Myrmicinae*.

Table 3.3. Recruiting Pheromones

Compound	Structure	Insect[a]
Class U (4–6 C)		
Hexanoic acid	$CH_3(CH_2)_4COOH$	Termite
Class 2U (8–12 C)		
Geraniol	$(CH_3)_2C=CH(CH_2)_2C(CH_3)=CHCH_2OH$	Bee
cis,cis,trans-3,6,8-Dodecatrien-1-ol	$CH_3(CH_2)_2HC=CH-CH=CHCH_2CH=CHCH_2CH_2OH$	Termite[b]

Source: Adapted from reference 280.
Note: In the structures, CH=CH signifies a *cis* double bond and HC=CH a *trans* double bond.
[a] Bee = honey bee (*Apis mellifera*); termite = *Zootermopsis nevadensis*.
[b] Isolated from the fungus on which *Reticulitermes flavipes* feeds.

Table 3.4. Aggregating Pheromones

Compound	Structure	Organism[a]
Class 2U (8–12 C)		
2-Methyl-6-methylene-2,7-octadien-4-ol	$(CH_3)_2C=CCH(OH)CH_2C(=CH_2)CH=CH_2$	beetle 1
Frontalin	**4**	beetle 2
Class 3U (13–18 C)		
Methyl oleate	$CH_3(CH_2)_7CH=CH(CH_2)_7COOCH_3$	beetle 3
Cyclic AMP (3′,5′-Cyclic adenosine monophosphate)	**5**	slime mold

Source: Adapted from reference 280.
Note: In the structures, CH=CH signifies a *cis* double bond and HC=CH a *trans* double bond.
[a] The aggregating pheromones have been identified in the following organisms: 1, bark beetle (*lps confusus*); 2, bark beetle (*Dendroctonus frontalis*); 3, beetle (*Trogoderma granarium*); and slime mold (*Dictyostelium discoideum*).

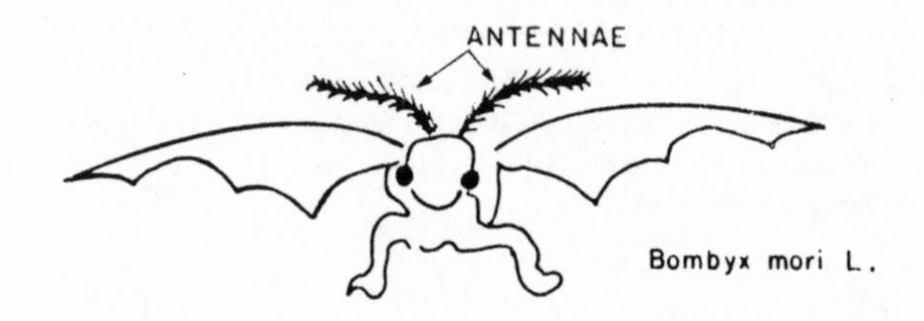

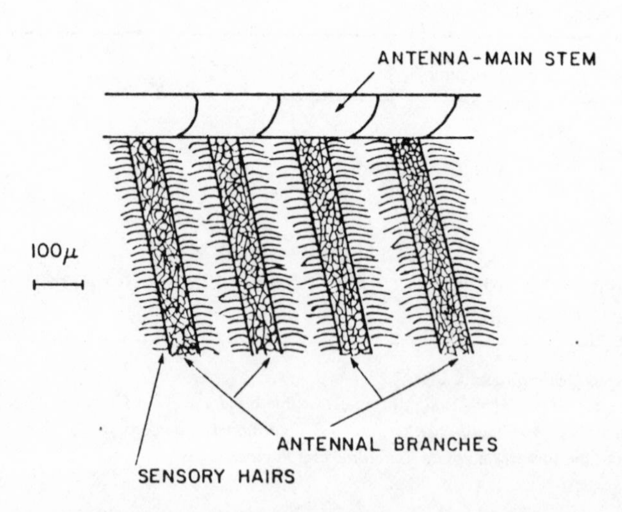

Fig. 3.2. Silkworm moth, *Bombyx mori* L. The antennae are thick, feathery structures located on the head of the moth. The antenna has branches from a main stem. Fine sensory hairs project from each branch.

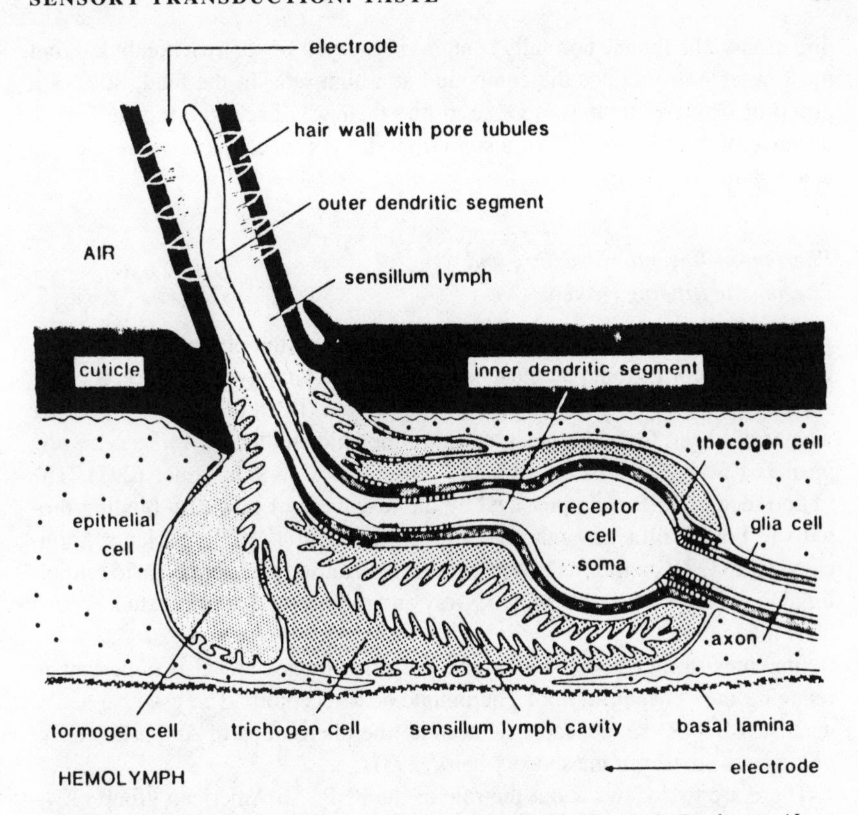

Fig. 3.3. Fine-structure of a sensory hair of the antenna of a silkworm moth, *Bombyx mori* L. The olfactory sensillum trichodeum shows one receptor neuron and three auxiliary cells. The hair is 5 μm thick and 300 μm long (reproduced, with permission, from the *Annual Review of Neuroscience*, vol. 9, copyright © 1986 by Annual Reviews Inc. [reference 295]).

each sensory hair, producing one impulse/second from the neurons. Since it is unlikely that all bombykol molecules would be delivered to the receptor sites, one may conclude that one molecule of bombykol is sufficient to stimulate one nerve impulse. The behavioral response (wing fluttering) was stimulated by 200–300 nerve impulses/second (ca. 1000 molecules/second deposited on the antenna) over and above the normal background traffic of 1600 impulses/second.

The effectiveness of the antennal detection system implies strongly that molecules that do not impinge directly upon a detector cell or site must travel along the antenna to the appropriate site.

The high sensitivity reported in these experiments corresponds to the well-documented long-distance detection of females by male moths. A female moth releasing 1 μg bombykol/second would produce a concentration of 1000 molecules $cm^{-3}sec^{-1}$ at a distance of 140 meters, assuming equal spread in all directions. The female normally contains between 0.01–0.1 μg bombykol, but must be able to produce the compound at a high rate. In the field, 40% of a group of Chinese moths were able to find their way back to a female from a distance of 4 km, and 26% of a similar group were successful when released at a distance of 11 km.

Pheromone Receptor Structure and Pheromone Binding Protein

In order to prevent overload and/or saturation of the pheromone receptor, mechanisms for its removal must be present in the vicinity of the receptor site. An abundant soluble pheromonal binding protein (PBP) of 15-kDa molecular weight has been found in the antennae of the wild silk moth, *Antheraea polyphemus* (296). The PBP is probably present as a 30-kDa dimer (297). The importance of PBP is emphasized by the discovery of olfactory binding proteins in bovine olfactory mucosa (298–301) and frog (302), the latter implicated as an OBP on the basis of its similarity in amino acid sequence to retinol-binding protein (RBP), for which X-ray and molecular dynamics studies have been made (303).

Enzymes that catalyze the removal of pheromones are also important in reducing the concentration of pheromone near receptors. The esterase active against (6E,11Z)-hexadecadienyl acetate (the pheromone of *Antheraea polyphemus*) is present in the sensory hairs (293).

There are two views about the role of the PBP. An American group (304–307) believes that the PBP acts both to sequester and to transport the pheromone to the receptor (carrier theory). A German group (295) suggests that the PBP (and the enzymes) act to lower concentration after the pheromone has acted on the receptor (removal theory).

Although both views are quite logical, two points favor the PBP as a carrier. First, the pathway for the pheromone from the tubule to the nerve receptor must involve the aqueous fluid surrounding the nerve; a binding protein would clearly solubilize the normally insoluble pheromone. Second, the PBP could enhance the specificity of the pheromonal recognition, and account for the exquisitely well-defined structural recognition by the receptor. The point proposed here may apply to other carrier proteins. The reversible association of PBP with a pheromone (ph) is shown in Eq. (3-1). The delivery of the ph to the receptor is a further equilibrium step (Eq. [3-2]). The binding of a non-pheromone (no) is expressed with Eq. (3-3). I propose that the conformation of the PBP.ph complex is different from that of the complex with the non-pheromone (PBP.no), even though the binding constants may not be very different. Genetic variants in the binding protein from *Antheraea polyphemus* may be useful in studying the binding mechanism of pheromones (308).

(3-1) PBP + ph $\rightleftharpoons$ PBP.ph
(3-2) PBP.ph + PR $\rightleftharpoons$ PBP + PR.ph
(3-3) PBP + no $\rightleftharpoons$ PBP.no
(3-4) PBP.no + PR $\rightleftharpoons$ PBP + PR.no

where PBP = pheromone binding protein; PR = pheromone receptor; ph = pheromone; no = nonpheromone.

The conformation of the PBP.ph must be appropriate for combining with the ph-receptor and delivering the ph molecule to the binding site. The PBP.no has a conformation that is much less appropriate for combining with the receptor, and thus does not deliver the nonpheromone to the binding site. (Eq. [3-4]).

There are two reasons for believing that pheromone receptors belong to the G-protein receptor family (Chaps. 4, 5). First, olfactory receptors might well be related to pheromone receptors, since the stimuli are delivered in similar ways. Second, of the two superfamilies of ligand-gated receptors, the acetylcholine receptor (AChR) superfamily and the G-protein superfamily, the AChR family may be excluded on the grounds that it requires two molecules of ACh to fully activate the receptor. The low concentrations at which pheromones are normally active would exclude a gating mechanism requiring a "high" ligand concentration.

These conclusions suggest that a careful search for the G-proteins of the hair might be worthwhile.

Environment, Pheromones, and Biological Evolution

A change in pheromones related to a change in food in a new environmental niche might be a proximate cause for formation of a reproductive isolate from an otherwise stable population. The organisms that enter the new niche may carry a variety of neutral and/or adaptive mutations. The isolate may or may not diverge substantially from the originating species, showing in general small changes. Abrupt changes ("punctuated equilibrium") are also possible within the context of modern evolutionary theory (309). In terms of the mechanism formulated previously, a change in pheromonal composition would change the effectiveness of the PBP-receptor activation, yielding a system that might then evolve to accommodate the changes in pheromone.

The formation of reproductive isolates is a central element in biological evolution (310, 311). However, the mechanisms of isolation, closely connected to the question of species number (311), are unknown. The usual proposals are restatements of the facts: (a) Accumulation of genetic differences leads to a divergence within groups. It is difficult to see how such differences might lead individuals to participate in particular breeding groups. (b) Mutations make the mutants averse to mating with the originating group. Intraspecies reproductive activity is normally mediated by pheromonal attraction and other semiochemical aids to copulation and impregnation. In the extreme, a mutation that produces aversion would be suicidal genetically, or must be produced simultaneously in many individuals within a group.

The macroscopic evolutionary problem is to connect occupancy of a new environmental niche with the formation of a reproductive isolate. The prospective breeders may carry a variety of genetic changes (in ''quasi-species''? [311]), including point mutations and chromosomal translocations, which may or may not be neutral (312, 313). A new niche is essentially a new food source. The chemical composition of the food taken by the individual organism may differ within the new niche from that in the old niche.

I now propose that the change in food composition derived from the new niche may lead to a change in the pheromones produced. The breeding groups (the reproductive isolates) will then succeed or fail within their new niche for a variety of reasons such as climate, luck, effectiveness of pheromonal responses, and genetic compatibility for reproduction. The development of races, subspecies, and species is obviously possible in this scheme; the rhythm of evolutionary development can correspond to almost any rate from low (morphological stability) to high (rapid change), influenced in part by pheromonal interactions (311).

The proposal requires that there be (a) conversion of food substances into pheromones, (b) changes in the pheromonal responsiveness of a species in a new environmental niche, and (c) a link between genetics and pheromonal communication in mating behavior.

(a) *Food conversion*. Pyrrolizidine alkaloids ingested by various species of *Lepidoptera* (314) function as a chemical defense against predation, a state often associated with aposematic features such as bright colors or characteristic patterns. Some butterflies (*Danainae* [*Euploea*]) (315, 316) and moths (*Creatonotos* and *Utetheisa*) (317–319) produce an ''aphrodisiac'' pheromone, R(−)-hydroxydanaidal (**6**) derived from the pyrrolizine alkaloid

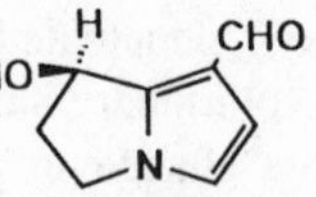

H CHO
HO
N

R(-)-hydroxydanaidal
(6)

present in the plant on which they feed. For *Creatonotos*, the plant alkaloid controls development of the male scent organ; if the plant substrate lacks alkaloids, an atrophied organ containing little or no pheromone appears. The female uses several C_{21} derivatives as attractants (320). The molecules have three *cis* double bonds (or two *cis* double bonds and an epoxide) (321) and so would fit into the ''3U'' category noted under pheromones.

Certain species of orchids in the American tropics (Mexico, Panama, Ecuador) are pollinated exclusively by euglossine bees. An extremely careful gas-chromatographic–mass-spectrometric study has shown that species-specific fragrances attract particular bee species. Since the bees do not acquire food from the orchid flowers, the fragrance compounds may be used as sexual pheromones. Studies of transformation of ^{14}C-labeled fragrance compounds within the bee are still needed to define the specific role of the fragrances within the bee (322). Nevertheless, semiochemically mediated reproductive isolates are clearly present in the orchid-bee system.

(b) *Pheromonal changes*. The fruit fly, *Dacus dorsalis*, was introduced into Taiwan (1911) earlier than into Hawaii (1944). Although both groups respond strongly to the natural feeding stimulant, methyl eugenol (3,4-dimethoxy-1-allylbenzene), the response of the Taiwanese flies is much weaker than that of the Hawaiian flies to such compounds as 2-amino- or 2-methylthio-1-methoxybenzene. Benzyl acetate, $C_6H_5CH_2OOCCH_3$, is an attractant for the male Oriental fruit fly, *Dacus dorsalis*, and the male melon fly, *Dacus cucurbitae*. Variation in the substitution on the benzene ring diminished the response of one species without affecting that of the other (323).

From a broader perspective, the responses of the South Pacific *Dacinae* to two different feeding attractants segregate substantially according to morphologically determined taxonomic divisions, 23 species being attracted to methyl eugenol and 56 species being attracted to 1-(4-hydroxyphenyl)-3-butanone. The attractants for male *Dacinae* act as oviposition stimulants for female *Dacinae*, suggesting that changes in feeding attractants may be paralleled by changes in pheromones for reproductive behavior (324, 325).

(c) *Genetic influence on mating behavior via pheromonal communication*. The genetic matching of the MHC (major histocompatibility complex [HLA, human; H-2, mouse]) is central to the success of organ transplantations (326). Thomas had an intuition that olfactory recognition might have a connection to immunological recognition (327). His colleagues discovered that each of two groups of congenic mice differing in H-2 genes preferred the other group in mating, the preference being communicated by pheromones (328). The pheromones are present in the urine (329), are readily dialyzable, and survive at least one hour heating at 100°C in water (327).

The precise pathway from the H-2 genes to the control of pheromone composition is unknown, with no obvious link between MHC function and metabolism (330–332). In addition to the H-2 locus on chromosome 17, both X and Y chromosomes add individuality to the mouse urine scent (333). Steroids are a likely choice as pheromones (329), and testosterone levels, possibly related to organ development (334), are influenced by the H-2 genes. Transplantation of congenic marrow, from mice of a different H-2 type, changes the urine odor and the mating preference. The pheromones seem to be influenced by the lymphocytes, although the lymphocytes are not the major steroidogenic tissue within the organism (335). Mice can distinguish between the urine of a parent strain and that of a mutant differing by only 3 amino acids in a cell surface glycoprotein arising from the H-2K gene. Two class-I MHC antigens in urine can be distinguished by mice (336). Early pregnancy termination (337) by exposure of the pregnant female mouse to congenic males or females with different H-2 may also be a manifestation of genetic influences on the formation of reproductive isolates via pheromones.

Molecular transformations connect food composition (or any extrinsic property of the new niche, for example, a change in light intensity that could influence hormone levels) to pheromone composition.The chemical changes might be catalyzed by a "normal" enzyme operating on new substrate or by new enzymes or enzyme systems (338). The development of reciprocal pheromonal systems (i.e., those that involve both breeding partners) would be an interesting field of study. A further interesting finding connected to evolution is the chemical mimicry practiced by the bolas spider, which emits female sex attractants of several different species, and successfully lures male moths to their doom (339).

The formation of breeding groups mediated by a change in pheromonal composition can accommodate a variety of genetic changes (mutations, movement of transposons, insertions, gene doublings, or those stemming from a postulated protein "success" monitoring system [340], etc.). Small changes in species as well as larger changes, adaptation and "exaptation" (341, 342) are possible. The influence of food composition on pheromonal composition is clearly open to experimental test.

Chemotactic Responses in Higher Vertebrates

Chemical stimuli play a role in the behavior of all higher vertebrates. In some cases, the effects are dramatic, as in the fetal resorption stimulated in pregnant mice by the odor of a strange male. The composition of the volatile metabolites in urine appears to control mating preferences among genetically defined mice. The composition is associated with the genotype of the major histocompatibility gene complex (H-2) (343, 344). Although a human pheromone to produce fluttering of the heart has not yet been found and tested (in analogy to the moth sex attractant), there are many qualitative indications that human sex pheromones exist (345). The stimuli for humans that correspond to the pheromones listed in Tables 3.1–3.4 (sex, alarm, recruiting, and aggregating) are sex, perspiration (fear), floral, and the smell of cooking food. A nonvolatile proteinaceous aphrodisiac pheromone active on the vomeronasal organ (see Chap. 4) has been found in hamster vaginal discharge (346).

The lives of vertebrates are guided in part by olfactory (smell) and gustatory (taste) cues. Scientific, cultural and economic considerations have stimulated much research into the identification, classification, and re-creation of these chemical stimuli. The molecular mechanisms by which the responses to the stimuli are created are unknown. The classic method of searching for analogous structural patterns in the stimulant molecules is still the most widely used approach to finding new stimulant molecules.

The Gustatory (Taste) System

Gustation is the art or faculty of tasting (Latin, *gustus*, taste). The structures that respond to chemical stimuli on contact with a liquid or solid may occur on the head and barbels of aquatic animals like catfish, but are mostly found in special areas like the tongue and oral cavity in man and other vertebrates. Taste sensation might include responses to texture in mechanoreceptors, temperature with thermoreceptors, as well as to chemical stimuli of olfactory and taste chemoreceptors. Only taste chemoreceptors will be treated in the present chapter.

Taste research began with the Greek philosophers Democritus and Aristotle. Democritus thought that the shapes of particles (molecules) determined their taste, the equivalences being angular = sour, spherical = sweet, hooked spheres = bitter, while Aristotle introduced the "classical" qualities—sweet, bitter, sour and salty (347). The phylogeny of taste receptors from fish to man reveals a basic similarity in bud structure, except for frogs (348). Careful analysis of taste responses at all levels between the taste cell and the cerebral neuron suggests only minimal specificity in taste receptors (349, 350). Although the discussion centers around the "classical" qualities, I have concluded that there are no specific receptors corresponding to these qualities. The molecular basis of taste reception remains to be established.

Taste Receptors

The taste receptors in man are located on the tongue, and to a lesser extent on the soft palate, the pharynx, and the larynx. Although there are chemoreceptors associated with "free" nerve endings throughout the oral cavity, the receptor cells are mostly grouped together into small structures called taste "buds." The buds are found on papillae (finger-like projections), which vary in appearance according to their location on the tongue (Fig. 3.4). There are four types of papillae on mammalian tongues, of which three—the fungiform, the foliate, and circumvallate—carry taste buds (351). A single taste bud is found at the top of a rat fungiform papilla. Single human taste buds respond to all four "basic" taste stimuli (salt, sweet, bitter, and sour) (349, 350, 352).

There are 5–15 receptor cells out of a total of 30–80 cells within each taste bud. The cells have a relatively short lifetime, and are replaced within 10–15 days. Adult humans have about 9000 taste buds, and the number decreases with increasing age.

There are four cell types in a taste bud, of which only one (type III) appears to be a receptor cell (351). The cells are separated from the medium that carries the chemical stimuli by a "dense substance," through which microvilli

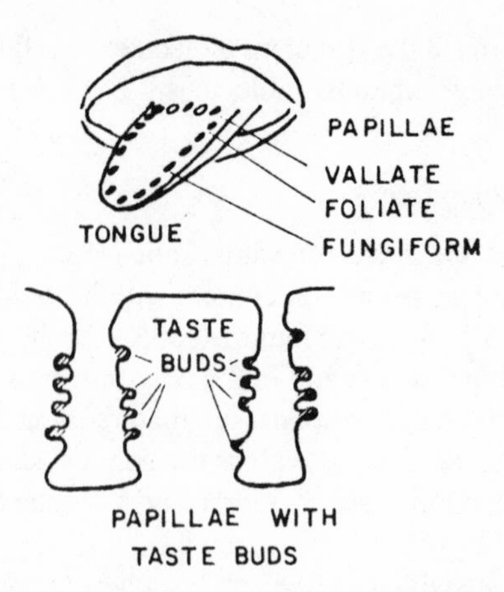

Fig. 3.4. The location of taste buds on a circumvallate papilla of the tongue. The upper diagram indicates the distribution of the various types of taste buds on a tongue.

project into a pore. The receptor cells are secondary sensory cells (cells without an efferent axon) and make synaptic contacts with first-order sensory neurons. Innervation of taste receptor cells is supplied by three distinct sensory ganglia: the geniculate ganglion of the facial nerve, the petrous ganglion of the glossopharyngeal nerve, and the nodose ganglion of the vagus nerve. The nerves are connected to different parts of the oral cavity. Between 10 and 50 taste cells are innervated by a single taste nerve fiber (353, 354).

Identification of Gustatory Cell

To identify a gustatory cell, both electrophysiological methods and electron microscopy are used. Penetration of a taste bud cell by a microelectrode shows a typical shift to a negative resting potential. Entry into a taste receptor cell (-50 mV for water-adapted cells) is affirmed by the following criteria: (a) change in the resting potential on application of a chemical stimulus; (b) change in cell membrane resistance during chemical stimulation; (c) return to initial state on removal of the chemical stimulus; (d) disappearance of potential difference on withdrawal of the electrode. Only a limited number of taste bud cells respond in a way that marks them as gustatory receptor cells.

Of the four cells types in the bud, only a type-III cell can be identified in an electron micrograph as a gustatory cell through accumulation of labeled catecholamine precursors and the presence of large dense-cored vesicles in synaptic contact regions (351). Although a single cell can be stained iontophoretically with dye through the microelectrode, unequivocal matching of the behavioral and anatomical approaches has not yet been possible. The detailed arrangement of the cells in a taste bud is shown in Figure 3.5.

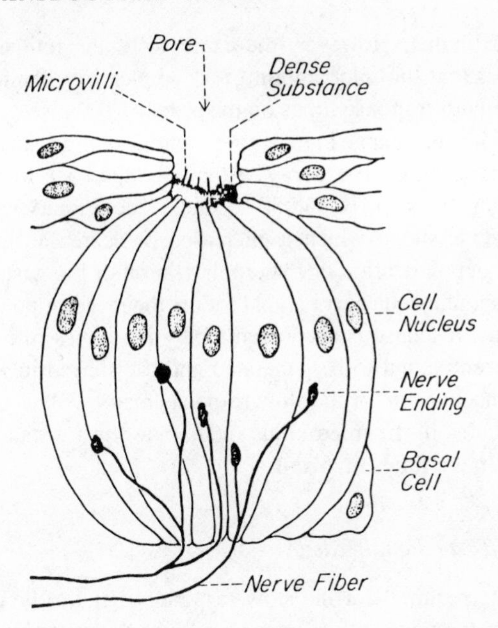

Fig. 3.5. Longitudinal section through a taste bud on a circumvallate papilla. Microvilli within an impermeable dense substance interact with stimuli at the entry to the group of taste receptor cells. The signal is transferred within the receptor cells to the neurons, which produce the signals for the nervous system.

Nature of Cell Response

The potential within a gustatory cell undergoes a change in level in response to a pulsed stimulus applied to the taste bud, its course characterized by a plot of the value versus time. The responses obtained to the four major taste stimuli, exemplified by NaCl, HCl, sucrose, and quinine hydrochloride, are typical of the experiments (351, 355). The responses depend on the initial state of the taste bud, whether adapted to water or to 41.4 mM NaCl (saliva salt level). Responses have been classified into three types: depolarizing; hyperpolarizing, then depolarizing; and hyperpolarizing.

Time Course of Gustatory Cell Response to Taste Stimuli

There are fast and slow phases in the response of a gustatory cell to a taste stimulus. The times involved in the fast response have been estimated, using stimulus solutions, by careful measurements of rat licking and human sipping times (monitored in the latter by electromyograms). The times, corrected for a 25-msec latency in the gustatory nerve response, vary between 150 msec for NaCl and 500 msec for sucrose (356, 357). One plausible possibility is that an electrical message is transmitted from the receptor region to the gustatory neuron. The permeability of one or more types of ion channel must then be affected by the stimulus. However, the example of the retinal rod response (Chap. 5) shows that the intervention of a complex biochemical pathway is still compatible with response times on the order of 100 msec.

The rise time for the change in receptor potential after a taste stimulus varies between 5 to 50 seconds. The decay of the potential back to the initial state takes even longer, between 10 and 80 seconds. The primary nerve cell transmits a spontaneous signal which is increased or decreased in frequency in response to the depolarization or hyperpolarization of the gustatory cell. The changes in the spontaneous nerve signal follow the receptor potential changes. Although the fast response is accounted for by ion fluxes, the slow response requires a different signal to be generated and transferred in response to the stimulus. The mechanism for the slow response may involve some lateral diffusion of molecules in the microvillus membrane for a distance of not more than 5 μm, given the times involved.

Nature of Slow Taste Signal: Tastin

The long time required for the slow response signal to be developed and transferred from the microvilli receptor region to the primary nerve cell suggests that a carrier is involved. The length of the taste bud is 50–80 μm, with a type-III cell occupying 44–70 μm. Innervation may occur over the lower 2/3 of the cell, from which the average distance from microvilli to gustatory-cell–neuron synapse may be estimated as 30 μm. Taking 10 seconds as the typical rise time for depolarization, a signal diffusion coefficient may be derived from the equation $D = (\Delta x)^2/2t$, as 4.5×10^{-7}cm²/sec. Proteins, such as lysozyme (Mr 14.5 kDa, 10.6×10^{-7}cm²/sec) and serum albumin (Mr 65 kDa, 6.7×10^{-7}cm²/sec) (358), have diffusion constants in the required range. The signal is almost certainly a protein, most likely a G-protein, for which I suggest the name *tastin*. A less likely possibility for explaining the slow potential change would be lateral diffusion of a very large protein complex in the microvillus membrane.

"Slow" changes in resting potential are probably due to the effect of the signal protein, tastin, on the permeability of channels. A plausible mechanism could involve phosphorylation of the channels, which might lengthen the open time and increase depolarization. If tastin were a channel phosphorylating protein, then loading this protein with a phosphoryl group would increase channel phosphorylation and thus depolarization. Tastin might be a protein analogous to transducin (Chap. 5) or the olfactory G-protein (Chap. 4).

One may suggest that the stimuli combine with receptors which affect (a) the operation of ion channels and (b) the availability of protein like tastin. A change in the permeability of ionic channels occurs in response to the changed level of tastin. A similar mechanism is well worked out for the case of the retinal rod cell in which light produces, through several biochemical cycles, an amplified (negative) chemical signal which interferes with a spontaneous current in the cell (Chap. 5).

The way in which the taste receptors cause a change in the level of active signal protein is unknown and requires, for the moment, an extensive empirical approach. The relatively low complexing constants (high concentrations of taste stimulus molecules are required for detection) and the modest specificity make ineffective the usual techniques of identifying receptors. Since the structural requirements for a stimulant molecule are quite loose, it is difficult to develop a reliable labeling procedure.

Classification of Taste Stimuli

Food contains significant quantities of substances like sugars, salts, and acids which stimulate taste receptors (359). The subjective nature of reports on taste sensations, being dependent on the ability of an individual to describe what he or she feels, makes objective classification of taste stimuli difficult. Drugs and disease can alter taste perceptions (360). Cooperative effects between different substances, time-dependent responses, and saturation of particular responses complicate further the problem of classification. In addition, there are genetic variations in taste responses.

Although some have held the viewpoint that there are specific cells for specific tastes, the neurophysiological evidence now suggests that taste sensations are neural composites of the responses of different kinds of receptors (351, 361, 362).

Confusion about exactly what is meant by a taste category (363, 364) is exacerbated by the lack of a molecular connection, i.e., proteins that function as the purported taste receptors. For psychophysical and neurophysiological studies, an empirical classification is a necessity. The four "classic," generally accepted categories are salt, sweet, sour, and bitter. To these can be added a fifth category, *umami* (Japanese, deliciousness) (365). Duplicates of three of these classes with different specificities are reported to occur in different regions of the tongue, and there also may be receptors for tastes described as "pleasant," "astringent," "pungent," and "metallic." Since the "duplicate" receptors may function in a way similar to that of the "original," and since the other claimed tastes may involve inputs from different types of receptors or are too subjective for detailed study at this time, our discussion will utilize only the five categories noted above. The possibility that there may be different receptor subtypes (i.e., more than one kind of sweet receptor [362]) will not be taken into account.

The reconstitution of snow crab flavor (a prized food in Japan) was achieved with a mixture of 45 compounds. The primary attributes of the crab taste could be produced by a limited group of compounds including glycine, glutamic acid, arginine, adenosine monophosphate (AMP), cytosine monophosphate

(CMP), and sodium chloride, the flavor being further enhanced by alanine, glycine betaine, and potassium phosphate. Removal of chloride ion made the synthetic mixture almost tasteless; removal of sodium ion decreased greatly the umami and sweet tastes (365). Another example of the significance of trace components is shown by the effort to add the carotenoid component, astaxanthin, to the flesh of tank-raised shrimp, in order to obtain the characteristic pink color and characteristic flavor. The pigment gene has been engineered into the yeast fed to the shrimp (366).

One Kind of Taste Receptor Explains Four Taste Categories

Before considering the phenomenology observed for various taste stimuli, I now point out that the four apparent taste categories can be explained by a single receptor model. (Appropriate neurophysiological data for the umami flavor are not available.) Taste detection involves fast and slow neural signals arising from weak combinations of stimuli with a receptor. I propose that the receptor is directly connected to an ion channel and contains, on the interior of the membrane, a tastin binding site. The receptor can respond in four ways (at least) to a stimulus: (a) ion flow leading to depolarization (salt taste, NaCl); (b) partial or full blockage of the ion channel (bitter taste, as from quinine hydrochloride); (c) interference with ion channel through protonation (sour taste, as for HCl); and (d) combination with a binding site of modest specificity (the "sugar binding site") in the receptor which undergoes a conformational change so as to block ion flow. The sugar-loaded receptor might block the channel directly or via a set of biochemical reactions. The intracellular signals are of two types, fast and slow, the latter involving release or nonrelease of phosphorylated tastin.

Salt Taste

The taste of salts depends upon both the cation and the anion as well as the concentration. Salts are far from uniform in taste and have been divided into groups by taste. The taste groups appear to be related to sums of the ionic radii of the salt components: **Salty** (small or medium sized cation, small or medium sized anion), **Salty** + **Bitter** (medium sized cation, medium or large sized anion), and **Bitter** (medium or large sized cation, medium or large sized anion). The **S** salts include NaCl, KCl, NH_4Cl, LiCl, RbCl, NaBr, NH_4Br, LiBr, NaI, and LiI. Some **S** + **B** salts are KBr and NH_4I. **B** salts are CsCl, RbBr, CsBr, KI, RbI, and CsI. The fact that bitter-tasting salts have large ions may mean that these ions block the ion channel, as suggested for quinine hydrochloride.

A salt like NaCl can be detected at concentrations as low as 3×10^{-3} M, and metal ions such as Fe^{+3} and Mn^{+2} can be detected at concentrations as low as 2×10^{-5}M. Wild animals (kangaroos and rats) in the Snowy Mountains of Australia will choose NaCl over KCl, $MgCl_2$, or $CaCl_2$, indicating that salt detection may be sufficiently precise to induce behavior conducive to survival. The well-documented search of herbivores for NaCl sources (''salt licks'') arises from the sodium-deficient nature of a plant diet. Human preoccupation with salt is a consequence of a largely vegetarian diet. Roman soldiers received a portion of their pay in salt (Latin, *salarium*, of salt), and the salt trade and salt taxes were for a long time significant factors in history.

Sweet Taste

The sweet taste is important in the perception of food and therefore of prime economic and social importance. Sugar in the form of sucrose is an important article of commerce, more than 100,000,000 tons being produced per year. The response to sweet foods is a common chemosensory element in evolution.

Both natural and synthetic substances are used as sweeteners. The social and medical conditions suggesting the use of non-nutritive sweeteners include obesity, heart disease, dental caries, hyperlipemia, and diabetes. Since sweetness enhances the attractiveness and palatability of foods, control of calorie intake can be accomplished more pleasantly with a nonnutritive sweetener than by strict dietary supervision.

A sweet-taste sensation is produced by a large variety of substances. The molar concentrations required to be equivalent to a sucrose solution may differ by as much as a factor of 10^5 (367, 368). In order to understand the molecular basis for the sweet response, one should obtain sequence and structural information on the taste receptor. Although one valuable approach is to label the receptor, one lacks effective means for such labeling and so must approach the problem of receptor identification through empirical classification of the myriads of sweet-tasting compounds. The hope is that inferences can be made about the space occupied by ''sucrophoric'' structures and groups (''sucrophoric'' means sweet-taste producing, by analogy to ''pharmacophoric,'' medicinally active). Although inferences about the structure of the receptor and the dynamics of its interaction with taste stimuli are premature, molecular conformations for sweet- and bitter-tasting molecules have been suggested. These results and other considerations lead to a mechanism of taste. An empirical approach has an intrinsic utility for research on sweet-tasting compounds, by providing models for further synthetic targets. However, an amino-acid sequence for the receptor is needed, along with information about molecular shape, in order to define the mechanism of action on a molecular level.

The classes of sweet-tasting compounds are listed in Table 3.5. Representative examples of sweet-tasting compounds illustrate their diverse character, among which are cyclic sulfonamides (**7**, **8**, **10**), ureas (**9**), oximes (**11**), dipeptides (**12**, **25**), amino acids (**13**), chalcones (**14** and **15**), chlorides derived from sugars (**16**), anilides (**17**), nitroarylamines (**18**), sugars (**19**–**22**), polyols like $HOCH_2CH(OH)CH_2OH$ (**23**) and $HOCH_2CH_2OH$ (**24**), terpene glyco-

Table 3.5. Classes of Sweet-tasting Compounds

1. Saccharides: a. Monosaccharides; b. Disaccharides
2. Polyols
3. Amino acids
4. Other natural products
5. Synthetic substances: a. Sulfone derivatives; b. Peptides; c. Others

Table 3.6. Taste Intensity of Sweet Substances

Compound	Relative Sweetness (Sucrose = 1)
Saccharin **(7)**	500[a]
Cyclamate **(8)**	30[a]
Dulcin **(9)**	100
OTD[b] **(10)** (Sunnett)	130[a]
SRI Oxime V **(11)**	450
L-Aspartyl-L-phenylalanyl methyl ester (Aspartame) **(12)**	100–200
D-6-Chlorotryptophan **(13)**	1,000
Neohesperidin[c] dihydrochalcone **(14)**	1,500
4′,1′,4,6′-Tetrachloride of galactosucrose[d] **(16)**	2,200
Trifluoroacetyl-L-glutamyl-N-cyanoanilide **(17)**	3,000
2-Amino-4-nitro-1-propyloxybenzene **(18)**	4,000
Sucrose[e] **(19)**	1
Fructose[f] **(20)**	1.7
Glucose[g] **(21)**	0.7
Lactose[h] **(22)**	0.2
Glycerol **(23)**	1
1,2-Ethanediol **(24)**	1.3
L-Aspartyl-DL-aminomalonic acid monomethyl monofenchyl diester **(25)**	60,000
Monellin (Protein; MW 10,500)	90,000
Thaumatin (Protein; MW 21,000)	100,000
Stevioside **(26)**	300[a]
Rebaudioside A **(27)**	450[a]

[a] Bitter and metallic aftertaste.
[b] 6-Methyl-1,2,3-oxathiazin-4(3H)-2,2-dioxide.
[c] Neo = 2-O-α-L-Rhamnosyl-β-D-glucosyl-**(15)**.
[d] α-D-Galactopyranosyl-β-D-fructofuranosidyl 4,1′,4′,6′-tetrachloride (368).
[e] α-D-Glucopyranosyl-β-D-fructofuranoside.
[f] β-D-Fructofuranose.
[g] α-D-Glucopyranose.
[h] 4-O-(β-D-Galactopyranosyl)-D-glucopyranose.

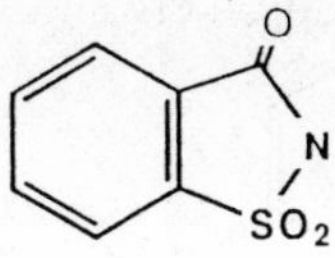

O
NH
SO_2
7

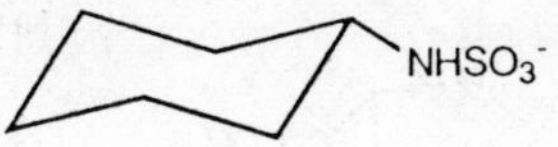

$NHSO_3^-$
8

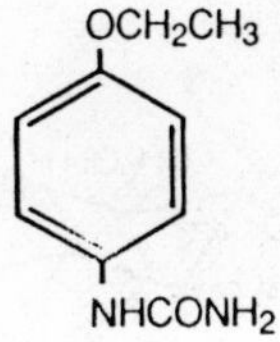

OCH_2CH_3
$NHCONH_2$
9

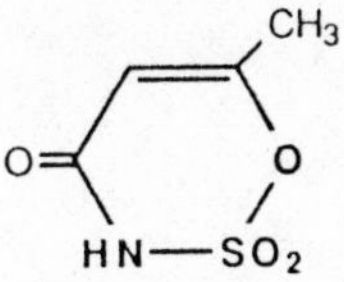

CH_3
O
O
HN—SO_2
10

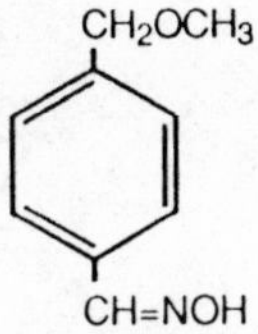

CH_2OCH_3
CH=NOH
11

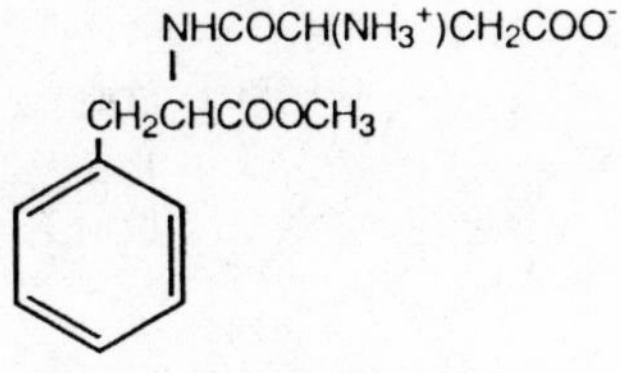

$NHCOCH(NH_3^+)CH_2COO^-$
$CH_2CHCOOCH_3$
12

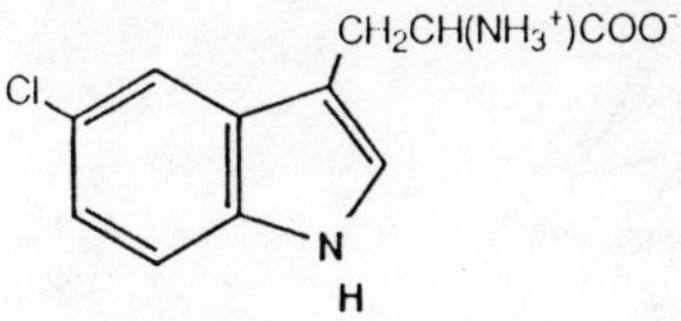

$CH_2CH(NH_3^+)COO^-$
Cl
N
H
13

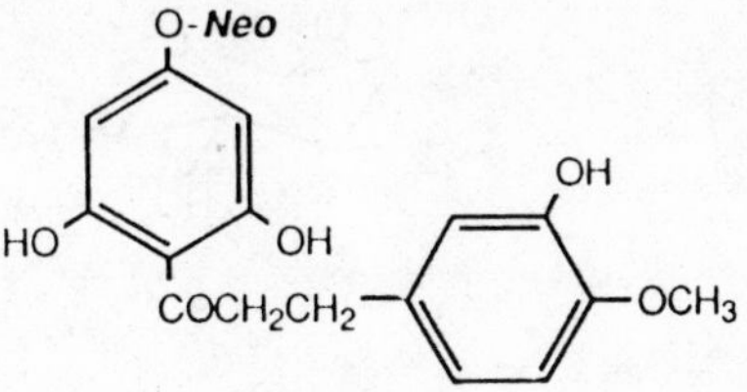

O-*Neo*
OH
HO
OH
$COCH_2CH_2$
OCH_3
14

Neo = 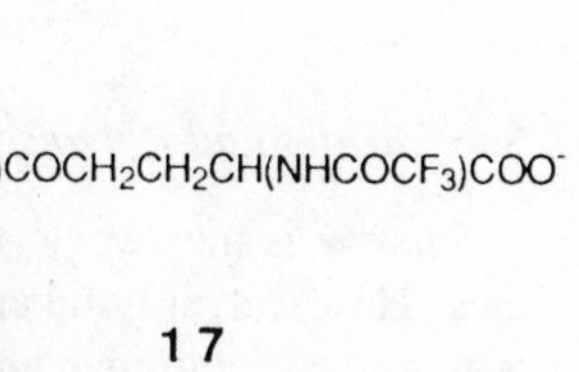
15

CH_2OH
O
HO
HO
HO
O
O
O
CH_3
HO
OH

Cl
CH_2Cl
O
CH_2Cl
O
HO
HO
O
CH_2Cl
OH
OH
16

$N(CN)COCH_2CH_2CH(NHCOCF_3)COO^-$
17

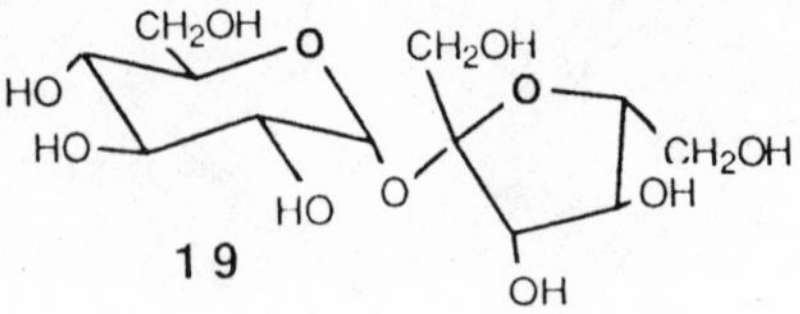

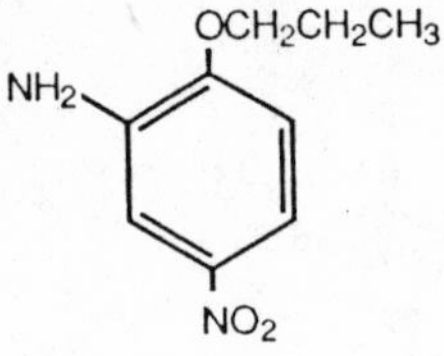

$OCH_2CH_2CH_3$ NH_2 NO_2

18

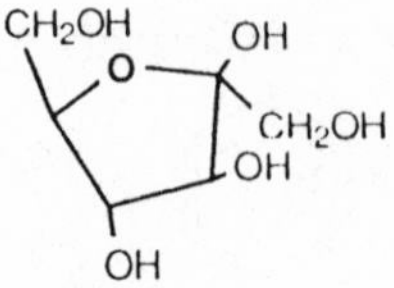

CH_2OH O CH_2OH HO HO O CH_2OH HO OH OH

19

CH_2OH O OH CH_2OH OH OH

20

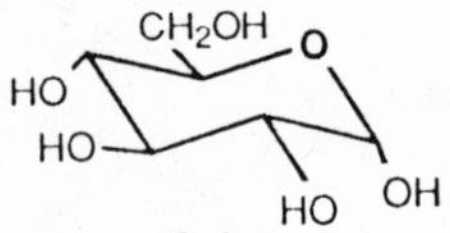

CH_2OH O HO HO HO OH

21

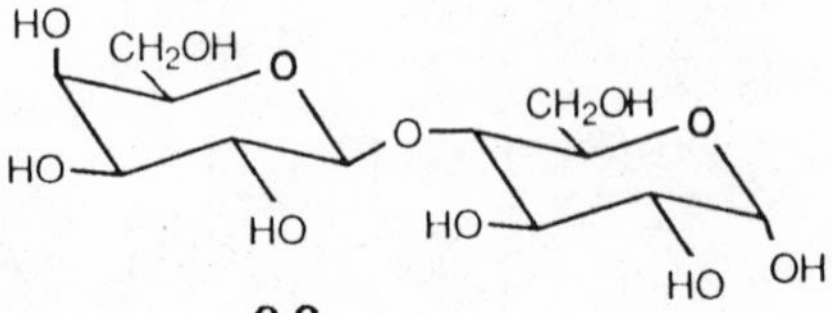

HO CH_2OH O HO O CH_2OH O HO HO HO OH

22

$OOCCH_2CH(NH_3^+)CONHCH(COOCH_3)COO$ CH_3 CH_3 CH_3

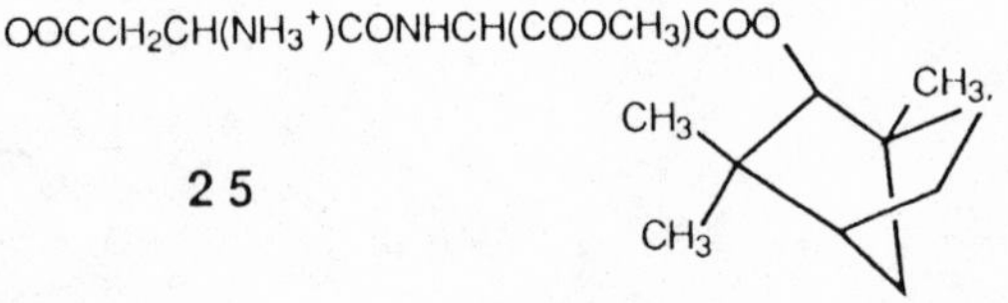

25

sides (**26**, **27**), and proteins such as monellin and thaumatin. The sweetness of these compounds is compared to that of sucrose in Table 3.6 (369).

Saccharides and Polyols

Sucrose is the sweetest disaccharide for humans and for most animal species. However, a number of compounds like Aspartame or thaumatin (349) that taste sweet to man are not attractive to many primates. Fructose, a monosaccharide, is almost twice as sweet as sucrose. Monosaccharides and derivatives with fewer hydroxyl groups (e.g., methyl glycosides or deoxysugars) are usually less sweet than sucrose. Some significant exceptions are a number of compounds in which chlorine has replaced hydroxyl groups, these being considerably sweeter than sucrose. The tetrachlorogalactosucrose (four Cl in place of OH, Table 3.6) is 2200 times sweeter than sucrose. There is considerable latitude in the arrangement of the hydroxyl groups required for stimulating sweet taste. An increase in hydrophobicity (C-Cl groups are compatible with nonpolar groups) increases sweetness. Polyols (1,2-dihydroxyethane, glycerol) are less sweet than sucrose. Five-, six-, and seven-carbon polyols (D-ribitol, D-sorbitol, *myo*-inositol, perseitol) are similar to sucrose in sweetness.

Amino Acids

D-amino acids are sweet-tasting. D-6-chlorotryptophan is an extreme example, being 1000 times sweeter than sucrose, while glycine is somewhat sweet. The rat is similar to man in amino acid-taste responses, but flies respond differently to sweet materials than do mammals.

Other Natural Products

The mystique that "natural" products are normally safe for human consumption rests upon a philosophical outlook derived from Rousseau and Thoreau, and human tradition based on successful use over many generations. Honey, for example, was used in the Neolithic period (10,000 years ago), as shown in cave paintings (376). A volatile, sweet sesquiterpene, hernandulcin [3-methyl-6-(1′-hydroxy-7′-methyl-4-hexen-1-yl)-cyclohex-2-en-1-one], 10^3 times sweeter than sucrose, was isolated from a plant described in 1576 as sweet (Nahuatl name: *Tzonpelic xihuitl*; Latin, *Lippia dulcis* Trev. [*Verbenaceae*]) (370). However, considering only the range of toxins available in nature, from curare to cholera toxin, one may dispute the idea that Nature is entirely benign.

Stevioside and Its Aglycone, Steviol

The natives of northeastern Paraguay had discovered that the leaves of the shrub *Stevia rebaudiana* Bertoni (Compositae) were useful as a sweetener. The active compound is stevioside (**26**, R_1 = β-D-glucopyranosyl, R_2 = 2-O-β-D-glucopyranosyl-β-D-glucopyranosyl [β-D-sophorosyl], R_3 = H), a diterpenoid derivative which is about 300 times as sweet as sucrose. Intensive cultivation of the plant occurs in China, Korea, and Japan. The product is approved for human use in Japan, and is used on the scale of many tons per year. The compound may also be active as a hypoglycemic agent for individuals with high blood sugar and is sold for this purpose in Brazil.

Although stable under the usual conditions of food processing, stevioside is converted to steviol (**26**, $R_1 = R_2 = R_3$ = H) by enzymes found in rat intestinal microflora. Steviol has been found to be converted to a potent mutagen for *S.typhimurium* by cytochrome P-450. Other toxic effects of steviol have been reported. A stevioside derivative less sensitive to metabolic breakdown

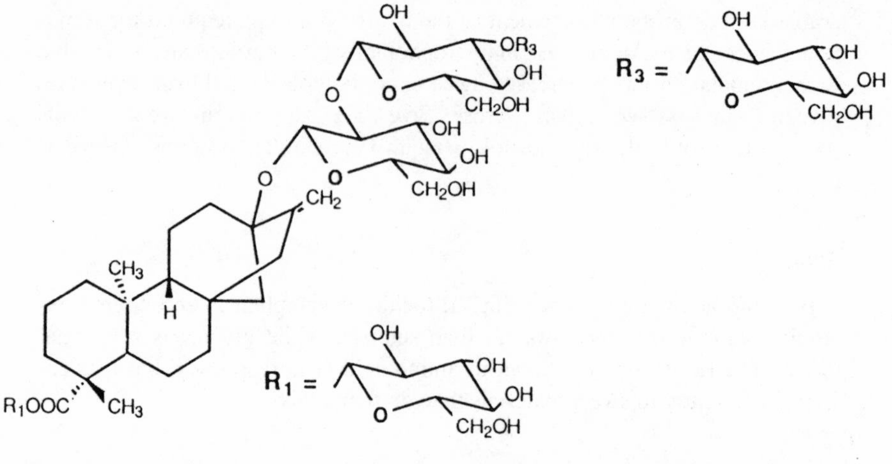

26, R_3 = H

Stevioside (**26**)

27, R_3 = R_3

Rebaudioside A (**27**)

to steviol was sought. Such a compound might be less toxic, and have less of the bitter aftertaste characteristic of stevioside. A somewhat more hydrophilic derivative, a sulfopropyl ester (SPS) (**26**, R_1 = $CH_2CH_2CH_2SO_3^-Na^+$; R_2 = 2-O-β-D-glucopyranosyl-β-D-glucopyranosyl [β-D-sophorosyl] R_3 = H), was found to have excellent sweetness characteristics (about equivalent to Aspartame) (371–373). Rebaudioside A (**27**, R_1 = β-D-glucopyranosyl, R_2 = 2-O-β-D-glucopyranosyl-β-D-glucopyranosyl [β-D-sophorosyl], R_3 = β-D-glucopyranosyl) has a less bitter aftertaste but is only present as a minor component (3%) in the plant product (373).

The hydrophilic-hydrophobic balance in the molecule affects the relative contributions of sweetness and bitterness to the overall taste. The change in the sweetness-bitterness ratio could arise from a modest change in the nature of binding to the receptor.

Synthetic Sweeteners

In 1879, the sweet taste of saccharin was noted by the chemist who had made it. The commercial potential of a synthetic sugar substitute was recognized and myriads of compounds prepared and tasted (374). Very few compounds have gained wide distribution and commercial success. One such material, cyclamate, was later withdrawn on the grounds that a metabolic product, cyclohexylamine, was said to be carcinogenic. Some of the biological and physical constraints on molecular properties of sweet substances have been reviewed in a description of the discovery of the oxathiazinone dioxides (375). These include, quite naturally, lack of both short- and long-term toxic effects like teratogenicity (effects on embryo development), mutagenicity (changes in genetic messages), carcinogenicity (cancer-causing effects), and effects on development as well as stability to the pH and temperature ranges used in the processing and storage of food.

Peptides and Proteins

The sweet taste of the most widely used synthetic sweetener, L-aspartyl-L-phenylalanyl methyl ester (Aspartame), was found by accident. Much controversy arose about its potential effect on individuals who lack the hydroxylase enzyme required for conversion of phenylalanine to tyrosine (the genetic defect is present in phenylketonuria). Formation of D-amino-acid residues through racemization at 100 °C ($t_{1/2}$ 13 and 23 hours for asp and phe residues at pH 6.8 [376]) can yield a bitter-tasting diastereomer (see below). The success of Aspartame has engendered intensive research on di-, tri-, and tetrapeptides. Peptides have well-understood structures, are easily accessible through synthesis, and have conformations that may be probed by spectroscopic techniques with relative ease. These factors have encouraged the use of polypeptides in studies designed to elucidate the molecular requirements for the sweetness taste receptor.

A derivative of aminomalonic acid, L-aspartyl-DL-aminomalonic acid monomethyl monofenchyl diester (**25**), is worth special mention as the sweetest dipeptide yet discovered.

Many peptides derived from 1,1-diaminoethane (so-called reverse or "retro-inverso" peptides) are sweet. Reverse peptides may not be metabolized easily and might be non-nutritive. Interesting examples are N-L-aspartyl-N′-(2,2,5,5-tetramethylcyclopentanyl)carbonyl-(R and S)-1,1-diaminoethanes (**A** + **28**). The R-isomer is 800–1000 times as sweet as sucrose

$^{-}OOCCH_2CH(NH_3^+)CONHCH(CH_3)NHCOR'$ (A)

$^{-}OOCCH_2CH(NH_3^+)CONHCH(CH_3)CONHR'$ (B)

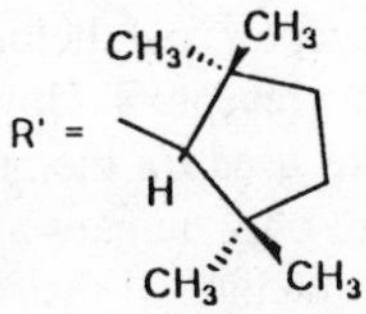

R' = [2,2,5,5-tetramethylcyclopentyl: CH_3, CH_3, H, CH_3, CH_3]

28

$^{-}OOCCH_2CH(NH_3^+)CONHCCOOCH_3$ with $(CH_2)_n$ ring

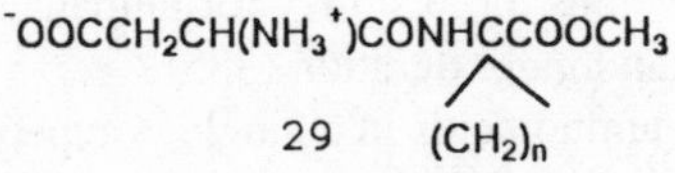

29

Cycloalkane (n + 1)	Taste
cyclopropane	sweet
cyclobutane	sweet
cyclopentane	sweet
cyclohexane	bitter
cycloheptane	very bitter
cyclooctane	tasteless

while the S-isomer is 600–800 times as sweet (377). The diastereomer of Aspartame, L-aspartyl-D-phenylalanine methyl ester, is bitter. A pair of derivatives topologically similar to **A** are N-L-aspartyl-(L or D)-alanyl-2,2,5,5-tetramethylcyclopentanylamide (**B** + **28**) (378, 379). The L,R-, L,S- (**A**), and L,D- (**B**) peptides are sweet, while the L,L- (**B**) derivative is bitter. Energy minimizations for evaluating the peptide conformations suggest that the sweet peptides are L-shaped, resembling Aspartame in the crystal and solution, whereas the bitter peptide is more compact.

An interesting series of L-aspartyl-α-aminocycloalkanecarboxylic acid methyl esters (**29**) range in flavor from sweet through bitter to tasteless, as noted above (380, 381). The changes within the series suggest that compounds bound to a particular site on the sweetness receptor can elicit different taste sensations in a manner dependent on the details of the binding. The geometric arrangements in the various stereoisomers of the reverse peptides have been examined to elucidate the factors that favor bitter or sweet taste (378).

Sweet-tasting Proteins

Two intensely sweet proteins have been isolated from plants well known to the natives of West Africa. Monellin is isolated from the fruit of the serendipity berry bush (*Dioscoreophyllum cumminsii*) (Ibo, *nmimimi*, *nwambe*; Yoruba, *ito-igbin*, *ayun-ita*). Thaumatin (382) is isolated from the berry (''miraculous fruit or berry,'' names also used for the berry from which the taste-modifying protein miraculin is obtained) of an unrelated bush (*Thaumatococcus daniellii*) (Yoruba, *anwuram-asie*, *owuramsie*).

Monellin is a protein containing 94 amino acids in two nonidentical polypeptide chains which associate in a noncovalent fashion. Monellin can be methylated (reaction with HCHO followed by reduction with $NaBH_4$ or $NaBT_4$) and retains its sweetness after this limited chemical change. Radioactively labeled monellin has been used in binding studies for the ''sweetness'' receptor, paralleling the technique so successfully applied to hormone and opiate receptors. However, nonspecific binding is too prevalent under the conditions used for the presumed receptor protein to be identified (383). Furthermore, only humans and Old World monkeys are responsive to the sweet-tasting proteins, so that the most convenient experimental material, bovine tongue, cannot be used to study the binding of monellin or thaumatin. Thaumatin by itself is not detected but enhances the taste of sucrose for the rat (384). The sweet taste of the proteins develops more slowly for humans than that of sucrose but is of higher intensity and longer duration (385).

Thaumatin is a protein containing 207 amino acids in a single polypeptide chain with 8 disulfide bridges. The sweet taste of thaumatin disappears after acetylation of four lysine ε-amino groups but reductive methylation (HCHO, $NaBH_4$) of six lysines does not change the taste. Reduction of the disulfide bonds with cysteine leads to loss of sweet taste. Reaction of the guanidino groups of arginine with 1,2-cyclohexanedione has almost no influence on the intensity of the sweet taste. Radioactively labeled methylated thaumatin also gives rise to a high degree of nonspecific binding on homogenates of monkey tongue.

Only a limited group of exactly homologous regions may be found by a comparison of the amino-acid sequences of the two proteins monellin and thaumatin, as shown in Table 3.7. Nevertheless, antibodies raised against thaumatin cross-react with monellin (386, 387).

Table 3.7. Common Sequences for Sweet-tasting Proteins, Thaumatin and Monellin

Sequence	Thaumatin	Monellin
I	92 leu-asn-*glu-tyr-gly*-lys	26B ile-gly-*glu-tyr-gly*-arg
II	99 tyr-*ile-asp-ile*-ser	5B ile-*ile-asp-ile*-gly
III	100 ile-*asp-ile-ser*-asn	21A ala-*asp-ile-ser*-glu
IV	117 thr-*thr-arg-gly*-cys	28A lys-*thr-arg-gly*-arg
V	127 ala-*ala-asp-ile*-val	20A arg-*ala-asp-ile*-ser

Source: Adapted from reference 382.
Note: Monellin is composed of two noncovalently associated chains (A and B).

Crystal Structure of Thaumatin

The three-dimensional structure of thaumatin I has been determined to 3.1-Å resolution (Fig. 3.6). Thaumatins I and II differ in 5 amino acids (46, 63, 67, 76 and 113) out of 207 (388). The structure consists of a flattened β-barrel and some disulfide-stabilized loops. The latter resemble those found in snake neurotoxins; inquiries among zoologists familiar with neurotoxins did not turn up any anecdotal information concerning the flavor of such toxins. There was an understandable reluctance to test the suggestion that these might be sweet.

A plausible biological function of thaumatin as an inhibitor produced by the plant against enzymes of predators has been suggested on the basis of the high homology (52%) of the amino-acid sequence to the sequence of the maize protein inhibitor (MPI) of the α-amylase of the beetle *Tribolium castaneum*. The MPI in turn is highly homologous (57%) to the PR-protein produced in tobacco after infection by tobacco mosaic virus. The sweet taste of the thau-

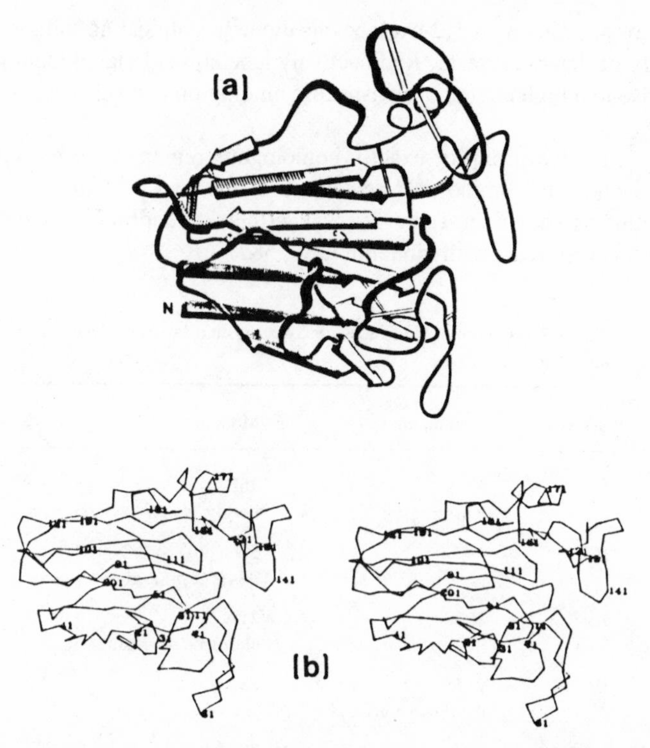

Fig. 3.6. X-ray crystallographic structure of thaumatin I. (a) The backbone structure shows a flattened β-barrel with disulfide-containing loops projecting from the side (b) Stereo drawing of the backbone structure of thaumatin I (adapted, with permission, from reference 388).

matin may be an evolutionary accident, since MPI does not have an obvious sweet taste (389, 390).

Crystal Structure of Monellin

A crystal structure of monellin (3Å resolution) reveals a five-stranded antiparallel β-sheet with three strands from the 44-amino-acid A-chain and two strands from the 50-amino-acid B-chain. An α-helix in the B-chain runs perpendicular to the β-sheet. Although monellin and thaumatin are immunologically cross-reactive, the amino-acid sequences are not homologous and the three-dimensional structures are substantially different. However, there are at least five pairs of homologous tripeptide sequences which may account for common antigenic sites, assuming that antigenic determinants (epitopes) are contiguous (Table 3.7). The structures of monellin and thaumatin are compared in Figure 3.7. The locations of the homologous tripeptides are marked in the figure (391, 392).

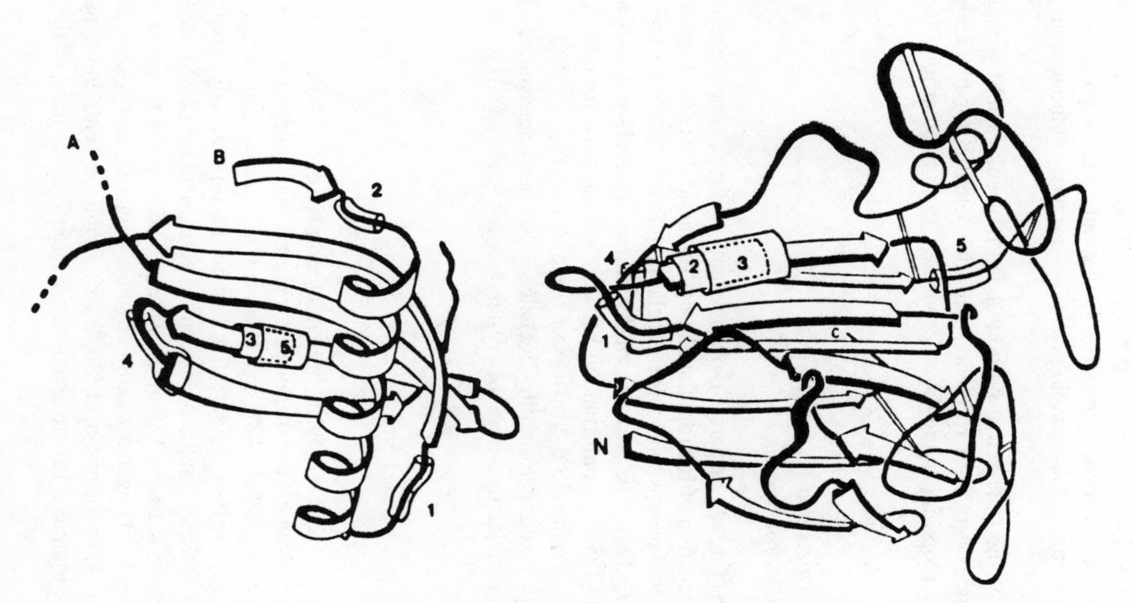

Fig. 3.7. Backbone structures of monellin (left) and thaumatin (right). The ''tubular'' sections represent the homologous tripeptides listed in Table 3.7. The broken lines show regions of disorder in the electron density maps (adapted, with permission, from reference 391).

The in-vitro immunological cross-reactivity may be of some use in searching for the molecular configuration responsible for the sweet taste. Antibody-protein complexes are not sweet. Sweet compounds like Aspartame compete for the binding sites of the antibodies, but nonsweet derivatives of Aspartame do not. Monellin and thaumatin appear to be the only pair of proteins with known crystal structure and immunological cross-reactivity without sequence similarity (392).

Identification of the taste-active portion of the molecule would stimulate research on peptides active for receptors, with the potential for blocking or activating the receptors, including those for viruses, or in photoaffinity labeling of receptors.

Structure of Aspartame and Neohesperidin Dihydrochalcone

The crystal structure of L-aspartyl-L-phenylalanyl methyl ester (Aspartame) (393) shows segregation of hydrophilic and hydrophobic domains, formation of columns of phenyl groups, and perpendicular neighboring phenyl groups as found in proteins (394). The structure of Aspartame hydrochloride has also been determined (395). A conformation has also been derived from the nmr spectrum of an Aspartame-18:6-crown ether complex (396).

The structure of neohesperidin dihydrochalcone has been determined. The arrangement of the molecule within the crystal does not lend support to the A-H· · ·B "theory" (397).

Taste Modifiers

In 1934, 150 of the 250 biologists present at a dinner of the American Association for the Advancement of Science reported that water tasted sweet shortly after eating artichokes. Although *Cynara scolymus* had been known to man since 700 B.C., the taste modification had not been previously reported. Artichoke extract (2.75%) produces a sweetness equivalent to 0.15 M sucrose (e.g., 2 teaspoons of sugar in 125 ml of water). The effect is transient, appearing in 15 seconds, rising to a maximum within a minute, and disappearing after 4–5 minutes. Chlorogenic acid (**30**, 3-caffeoylquinic acid) and cynarin (**31**, 1,5-dicaffeoylquinic acid) are apparently responsible for most of the effect (398).

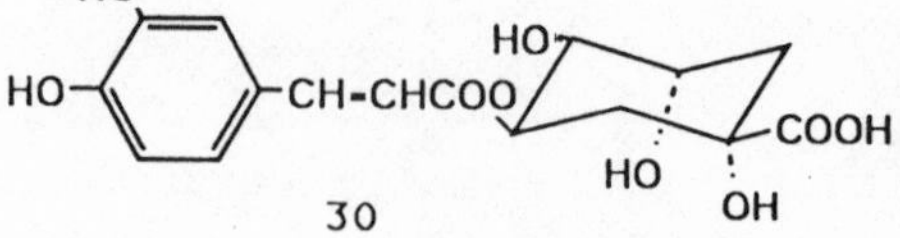

HO
HO–⟨benzene⟩–CH=CHCOO HO COOH
HO OH

30

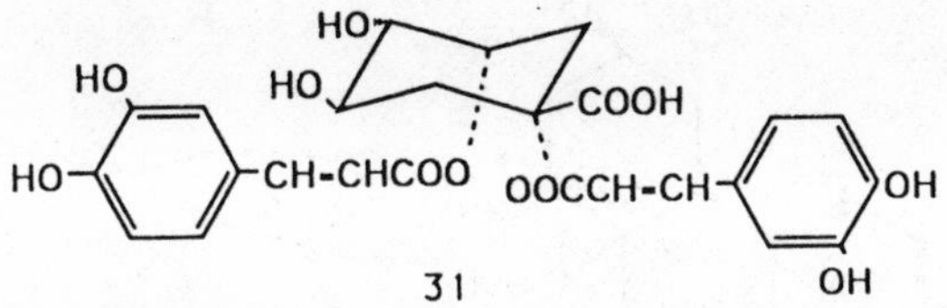

31

Gourmets have disputed whether or not wine should be served with artichokes (would the taste of Chateau d'Yquem be affected?). The use of taste modifiers as non-nutritive sweeteners has been suggested. A more serious question is whether or not the idea of taste modification is valid. A sweet taste is a sweet taste, and a taste is transient for some molecular reason. The stimulant combines with the taste receptor without activating the conformational change that leads to depolarization. Addition of water decreases the affinity of a portion of the taste-modifying molecule to the receptor and allows execution of the "correct" conformational change. The transient nature of the effect is due to the fact that the supply of combined molecules is limited by the number already bound to the receptor, since the source (the artichoke) has been removed.

Gymnemic acid (**32**), isolated from the leaves of *Gymnema sylvestre*, an Indian plant related to milkweed, inhibits detection of sweet substances (including the sweetness induced by artichoke). The effect of the plant extract was first reported in 1847 (399), but the characterization of the active substances was not completed until 1981 (400). After tasting an extract containing gymnemic acid, "grains of sugar will taste like grains of sand" (401). Another triterpene saponin, ziziphin (**33**), blocks only the sweet-taste response without suppressing the salty taste of NaCl, the sour taste of HCl, or the bitter taste of quinine (402). An apparently minor change in ziziphin, that of removing two acetyl groups (see formula), yields a compound without effect on the sweet taste. Another saponin, hodulcin, is probably closely related to ziziphin and has been isolated from *Hovenia dulcis* (403).

Our model suggests that gymnemic acid combines with the receptor and inhibits the "correct" conformational change, one which would block the channel and promote hyperpolarization. Gymnemic acid has no effect on the neural signal stimulated by sucrose in the *chorda tympani* nerve of gerbils. In fact, of all inhibitors tested, only DCG (**34**) affected the response (404, 405). A behavioral study on the tasting of sucrose by 22 primate species in comparison to man showed that gymnemic acid had an effect only on humans. Some effects have been reported on hamsters and dogs (406).

Miraculin, a Taste-modifying Protein

In 1852, Daniell described a small, red "miraculous berry" from the West African shrub *Synsepalum dulcificum*, which changed a sour taste into a sweet

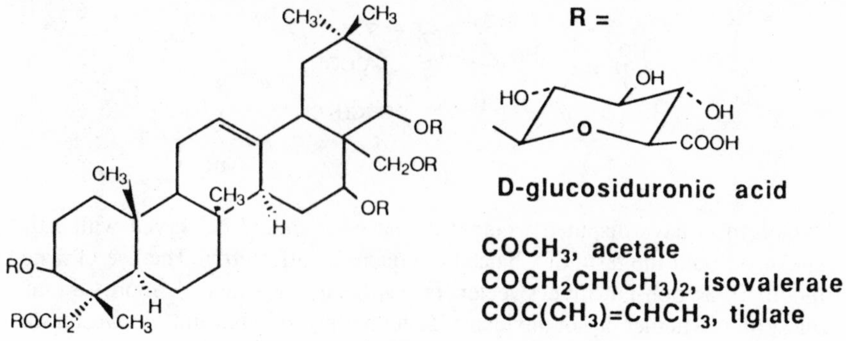

R =

D-glucosiduronic acid

$COCH_3$, acetate
$COCH_2CH(CH_3)_2$, isovalerate
$COC(CH_3)=CHCH_3$, tiglate

32

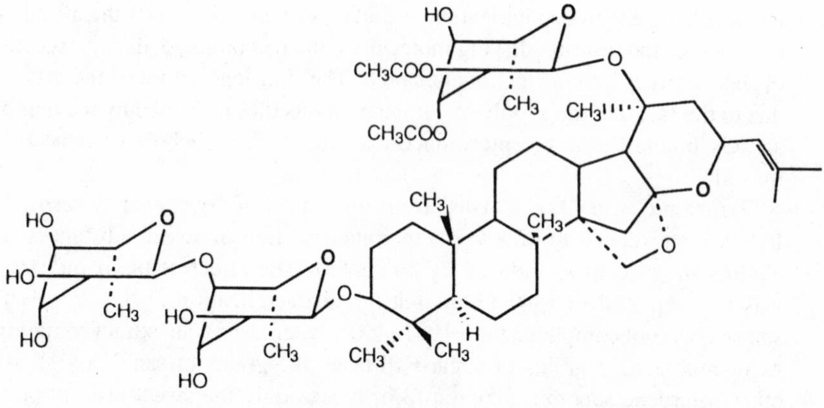

33

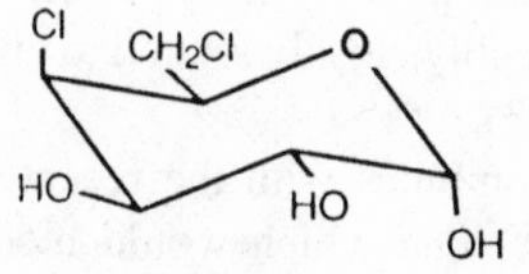

34

taste. The natives used the berry (Yoruba, *agbayun*; Twi, *asaa, asewa*) to sweeten the sour taste of bread and palm wine. In 1968, a glycoprotein, miraculin, was isolated using base with an M_r of 44,000 and a carbohydrate content of 7–20% (407). Extraction with 0.5 M NaCl at pH 4 led to a colorless solution with high physiological activity from which a pure protein, M_r 28,000, containing 13.9% carbohydrate (glucosamine, mannose, galactose, xylose, and fucose) was isolated (408). The plant was introduced into Florida (USA) with the idea of producing miraculin commercially, but the Food and

Drug Administration did not approve the use of the protein as a food additive and the project was abandoned.

Miraculin changes the sour taste of dilute acids into a sweet taste which persists for several hours. The taste of a mixture containing dilute acid and miraculin is sour at first but changes to sweet after about one minute. Is the nerve signal produced by acid modified by miraculin? Its amino acid sequence shows little homology with those of thaumatin and monellin.

Bitter and Other Tastes

The taste receptor is most sensitive to bitter substances. Highly toxic alkaloids (brucine, strychnine, solanidine) are intensely bitter (detectable between 10^{-6} and 10^{-5} M), and one may speculate that an organism that detects these substances at the lowest possible level has a better chance for survival. Intensely bitter peptides can be formed by bacterial degradation of the milk protein, a fact of possible economic importance (409–412). Organisms at all levels of the evolutionary scale have an aversion for bitter substances, although pigeons do not respond to quinine. Quinine (**35**), used in substantial amounts to cure malaria, is detectable at 4×10^{-5} M and is the flavoring agent in "quinine water." The bitterest substances known are the cucurbitacins (**36**, cucuB), detectable at 1 ppb (ca. 10^{-9} M). A strongly bitter-tasting substance, N-2-[2,6-dimethylphenylamino]-2-oxoethyl-N,N-diethyl-benzene-methanaminium benzoate (**37**, "Denatonium benzoate") is added as a warning substance to toxic materials (413). A small subset of rat taste cells release intracellular Ca^{++} in response to the material (414), with various controls suggesting a second messenger mechanism in taste cells (415). Intracellular injection of cGMP and cAMP into mouse taste cells decreases K^+ conductance, as does application of sucrose, thus depolarizing the cell (416). A similar effect is found for cAMP on frog taste cells (417). Disaccharides promote the formation of cAMP via a GTP-mediated activation of adenyl cyclase (418, 419).

The cucurbitacins are kairomones (feeding stimulants) for a family of beetles of the Luperini tribe (subfamily *Galerucinae*, family *Chrysomelidae*). The cucurbitacins are highly toxic to mammals (Cuc B i.p. LD_{50} 1.0 mg/kg in mouse), but the basis for the toxicity is unknown. The α,β-unsaturated ketone system may bind strongly to a nucleophilic thiol group. Noteworthy is that compounds that evolved as a defense for plants against herbivores stimulated the evolution of predators that are specifically attracted to the toxic compounds (420).

Picric acid (**38**, 2,4,6-trinitrophenol) has a bitter taste (Greek, *pikros*, bitter) which overwhelms the expected sour taste of a strong acid. 5-Nitrosaccharin (**39**) is very bitter, in contrast to the sweet taste of saccharin, which has, however, a bitter aftertaste. Thiourea (**40**) and other thioamides are very bitter.

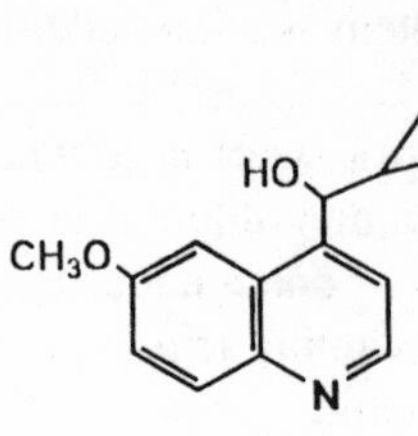

35

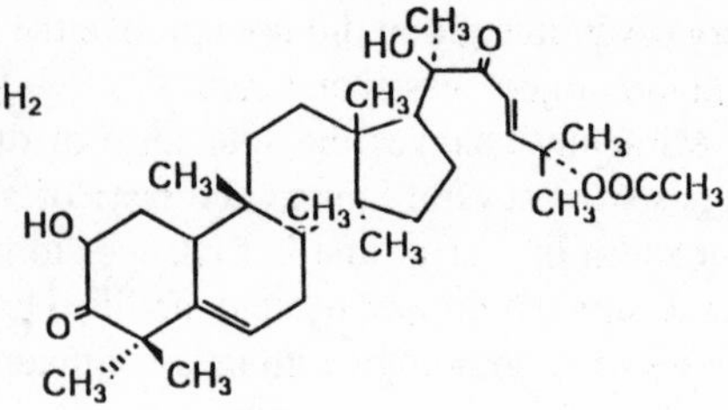

36

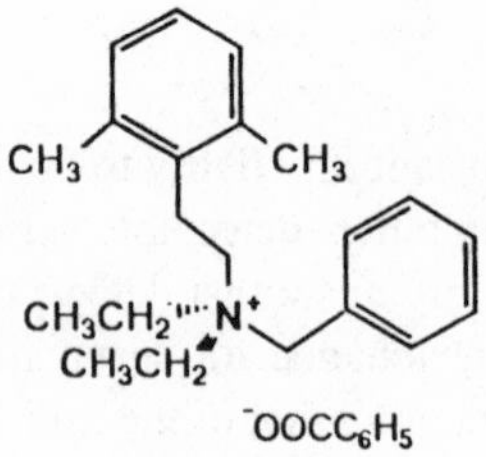

37

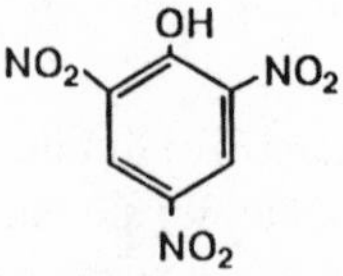

38

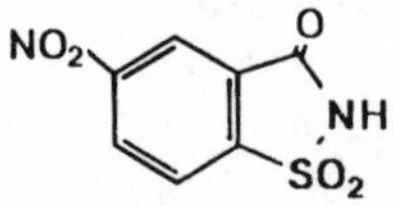

39

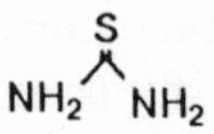

40

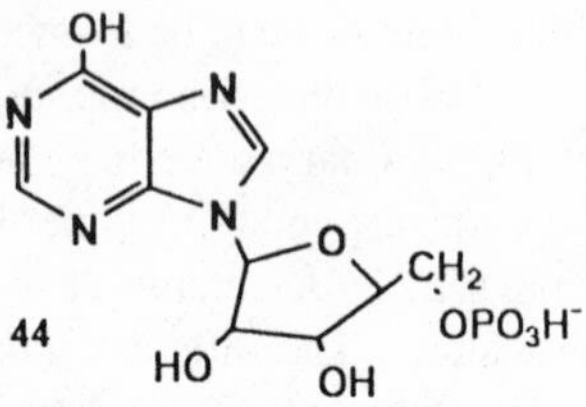

44

$Na^+ \, ^-OOCCH_2CH_2CH(NH_3^+)COO^-$

42

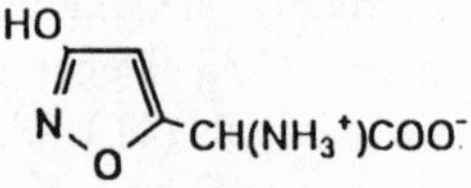

43

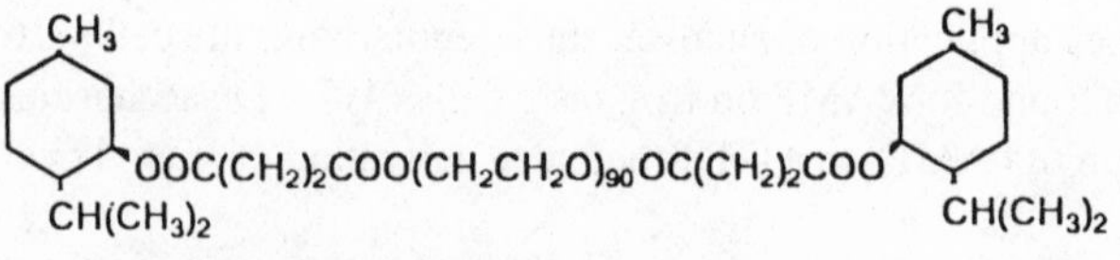

41, DP-90 bis-menthyl ester

The study of the genetics of taste depended on the bitter-tasting phenylthiourea. A 1:2 complex of diketopiperazines and theobromine is responsible for the bitter flavor of roasted cocoa, the complex being much more bitter than either of the bitter-tasting components. An attempt to create a polymeric material (**41**, DP-90) bearing "minty" flavor resulted in a bitterness "so strong and vile that one panelist threatened to quit." Some caution is thus advised in the development of water-soluble, nonabsorbable (and therefore non-nutritive) polymers bearing "sweet-tasting groups," since metabolism or unexpected reactions with other food components may produce surprising and unpleasant results (421).

Sour Taste

The sour taste is clearly associated with acids; responses to acids seem to occur in organisms throughout the evolutionary tree. A sour taste can be detected in a pH 3.5 solution of hydrochloric acid, but the taste may be modified by the presence of different anions and other substances.

Umami Taste

The flavor-enhancing qualities of soy sauce have long been known to the Japanese, the enhancement being due mainly to monosodium glutamate (**42**, MSG). The special taste quality, *umami*, is associated with MSG and other compounds (**43**, **44**) (422). The meaty flavor of soups is particularly enhanced, and the palatability, taste intensity, and impact of food flavors are favored by MSG. Ribonucleotide monophosphates such as inosine monophosphate (**44**, IMP) and guanosine monophosphate (GMP) also elicit an umami response somewhat weaker than that of MSG. A synergistic effect is found for a mixture of MSG and IMP. Compounds that are glutamic acid agonists (i.e., stimulate the same receptors as does glutamic acid) lead to an umami sensation. L-Ibotenic acid (**43**) is 5–30 times as strong as glutamic acid in producing the umami response, whereas L-aspartic acid is about 0.08 as strong. Two peptides with a "delicious taste" have been isolated from papain-digested beef. The two glutamic acid residues in peptide A (composition glu_2, asp, ala, leu, val) supposedly make the peptide more delicious than peptide B (glu, asp, leu_2, ile, ala_2, val) (423).

Mechanism of Taste

Phenomenology, correlation analysis, and enormous effort over many years have not led to an understanding of the mechanism of taste, although the G-protein pathway seems likely (415–418). Pending the isolation and structural determination of the taste receptor, a heuristic approach must be used. The following ideas might be used to construct a plausible mechanism of taste. The molecular details must be left as a problem for the future.

1. There is only one taste receptor. The same idea has been proposed on the basis of models for enantiomerically related molecules which yield either bitter or sweet responses (424). A variety of channels might contribute to the net response (425).

2. (At least) four general types of signals may be produced within the taste receptor. These may be somewhat modified according to the adapting solution used for the tongue, 41.4 mM NaCl or distilled water. For the latter medium, depolarization is the most frequent result.

3. Salt application always produces depolarization of receptor cells, while bitter, sweet, and sour taste are more variable mixtures of hyperpolarization

and depolarization. Sucrose, like cGMP, leads to closure of K^+ channels via protein phosphorylation, accompanied by a modest depolarization (416). This is in agreement with the suggestion that a voltage-sensitive K^+ conductance contributes to taste-cell response, with the channels restricted to the apical membrane of the taste cells on the tongue (425).

4. The early responses in certain fibers of the *chorda tympani* nerve of the hamster (426) are (a) salt (very rapid bursts for 1 sec followed by rapid firing), (b) bitter (400-ms burst followed by relative silence), (c) sweet (decrease in spontaneous signals), (d) sour (irregular rapid bursts for 400 ms followed by silence).

5. Sweet, bitter, and even "tasteless" responses are closely related, with small changes in the molecular structure of the stimulus molecule resulting in large changes in the taste sensation (see point 1).

6. The taste receptor cannot be highly specific in view of the wide variation in molecular structures and molecular types that produce the responses. A pseudo-β-DL-fructopyranose which has a CH_2 in place of an ether O in β-D-fructipyranose is as sweet as D-fructose (427).

7. The simple, "classic" AH,B criterion for sweet compounds (428, 429) is met by many compounds that are not sweet. The postulate that a hydrophobic domain must also be present does not remove this difficulty, but does suggest some general properties of the site.

8. Some of the most bitter substances (e.g., quinine, picric acid) fit the "AH,B, hydrophobic tripartite" criterion, but others (strychnine, solanidine) are probably best accommodated by either the AH, hydrophobic portion or weakly (thiourea) by the AH,B portion of the receptor.

Genetics of Taste

The common phrase "*Chacun a son gout*" (every man to his own taste) has some basis in biology, since there are substances that are unpleasant or bitter to some individuals and are not detected (or only weakly detected) by others. Relatively few studies of the genetics of taste have been made, although there are many self-selected expert tasters (wine, brandy, coffee, tea, cheese) who would be good subjects for comparison with others chosen at random.

Phenylthiocarbamide (**45**, PTC) was once widely used to demonstrate genetic differences in taste, a use terminated by reports of carcinogenicity. Some 60–70% of North Americans tested found PTC bitter in taste, others calling it innocuous or weak. The proportion of nontasters among American Indians is only about 6%. Many thioamides are bitter to tasters, including allyl thiocarbamide (**46**), thioacetamide (**47**), and thioacetanilide (**48**). Allyl thiocarbamide produces a momentary faint bitterness in nontasters, suggesting that the receptor is slow in responding.

Detection of the bitter taste that accompanies the sweet taste of saccharin has been correlated with detection of 6-*n*-propylthiouracil (**49**), a compound that tastes bitter to the same people who find PTC bitter (430, 431). The bit-

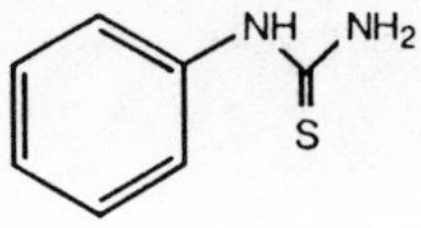

45 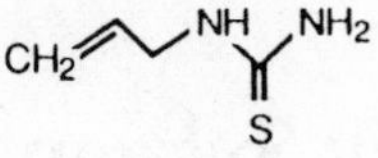 46 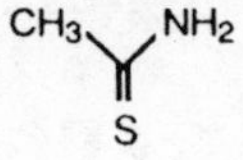 47

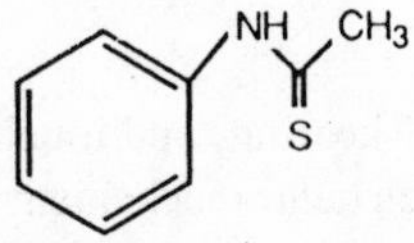

48 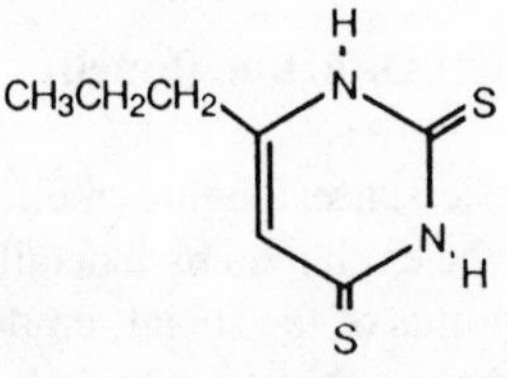 49

terness becomes appreciably stronger to tasters at concentrations (10^{-3} M) used to sweeten commercial products. If somewhat lower concentrations of saccharin were used in conjunction with another nonbitter (but more expensive) sweetening agent, the bitter taste that affects the commercial acceptance of a product might be avoided. Another compound, sodium benzoate, used as a preservative in carbonated drinks, is tasteless to many, but bitter-sweet to some. A single genetic locus has been implicated in determining the difference between two strains of mice with respect to the ability to detect the sweet amino acid, D-phenylalanine (432).

A very rare condition, thus far reported in only two patients also afflicted with congenital hypoparathyroidism, is aglycogeusia—inability to taste sweet substances. Sweet substances produced either a bitter taste or a sour taste. A gene mutation that changes repellency from concentrated NaCl to attraction has been produced and characterized in *Drosophila* (433).

Hypogeusia

Thiol drugs, like the D-penicillamine used in the treatment of Wilson's disease to aid in the removal of copper ion from the body, diminish the taste acuity in a substantial number of patients (> 30%). Total loss of taste occurred in a patient treated with 5-thiopyridoxal, but acuity returned after drug treatment ended. The loss of taste was less severe in patients with Wilson's disease, and it was thought that the hypogeusia was related to lowered metal-ion levels in the bloodstream. Administration of metal ion (zinc or copper) was claimed to return taste acuity to normal (434). Unfortunately, the limited amount of information available on hypogeusia is insufficient to support a theory.

4 Olfactory System

''A rose is a rose is a rose.''
— Gertrude Stein

Sensory Transduction: Olfaction (Smell)

There are five classic senses: sight, smell, taste, hearing, and touch. Only the sense of sight can be explained by a detailed molecular mechanism (Chap. 5). Even with the stimulus of the social, emotional, and economic importance of applied olfactory theory, the effort to understand olfaction was not successful until developments in biochemistry and the mechanism of vision provided useful guidelines (435–445). The sense of smell or olfaction (Latin, *olfacere*, to smell) is closely related in mechanism to the sense of sight, and a theory for smell has begun to develop. Using the new prototheory and analogies between rhodopsin, the light-receptor protein, and the muscarinic acetylcholine receptor protein, a molecular-level mechanism for the action of odorants can be developed. Present theory of the sense of taste is more qualitative (Chap. 3).

A number of proposals have made to define the role of the odorant or olfactant. One important and attractive theory relating molecular shape and smell (440–442) leaves unanswered important questions, such as how the olfactory nerve signal is produced. Other theories are less instructive. The ''infrared detection'' mechanism (446) does not allow for the fact that enantiomeric molecules smell different (447, 448). The ''penetration and puncturing'' idea (449) is a vivid and acknowledged modernization of the suggestion made in 47 B.C. by Lucretius that pleasant aromas are composed of smooth round atoms, and that harsh, bitter smells are made of hooked particles which tear their way into our senses. The theory claims that odorant molecules open ion pathways through the hydrophobic bilayer of cell membranes. However, ion flow in and out of cells is normally via proteins and protein channels (Chap. 7), and it is difficult to imagine how lipid perturbation could give rise to molecule-specific signals.

The Olfactory System

Molecules present in the external medium (respired air or water) act as stimuli for the olfactory system. The mammalian olfactory system consists of two organs, the olfactory mucosa where nerve signals are generated in response to odorants, and the olfactory bulb of the brain where the signals are processed.

The mucosa are located in the nasal cavity directly beneath the bulb within the skull (Fig. 4.1).

The detector organ is spread over about 10 cm^2 in humans, and consists of a tissue in which there are nerve cells, secretory cells, and other epithelial cells (450). The nerve cells end in cilia which are bathed in a mucopolysaccharide solution secreted by Bowman's gland. The ciliated nerve cells are surrounded by two types of supporting cells on the exterior side and are adjacent to basal cells on the interior side (Fig. 4.1). Basal cells serve as the source for new olfactory nerve cells, turnover occurring within less than three months (451), but fully differentiated mouse olfactory receptor cells may last longer than 12 months (452). Severing olfactory axons or treatment with zinc sulfate causes the loss of the olfactory neurons, which are replaced by a functional epithelium within 30–60 days (453).

The axons of the olfactory neurons are unmyelinated and difficult to probe individually because of their small size (0.1–3 μm) (454). The axons are gathered into bundles within or close to the olfactory epithelium. The olfactory axons do not branch, but pass through the skull to olfactory macroglomeruli, 100–200 μm-diameter clusters of synaptic connections surrounded by a single layer of glial cells. The terminals of the olfactory neuron make contacts with the ramified dendrites of periglomerular short-axon cells (PG), tufted cells (T), and mitral cells.

The olfactory bulb (454) has a number of cell layers, among them the olfactory glomeruli, the external plexiform layer (EPL), the mitral body layer, and a granule layer (GL). For approximately 10^8 olfactory axons of the rabbit, there are 10^5 mitral cells with contacts being made within ca. 4000 glomeruli. The major output from the olfactory bulb is via the mitral cell axons to the lateral olfactory tract. There are contacts with nerve axons from other parts of the brain in both the EPL and the GL, the latter containing a "deep" short-axon cell and granule cells (G) which are different from the granule cells that give rise to the parallel fibers of the cerebellum. There are many G (200:1) and numerous PG (20:1) for each mitral cell. The G, T, and PG have complex dendritic links with the other cells. The large number of olfactory neurons and other cells per mitral cell suggests that considerable processing of the sensory input occurs in the olfactory bulb.

The details of the olfactory epithelium and olfactory bulb are shown in Figure 4.1 together with an overall drawing of the olfactory area.

Olfactory Sensory Neurons

The sensory cells are bipolar neurons divided into a dendritic region (olfactory rod), a cell body, a transmitting axon, and axon terminals (Fig. 1.4). Rods of olfactory neurons often form tight junctions with one another and with neighboring supporting cells. The junctions seal the epithelium against pene-

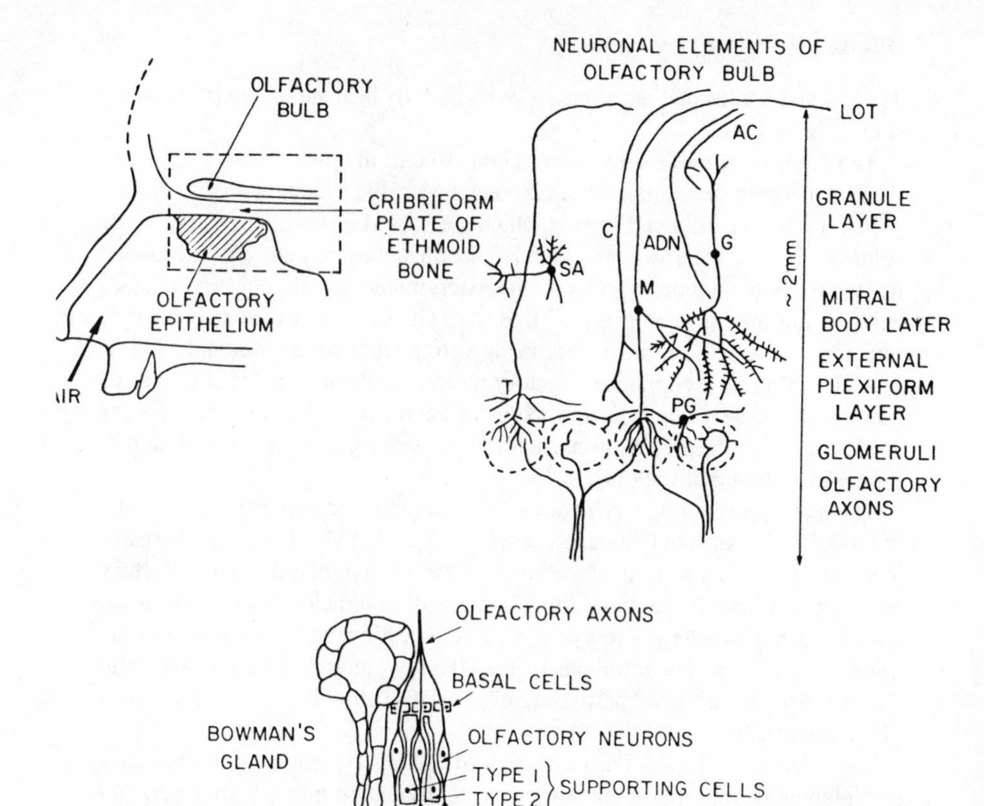

Fig. 4.1. Elements of the olfactory system. The system consists of a portion exposed to the exterior environment and a portion at the front of the brain, as denoted in the side view of part of a human skull. The olfactory epithelium is at the upper end of the nasal cavity. Axons from olfactory neurons extend through the skull (at the cribriform plate of ethmoid) to the olfactory bulb, a portion of the brain in which preliminary processing is carried out on the signals from the olfactory neurons. Each of the two major sections of the olfactory system is shown in greater detail. The elements of the olfactory epithelium are the olfactory neurons, two types of supporting cells, basal cells, and the cells that constitute the mucus-secreting Bowman's gland. The neuronal elements of the olfactory bulb are the glomeruli, the external plexiform layer (EPL), the mitral body layer, and the granule layer. The glomeruli are complexes of many synapses between the olfactory neurons and mitral cells (M) as well as tufted cells (T) and periglomerular short-axon cells (PG) surrounded by a monolayer of glial cells. The EPL contains synapses mainly between mitral cells and granule cells, and a few contacts between fibers from the nucleus of the horizontal limb of the diagonal band, a region at the base of the brain. More numerous contacts between distant neurons and those in the olfactory bulb are made in the granule layer with a portion of the olfactory cortex, the anterior olfactory nucleus (AON) on the same side of the brain, and with the same nucleus on the contralateral side through the anterior commissure (AC). Some interaxon contacts are mediated by a ''deep'' short-axon cell (SA) in the granule layer. The main output of the olfactory bulb from the mitral cells, modified in unknown ways by all of the contacts noted above, leads to the lateral olfactory tract (LOT) in the olfactory cortex. Further and more complex relationships between neuron groups exist in the olfactory cortex (references 24, 455).

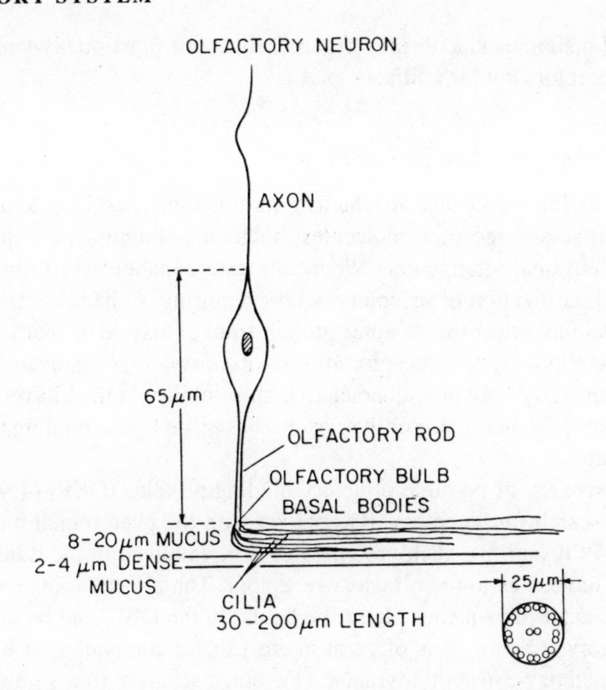

Fig. 4.2. The olfactory neuron. The dendrite (olfactory rod) of the olfactory neuron ends in the olfactory bulb at the interface of the cell with the mucus layer. Cilia (0.25 μm diameter, 30–200 μm length) emerge from the bulb and lie within a mucus layer, the latter a viscous aqueous solution of mucopolysaccharides. A tangled mass of cilia is present in areas with a high population of olfactory neurons. A cilium begins at a basal body and has a typical cilial structure in the case of the macaque monkey, with nine pairs of tubules surrounding a central doublet.

tration by proteins (455) and bilayer-soluble fluorescent lipids (456) and perhaps smaller odorant molecules. The cell bodies are separated by supporting cells. The dendrite has a bulbous area (''olfactory knob'') at the mucosal interface. The vesicle carries 8–20 cilia and microvilli, which expand the mucosal surface area by a factor of 100–1000. The cilia form a dense meshwork over the olfactory epithelium and contain the olfactory receptor molecules. The evidence (450) is (a) the electrophysiological response on the surface (the electroolfactogram or EOG) is greatest at the outside; (b) the olfactant-stimulated nerve signal disappears after removal of the cilia; (c) cilia preparations bind odorants quite well (the caveat with respect to nonspecific binding must be kept in mind); (d) a high density of intramembranous particles (IMP) are found in freeze-etched cilia by electron microscopy (457); and (e) after removal, the cilia regenerate and a signal may eventually be recorded from olfactory nerves after stimulation with an odorant. The olfactory neuron and ciliary structure are shown in Figure 4.2.

There are olfactory regions, the vomeronasal receptors in amphibians, reptiles, and mammals (Jacobson's organ), which are functionally similar to olfactory receptors but lack cilia.

Olfactory Receptors and Binding Proteins

The basis for a molecular mechanism for olfaction must be a detailed structure of olfactory receptor molecules, including amino-acid sequence and three-dimensional arrangement. Sequence determination will follow the isolation and purification of receptors, a task requiring a "handle" by means of which one may select the receptor protein from a mixture of many other proteins. The relatively low receptor affinity of most odorant molecules coupled to their generally lipophilic character (high solubility in the bilayer and other nonreceptor protein regions of the system) has made ligand binding techniques problematic.

The discovery of possible olfactory binding proteins (OBP) (458) complicates the search for receptors. By analogy with the pheromonal binding proteins (PBP) (Chap. 3), such proteins might be involved in the transfer of the olfactant molecules to the olfactory receptors. The olfactant molecules in the air would enter the aqueous phase, be bound to the OBP, and be delivered to the olfactory receptor. It is of great interest that a component of bell pepper odor, 2-methoxy-3-isobutylpyrazine (**1**), binds strongly to a soluble 19-kDa protein (native form 38 kDa) localized more in olfactory rather than respiratory epithelium (300, 459). Binding was followed with ^{3}H-methoxy-**1**; two binding sites, K_d, 10^{-8} M and 3×10^{-6} M, are found for the bovine protein. Its solubility and molecular weight suggest that the protein might be an OBP rather than an olfactory receptor protein. Rat OBP is homologous to α2-microglobulin and may be related to a supergene family that includes retinol-binding protein (460). The olfactory receptor protein should be membrane-bound and of higher molecular weight. There is no immunological cross-reactivity of the pyrazine OBP to anti–rat olfactory marker protein (OMP) antibody (459), an intracellular olfactory neuron protein of molecular weight similar to that of OBP (461–463). Surprisingly, mouse and rat OMP antibodies bind to a number of neuronal areas in mouse and rat brains (464).

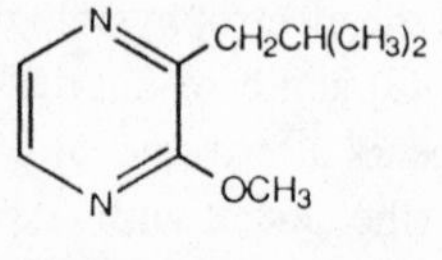

N, N, $CH_2CH(CH_3)_2$, OCH_3

2-Methoxy-3-isobutylpyrazine

(1)

Frog olfactory cilia contain three major (lectin-binding) glycoproteins (465), of which one (gp95) is present in substantial amounts and fits the properties expected for a transmembrane receptor (466). The integral membrane character of gp95 is suggested by its failure to be solubilized in 0.1 M Na_2CO_3 and its ready incorporation into Triton X-114 micelles. About 30% of the 95-kDa mass is cleaved by endoglycosidase-F (which attacks both high mannose and N-linked oligosaccharides), diminishing greatly its binding to the lectin, wheat germ agglutinin (WGA). The gp95 has been purified about 50-fold on a WGA column. Toad, cow, and rat olfactory cilia contain proteins in the 85 kDa range. However, a gp56 appears to be the best receptor candidate in bovine olfactory cilia (467). A camphor and decanal binding glycoprotein subunit of 88 kDa has been isolated from rat olfactory epithelia (468). Non-sensory cilia are almost devoid of glycoproteins. Monoclonal antibodies to gp95 (469) may make possible the further purification and eventual cloning of the protein.

A structural model for the olfactory receptor protein will be presented in a later section.

Classification of Olfactory Stimuli

Direct recording of neurophysiological changes in single olfactory receptor cells is not technically feasible, nor has a reconstituted receptor complex been prepared. One important feature of olfactory receptors is the wide variety of odorant molecules that stimulate an olfactory neural signal. Identification of odorant classes might cast light on the number of receptor variants to be expected. An appropriate classification method might also help in the interpretation of detailed mechanism and could be of considerable practical interest. With respect to the theory that relates odor to molecular shape, it is important to note that the shape of optical enantiomers is exactly the same and yet their odors can be quite different.

Substances were grouped into odor classes according to their smell by experienced humans (440, 441). The number of recognizable classes is small, but varies with the criteria used to define a class. There may be as many as 30 to 40 classes (470–472). The steric dimensions of the molecules within certain classes were then measured using models and shadowgraphs, giving rise to the shape functions illustrated in Figure 4.3.

The classes are named according to familiar descriptions of the odor type, such as ethereal, camphoraceous, musky, floral, minty, pungent, putrid, almond, aromatic, and anise. Examples of molecules (**2–25**) in each of these classes are listed in Table 4.1. Classes for which specific anosmias have been recognized are marked.

The power of the approach is illustrated with a variety of chemically different, but similarly shaped molecules (**3**, **26–30**). 1,4-Dichlorobenzene (**3**), methyl isobutyl ketone (**27**), hexamethylethane (**28**), hexachloroethane (**29**), and cyclooctane (**30**) all have odors resembling that of camphor (**26**) (''camphoraceous'').

$ClCH_2CH_2Cl$

2

Cl Cl

3

O

4

$CH_2CH_2C(CH_3)CH_2CH_2CH_3$

OH

5

O

CH_3 $CH(CH_3)_2$

6

CH_3COOH

7

CH_3SSCH_3

8

CHO

9

CH_3O

10

I N=C=S

11

N

12

CH2=CHCH2N=C=S

13

$CH_3CH_2CH(CH_3)CH_2OOCCH_3$

14

CH=CHCHO

15

CH_3 CH_3 O CH_3 CH_3

16

E-$(CH_3)_2C=CHCH_2CH_2C(CH_3)=CHCHO$

17

$(CH_3)_3N$

18

CH_3CH_2 OH

19

$(CH_3)_2CHCH_2COOH$

20

CH_3 CH_3 O H

21

N=C

22

CH_3 N H

23

HCN

24

$(CH_3)_2CHCHO$

25

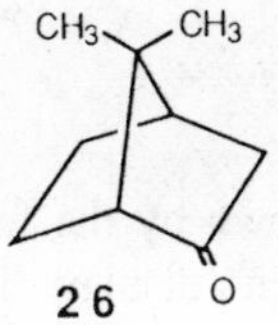

26 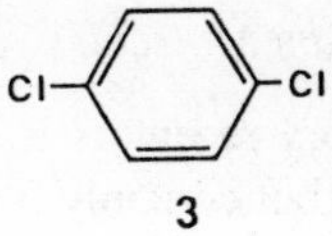 3 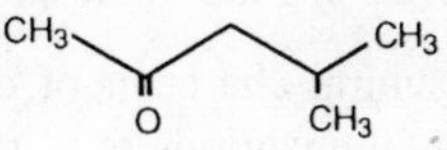 27

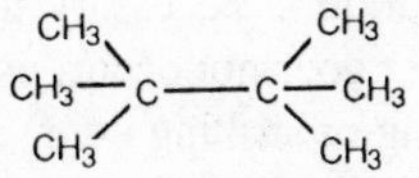

28 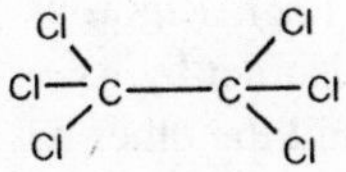 29 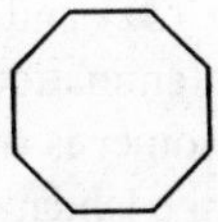 30

Table 4.1. Odorant Classes

1. Ethereal[a]	1,2-Dichloroethane **(2)**
2. Camphoraceous[a]	1,4-Dichlorobenzene **(3)**
3. Musky[a]	7-Hydroxypentadecanolactone **(4)** (''Exaltolide'')
4. Floral[a]	1-Phenyl-3-methyl-3-hexanol **(5)**
5. Minty[a]	Menthone **(6)**
6. Acid[a,b]	Acetic acid **(7)**
7. Putrid[a]	Dimethyl disulfide **(8)**
8. Almond	Benzaldehyde **(9)**
9. Aromatic[a]	Methoxybenzene **(10)** (Anisole)
10. Anise	4-Iodophenyl isothiocyanate **(11)**
11. Carrion[a,c]	1-Pyrroline **(12)**
12. Garlic[a]	Allyl isothiocyanate **(13)**
13. Fruity[a]	Isoamyl acetate **(14)**
14. Spicy[a]	Cinnamaldehyde **(15)**
15. Violet[a]	β-Ionone **(16)**
16. Citrous[a]	Geranial **(17)**
17. Fishy[a]	Trimethylamine **(18)**
18. Phenolic[a]	4-Ethylphenol **(19)**
19. Sweaty[a]	Isovaleric acid **(20)**
20. Urinous[a]	Androst-16-en-3-one **(21)**
21. Repulsive[a]	Phenyl isocyanide **(22)**
22. Fecal[a]	Skatole **(23)**
23. Cyanide[a]	Hydrogen cyanide **(24)**
24. Malty[a]	Isobutyraldehyde **(25)**

Source: Based in part on references 442, 470–472.

[a] Specific anosmias have been reported for these odorant classes (440–442).

[b] Non-olfactory responses may contribute (478).

[c] Also called spermous.

Properties of Enantiomeric Odorant Molecules

The chiral character of olfactory receptors is best shown by the fact that individual enantiomers of many (but certainly not all) pairs display different odorant properties. The enantiomers *d*- and *l*-camphor do not differ in odor. A vivid difference exists between enantiomeric pairs of two terpene derivatives, carvone (**31**) and carveol (**32**). Rigorous purification is necessary to ensure that contamination of one isomer by its enantiomer does not occur, especially if one isomer is odorless and the other has a strong or striking odor, as in the case of the labdane derivative (**33**) and its enantiomer, *ent*-labdane (**34**), which has a strong ambergris-like odor (473).

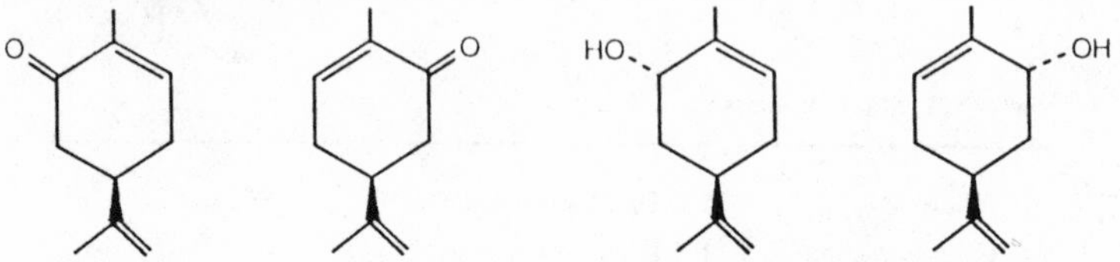

R(-) (31R) spearmint — S(+) (31S) caraway — carvone

R(-) (32R) minty — S(+) (32S) musty — carveol

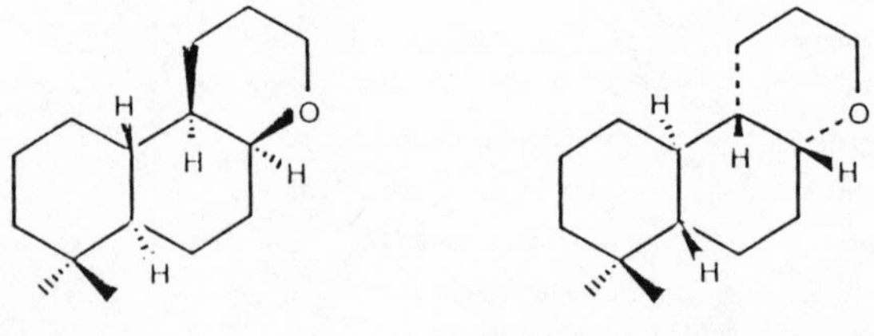

8β,13-epoxy-14,15,16-trinorlabdane (33) odorless

ent -8b,13-epoxy-14,15,16-trinorlabdane (34) strong ambergris

The odor differences between the enantiomers (Fig. 4.4) are of practical and theoretical importance, as shown by the examples (474) listed in Table 4.2. Highly stereospecific syntheses are required for synthetic derivatives in order to ensure the reproducibility and quality of perfume essences (438). The quality of information derived from experiments on the olfactory system might be dependent on the purity of the odorants used.

Intensities of Olfactory Stimuli

The names for odorant classes are frequently descriptions for the perceived character given in somewhat literary, familiar terms. Although there has been

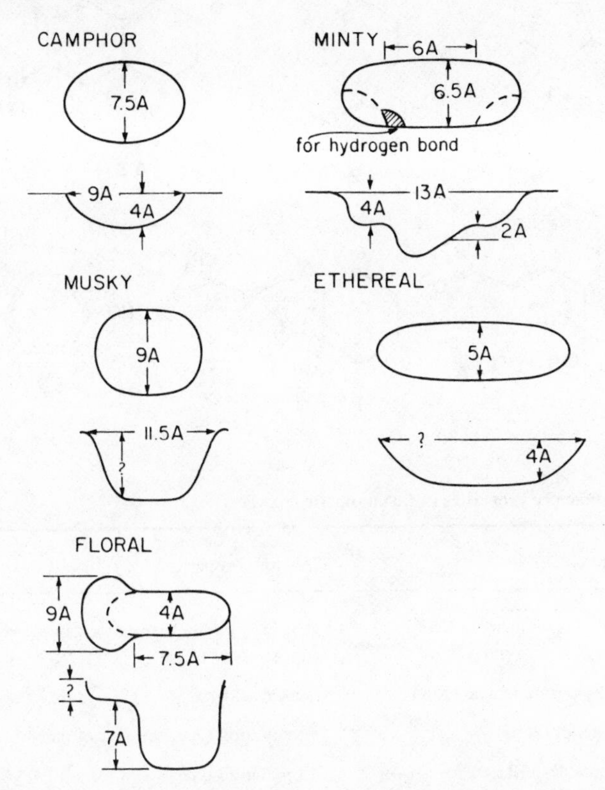

Fig. 4.3. Estimated dimensions for odorant receptor sites for five of the "primary" classes derived from estimations of the molecular shape of odorants (references 440–442).

an interest in classifying odors for several thousand years, there are difficulties in defining sensations (the perceptions of humans) that do not have readily accessible physical counterparts such as nerve signals. To deliver a prescribed amount of odorant per olfactory neuron is not possible, considering problems in flow, nose size, fluidity of mucus, rate of respiration, day of month (especially for females), time of day, and age (475). In addition, odors vary in their perceptibility, i.e., are detected with greater or lesser efficiency (476).

Relying on the response of human subjects, one may conclude that the odor intensities vary from very intense to barely perceptible (or odorless). A selection of threshold values (recorded as pT, a number related to the concentration presented for detection by a human being) is given in Table 4.3 (477). Some of the compounds are illustrated below (**41–45**).

Although some natural aromas are dominated by the odor of a particular substance, most natural aromas are highly complex mixtures in which many compounds contribute to an overall impression. Nonolfactory responses may also affect the sensation (478). Modern high-resolution gas chromatographs

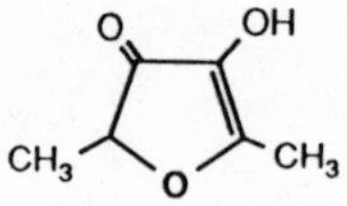

41 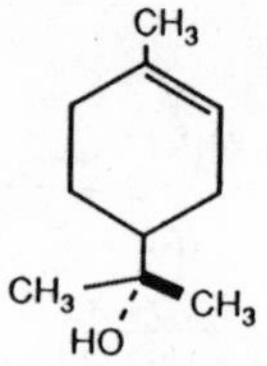 42 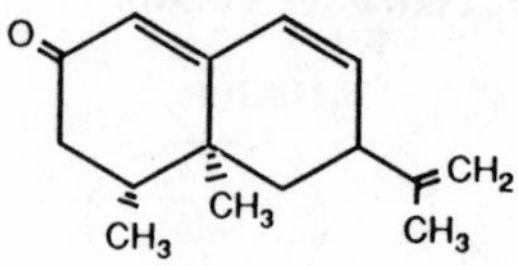 43

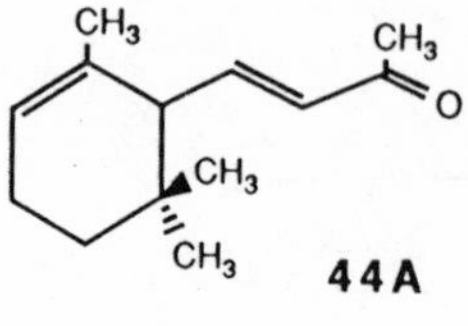

44A 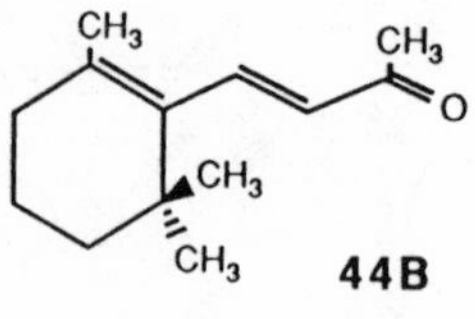 44B 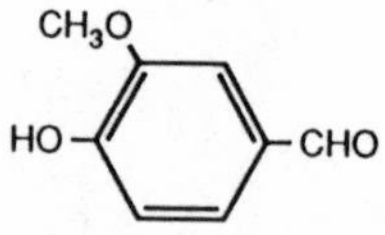 45

Table 4.2. Olfactory Properties of Enantiomeric Odorants

Enantiomers	Olfactory Properties
R(−)-4-Methylhexanoic acid (**35R**)	(goat-like [hircine], sweaty), caproic acid but more fatty; stronger of pair
S(+)-4-Methylhexanoic acid (**35S**)	Caproic acid
3S(−)-Citronellol (**36S**)	Geranium-oil type; stronger of pair
3R(+)-Citronellol (**36R**)	Citronella-oil type
cis-(+)-Rose oxide (**37+**)	Green, geranium-type odor; *cis* derivatives stronger than corresponding *trans* derivatives; (+)-compounds sweeter and less herbal than (−)-compounds. *Cis*-(−)-Rose oxide has the strongest odor of the four isomers.
cis-(−)-Rose oxide (**37−**)	
trans-(−)-Rose oxide (**38−**)	
trans-(+)-Rose oxide (**38+**)	
3R(−)-Linalool (**39R**)	Woody lavender-like odor (Ho oil); stronger of pair
3S(+)-Linalool (**39S**)	Sweet shade of lavender-like odor, like petit grain oil
S(−)-7-Hydroxydihydrocitronellal (**40S**)	Reminiscent of lily of the valley with a fresh green and minty shade
R(+)-7-Hydroxydihydrocitronellal (**40R**)	Strong, sweet lily-of-the-valley-like odor; stronger of pair

Source: Adapted from reference 474.

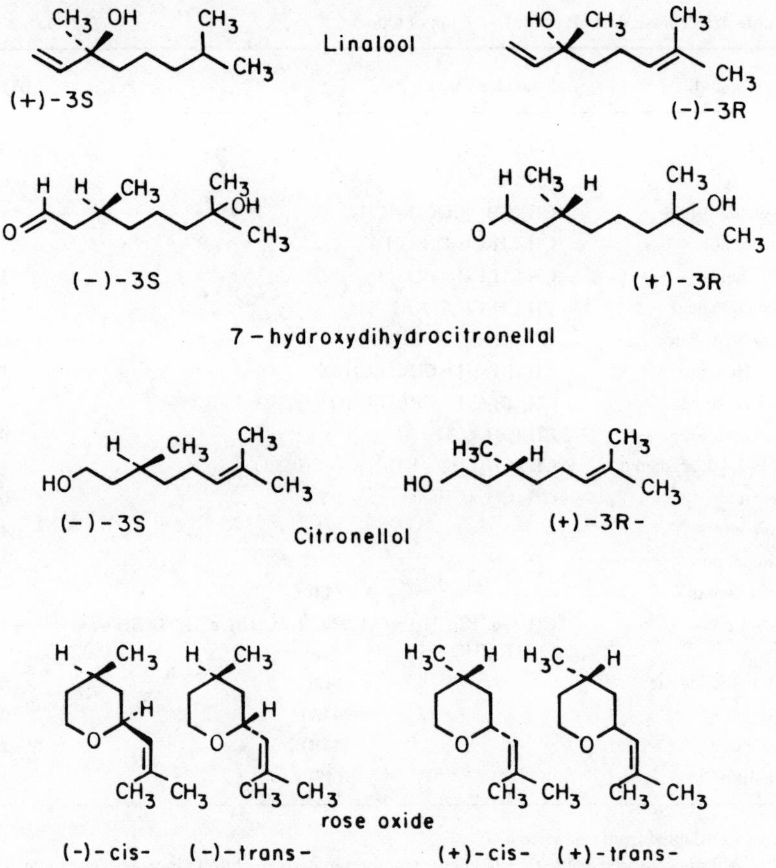

Linalool

(+)-3S (−)-3R

7-hydroxydihydrocitronellal

(−)-3S (+)-3R

Citronellol

(−)-3S (+)-3R-

rose oxide

(−)-cis- (−)-trans- (+)-cis- (+)-trans-

Fig. 4.4. Formulas of enantiomeric odorants. The enantiomers of linalool (**39S**, **39R**), 7-hydroxydihydrocitronellal (**40S**, **40R**), citronellol (**36S**, **36R**), *cis* (**37−**, **37+**) and *trans* (**38−**, **38+**) rose oxide are illustrated. The odor differences are listed in Table 4.2. The formulas for the enantiomers of 4-methylhexanoic acid (**35S**, **35R**) are not shown.

can resolve almost all compounds present in different mixtures as separate peaks, and mass spectrometry and infrared spectroscopy often make possible the identification of the very small quantities obtained by chromatographic separations. Nevertheless, the picture is not yet complete, as indicated by the fact that the number of known components in various aromas is still low in comparison to the total number of detectable components: beef (270/520); cheddar cheese (150/280); coffee (468/>968); raspberry (95/215); rye crispbread (92/122); bell pepper (43/106) (477). Techniques such as gas chromatography/mass spectrometry and high-pressure chemical ionization with tandem mass spectrometry (MS/MS) will lead to more complete information on the nature of the compounds present in natural mixtures (479).

Table 4.3. Threshold Values for Detection of Odors

Compound	Structure	pT[a]
Acetone	CH_3COCH_3	5.7
Diacetyl	$CH_3COCOCH_3$	0.6
1-Octen-3-one	$CH_3(CH_2)_4COCH=CH_2$	−1.0
1-Butanol	$CH_3CH_2CH_2CH_2OH$	4.0
Butanal	$CH_3CH_2CH_2CHO$	1.8
Butanoic acid	$CH_3CH_2CH_2COOH$	3.5
Ethyl Butyrate	$CH_3(CH_2)_2COOCH_2CH_3$	−1.0
Z-3-Hexenol	$CH_3CH_2CH=CHCH_2CH_2OH$	1.8
Z-3-Hexenal	$CH_3CH_2CH=CHCH_2CHO$	−0.6
Hexanal	$CH_3(CH_2)_4CHO$	0.7
E,E-2,4-Decadienal	$CH_3(CH_2)_4HC=CH(H)C=CHCHO$	−1.2
Decanal	$CH_3(CH_2)_8CHO$	−0.1
Furaneol	**(41)**	−1.4
Limonene[b]		1.0
α-Terpineol	**(42)**	2.5
α-Sinsensal	$CH_2=CHC(CH_3)=CHCH_2CH=C(CH_3)CH_2CH_2CH=$ $C(CH_3)CHO$	−1.3
(+)-Nootkatone	**(43)**	0.0
α-Ionone	**(44A)**	−0.4
β-Ionone	**(44B)**	−2.2
Vanillin	**(45)**	1.3

Source: Adapted from reference 477.

[a] Logarithm of the detectable threshold concentration in parts per billion in water. The molecular weights vary by less than a factor of about three for most of the compounds. Thus, pT is proportional to the vapor-phase concentration above the solution provided that the vapor pressures are proportional to the solute concentration. The values listed reflect a human perception and are only a qualitative indication of the degree of physical interaction of the detected molecule with the olfactory receptor.

[b] See formula **3**, Chap. 3.

Genetics of Olfactory Response

If specific receptor proteins are responsible for well-defined olfactory responses, a genetically determined lack of one of these proteins should produce an olfactory deficit. In certain cases, this may manifest itself as a specific anosmia, the inability to smell a particular class of odorants. The counting of specific anosmias may help in defining the number of olfactory classes and, possibly, the number of receptor types. Genetic variation in olfactory binding proteins could also be responsible for some anosmias.

Some substantial barriers stand in the way of gathering information on the genetics of olfactory response. These include the variation in individual characteristics; the problem of universal odor descriptions (480, 481) except for experts (wine-tasters, perfumers, organic chemists [482, 483]); the difficulty in finding families with pedigrees in cultures in which individuals are highly mobile; and the irreproducibility of odorant delivery.

To test a sample compound, one must (a) disperse the material in the gaseous state in a simple and reproducible way, (b) know the concentration in the gas, (c) deliver that gas under well-defined conditions to the olfactory mucosa which reside at the end of a short but ill-defined nasal passage lined with hairs, mucus, and various types of other cells, (d) avoid fatigue of the response (or adaptation, which changes the intensity and the nature of the response), (e) correct for the physical and psychological state of the percipient, including age (484), and (f) record the results using terms that are crude, imprecise, and descriptive, and, further, may not mean the same thing to each individual tested. Thus, relatively crude and empirical data on odor quality, intensity, and persistence is what must be used in order to probe the molecular nature of the olfactory receptor.

Anosmias have been reported for 92 compounds (442), but only four of these have been supported by serious genetic study. Five anosmias are listed in McKusick's compendium of genetic deficiencies (485–493). The odor classes associated with anosmias are urinous (androst-16-en-3-one), cyanide (hydrogen cyanide), musk (ω-pentadecanolactone), and putrid (butanethiol), with anosmia to Freesia flowers as another possible class. Other classes for which anosmias have been reported without family studies are sweaty, spermous, fishy, malty, minty, and camphoraceous (441, 442).

The olfactory sensitivity of monozygotic and dizygotic twins to acetic acid, isobutyric acid, and 2-*sec*-butylcyclohexanone could not be correlated with the genetic relationship (494). A similar study using androst-16-en-3-one as the stimulant indicated that identical twins were more similar in their response than fraternal twins (486). An extensive search using isovaleric acid as the stimulus revealed only a few individuals who were unresponsive and caused considerable olfactory discomfort to all those who worked in the same building where the tests were carried out.

The odorant used by the skunk to ward off unwanted intruders is a mixture of 3-methylbutane-1-thiol, *trans*-2-butene-1-thiol, and *trans*-2-buten-1-yl methyl disulfide (476, 495, 496). The related 1-butanethiol was tested in 1948. Of 4030 individuals tested, 17 failed to note the odor of a 10^{-3} M solution. Of the 17, 11 had generalized anosmias. Of the remaining 6, 4 could not detect a 10^{-2} M solution! On the basis of perfunctory inquiries into the olfactory abilities of the families, it was concluded that 1-butanethiol anosmia might be genetic in origin and recessive in character (487).

Hydrogen cyanide (HCN) anosmia is relatively common, 10–16% of all individuals being unable to detect the smell. This anosmia is dangerous for those who work with cyanides (chemists, electroplating industry workers).

Three different thresholds (low, medium, and high) for the detection of HCN were found using aqueous solutions of KCN as the source. No simple genetic pattern could be established in several independent studies (490).

In 1968, a specific anosmia for the odor of Freesia flowers was noted among 5–8% of Oxford University undergraduates tested. Homozygous recessives for a single autosomal gene were anosmic (492).

Chemical Probes of the Olfactory Response

Reactive agents are potentially useful in the study of the olfactory system, by labeling molecules or cells in association with electrical measurements of olfactory function. N-Ethylmaleimide (NEM, **46**) was shown to inhibit the generation of an electroolfactogram (EOG) in the frog olfactory epithelium in response to various stimulants. The fruity odorant ethyl butyrate prevented the inhibition (497). Very low concentrations of NEM (10^{-6} M) induce an EOG and an olfactory nerve signal similar to that produced by N-ethylsuccinimide (**47**). Higher concentrations change the EOG in ways that are partially prevented by the odorant isoamyl acetate. NEM reacts most rapidly with dissociated thiol groups, and the results suggest that thiol groups are exposed when the receptor is unoccupied by an odorant (498).

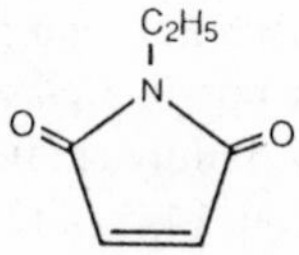

46 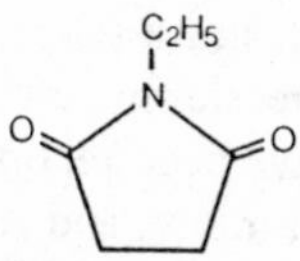 47

In an effort to increase the specificity of the response, a reactive odorant was used. A fruity odorant (also a tear gas!), ethyl bromoacetate was then shown to irreversibly inhibit the EOG response of the frog to isoamyl acetate and isoamyl sulfide, but had much less effect on the response to isoamylamine (499, 500). Isoamyl acetate protected the epithelium against inactivation by ethyl bromoacetate, but isoamyl sulfide did not. Other "war gases" have characteristic odors, "sulfide odor reminiscent of garlic" (mustard gas, $ClCH_2CH_2SCH_2CH_2Cl$), "fishy" (nitrogen mustard, $ClCH_2CH_2N(CH_3)CH_2CH_2Cl$), and "geranium" ("Lewisite," chlorovinylchloroarsine, $ClCH{=}CHAsCl_2$), and have been suggested as interesting olfactory labeling agents. However, the odor qualities are not universally agreed upon (374, 413) and further efforts to find specific chemically reactive odorants have not been successful (501). Chemical modification using a mercury derivative or iodine reduced EOG amplitudes from rat olfactory mucosa using odorants from many classes, but attempts to protect against modification with a number of the odorants failed (502). Treatment of tiger salamander olfactory epithelium with ethyl acetoacetate ($CH_3COCH_2COOCH_2CH_3$) followed by sodium cyanoborohydride ($NaBH_3CN$) diminished by ca. 30% the response to cyclopentanone but did not affect response to dimethyl disulfide. The reagent combination is known to convert free amino groups to alkylated derivatives via an imine (Schiff base) intermediate (503). A photochemically activatable labeling odorant has been used on the frog olfactory epithelium (504). Irradiation of 1-azidonaphthalene diminishes the EOG after irradiation on frog olfactory mucosa (505).

Some Applications of Human Olfactory Stimulants

Olfactory stimulants are molecules present in the immediate environment of an individual which convey information that may be important for survival. Broadly speaking, olfactory stimuli produce psychological changes in the individual, changes which frequently lead to some response on his or her part. Odors guide mammals in their relationships with one another and with the environment (506, 507).

On the attractant side, the smell of food may stimulate appetite and movement in the direction of the intensity gradient, perfumes and other pheromonal substances may produce appropriate sexual responses, and various smells enhance the attractiveness of objects (''new-car smell,'' ''seashore smell,'' etc.) which might lead to purchases or visits. The Japanese have a clock (Koban-Dokei) with which the time is indicated by a particular incense smell, each source burning for about 1 hour. On the other hand, anosmics of many ages and types have normal food intakes, but feel deprived by their failure to smell the food (360, 508).

Repellent odors have uses. Olfactory detection of chemical warfare agents (nerve gases in particular) utilizes the formation of the vile-smelling, but not very toxic, alkyl isocyanides. Nerve gases and many useful insecticides (parathion) inhibit acetylcholinesterase, which hydrolyzes acetylcholine. Reaction of the agent (4-toluenesulfonyl chloride is taken as a model) with a complex of zinc chloride and N-methylformamide yields CH_3NC, for which the detection limit is 7×10^{-9} g/L (Eq. [4-1]). Adaptation and interference from other odorants such as NH_3 and 2-methoxyphenol (guaiacol) were noted (509).

$$\text{(4-1)} \quad CH_3NHCHO.ZnCl_2.H_2O + 4\text{-}CH_3C_6H_4SO_2Cl \rightarrow$$
$$CH_3NC + ZnCl_2.H_2O + 4\text{-}CH_3C_6H_4SO_3H + HCl$$

The most common warning agent is the pentanethiol mixture added to odorless alkanes to make them detectable (the ''smell of gas''). The highly toxic H_2S smells like ''rotten eggs,'' and few people care to remain in the vicinity of even a modest concentration. Food that has been attacked by microorganisms often has an unpleasant smell (''rotten,'' ''moldy,'' etc.), but there are foods that are esteemed in spite of their nominally bad odor (e.g., limburger cheese, Germany).

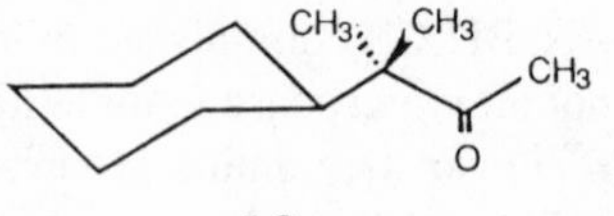

48

A "malodor counteractant," 4-cyclohexyl-4-methyl-2-pentanone (**48**), has been reported to inhibit the perception of thiols, amines, and short-chain carboxylic acids (510). No apparent commercial use has been made of the discovery, possibly because repellent warning signals would be neutralized. Synthetic molecular sieves, which are particularly active in sequestering odorants, are now used in personal care and household products (511).

Olfactory Neural Signals

If recognizable neural responses could be associated with particular odorants, a neural code for olfactants could be constructed. However, electrophysiological approaches have not yet led to an olfactory code, assuming one exists. Olfactory neurons are so small that direct recording from microelectrodes inserted into the cell has been used only by a few highly skilled investigators (512). Olfactory neurons of the tiger salamander (*Ambystoma tigrinum*) have resting potentials between 40 to 60 mV, and generate action potentials (>60 mV and 5 msec long) in response to a depolarizing current pulse. The frequency increases from 25 Hz to 35 Hz with larger current pulses (513). A fluorescent dye introduced through the microelectrode used for the electrophysiological measurements made the cell visible and allows correlation of the morphological type with electrical behavior. Single-cell responses to odorants have been recorded in one case (514).

Olfactory neural signals are detected after stimulation, using electrodes placed at appropriate points in the olfactory system (515) (Fig. 4.5). The signal varies in detail or content depending on whether groups (signals a, d, e, and f) or single cells (signals b and g) are being probed. A pulse of odorant produces a signal (electroolfactogram, EOG) from an electrode in the mucus adjacent to the olfactory mucosa, the potential decreasing as ions flow into the cell. The EOG sums events over many cells and a myriad of receptors. Neural signals from the olfactory neuron can be detected as spike activity (1–20 spikes/sec) (455) with an electrode placed next to an axon immediately below the olfactory epithelium. The individual spike (signal c) has a shape opposite to that of a signal obtained from an intracellular electrode, as expected. Certain odorants inhibit the neural signal.

The EOG is usually negative and has a rise time of about 1 second after an odorant stimulus. The maximum amplitude varies with the nature and the concentration (over several orders of magnitude) of the odorant, and returns to the

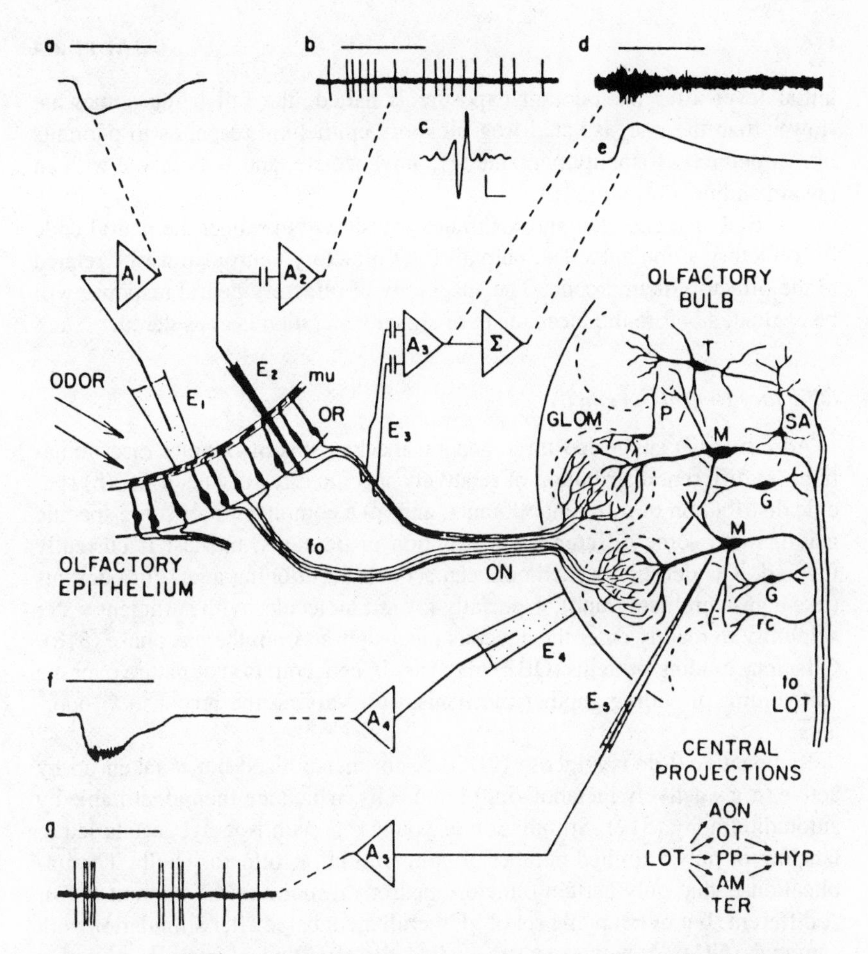

Fig. 4.5. Electrical signals detected at various locations in the olfactory system after stimulation by odorants. Electrode E1 is in the mucus (mu) bathing the cilia of the olfactory neurons (OR) in the olfactory epithelium. In response to an odorant (arrows), a change in potential over seconds is detected (the electroolfactogram [EOG], signal *a*). A metal microelectrode (E2) records the signals from single OR (signals *b* and *c*). A pair of electrodes E3 shows asynchronous spike activity in a bundle of nerves (*fila olfactoria*, fo) or the whole nerve (ON) as signal *d*. A smoothed summation of evoked activity (signal *e*) can be obtained after rectification. The axons within the ON terminate within glial-cell-bounded neuropilar complexes (glomeruli, GLOM) with processes from mitral (M), tufted (T), and periglomerular (P) cells. The axons of M and T project to the lateral olfactory tract (LOT). Other connections are made with granule (G) and short-axon (SA) cells. Recurrent collaterals (rc) from M are also present. Surface potentials at the olfactory bulb (with a saline-filled pipette, E4) resemble the sum of the EOG and signal *d* (signal *f*). A microelectrode E5 near M gives signal *g*. The line near signals *a*, *b*, *d* (*e*), *f*, and *g* represents five seconds, the length of the odorant delivery period. The voltages are positive upward. Signal *c* has calibration marks for 200 μV and 5 msec. Amplifiers are single-ended, except for A3, which is differential.

The LOT projects to the anterior olfactory nucleus (AON), the transitional entorhinal complex (TER), the olfactory tubercle (OT), the prepyriform complex (PP), and the amygdaloid complex (AM), the latter three projecting further to the hypothalamus (HYP) (reproduced with permission from reference 515).

initial level after the odorant exposure is halted, the fall being somewhat slower than the rise. Isolated frog olfactory epithelium responds to odorants like 2-methoxy-3-isobutylpyrazine (**1**), amyl acetate, and 1,8-cineole with an inward sodium current (516).

The EOG is generated and lost much too slowly to reflect the neural code for olfactory stimulants. The output of the olfactory neurons must be related to the olfactory neural code. The specificity of olfactory neural response will be evaluated before the mechanism of signal generation is considered.

Olfactory Neural Specificity

Adrian (517) suggested three general mechanisms of odorant discrimination: (a) different distribution of relatively specific olfactory neurons, (b) specific distribution of different odorants, and (c) a combination of some specific neurons and some differential distribution of odorants; the last is currently favored. The densely packed cilia can act as a partitioning agent for odorants (like a gas chromatograph), especially toward molecules with sufficient water solubility to readily enter the aqueous mucus phase from the gas phase (518). Olfactory binding proteins (OBP) may vary in concentration or nature over the epithelium, providing another mechanism for varying the response to odorants.

Radioactive 2-deoxyglucose (2DG) is not metabolized but is taken up by active (e.g., actively metabolizing) brain cells, which are then identifiable by autoradiography (519). Stimulation of young rats with isoamyl acetate led to labeling of only a limited number of glomeruli in the olfactory bulb. The implication is that only certain olfactory neurons respond to the odorant (520). A different, but overlapping set of glomeruli are labeled after stimulation with camphor (521). A new structure, called the modified glomerular complex (MGC), is found by 2DG labeling after stimulation by the suckling stimulant in rat pups and is adjacent to the accessory olfactory bulb (AOB). The AOB is quite active in the 1–2 days preceding birth (521). Embryonic mouse olfactory neurons are as sensitive to odorants as adult olfactory neurons but less specific (522).

A topographically distinct group of olfactory neurons is associated with the MGC. Horseradish peroxidase–wheat germ agglutinin (HRP-WGA) iontophoresed into the MGC is transported in a retrograde fashion to a limited group of neurons. Thus, a distinct set of neurons is associated with a distinct set of glomeruli (523). The fact that the suckling cue is involved (perhaps a genetically predetermined set of receptor molecules are the targets) precludes too wide a generalization of the finding that some topographic specificity of organization exists in the olfactory system. An anatomically analogous glomerular complex in insects is preferentially responsive to an important pheromone (523–525). Two classes of olfactory receptor neurons can be distinguished by a monoclonal antibody (526).

Origin of the Olfactory Neural Signal

Interaction of an odorant with an olfactory receptor leads to depolarization of the dendrite (cilia, olfactory rod) and the cell body of the olfactory neuron. If the depolarization is sufficiently large at the trigger region (axon hillock) of the axon origin, a neural signal is initiated which travels down the axon to activate transmitter release at synapses in the appropriate glomerulus of the olfactory bulb. The EOG response for many odorants implies that the olfactory neuron is depolarized by many different molecules and that the depolarizations are associated with neural signals (Fig. 4.5). Direct depolarization of the olfactory neuron with an intracellular microelectrode leads to neural signals.

At least two main questions must be answered: (1) how does the odorant produce depolarization? (2) how is the depolarization modulated to produce a signal characteristic of the odorant?

The mechanism by which the odorant leads to depolarization is not completely understood, but some significant points have been established. Most important is that the olfactory receptor is associated with a "G-protein," which in turn promotes the synthesis of cyclic adenosine 3′,5′-phosphate (cAMP) (**5**, Chap. 3) (527, 528). The flow of information is shown in Eq. (4-2).

(4-2) Odorant → Receptor → stimulatory G-protein (G_s) → adenylate cyclase → cAMP → depolarization

Odorant-sensitive Adenylate Cyclase

The cAMP is formed from adenosine triphosphate (ATP) by the enzyme adenylate cyclase (Adc), which is present at a substantial level in olfactory epithelium (529). The cilia may contain as much as 40% of the epithelial Adc in only 3% of the total protein (530). Rat olfactory epithelia also contain odorant-stimulated Adc (531, 532).

Frog olfactory Adc is activated by odorants and guanosine triphosphate (GTP) or the GTP analogues, guanosine 5′-(β,γ-imido)triphosphate (GppNHp) or guanosine 5′-O-(3-thio)triphosphate (GTPγS) as well as fluoride ion and forskolin. It is inhibited by the GDP analogue, guanosine 5′-O-(2-thio)diphosphate (GDPβS). These responses are characteristic of G-protein-activated systems (530–533) (see Chap. 5). Cyclic AMP is converted to adenosine 5′-phosphate by a phosphodiesterase; several inhibitors of the enzyme diminish by 40–80% the EOG induced by odorants from six different classes (floral, fruity, camphoraceous, minty, musky, and putrid) (534).

An early report that cAMP strongly excites the olfactory EOG and that PDE inhibitors prolong the EOG (535) has been confirmed in part. Patch clamp measurements on excised olfactory ciliary membrane suggest that both cAMP and cGMP (cyclic 3′,5′-guanosine monophosphate) but not cCMP (cyclic 3′,5′-cytidine monophosphate) act directly to produce an increase in membrane conductance (536). Such direct action of cyclic nucleotide on conductance has been found only for rod and cone membranes.

Fruity, floral, minty, and herbaceous odorant stimulation of Adc has been demonstrated. The stimulation is correlated within a particular series of odorants with hydrophobicity as measured by the estimated distribution coefficient between octanol and water. Several hydrophobic and normally agreeable odorants do not stimulate Adc activity. Interestingly, putrid odorants (isovaleric acid, triethylamine, thiazole) do not affect Adc activity, implying that another mechanism of odorant detection might exist (537). There is conflicting evidence (527, 537) on the activation of phospholipase C, which hydrolyzes phosphatidylinositol 4,5-bis-phosphate to inositol triphosphate (InP_3) and a diacylglycerol, the latter stimulating protein kinase C (538–540). One complication in using the isolated preparation is that olfactory binding proteins (OBP) might be lost, much as sugar-binding proteins can be released from *E.coli* by osmotic shock (Chap. 2). The OBP might be essential in transferring the stimulus molecule (odorant) from the mucus to the receptor, as well as from the air to the mucus (300).

According to eq. (4-2), raising the concentration of cAMP should increase the response of the olfactory neuron. One should observe an increase in depolarization after addition of cAMP, a situation which is achieved in part by inhibiting the hydrolysis of cAMP by phosphodiesterase with inhibitors, or more easily by addition of $N^6,O^{2\prime}$-dibutyryl-cAMP (533). In fact, the EOG response is decreased, opposite to the expected result. Free cAMP prevents depolarization, contradicting the results from the patch clamp measurements. A partial explanation suggests that depolarization must be promoted by cAMP formed within a protein complex, as schematically illustrated in Figure 4.6.

The odorant (O) interacts with the receptor (R) to yield a complex (OR) which undergoes a conformational change to yield the modified complex, OR′. The OR′ catalyzes the replacement of GDP in the olfactory G-protein (G) by GTP, giving G.GTP. The olfactory cyclase, Adc, and a kinase (K) form a complex (Adc.K) which gives rise to Adc.cAMP.K under the influence of G.GTP in a reaction with ATP. The cAMP must alter the conformation or reactivity of the kinase. Replacement of the cAMP by ATP followed by reaction between ATP and K leads to an active phosphokinase (K*). The latter affects ion channels in some way, possibly through phosphorylation of a K^+ channel and its activation. If dissociation of the cAMP is reversed through raising its concentration, the ion channel would be less affected and depolarization would be decreased. The results explained by Scheme 4.1 (below) need to be reconciled with patch clamp experiments, which show that cAMP and cGMP raise conductance through olfactory ciliary membrane.

Olfactory G-Proteins

The fact that GTP promotes cAMP formation suggests that a G_s-protein has a role in the olfactory response. Support for this conclusion comes from human genetics. Disturbances in the sense of smell were noted years ago in cases of pseudohypoparathyroidism (541), and such olfactory dysfunction has been

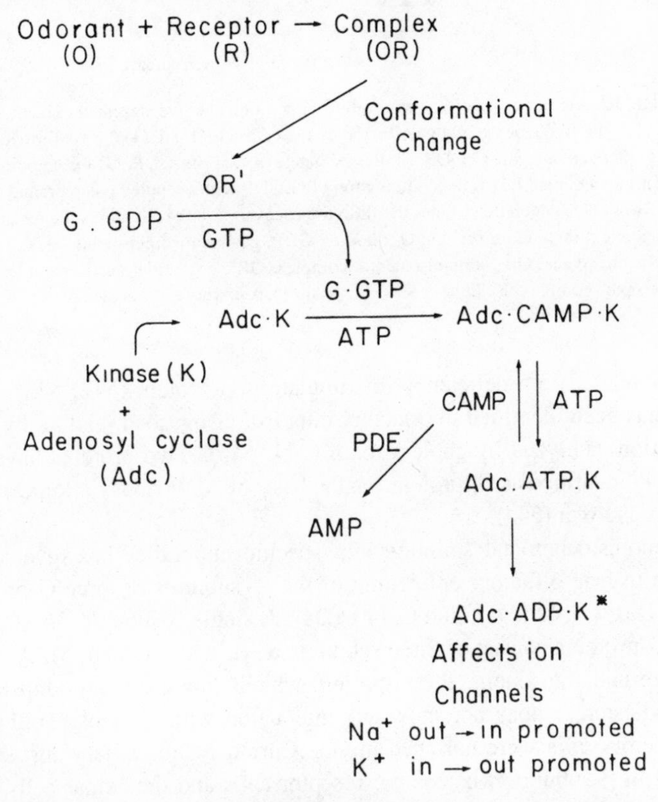

Fig. 4.6. Overall molecular scheme for olfactory response. The odorant (O) interacts with the receptor (R) to yield a complex (OR) which undergoes a conformational change to yield the modified complex, OR′. The OR′ catalyzes the replacement of GDP in the olfactory G-protein (G) by GTP, giving G.GTP. The olfactory cyclase, Adc, and a kinase (K) form a complex (Adc.K) which gives rise to Adc.cAMP.K under the influence of G.GTP in a reaction with ATP. The cAMP must alter the conformation or reactivity of the kinase. Replacement of the cAMP by ATP followed by reaction between ATP and K leads to an active phosphokinase (K*). The latter affects ion channels, possibly by activation of a K^+ channel through phosphorylation. If dissociation of cAMP is inhibited through a rise in its concentration, less active kinase would form per unit time, the ion channel would be less affected, and depolarization would be decreased. (Symbols: R, receptor; G, guanosine nucleotide binding protein; O, odorant; GDP, guanosine diphosphate; GTP, guanosine triphosphate; OR, ordorant-receptor complex; OR′, conformationally modified odorant-receptor complex; K, kinase; K*, phosphate-loaded kinase.)

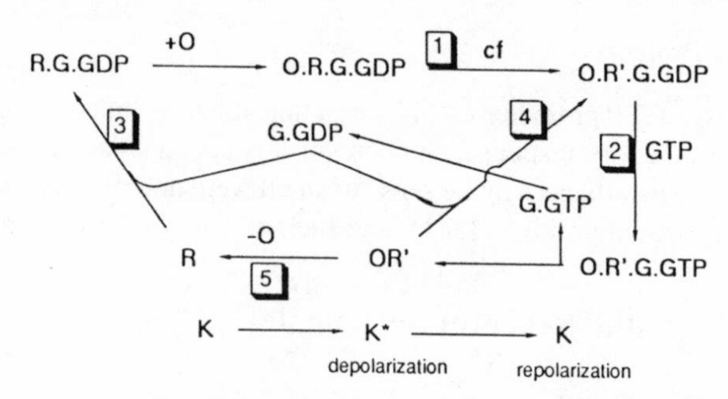

Scheme 4.1. Kinetic factors in olfactory receptor operation. The five significant kinetic steps through which the frequency might be affected by the odorant (O) are (1) the conformational change (cf) that activates the G.GDP; (2) the exchange of GTP for GDP; (3) the combination of G.GDP with unoccupied R; (4) the combination of G.GDP with occupied receptor; and (5) the loss of O from OR′, which determines the ratio of processes 3 and 4. Symbols: R, receptor; G, guanosine nucleotide binding protein; O, odorant; GDP, guanosine diphosphate; GTP, guanosine triphosphate; OR, odorant-receptor complex; OR′, conformationally modified odorant-receptor complex; K, kinase; K*, phosphate-loaded kinase; cf, conformational change.

carefully related to a deficiency in stimulatory G-protein (542, 543). A G_s-protein has been identified in olfactory cilia from frog, toad, and rat by ADP-ribosylation catalyzed by cholera toxin (544, 545). Two proteins have been tentatively identified as G_i-proteins on the basis of ADP-ribosylation catalyzed by pertussis toxin (544).

Immunohistochemical staining with specific antibodies has revealed the presence in frog olfactory epithelium of the α- subunits of three G-proteins, G_s (45 kDa), G_i (42 kDa), and G_i (40 kDa). A single β-subunit (36 kDa) was found. G-proteins consist of three subunits, α, β, and γ (Chap. 5). Antibody against retinal transducin, the G-protein which "connects" rhodopsin to a phosphodiesterase does not show any interaction with olfactory epithelium. The olfactory cilia were enriched in $G_{s\alpha}$. Curiously, the ciliary surface was enriched in β-subunit, whereas the receptor cells and the acinar cells of the mucus-secreting gland (Bowman's gland) showed the most interaction with antisera to a peptide against a peptide sequence common to all G-protein α-subunits (546).

Kinases and Other Components of the Olfactory System

The nature and role of the kinase indicated in Figure 4.6 is speculative, and based on knowledge about the phosphorylation of membrane proteins, including the nicotinic acetylcholine receptor (547).

The Olfactory Code

The question of how the odorant produces depolarization is answered in part with the idea that the odorant catalyzes the formation of cAMP, a "second messenger," via a G-protein. Pace and Lancet (527) have noted a number of other possibilities for depolarizing the cell, such as (a) direct action of the odorant on a channel; (b) other "second messenger" systems; (c) direct action on the G-protein system—a mechanism suggested for thiols, which they believe to have similar smells. However, no one with a sense of smell could confuse the unpleasant stench of thiophenol with the unpleasant odor of methanethiol. One is reminded of the young Gergel, who synthesized quantities of 1-butanethiol without realizing that its smell was to be avoided (548).

A more difficult task is to answer the second question of how the depolarization is modulated to produce a signal characteristic of the odorant. Among the potential odorant discriminators are olfactory binding proteins, olfactory receptors, and olfactory receptor neurons. At least some olfactory neurons can be differentiated by their surface proteins, probably glycoproteins. The lipid and protein portions of the membrane might influence the ease with which the olfactory receptor undergoes conformational change and thereby activate the G-protein. Some of these points are considered in the following sections.

Neural Frequencies

Information about the nature of the odorant (its class and its specific identity) must be carried in the frequencies of the neural signal produced by olfactory neurons. How can the odorant produce a characteristic frequency in a given neuron? The physical basis must lie in some property of the receptor and the associated proteins. Scheme 4.1 below is patterned after the general scheme for chemotaxis (Scheme 2.2) and shows how the speed of depolarization and repolarization might be connected to the action of the odorant on the olfactory receptor.

Depolarization leads to the firing of the nerve cell. Repolarization will prepare the cilia and dendrite to initiate another signal through depolarization. There are five kinetic steps in the scheme through which the frequency might be affected by the odorant (O): (1) the conformational change (cf) which activates the G.GDP, (2) the exchange of GTP for GDP, (3) the combination of G.GDP with unoccupied R, (4) the combination of G.GDP with occupied receptor and (5) the loss of O from OR′, which determines the ratio of processes (3) and (4). The rate of dissociation of G.GTP from its receptor complex and the rate of hydrolysis of GTP are probably independent of the nature of O. The rate of combination of O with R is likely to be too fast to control the frequency. The physical basis (distribution in time and space) for the rate of arrival of the odorant at R has been carefully analyzed (450), and seems to vary too much with conditions to provide a reliable code. After the odorant molecules arrive at the olfactory region, there is a "latency" period of 150–400 msec, depending on odorant concentration, for diffusion through the mucus layer to the receptors (450, 549). Olfactory binding proteins could modify behavior at this point. The rate of formation of K* could be determined by one or more of these five steps. More than one stable frequency might exist for such a system. On the molecular level, the interaction of the odorant with the receptor would control the rates.

Origin of Specificity of Neural Connections

Olfactory neurons are connected to specific glomeruli. The cells are replaced within 3 months by cells (cell longevity may be greater [452]), which presumably maintain whatever specificity existed. The most economical hypothesis for the specificity is that the receptor carries the specifying message, probably as glycosyl sequences. The receptor is synthesized in the cell body and migrates in either direction, either within the cell or within the membrane. Post-translational modification by glycosylation places the specifying message on the protein. At the axon end, recognition by the dendrite of a target cell (mitral or tufted) leads to the formation of synapses. The receptors in this region are then excluded by the presynapton components. Membrane carrying receptors in contact with basal cells could control (a) the nature of the message expressed by such cells, and therefore the specificity of the olfactory neurons for which they are the progenitors, and (b) the growth of the basal cell, which does not occur until the olfactory neuron fails to generate messages (550).

The variation in the specifying message could be due to a variation in an enzyme (as in the case of the ABO blood groups [551]), or in the receptor protein as the target for glycosylating enzymes. The latter seems more plausible in view of the relatively weak interaction of the odorant with the receptor.

The receptor specification hypothesis is closely related to the idea that a recognition molecule is coexpressed (552). An alternative idea is that the target neuron (mitral or tufted cell) would specify receptor expression (443).

Origin of Receptor Specificity

The wide variety of molecules that may be detected, the indications of differential sensitivity to odorants, the existence of various anosmias as well as the diverse sensations reported by humans, suggest that there are a number of olfactory receptors. These must exhibit the following properties: specific interaction with small molecules (HCN, H_2S, H_2Se); similar interactions with related molecules; highly characteristic responses to individual molecules; some genetic variability.

Without sequence and detailed structural information on the receptor molecules, a physical organic approach can be applied to the problem of receptor specificity. How can one bind a small molecule in a noncovalent, reversible fashion to a protein? Hydrogen cyanide (HCN) is a weak acid ($pK_a = 10$), which would be bound in the undissociated form. Its large dipole moment (2.9 D) could interact with an appropriate dipole or pair of dipoles of opposite sign. Among the candidate binding groups (receptor polypeptide side chains), imidazole (histidine) or amide (asparagine/glutamine) have appreciable group dipoles. A hydrogen cyanide molecule fits easily into the space between the two imidazole rings (amide groups). An Im.HCN.Im complex would have dipole-dipole attractions similar to those in liquid HCN (high dielectric constant, D = 110). In the case of an H_2S receptor, two SH groups (cysteines) seem appropriate on the grounds that hydrogen sulfide is not highly polar, and requires two polarizable groups for interaction. A SH· · ·S bond might also be involved.

Receptor binding sites may be modeled as follows: (1) Groups like those in the odorant are chosen for the primary interaction. (2) Groups are selected from one of two types: (A) flat, and (B) small; each type includes (a) charged, (b) dipolar, and (c) nonpolar groups. A list of potential primary receptor groups is given in Table 4.4.

A set of primary olfactory receptors can be derived from the list. Charged groups are not excluded, in view of the bilayer groups in adrenergic receptors and rhodopsin (Table 4.5).

A Structural and Dynamic Model for the Olfactory Receptor

The exciting discovery that the β-adrenergic receptor (βAR) (553, 554) and the muscarinic acetylcholine receptor (mAChR) (555–557) (Chap. 8) are homologous to rhodopsin (Chap. 5) raises the possibility that many, if not all, G-protein receptors are related to rhodopsin (558, 559). The important dopamine receptors fall into at least two classes, D-1 activating Adc and D-2 possibly inhibiting the enzyme (560, 561). The olfactory receptor is also a G-protein-linked receptor. The polypeptide gp95 has been identified as a plausible receptor candidate (465–466). The molecular weight of sugar-stripped gp95 (ca. 65,000) is not far from that of mAChR (55–60,000). A plausible structural arrangement for the olfactory receptor is then one in which there are seven transmembrane helices, arranged in an ellipse like those proposed for rhodopsin (562–564).

The proposed binding-site groups can be present on different helices of the interior of the ellipse within the bilayer. If the groups are on helices 3 and 6, and the odorant enters the region between these groups for binding, helices 3 and 6 must be pushed apart by a distance equal to one of the dimensions of the odorant. A change in the shape of the olfactory receptor molecule must ensue, its nature depending on the shape of the odorant and the interactions between the other helices. The interaction between the receptor and a G-protein on the

Table 4.4. Candidate Receptor Groups

A. Flat

a. Charged: $CH_2(CH_2)COO^-$(glu, asp); $(CH_2)_3NH_2C(=NH_2^+)NH_2$(arg)

b. Dipolar: $ImCH_2$(his); $CH_2(CH_2)CONH_2$(gln, asn); $HOC_6H_4CH_2$(tyr)

c. Nonpolar: $IndoleCH_2$(trp); $C_6H_5CH_2$(phe)

B. Small

a. Charged: $(CH_2)_4NH_3^+$ (lys)

b. Dipolar: $HOCH_2$(ser); $HOCH(CH_3)$ (thr)

c. Nonpolar: $HSCH_2$(cys); $CH_3SCH_2CH_2$(met); $(CH_3)_2CH_2$(leu); $CH_3CH_2CH(CH_3)$(ile); $(CH_3)_2CH$(val)

Table 4.5. Primary Olfactory Receptor Groups

Receptor Pair		Possible Interactants
Im	Im	HCN,-CHO
$CONH_2$	$CONH_2$	HCN,-CHO
HOC_6H_4	HOC_6H_4	RCOOR
C_6H_5	C_6H_5	ArH
HO	HO	ROH
HS	HS	H_2S
$(CH_3)_2CH$	$(CH_3)_2CH$	*n*-pentane
$CH_3CH_2CH(CH_3)$	$CH_3CH_2CH(CH_3)$	isopentane

Note: These groups are attached to β-CH_2.

interior of the bilayer (within the cilium) would be altered, and thus make possible the exchange of GTP for GDP.

A model olfactory receptor in the shape of rhodopsin, an ellipsoidal arrangement of seven helices, is shown in Figure 4.7. (Only the portion of the receptor within the bilayer is shown.) The interior is the binding site. In the model, each helix has a valine on the uppermost turn directed toward the center of the ellipse, making the binding site very nonpolar. The binding site is loaded with an *n*-pentane molecule to suggest how well this arrangement accommodates the shortest straight-chain hydrocarbon with an odor. The model corresponds to the conformationally altered odorant-olfactory receptor complex (OR′).

An HCN molecule inserted as a hydrogen-bonded link between the N and NH of the imidazoles of histidines would increase the separation between the rings by ca. 3 Å. A sulfur atom between two cysteines would also increase the

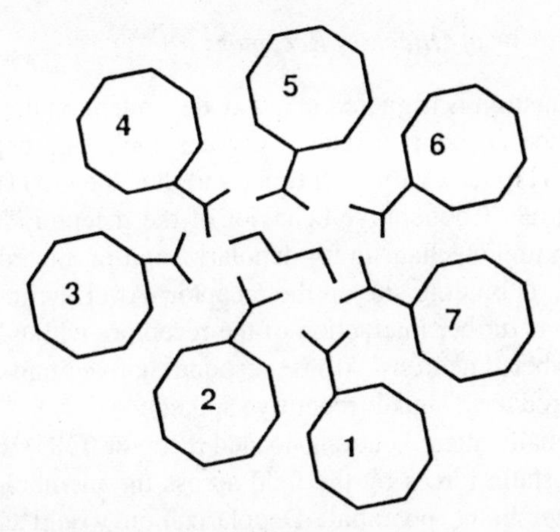

Fig. 4.7. An olfactory receptor model. Seven helices are arranged in an ellipse like that proposed for rhodopsin or known for bacteriorhodopsin. The links between the helices are not shown, but they presumably are extracellular between helices 2–3, 4–5, and 6–7. The side chains directed at the interior at the first turn of the helix are all isopropyl, part of the amino acid valine. The receptor is shown as loaded with *n*-pentane, which fits nicely into the model site. Inspection of the examples shown in Table 4.1 shows that the largest olfactants (steroids, musks) reported could be accommodated. If two helices had one variable amino acid each, 16 × 16 or 256 different receptors could be produced. An alternative binding mode would be one in which the pentane was aligned parallel to the bilayer helix direction.

separation by ca. 3 Å. If these changes were transferred directly to the G-protein GDP binding site, preference for GDP might be changed to a preference for GTP through increased separation between binding groups.

Although a simple picture like that just described is unlikely, the notion of specific geometric molecular changes is introduced.

The rate at which odorant is dissociated from the conformationally altered complex might be the property that most tightly couples the behavior of the system to odorant type. The lifetime of the OR′ complex determines how much G.GTP is formed, and therefore how much depolarization occurs.

Attributes of Olfactory Receptor Model

Many properties of the olfactory system can be explained with the olfactory receptor model, as crude as it is in the absence of sequence and structural information. Among the features that can be approached on the basis of the model are (1) cooperative behavior of the olfactory receptors, which seems necessary for multiple-spike generation; (2) the qualitative success of the shape theory for classifying odorants; (3) the dependence of odor sensation on odor concentration and the presence of other odorants; (4) some genetic aspects of the olfactory system, such as the apparent multiplicity of receptors and the occurrence of anosmias.

Cooperative Behavior of Olfactory Receptors

The olfactory neuron is triggered in a way dependent on the structure of the odorant to produce an odorant-related nerve signal. The depolarization induced by the G.GTP must vary with time, and thus the G.GTP concentration must vary with time. Cooperative behavior of the receptors is required. The model offers a natural mechanism for depolarization-promoted dissociation of the odorant from its binding site on the receptor. After the initial depolarization by an odorant, further interaction of the receptors with odorant would be cooperative. ''Labeled neurons'' (those responding to certain classes of odorants) must still produce a signal (repetitive spikes).

A conformationally altered odorant-loaded receptor (OR′) would be subject to certain electrostatic forces by the field across the membrane. The field is related to the membrane potential. Depolarization would change the field across the membrane, and could certainly alter the binding constant for the odorant. It is reasonable that the dissociation be promoted in the absence of the field, thus opening all of the receptors in a given cilium at once. This should lead to cooperative behavior. A schematic view of the receptor shows conformational changes in response to odorant loading (Fig. 4.8).

Shape Theory and Olfactory Receptors

The model receptor could accommodate a wide variety of molecules of the hydrophobic type. Replacement of one or two amino acids would change drastically the specificity of the receptor without changing the mechanism through which G-protein is activated. Shape theory combined with knowledge of binding-site groups (Table 4.5) might well be useful in designing new odorants. The receptor site is chiral so that distinctions among enantiomers can be understood. Specific proposals about binding in terms of the importance of stereochemistry and the precise positions of weakly polar groups (473, 474, 565–570) can be treated in terms of the model receptor site. The model receptor site opens up consideration of both odorant and receptor site structure.

Odor-sensation Dependence on Odor Concentration and Purity

The signal (spike) frequency from a single cell increases as the odorant concentration rises (455). In the model, an increase in odorant concentration would increase the concentration of G* and increase the rate at which depolarization occurred. An extensive study of odorants of overlapping and different classes showed that odorants of the same class (Table 4.1) gave more similar spiking frequencies than odorants of different classes. However, no two odorants gave the same result (571). The model suggests how rather subtle differences between molecules might appear as differences in frequencies.

The concentration of odorant influences the magnitude of the generator potential and the number of action potentials produced (572).

Human responses in the olfactory bulb to odorants have been studied in the course of brain operations involving regions at or near the olfactory lobe. Sets of signal frequencies appeared for particular classes of odorants. However, only eleven patients could be investigated, so the assignments must be regarded as tentative (573). Frequency components have been reported in the olfactory responses of rainbow trout (*Salmo gairdneri*) (574) and himé salmon (*Oncorhynchus nerka*) (575).

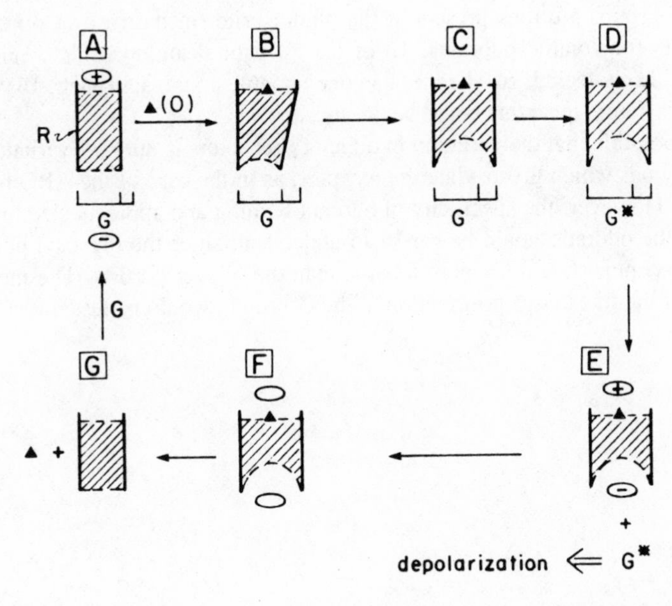

Fig. 4.8. A schematic side view of the olfactory receptor model (Fig. 4.7) during operation. Stages are marked with letters inside squares. The shaded area represents the nonpolar receptor interior. The field is produced by the charges shown in ellipses, with a blank indicating no relative difference in charge, and thus no field. In stage *A*, the receptor is unoccupied. Combination with an odorant (black triangle) leads to stage *B*, which expands in a conformational change to stage *C*, that of an activated odorant-receptor (OR′) complex. The binding arrangements of the G-protein and receptor change as a result of the conversion of OR to OR′, with G-protein becoming activated for the exchange of GTP for GDP (stage *D*). The membrane potential is appreciable, which must produce a considerable field across the low-dielectric portions of the receptor or its complexes. A horizontal expansion of the interior could lead to a shrinking of the vertical interior as hydrophobic groups rotate upward (SGR, Chaps. 6, 7) through stages A–D. Dissociation of the G.GTP complex yields an activated G* (stage *E*), which catalyzes a sequence that ultimately affects ion channels and produces depolarization. The receptor-odorant complex (stage *F*) is then subjected to a much smaller field, after depolarization, and the odorant dissociates. All of the receptors would be coordinated by this circumstance, and subsequent depolarizations would occur at a rate determined by the competition between the dissociation of the odorant (stage *F* → stage *G*) and the reformation of the complex with G (stage *F* → stage *C*).

The purity of odorants will clearly affect the sensations provoked in humans. It seems likely that musks occupy the largest binding sites, and possibly change sensations by preventing occupancy of such sites by smaller odorant molecules.

Some Genetic Aspects of the Receptor Model

The model leads to a simple picture for anosmias. In the case of fish (rainbow trout, *Salmo gairdneri*), amino acids are detected by the olfactory system (576) in a pH-dependent fashion (577) and bound by the cilia (578); charged binding groups are thus present at the binding site. In the case of olfactory receptors for volatile odorants, 16 of the 20 natural amino acids would be reasonable choices. If two helices had one variable amino acid each, 16×16 or 256 different receptors could be produced.

It is possible that the variation in olfactory receptors is due to a variation in the enzymes which glycosylate the receptor, as in the case of the ABO blood groups. However, the specificity of odorant binding and apparent size limitation in the odorant would be harder to understand since the glycosyl portion of the receptor should be more flexible than the bilayer portion. The mechanism for the transfer of information to the G-protein would be less direct.

5 Visual System

''Seeing is believing''

— Folk saying

''None so blind as those who will not see''

— Folk saying

Information about the external world is needed for most learning, even if one takes into account genetically programmed predispositions to respond to certain stimuli. Humans acquire most of their information about the external world via their eyes, the receptor organs of the visual system (579–583). In the first chapter, I noted that level assignments could aid us in understanding relationships within complex systems. A level analysis for the visual system is presented in Table 5.1. It begins with family groups and ends with the molecular fragment responsible for the absorption of light, and thus for initiating the sequence that leads to the perception of light. The components at each level are indicated together with a global description of function. The ultimate purpose of this chapter is to explain in molecular detail the conversion (''transduction'') of photons into neural signals.

The space from which an eye receives light is called a visual field. The two eyes of mammals have receptive fields covering overlapping visual fields. Light from a particular part of a visual field passes through the pupil and the lens and then excites a particular group of cells in the retina. The excitation produced in the retina is passed on through a number of intermediate cells to a portion of the brain concerned with visual inputs, the visual cortex. Different parts of the receptive fields in the retina excite different groups of cells in the visual cortex. The neural transfer pathways via the optic nerve and a lateral geniculate body (LGS) are illustrated schematically in Figure 5.1. Light from the receptive field of the right eye is passed in substantial measure to the left LGS. Signals from the right LGS are transferred to the visual cortex of the right occipital lobe; signals from the left LGS are transmitted to the visual cortex of the left occipital lobe. The operation of the visual system can be described by the general sequence, stimulus → receptor organ (eye) → neural processing (retina) → brain (lateral geniculate bodies, visual cortex).

Organ: Human Eye

The human eye is located in the head, immediately in front of the brain. Most striking is a dark pupil, an opening controlled in size by the surrounding

Table 5.1. Level Analysis of the Visual System

Level	Description	Program
14	Family groups: Human families	
		Language: Eye contact
13	Organisms: Humans	
		Organism organization: Sensory systems
12	Organ systems: Visual	
		Organ system organization: Vision mapping to cortex
11	Organs: Eye	
		Organ structure: Image focusing through lens on detector
10	Tissues: Retina	
		Tissue organization: Signal detection and processing
9	Cells: Rods and cones	
		Cell segments: signal modulation
8a	Organelle system: Outer segments	
		Segment structure: Signal transduction
8	Organelles: Rod disks	
		Organelle assembly: Signal detection
7	Biochemical systems: Rhodopsin complex	
		System organization: Photon transduction
6	Polymolecules: Rhodopsin	
		Polymolecular structure: ''Photoswitch''
5	Molecules: 11-*cis*-Retinylidene-ϵ-lysine	
		Molecular structure: Light absorption
	Visual system specificity lost below this level	

Note: Level numbers are those specified in Table 1.3.

pigmented iris muscle (colors: blue, brown, hazel, grey). The iris regulates the intensity of the light admitted into the eye. Behind the pupil, a capsulated lens attached to fine collagen fibers (*zonula*) is controlled in focal length by the ciliary muscles. The primary focus of the eye is on the foveal region of the retina. The eye is surrounded by the *sclera*, which is transparent in front (cornea) and white over most of the eye. The anatomy of the eye is illustrated in Figure 5.2.

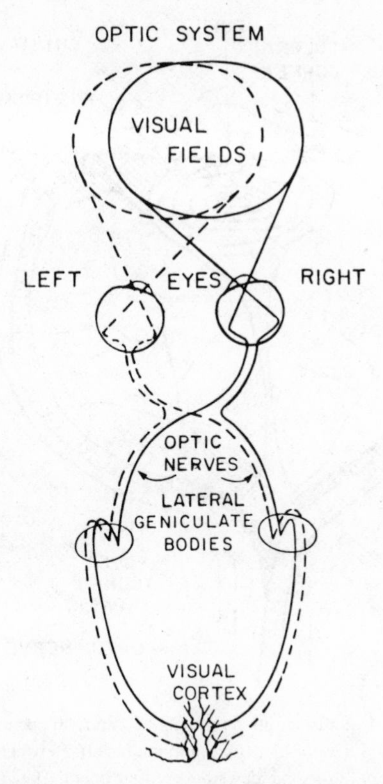

Fig. 5.1. The human visual system. The neural connections of the eye to the visual cortex of the brain are shown in schematic form. The optic nerve connects the eyes to the lateral geniculate bodies (LGS). Nerve signals are transferred for further processing to both the left- and right-hand portions of the intermediate (LGS) and final (visual cortex) portions of the system.

Transparency is critical for efficient passage of light through the lens, an organelle composed of layers of transparent cells built up in "onion-skin" layers. The major proteins are "crystallins." Transparency is lost in several ways (protein polymerization, galactose precipitation), with clouding of the lens (cataracts). The lens has the highest glutathione (GSH) concentration (ca. 10^{-2} M) of all tissues, presumably to intercept free radicals which may cross-link and thus lead to the polymerization of the lens proteins.

Lens cells are not replaced but added from the periphery. The lens loses flexibility with time, probably due to cross-linking of the crystallins. Individuals past the age of 40 or 50 normally require artificial lenses to focus light on the retina.

Beyond the lens is a hemispherical cavity filled with a hyaluronic acid-rich liquid ("vitreous humor") and lined with the cells of the retina. The portion of the retina adjacent to the liquid forms nerve processes which lead to the

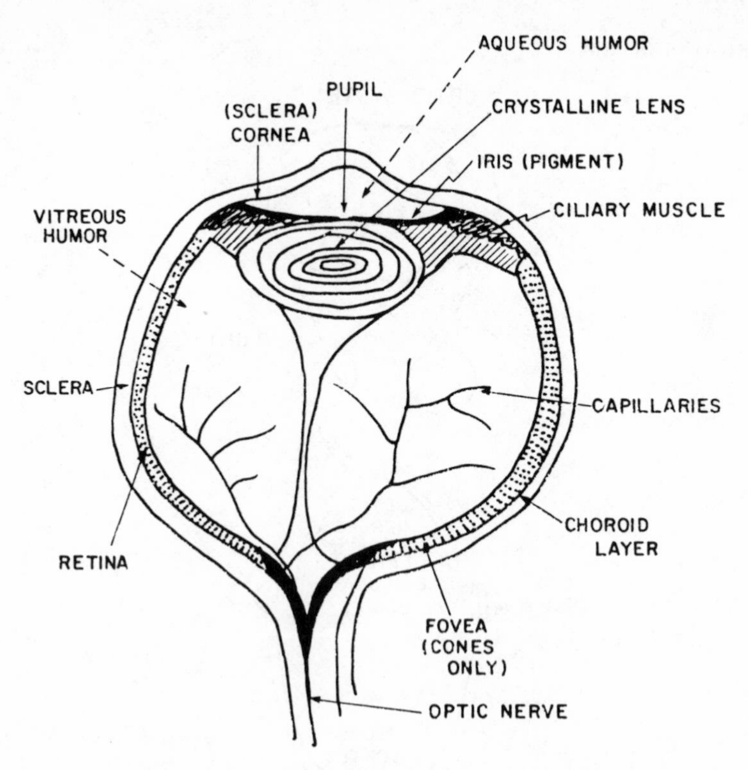

Fig. 5.2. The human eye. Light is admitted through the pupil, an opening controlled by the characteristically pigmented iris muscles (blue, brown, hazel). Between the iris and the cornea is a liquid (''aqueous humor'') that is renewed every four hours. The cornea is a transparent extension of the sclera, which covers the whole eye, has a convex shape, and is the major focusing element of the eye. The crystalline lens is built of layers of transparent cells, and serves as a weak focusing element for the light entering the pupil.

The light is focused on the retina through the viscous liquid (''vitreous humor'') located between the lens and the retina. The optic nerve arises from the nerve cells located on the inner side of the retina and extends to the lateral geniculate bodies of the brain. The detector cells in the retina are either rod-shaped cells (''rods'') or cone-shaped cells (''cones''). One small section of the retina (the fovea) contains only cone cells.

Between the retina and the sclera is a protective (and, in some species, reflective) region called the choroid layer. In the cat's eye, a reflecting layer (''*tapetum lucidum*'') augments the light-gathering capacity of the eye and accounts for the reflections observed in eyes of cats at night. The choroid layer of a small mammal, the bush baby, contains a pigment that fluoresces under ultraviolet illumination, enabling a response to uv light without change in the detection apparatus.

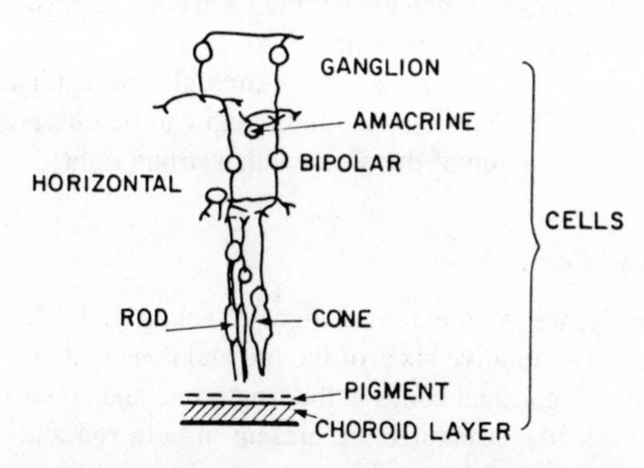

Fig. 5.3. Structure of the human retina. Rod and cone cells are located on the side of the retina farthest away from the light that enters the eye. The cells form synapses with bipolar cells, the latter in turn being connected to ganglion cells. The ganglion cells give rise to the fibers that combine into the optic nerve. Horizontal cells regulate the interaction between rod and cone cells, and amacrine cells the interaction between ganglion cells.

optic nerve. Immediately outside the retina is the choroid layer, which is lined with a protective-pigment epithelium.

Tissue: Retina

The retina is a tissue composed of light-detecting cells, nerve signal processing cells, and cells that have various protective and recovery functions. The general structure of the retina is shown in Figure 5.3.

The light-detecting cells, rod-shaped cells (“rods”) and cone-shaped cells (“cones”), are located immediately above the pigmented protective cells of the choroid layer. It is surprising that light must pass through the other cells of the retina before encountering the detector cells. Approximately 10^8 rods and cones are found in the retina, but there are only 10^6 optic fibers. Considerable signal processing occurs within the retina before transmission to the LGS. The ratio between outgoing fibers and detector cells is much closer to 1:1 for the cells in the *foveal* region of the retina. All of the cells in the *fovea* are cone cells.

The rod and cone cells form synapses with bipolar cells, which in turn form connections with the ganglion cells. Axons of the latter cells make up the optic nerve. Light-induced nerve signals (increased hyperpolarization) are affected by horizontal cells through hyperpolarization graded by the nerve signal and through lateral inhibition, which enhances contrast between neighboring detector cells. Red and green cone signals are “opponents” in cone-connected horizontal cells. Amacrine cells are divided into a number of subsets according to neurotransmitter, and produce signal spikes which presumably influence the spike outputs of the various types of ganglion cells (584). Signals from rod cells with depolarizations much greater than 5 mV are clipped as a preliminary form of visual information processing (585).

The blood supply of the retina passes through fine capillaries below and above the retinal cells. Those in the light path can be observed within one's eye during an examination of the retina with a strong light.

Cells: Rods and Cones

The morphologies of the retinal light-detecting rod and cone cells are closely related. The relative sizes of the rod and cone cells vary with the species, rods being larger than cones in the frog retina and cones larger than rods in primates. Disk-like structures are present in both rod and cone cells, and are formed by infolding of the cell membrane. In rod cells, the disks separate from the cell membrane. In cone cells, the disk-like structures remain as infolds of the cell membrane.

Rod and cone cells are divided into four segments (Fig. 5.4). The "outer" segment contains the disks or infolds. Rod outer segments (ROS) are easily pinched off by mild shaking, and are a source of visual pigment-rich material.

The disk structures are formed by infolding (and pinching off for disks) at the beginning of the outer segment. The disks migrate over several days to the end of the cell, are shed and then phagocytized by the cells of the protective pigment epithelium. Shedding may be promoted by the pineal gland hormone, melatonin (N-acetyl-6-methoxytryptamine) (586).

The metabolically active second segment is packed with mitochondria. It is connected to the outer segment by several cilia, which have a typical 9 + 2 structure, possibly for fast transport of metabolites and proteins. The third segment contains the nucleus. The last segment forms synapses with bipolar and horizontal cells.

Rod and Cone Cells: Intensity and Wavelength Sensitivity

Approximately 10^8 rod and 10^7 cone cells are found in the retina. Rods are much more sensitive to light than cones. To obtain a curve of apparent brightness versus dark adaptation time, the eyes of a human subject are dark-adapted, then exposed to a light of controlled brightness. The apparent brightness of the light is noted for different adaptation times. The wavelength variation of the low-sensitivity response corresponds to that of cone cells, whereas the wavelength dependence of the high sensitivity response parallels that of rod cells. The rod cells are much more sensitive under optimum conditions. An hour or more in the dark might be required to attain the highest degree of dark adaptation (Fig. 5.5). The cone density varies by a factor of almost 3 in a small sample of human retina, a range which might account for individual

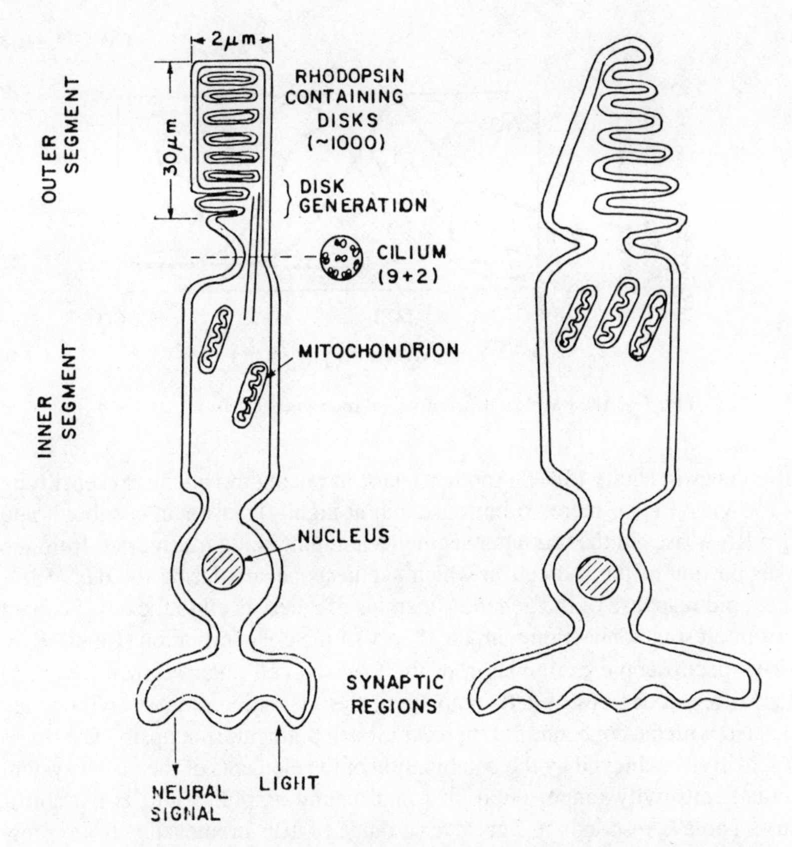

Fig. 5.4. Rod and cone cells. The light-detector cells of the retina are divided into two segments. The outer segments (OS) contain the light-absorbing pigments, rhodopsin or iodopsin. The inner segment is composed of three sections containing mitochondria and other regular cellular elements, the nucleus, and synapses to bipolar cells.

The rod cell outer segment (ROS) is about 30 μm long and about 2 μm wide. The rhodopsin pigment is located in 1000 disk-like structures, which are formed through pinching off from the cell membrane near the narrow neck between the outer segment and the inner segment. The corresponding structures in the cone cells do not pinch off in most species and contain iodopsin.

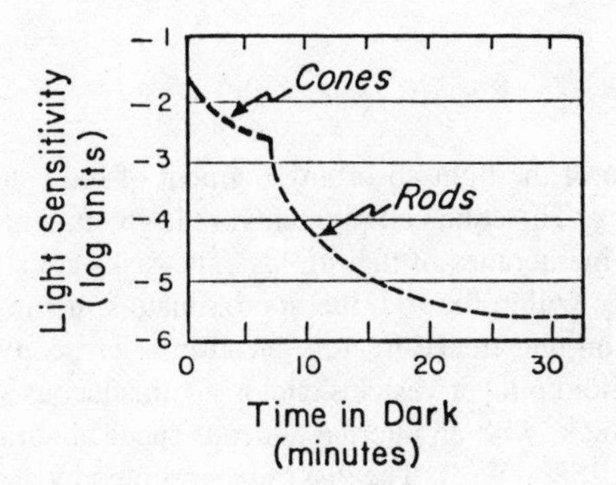

Fig. 5.5. Light sensitivity of the retinal cells of the eye after specified periods of dark adaptation.

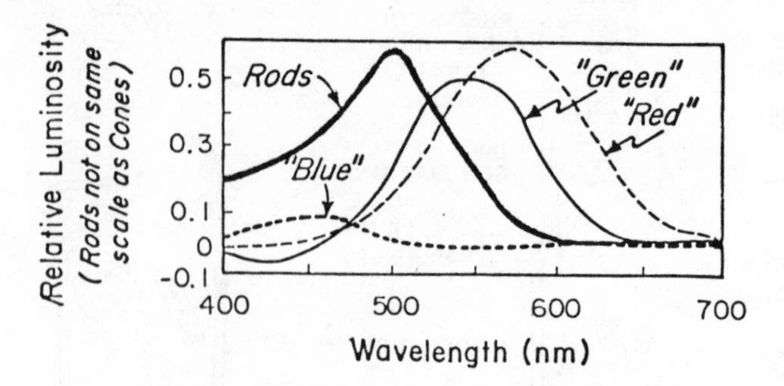

Fig. 5.6. The wavelength sensitivity of rod and cone cells.

differences in acuity (587). Another factor in raising the nocturnal sensitivity of rod vision is an increase in transducin at night. Transducin α-subunit and T_α mRNA rises in the rod outer segments at night, being transported from the inner portion of the rod cell in which synthesis occurs during the day (588). The rapid response of the genetic apparatus of a nerve cell to the environment is of interest in connection with the theory of memory formation (Chap. 9).

By spectroscopic examination of the frog rod cell outer segments excised after different degrees of dark adaptation, the adaptation process could be associated with the concentration of receptor cell pigment, rhodopsin. The range of sensitivity achieved by the combination of the elements of the visual system (retinal sensitivity range, pupil size, and neural amplification) is extraordinary. The eye responds to light over a range of 10^{10} in intensity, from a few quanta to bright sunlight.

The eye is sensitive to wavelengths between 400 and 700 nm. The wavelength response of rods parallels the absorption spectrum (λ_{max} 500 nm) of the rod pigment, rhodopsin. There are three types of cone cells: red-sensitive (λ_{max} 565 nm), green-sensitive (λ_{max} 535 nm), and blue-sensitive (λ_{max} 435 nm). The responses of the receptor cells as a function of wavelength are shown in Figure 5.6.

Organelles: Disks

A major portion of the light-absorbing pigment of rod cells, rhodopsin, is present in the disks. The cone cell pigments, red, green, and blue iodopsins, are located in the membranes of the infolds. The disks are stacked with considerable regularity within the rod; the approximate structure can be defined by X-ray diffraction and freeze-fracture electron microscopy. The disks are flattened phospholipid bilayer vesicles enclosing an aqueous space. The disks are about 150 Å thick, with an internal aqueous space about 70 Å wide (protein-free, ca. 25 Å [589, 590]). The disks are separated by about 160 Å.

Mammalian disk membranes have less protein than most mammalian membranes (44%, versus 57% in Golgi membranes) and a high proportion of unsaturated lipids (more than 50% of the acyl groups have four or more double bonds, and most have six [$C_{22:6}$]), the latter being essential for normal retinal function (591). Substantial amounts of longer (C_{24}–C_{36}) tetra-, penta-, and hexaenoic acyl derivatives of phosphatidylcholine are found in the rod outer segment of most species (592, 593). Diunsaturated phosphatidylcholines contain as much as 25% of the longer polyunsaturated acyl groups (594, 595). Present are all three major types of phosphatidyl lipids—phosphatidylethanolamine, phosphatidylcholine, and phosphatidylserine (ratios, 44:36:15)—along with small amounts of phosphatidylinositol (2%), sphingomyelin (<3%), and a relatively low amount of cholesterol (<10%) (596). Rhodopsin has a high lateral mobility in the disk bilayer; proteins are much more laterally mobile in bilayers made from unsaturated phospholipids than those assembled from saturated phospholipids.

The major part (90%) of the disk protein is the light-absorbing pigment, rhodopsin. Other proteins associated with the disk membrane include transducin, a 3′,5′-cyclic guanosine monophosphate (cGMP) phosphodiesterase, and a "rim protein," connecting disk edges to the ROS membrane. An ATP-dependent protein kinase phosphorylates conformationally altered rhodopsin, which may also be inhibited by complexation with "arrestin," the "48K protein" that may be responsible for an autoimmune disease (597), and an NADP-dependent retinol dehydrogenase to catalyze the conversion of retinol to retinal. A summary of important disk-related proteins is given in Table 5.2.

Table 5.2. Proteins Associated with Disks

Protein	Membrane Association	Mol. Weight (kDa)	Molar Quantity vs. Rhodopsin
Rhodopsin	Intrinsic	39[a]	1
Transducin ($T\alpha + T\beta + T\gamma$)	Peripheral	~80 (39 + 37 + 6)	0.1
"48K"[b]	Soluble	50	0.03
cGMP Phosphodiesterase + inhibitor(I)	Peripheral	~180 (88 + 84 + 13)	0.01
ATP-dependent rhodopsin kinase	Soluble	65	0.001
"Rim protein"	Intrinsic	240	0.003

Source: Adapted from Table 1, reference 583.

[a] Not including two polysaccharide chains.

[b] Antigen responsible for experimental autoimmune uveoretinitis, reference 597 ("arrestin").

Proteolysis of rhodopsin (598) and sequence information suggest that the rhodopsin molecule crosses the bilayer seven times. Electron micrographs of disk-derived vacuoles labeled with ferritin-labeled concanavalin A indicate that the carbohydrate portion of rhodopsin is located on the interior side of the disk bilayer. Concanavalin A binding (one monomer per retinylidene group) is inhibited by α-methyl-D-mannoside.

Biochemical System: Rhodopsin-Transducin System

Disk and cytoplasmic proteins are involved in the translation (''transduction'') of the light signal (light absorbed by rhodopsin) into a biological signal (Eq. [5-1]).

(5-1) Rhodopsin → Rhodopsin-X → Transfer/Amplification Agents → Membrane

Rhodopsin-X (RX) is the biochemically active, conformationally altered form of rhodopsin resulting from changes following light absorption. Rhodopsin-X interacts with another protein, transducin, and the combination acts to produce a transfer agent and amplification of the light signal (599).

The need for the transfer agent is evident. First, rhodopsin is present in disks that are structurally separated from the cell membrane, and yet the signal that emerges from the rod is produced in the membrane of the outer segment (that part of the rod cell which contains the disks). One is reminded of the agent that transfers the signal from the chemotaxis receptor to the flagellar motor of bacteria (Chap. 2). Second, the number of rod membrane ion channels that are affected by the absorption of light is much larger than one, i.e., amplification of the light signal has occurred.

Two candidates for the transfer agent have been proposed, Ca^{++} ion and cyclic guanosine 3′,5′-monophosphate (cyclic GMP or cGMP) (600). Calcium ion does fulfill certain criteria for the role of transfer agent: (1) Na^+ channels in the cell membrane are closed by the introduction of Ca^{++} (601), and (2) some Ca^{++} is produced by light stimulation of the rod, as shown by light emission from aequorin in the horseshoe crab (*Limulus*) eye (602). However, the rod calcium concentration falls on illumination and recovers in the dark. Apparently, both sodium and calcium enter the rods by way of sodium channels. Light-stimulated closure of the channels does not prevent calcium efflux via a light-insensitive sodium/calcium exchanger, and the internal calcium thus decreases (603, 604). This result is opposite to what would be needed for calcium to act as a transfer agent, but in agreement with the possibility that a modulator protein complex with Ca^{++}, $M.Ca^{++}$, dissociates to yield M*. The M* promotes the formation of cGMP from GTP and leads to adaptation within the photoreceptor (605–608).

The second possible transfer agent, cGMP, is present in high concentrations (0.3 μM) in rod outer segments (ROS) and is turned over at a high rate (^{18}O-exchange) even in the dark. ROS contains a light-activated phosphodiesterase which catalyses the conversion of cGMP to 5′-GMP (Eq. [5-2]). cGMP depolarizes the ROS membrane and increases the time (latency) required for light-induced hyperpolarization in the rod. Depolarization by cGMP has been demonstrated with patch clamp techniques on excised patches of ROS membrane (609) and on intact ROS (610). The cGMP- dependent conductance behaves like a channel in the absence of divalent cations (611) after incorporation into planar bilayers (612). The mechanism by which cGMP keeps ROS channels open is not clear. Regulation by Ca^{++} could be important, perhaps under the influence of inositol triphosphate (538). Both Ca^{++} and Mg^{++} decrease conductance in the absence of cGMP (613). The identification, purification, and functional reconstitution of the cGMP-dependent channel from bovine retina (614) will allow the use of cDNA techniques to obtain the amino acid sequence. The channel is probably a tetramer of 63 kDa units, requiring 4 cGMP for sodium conductance in the reconstituted form. The monomeric unit is not far in molecular weight from the probable protogene (57 kDa for rat brain sodium channel I) which evolved into the rat sodium channel (Chap. 7).

One photoexcited rhodopsin (RX) catalyzes a process in which the nonhydrolysable GTP analogue, ^{3}H-GppNHp, displaces an equivalent amount (500/RX) of ^{3}H-GDP from the ROS membrane. This result suggests that a GTP binding protein (which was named transducin [T]) is present and forms a T.GTP (or T.^{3}H-GppNHp) complex (615, 616). The discovery that T.GDP is bound strongly to ROS membranes in the absence of GTP (617) led to a procedure for the isolation of transducin. The uptake of ^{3}H-GppNHp (71/RX) could be demonstrated with purified T (618). Kinetic measurements using infrared light scattering on magnetically oriented ROS show that RX catalyzes the transformation of T.GDP to T.GTP in 1 ms (583, 619). The T.^{3}H-GppNHp complex dissociates on chromatography to two polypeptide units, $T_{\beta\gamma}$(MW β:36,000, γ:10,000) and a T_{α}.^{3}H-GppNHp complex (MW 39,500). The complex activates a phosphodiesterase (PDE) by removing the inhibitory subunit (PDE_{γ} or I) (620, 621). The PDE has an extremely high catalytic rate, close to the limit set by diffusion-controlled combination of the enzyme and the cGMP substrate (621). Absorption of a photon leads to considerable turnover of cGMP, via hydrolysis, within 100 ms. (Eq. [5.2]).

The phosphodiesterase (PDE) is a peripheral membrane protein (MW 180,000) with three subunits, (α) MW 88,000, (β) 84,000, and (γ) 13,000. Early research on the light-regulated PDE has been reviewed (622). The isolated enzyme is activated (activity 2100 $M^{-1}sec^{-1}$) by the protease, trypsin, by attack on the γ-subunit, showing that an inhibitory constraint decreases the esterase activity. The γ-subunit strongly inhibits the trypsin-activated esterase, suggesting that the γ-subunit is removed or altered by the T_{α}.GTP com-

(5-2)

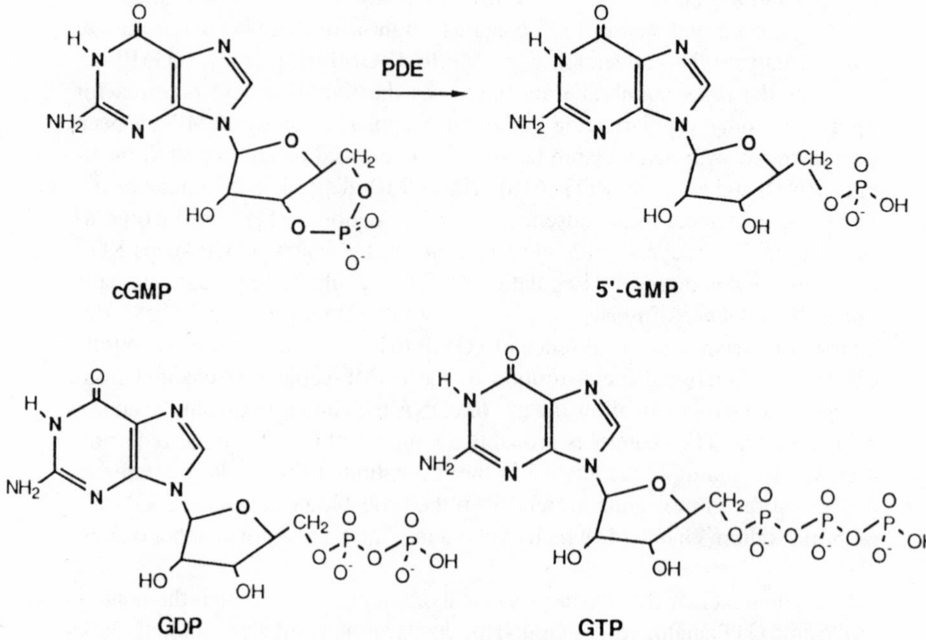

cGMP —PDE→ 5'-GMP

GDP GTP

plex. Alteration is favored (perhaps via formation of a $PDE_{\alpha\beta\gamma}.T_{\alpha}$.GTP complex) in view of the fact that the light-activated PDE is inhibited by purified PDE_{γ} much less (K_i 440 nM) than the trypsin-activated PDE (K_i 15 nM) (623). Studies of the effects of phosphodiesterase inhibitors used as drugs (624) with respect to rod and cone PDE would be warranted.

Since 500 molecules of transducin may be activated by one molecule of "activated" rhodopsin and each PDE hydrolyzes 500 molecules of cGMP, the overall biochemical result of the absorption of one light photon is the hydrolysis of 250,000 molecules of cGMP. A single amino acid substitution (lys 248 → leu) in rhodopsin (625, 626) prevents activation of transducin.

Biochemical controls limit the lifetime of the cycles. The transducin (α-subunit) has GTPase activity, bound GTP being hydrolyzed to GDP. Activated rhodopsin is inhibited by "arrestin" or phosphorylated by a kinase (627) and/or hydrolyzed to the protein, opsin, and the aldehyde, retinal, a combination no longer active in promoting the activation of T. Phosphoopsin is dephosphorylated by a specific phosphatase (627).

An overall scheme (Fig. 5.7) shows the complexities of the biochemical cycles and reactions for the amplification of the light signal and its conversion into a biochemical message. A schematic mechanism for the RX-catalyzed transformation of T.GDP into T.GTP is given later. The striking analogy between light and hormone amplification is described below.

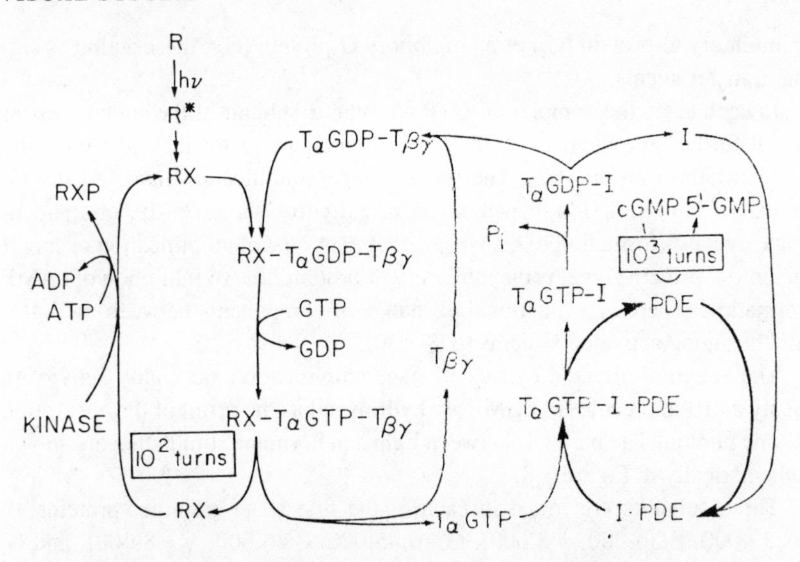

Fig. 5.7. Biochemical cycles related to light amplification. After absorption of a photon, rhodopsin (R) is excited to R*, which undergoes *cis-trans* isomerization and decays to the ground state of bathorhodopsin. The latter further isomerizes in two stages to metarhodopsin II (RX). The conformationally altered, ''activated'' rhodopsin (RX) combines with transducin. guanosine diphosphate (T.GDP) and catalyzes the exchange of GTP for GDP within the ternary complex, RX.T.GDP. The complex dissociates to RX, $T_{\beta\gamma}$, and T_{α}.GTP. The T_{α}.GTP extracts an inhibitor (I) from an inhibitor.phosphodiesterase (I.PDE) pair, releasing PDE. The PDE catalyzes the hydrolysis of cyclic 3′,5′-guanosine monophosphate (cGMP) to guanosine monophosphate (5′-GMP). The GTP in the transducin subunit inhibitor pair (T_{α}.GTP.I) is hydrolyzed to GDP, resulting in the release of I, which then inhibits the PDE. The hydrolysis of GTP occurs at a rate such that 1000 molecules of cGMP can be hydrolyzed by active PDE. The activity of the RX is terminated by a kinase, which catalyzes phosphorylation by adenosine triphosphate (ATP) to form phosphorylated rhodopsin (RXP), perhaps in combination with another protein, arrestin. About 100 molecules of T_{α}.GTP are produced by each RX. The overall amplification of the photon absorbed by a rhodopsin molecule in a disk membrane is large, with at least 10^5 cGMP hydrolyzed per photon. Many Na^+ channels are affected by the loss of the cGMP.

Parallels between Light and Hormone Amplification

A photon initiates the transformation of a receptor molecule (rhodopsin) into a form (RX) which amplifies the stimulus through forming many molecules of a protein, the transducin.guanosine triphosphate (T.GTP) complex. The RX catalyzes the loading of GTP into T which acts as a signal transfer agent.

A hormone (for example, epinephrine, adrenaline) initiates the transformation of a receptor molecule into a form (RH) which amplifies the stimulus by catalyzing the formation of many molecules of a G-protein.guanosine triphosphate (G.GTP) complex. The RH catalyzes the loading of GTP into either a stimulatory G-protein (G_s) or an inhibitory G-protein (G_i), these acting as signal transfer agents.

In each case, the complex of GTP with the α-subunit of the signal transfer agent further amplified the signal by another large factor through activating (or inhibiting) an enzyme. The second stage amplification involved the removal of inhibition of a phosphodiesterase (hydrolysis of cGMP) or an adenylate cyclase (formation of cAMP). A different type of amplification occurs if the initial protein signal is the inhibitory G-protein, that of inhibition of cAMP formation. There is a functional exchange of components between the light- and hormone-activated systems (628–630).

The enzymes affected by second stage amplification are highly active. As many as 10^5 molecules of cGMP are hydrolyzed as the result of the absorption of one photon. The parallels between light and hormone stimulation are shown schematically in Figure 5.8.

Three subunits are found in each of the first stage amplifier proteins, T (α:39,000, β:36,000, γ:8,000), G_s (α:45,000, β:36,000, γ:~8,000), and G_i (α:42,000, β:36,000, γ:~8,000).

Signal amplifying systems no doubt occur elsewhere in cells. There are significant homologies between the amino acid sequences of the T_α, the bacterial elongation factor, EF-Tu, and the p21 proteins produced by ras oncogenes (631). By using the X-ray structure of the GDP binding domain of EF-Tu (632) together with other information, a binding site for the GDP has been formulated (631). The crystal structure of human c-H-*ras* p21 lacking the last

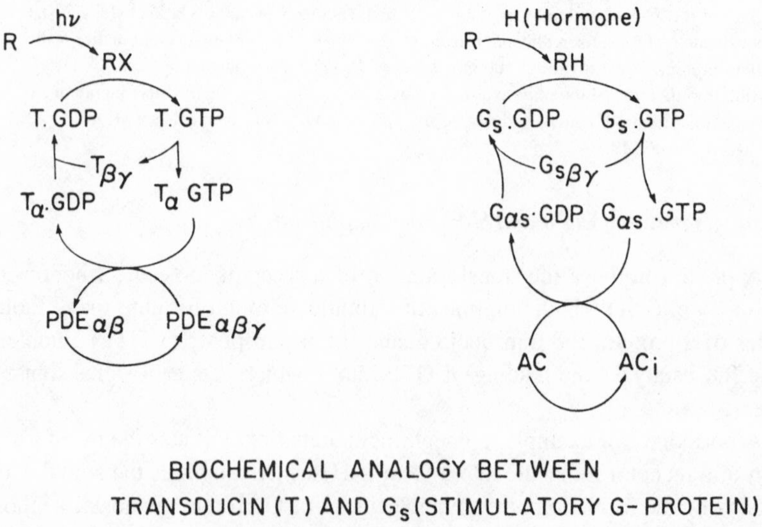

BIOCHEMICAL ANALOGY BETWEEN
TRANSDUCIN (T) AND G_S (STIMULATORY G-PROTEIN)

Fig. 5.8. Parallelism between hormone and light responses.

18 amino acids of the 189 in the oncogene protein provides a more precise picture of the relationship between GDP and the protein binding site (633, 634). The exchange of GTP for GDP presumably occurs at the binding site. A knowledge of the dynamics of the site with respect to the conversion of a GDP → GTP complex are central to an understanding of the mechanism by which the activated receptors catalyze the process and effect first stage amplification. Apparently p21 is not a regulatory component of adenylate cyclase (635). There is some disagreement about the amino acid sequence of bovine T_α (636–640). Desensitization of the β-adrenergic receptor depends on both an analogue of "arrestin" and a kinase (641).

Polymolecule: Rhodopsin

Rhodopsin contains only one molecular fragment (a "chromophore") which absorbs visible light. After the chromophore is excited by light, a series of changes occurs, ultimately leading to conformationally altered rhodopsin (RX) with catalytic activity. Neutron diffraction of magnetically oriented ROS has established that about 50% of the mass of the rhodopsin molecule is within the bilayer of the disk (583). The remainder is distributed equally in the intradiscal and cytoplasmic regions, after allowance is made for the complexation of transducin with the cytoplasmic face of rhodopsin.

Diffusion-enhanced energy transfer using terbium dipicolinate indicates that the retinylidene chromophore is located approximately in the middle of the bilayer. The technique depends on the diffusibility of the donor in the medium coupled with measurements of the fluorescence emission lifetimes of long-lived energy donors (642). The chromophore lies at an angle of 18° to the bilayer plane (643).

The chromophore of rhodopsin is an almost planar iminium ion for which the transition dipole for light absorption in the visible is parallel to the ground state dipole of the conjugated system. This has been shown for *cis*- and *trans*-retinals and for the imine and iminium ion of retinylidene N-butylamine (644). The dipole change on excitation is illustrated for an α,β-conjugated carbonyl system in Eq. (5-3). The electric vector of light propagating parallel to the rod axis is perpendicular to the rod axis. Since light is absorbed preferentially from a direction parallel to the rod axis, the chromophore must have the same orientation, which is almost parallel to the major portion of the disk surfaces. The relationship of the electric vectors of the light to the direction of maximum absorption is illustrated in Figure 5.9.

(5-3)

$$\mathrm{C{=}C{-}C{=}O} \xrightarrow{h\nu} \mathrm{C^{+}{-}C{=}C{-}O^{-}}$$

The bleaching of rhodopsin by light (formation of *trans*-retinal and opsin) was used to determine the rotational mobility of rhodopsin in the disk membrane. Rhodopsin molecules have chromophores parallel to the disk surface, but randomly oriented in the plane. Polarized light bleaches only those chromophores which have transition dipoles parallel to the electric vector of the light. With a 5-nanosecond flash, properly oriented rhodopsin molecules were bleached. By monitoring the recovery of light absorption by unbleached rhodopsin molecules, the time of rotational relaxation of rhodopsin in the disk membrane was measured as 20 μsec. The selection of the molecules and the recovery of absorption is shown in Figure 5.10 (645, 646).

The rotational relaxation rate has been estimated theoretically as consistent with a hopping process for phospholipid molecules (i.e., that a vacancy is created by a jump of a phospholipid molecule and the protein rotates into the vacancy) (647). The rotational rate is probably greater than that derived from the lateral diffusion rate for dipalmitoylphosphatidylcholine as expected on the basis of the high unsaturated acyl content of the phospholipids in the disk bilayer.

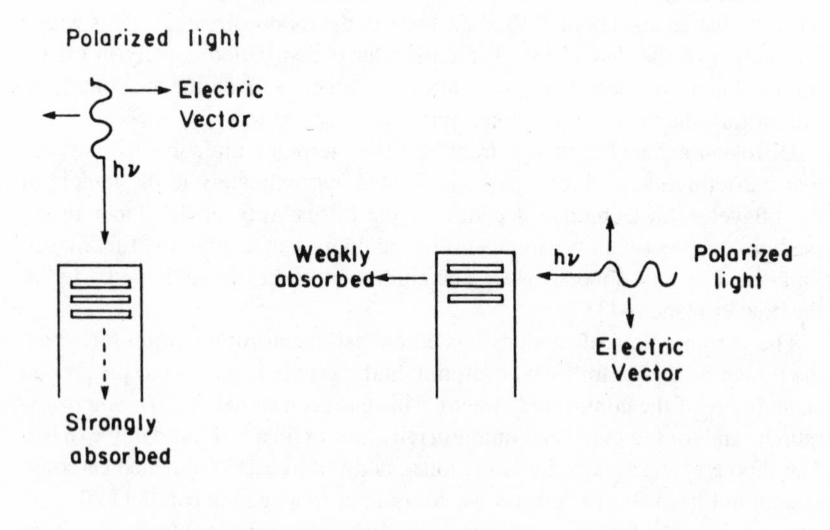

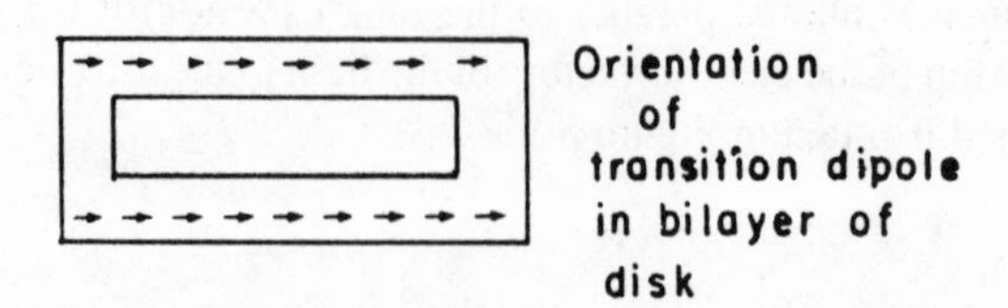

Fig. 5.9. Dichroic absorption of rhodopsin in retinal rod cells. Polarized light is absorbed strongly if the electric vector for the light is parallel to the transition dipole of the chromophore.

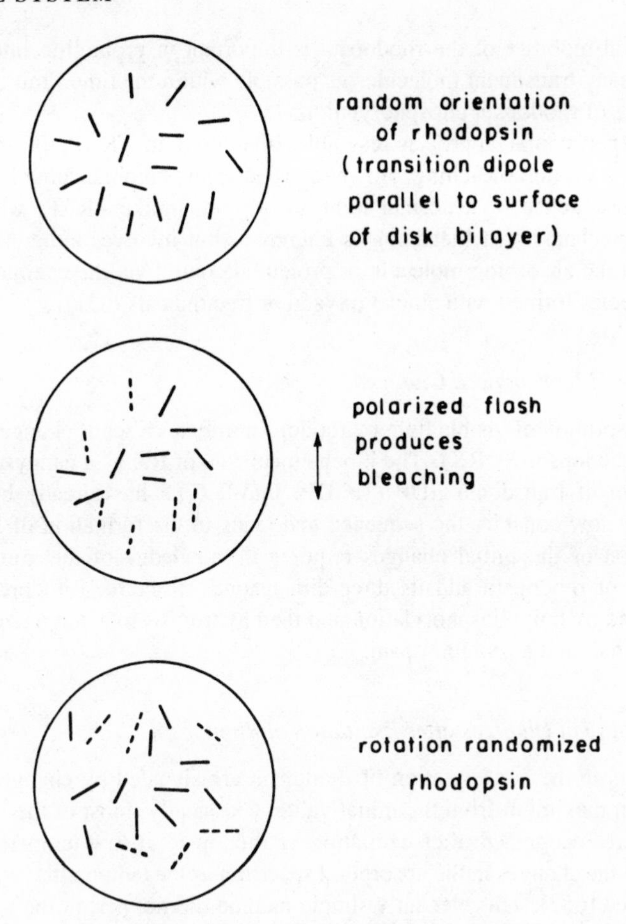

Fig. 5.10. Rotational relaxation of rhodopsin in rod disk membranes after bleaching with a flash of polarized light. After recovery of absorption, rhodopsin molecules with transition moments parallel to that of the polarized analyzing light are found.

The lateral diffusion rate for rhodopsin in the disk membranes of the frog, *Rana pipiens*, and for porphyrosin in the disk membranes of the mud puppy, *Necturus maculosus*, was determined by bleaching the rhodopsin in one half of the disk and following the return of rhodopsin from the other half of the membrane by lateral diffusion. The recovery times were 4 seconds in the frog (D 5.5×10^{-9} cm^2sec^{-1}) and 15 seconds in the mud puppy (D 4.7×10^{-9} cm^2sec^{-1}), faster than biochemical regeneration of rhodopsin (648, 649). An unexpected but useful result of the photobleaching experiments was that pigment molecules diffused laterally quite fast in cones but not around the ''hairpin'' turns which separate the two sides of an invagination. Thus, the cone structures are effectively separate entities, like disks. A diffusion constant of 3.3×10^{-9} cm^2sec^{-1} is found for rhodopsin in spherical vesicles (650). The high lateral mobility of the rhodopsin is important in promoting interaction with as many transducin molecules as possible within the time limit imposed by the rate of rhodopsin phosphorylation.

The experimental approach resembles that used in FRAP, fluorescence recovery after photobleaching. However, rhodopsin photobleaching is a "natural" consequence of a normal light absorption, unlike FRAP, where the detailed mechanism of bleaching is unknown, but involves either chemical change in the absorbing molecule or protein alteration via the chemically reactive species formed with singlet oxygen or free radicals (651).

Rhodopsin: Light-induced Changes

The absorption of visible light by rhodopsin initiates a set of changes which produce rhodopsin-X (RX). The biochemical role of RX as a catalyst for the conversion of transducin.GDP (T.GDP) into T.GTP has already been described. I now consider the sequence that leads to the formation of RX. Interpretation of the initial changes requires a knowledge of the amino acid sequence of rhodopsin and its three-dimensional structure. Rhodopsin-X is deactivated by polyphosphorylation and then hydrolysis to a conjugated aldehyde, retinal, and a protein, opsin.

Conformational Changes after Excitation of Rhodopsin

Changes in the conformation of rhodopsin are signaled by changes in the absorption maximum from the initial value. Classically, most of the intermediates were recognized after excitation of rhodopsin at low temperature by following the changes in the absorption spectrum as the temperature was gradually raised (652). This elegantly simple method did not permit the measurement of the rates of conversion of the intermediates into their respective successors. The formation or disappearance of intermediates could be followed spectroscopically after flash excitation at room temperature and below (653–657). The intermediates identified and their lifetimes are shown in Figure 5.11. The lack of interaction between spin-labeled stearic acids and a rhodopsin spin-labeled on the retinylidene moiety shows that the chromophore is completely sequestered from the lipid of the bilayer (658). Some lateral motion of the helices of rhodopsin is necessary, at least at later stages of the sequence shown in Figure 5.11. Replacement of unsaturated acyl groups in the phospholipid by saturated acyl groups leads to a less mobile bilayer microenvironment for the rhodopsin which greatly retards the meta I to meta II conversion (659).

The photochemical equilibrium between rhodopsin (11-*cis*-retinylidene opsin), isorhodopsin (9-*cis*-retinylidene opsin), and bathorhodopsin (11-*trans*-retinylidene opsin) can be established at temperatures too low for subsequent processes to occur (660, 661). The common intermediate, bathorhodopsin, is

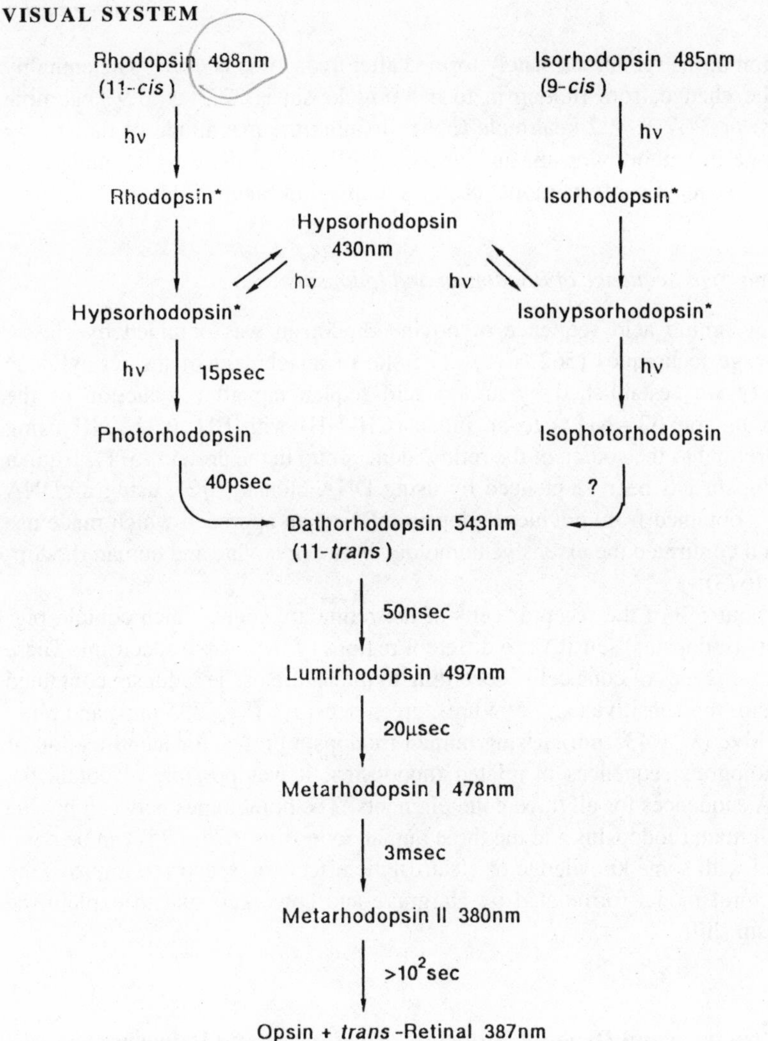

Fig. 5.11. Intermediates produced by light excitation of rhodopsin. The lifetimes measured at room temperatures are given for most of the intermediates.

11-*trans*, a conclusion supported by resonance Raman studies (662–664). An orientational change of 26° in the $\pi \rightarrow \pi^*$-transition dipole accompanies the transformation from rhodopsin to bathorhodopsin (665, 666). Another intermediate, hypsorhodopsin (657, 667), is detected only at very low temperatures and converted to bathorhodopsin on warming (654). The conjugated system of hypsorhodopsin is probably twisted, since the λ_{max} is 430 nm, a position reflecting less conjugation in the double bond system. Another short-lived intermediate, photorhodopsin, lies on the reaction pathway from rhodopsin to bathorhodopsin; isophotorhodopsin has not been reported. The conformation of the retinal ultimately formed after hydrolysis is *trans*. The enthalpy of the change from rhodopsin to bathorhodopsin is 32.2 ± 0.9 kcal/mole (668) or 34.7 ± 2.2 kcal/mole (669), defining the magnitude of the energy storage in bathorhodopsin, an "energy-rich" intermediate which undergoes the subsequent conformational changes outlined in Figure 5.11.

Amino Acid Sequence of Rhodopsin and Iodopsins

The amino acid sequence of bovine rhodopsin was obtained by classic cleavage techniques (562, 670). The site of attachment of the retinylidene moiety was established by amino acid sequencing after reduction of the iminium ion ($C{=}NH^+$) to an imine (CH–NH) with $BH_3.(CH_3)_2NH$ using [^3H]retinal as the source of the retinylidene group in the protein (671). Human rhodopsin has been sequenced by using DNA clones (672), using a cDNA probe obtained from bovine rhodopsin mRNA, an approach which made use of and confirmed the extensive homology between bovine and human rhodopsins (673).

About 6% of the receptor cells of the retina are cones which contain pigments (iodopsins) sensitive to different regions of the visible spectrum. There are three types of cone cells, according to the nature of the iodopsin contained therein: red-sensitive (λ_{max} 565 nm), green-sensitive (λ_{max} 535 nm), and blue-sensitive (λ_{max} 435 nm). Using human rhodopsin probes for identification of homologous sequences in related rhodopsins, it was possible to obtain the DNA sequences for all three cone pigments. The homologies between bovine and human rhodopsins and the three human iodopsins (674, 675) can be combined with some knowledge of electrostatic effects on spectra to improve the structural model formulated by Hargrave and coworkers and to explain the "opsin shift."

A Structural and Dynamic Model for Rhodopsins and Iodopsins

The absorption maximum of rhodopsin is at longer wavelengths than expected from studies on model iminium derivatives. The change in position is called the "opsin shift," for which an explanation has been vigorously sought. No homogeneous solvent can accurately reflect the complex environment of the chromophore in rhodopsin, but the magnitude of solvent effects on the maxima of model compounds is not large enough to account for the shift. It was suggested that specific interactions of particular groups, rather than some overall property of the retinylidene site in the protein, could explain the opsin shift. I have carried this idea forward by using genetic homologies to aid in the selection of the interacting groups. A semiquantitative understanding of the absorption maxima of rhodopsins and red, green, and blue iodopsins (cone pigments) is derived on the basis of a fairly detailed structural model for rhodopsins and iodopsins. Analysis of the model leads to an idea about switching on rhodopsin catalysis of transducin action.

Summary of Rhodopsin Model

The purpose in formulating a model for a protein is to enable analysis of the chemical and physical properties to proceed in the absence of a direct determination of the three-dimensional structure. In the case of rhodopsin and related proteins, it is necessary to make an assignment of the groups responsible for the ''opsin shift'' and the positions of the light absorption maxima.

A modified structural model of rhodopsin has been formulated and is described here (564). Seven α-helical segments of 24 largely hydrophobic amino acid residues are assembled with exobilayer connecting strands into an aligned set, using the sequences of human red, green, and blue iodopsins (cone pigments) and human and bovine rod rhodopsins. (Aligned set numbering based on the red-green iodopsins is used.) The inner region of the heptahelical hydrophobic domain includes one his-glu (asp) ion pair (red, green, rod) near the retinylidene moiety in addition to an iminium ion-asp 99 pair. The negative charges posited in the ''point-charge model'' to cause the shift of the retinylidene iminium ion light absorption to longer wavelengths in the protein (''opsin shift'') are asp 99 (red, green, rod), glu 102 (red, green), and glu 138 (rod). Blue iodopsin lacks both an ion pair and a counter charge to the iminium ion in the inner region, a fact that explains its absorption relative to rod rhodopsin. The presence of an iminium ion is proven by a 1662 cm^{-1} stretching vibration due to $C{=}NH^+$ (663). The spectroscopic difference between rod rhodopsin and the red/green iodopsins is due to the influence of glu 102 in the latter. The red-green difference is due to the net effect of seven OH groups around the chromophore, all such groups being found within one helix turn of the retinylidene location. The tryptophan that rotates as the retinylidene group isomerizes may be trp 142, trp 177, or trp 281.

The geometric changes (the rhodopsin ''photoswitch'') resulting from *cis-trans* isomerization in the first excited electronic state (S_1) ultimately lead to RX (photoactivated rhodopsin, metarhodopsin II) and result in a change of the activity of exobilayer groups. A possible RX binding site for transducin can be based on G-protein receptor homologies (Chap. 8). The active part is made of the two positively charged side chains from lys 83 and arg 85. These may compete for the two GDP negative charges in the transducin.guanosine diphosphate (GDP) complex. The site in RX becomes active after dissociation of the positively charged groups from the 263 and 265 carboxylate groups. Metarhodopsin II (RX) promotes the dissociation of GDP from T.GDP and formation of a T_α.GTP complex; a schematic mechanism for the T_α.GDP → T_α.GTP conversion is given.

Introduction to Model

Visible light (400–700 nm) is converted into a neural signal in animals by a series of processes beginning with light absorption by rhodopsin (R) in retinal rod cells (iodopsins in cone cells) (579, 676). The 11-*cis*-retinylidene chromophore rearranges to the 11-*trans* structure in the first excited electronic state (S_1), a species which then decays to photorhodopsin (677). Direct observation of the initial changes for *bacteriorhodopsin* using 6-femtosecond (fs) probe pulses after excitation for 60 femtoseconds showed that the primary ground state product appeared within 500 fs (678). A similar experiment for rhodopsin may resolve some of the questions about the early photophysics of rhodopsin (679). Photorhodopsin changes within 40 ps (677) to bathorhodopsin, an intermediate which is transformed via lumirhodopsin and metarhodopsin I to metarhodopsin II. Metarhodopsin II is the phototransformed protein, RX (680), which initiates the biochemical processes that temporarily decrease cGMP within the cell (616, 681). The initial biochemical step is the phototransformed rhodopsin (iodopsin)–promoted displacement of guanosine diphosphate (GDP) by guanosine triphosphate (GTP) from the G-protein (transducin, T) complex with GDP (558, 616, 681). The flow of information to the nervous system about a photon absorbed by a rod or cone may be expressed in the steps, photon + R → RX → T.GTP → (-cGMP) → hyperpolarizing signal to nervous system. A ''point-charge'' model (a negative charge within the protein that stabilizes the retinylidene excited state more than the ground state) has been proposed to explain the shift of the retinylidene iminium ion light absorption to longer wavelengths in the protein (''opsin shift'') (682–686).

On the basis of the alignment of the amino acid sequences of human rod rhodopsin and red, green, and blue iodopsins (674), I develop a modified structural model by specifying charge-charge interactions (562, 563, 670, 682). With this model, one can identify (a) the groups responsible for spectroscopic differences between the rhodopsins and iodopsins (consistent with the general idea of the ''point-charge model''), (b) the nature of the change in rhodopsin produced by the S_1 *cis-trans* isomerization, (c) part of the conformational change that produces RX, the catalyst for the formation of T_α.GTP, and (d) a schematic mechanism for the T_α.GDP → T_α.GTP conversion.

Rhodopsin-Iodopsin Model

Aligned sequences of human red-green-blue-rod and bovine rod rhodopsins (iodopsins) (562, 670, 672) are examined for hydrophobic character. The bilayer thickness is taken as 36 Å (24 amino acids in an α-helix). Hydrophobic segments can be selected without difficulty by inspection of the aligned set; more formal methods lead to essentially the same choices (562). The N-terminal end carries a known carbohydrate sequence (Fig. 5.12) (687, 688) and

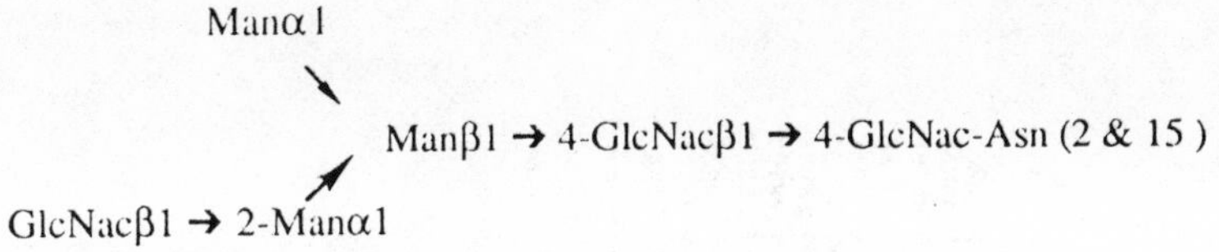

$$\begin{array}{l} \text{Man}\alpha 1 \searrow \\ \qquad \text{Man}\beta 1 \rightarrow \text{4-GlcNac}\beta 1 \rightarrow \text{4-GlcNac-Asn (2 \& 15)} \\ \text{GlcNac}\beta 1 \rightarrow \text{2-Man}\alpha 1 \nearrow \end{array}$$

Fig. 5.12. Glycosyl sequences bound to asparagines 2 and 15 (bovine rhodopsin numbering) near the N-terminal end of bovine rhodopsin.

is inside the disk (688–690) (outside the cone); an odd number of transmembrane segments puts the C-terminal end in the cell cytoplasm, as found (691). The connecting links within the disk will not be considered further; the links within the cytoplasm will be considered in connection with their interaction with transducin. A schematic structure for the rhodopsins based on these principles can be constructed using the single-letter code for amino acids (Figure 5.13). Other ungulate rhodopsins have amino acid sequences very similar to that of bovine rhodopsin; in particular, ovine rhodopsin has asp 99, glu 138, his 227, and lys 312 (692). Even the more distantly related *Drosophila* rhodopsins have the equivalent of asp 99 and lys 312 (rho_{1-6}: asp 96, lys 319 [693], rho_7: asp 104, lys 335 [694, 695], and rho_8: asp 103, lys 326 [696, 697]).

Several features are immediately apparent in the model. The retinylidene chain bound to lys 312 (671, 698) is close to the middle of the bilayer. A small number of charged groups is found well within the bilayer. It is proposed that these groups are functionally involved with the retinylidene group located in the inner region of the helices. Ion pairing will occur for oppositely charged groups in a moderately nonpolar environment. Four out of five opsins have a negatively charged group (asp) at position 99, which is therefore the counter-ion to the 312 iminium ion. (Most of the *Drosophila* rhodopsins have homologous ion-pairs, as noted above.) The same four opsins have another ion-pair, in addition to 99^-–312^+, in the center of the bilayer. The additional ion-pair in rod rhodopsins is composed of glu 138 and his 227; in red-green iodopsins, the pair consists of glu 102 and his 54. The glu 102–his 54 separation is between 7–9.5 Å, depending on conformation. Although the charge-charge separation is a bit large for an ion-pair, the choice of his 54 is validated by the occurrence of lys at the homologous positions in two *Drosophila* rhodopsins even though there is no identifiable intrabilayer counter-ion in the latter. The partners in the ion-pairs may vary and certain charged groups may not have partners. Given the angle of 18° between the chromophore transition moment and the bilayer plane (643), the retinylidene moiety should be located not far from the amino acid side chains at 102, 138, and 227.

The bilayer helices are numbered 1 to 7 from the N-terminal end of the protein. By assuming that ion-pairing occurs in a hydrogen-bonded arrangement within the cavity containing the retinylidene chain, I am able to orient

STRUCTURE: HUMAN RED-GREEN-BLUE IODOPSINS-HUMAN/BOVINE ROD RHODOPSINS

```
                                                                                    340 KKKGG         367    N
                                                                              339 KKGCC   341 VVAKK   366    P
                                                          255 EEEEE 256 SSSSS  338 GGCCC   342 DDMNN   365    G
                                                          254 KKQQQ   257 EEAAA  337 FFVIL   343 DDTPP   364 AAVAA
                                                          253 QQQQQ   258 SSTTT  336 LLMTT   344 GGDLL   363 PPQPP
                                                          252 QQQQQ   259 TTTTT  335 QQKTT   345 SSEEG   362 SSTAA
                                                          251 KKAAA   260 QQQQQ  334 LLMLV   346 EESDD   361 VVSVV
                                                          250 AAAAA   261 KKKKK  333 IIIMM   347 LLDDD   360 SSSQQ
          81 FFYHH            161 NNNNN 162 VVFFF          249 VVVAA   262 AAAAA  332 CCCCC   348 SSTEE   359 SSVSS
        80 KKRQQ   82  KKKKK  160 GGGSS   163 RRRRR      248 AAAEE   263 EEEEE  331 NNANN   349 SSCAA   358 VVTTT
        79 MMLVV   83  KKKKK  159 FFFMM   164 FFFFF      247 RRKLK   264 KKRKK  330 RRQRR   350 AASSS   357 SSSEE
        78 TTTTT   84  LLLLL  158 PPPPP   165 DDSGG      246 IILVV   265 EEEEE  329 FFFFF   351 SSSAT   356 SSVTT
        77 AAAVV   85  RRRRR  157 KKKKK   166 AASEE      245 AAATT   266 VVVVV  328 QQQQQ   352 KKQTT   355 VVEKK
        76 AAVYY   86  HHQTT  156 CCCCC   167 KKKNN      244 LLRFF   267 TTSTT  327 RRKKK    353 TTKVV 354 EETSS
__________________  75 LLLLL   87  PPPPP__ 155 VVIVV__ 168 LLHHH__   243 WWLVV__ 268 RRRRR__ 326 NNNNN________________
BILAYER         74 VVVTT   88  LLLLL  154 VVVVV   169 AAAAA      242 VVLLL   269 MMMMM  325 MMMMM
    |           73 LLMLL   89  NNNNN  153 LMIVV   170 IILII      241 QQQQQ   270 VVVVV  324 FFFMM
    |           72 GGAFF   90  WWYYY  152 WWYYY   171 VVTMM      240 LLTGG   271 VVVII  323 VVCII
    |           71 NNNNN   91  IIIII  151 RRRRR   172 GGVGG      239 YYYYY   272 VVVII  322 YYYYY
    |           70 TTLII   92  LLLLL  150 EEEEE   173 IIVVV      238 CCSCC   273 MMMMM  321 IIIII
    |           69 FFPPP   93  VVVLL  149 WWFII   174 AALAA      237 LLFFF   274 IVVVV  320 VVIVV
    |           68 VVFFF   94  NNNNN  148 SSAAA   175 FFAFF      236 MVCFF   275 FLGII  319 PPPPP
    |           67 SSGGG   95  LLVLL  147 IILLL   176 SSTTT      235 IIIII   276 AASAA  318 NNNNN
    |           66 AAILL   96  AASAA  146 IIFVV   177 WWWWW      234 IILIV  •277 YFFFF  317 YYYYY
    |          •65 TILVM   97  VVFVV  145 AAAVV   178 IITVV     •233 ASSII   278 CCCLL  316 IIIIV
    |           64 VVFII   98  AAGAA  144 LLLLL   179 WWIMM      232 LLLML   279 VFVII  315 TTCAA
    |           63 VVVLL  -99  DDGDD  143 SSSSS  •180 SAGAA      231 PPPPP   280 CCCCC  314 AAAAS
    |           62 FFTLL  100  LLFLL  142 WWWWW   181 AAILL     •230 ITVII   281 WWYWW  313 SSSST
    |           61 IIGFF  101  AALFF  141 LLGLL   182 VVGAA      229 IIITI   282 GGVVL +312 KKKKK
    |           60 MMMMM -102  EELMM  140 GGTAA   183 WWVCC      228 CCFFF   283 PPPPP  311 AASAA
    |           59 WWFYY  103  TTCVV  139 TTVII   184 TTSAA     +227 CCCHH   284 YYYYY  310 FFFFF
    |           58 VVAAA  104  VVILF -138 IILEE   185 AAIAA      226 TTFVV  •285 TAAAA •309 YFFFF
    |           57 SSAAA  105  IIFGG  137 GGGGG   186 PPPPP      225 VVIVV   286 FFASG  308 AASAA
    |           56 TTQLL  106  AASGG  136 CCAGG   187 PPPPP      224 MMFFF   287 FFFVV  307 PPPPP
    |           55 LLLMM  107  SSVFF  135 LLVLL   188 IIFLL      223 LLLMM   288 AAAAA  306 LLIII
    |          +54 HHYSS  108  TTFTT  134 SSTTT   189 FFFAV      222 VVFYY   289 CCMFF  305 AATTT
    |           53 YYFFF  109  IIPST  133 VVGAA   190 GGGGG      221 IIWII   290 FFYYY  304 AAVMM
    |           52 VVAQQ  110  SSVTT  132 TTLFF   191 WWWWW      220 MMTVV   291 AAMII  303 MMLFF
```

```
                                51 WWWWW__  111 IVFLL__  131 YYFFF__  192 SSSSS __ 219 YYYFF__  292 AAVFF__  302 LLRII
 27 IIEVV 28 FFFPP              50 RRVPP    112 VVVYY    130 GGGGG    193 RRRRR    218 SSSSS    293 AANTT    301 PPLPP
26 SSEYY     29 TTYFF           49 PPPEE    113 NNATT    129 EEEEE    194 YYFYY    217 QQEEE    294 NNNHH    300 HHDGG
25 SSEFF     30 YYLSS           48 AAAAA    114 QQSSS    128 LLLLL    195 WWIII    216 VVSNN    295 PPRQQ    299 FFLFF
24 QQSNN     31 TTFNN           47 IIILL    115 VVCLL    127 VVANN    196 PPPPP    215 GGRNN    296 GGNGG    298 APGND
23 TTMPP     32 NNKAK           46 HHHYY    116 SYNHH    126 CCCCC    197 HHEEE    214 PPYVT          297 YYHSS
22 SSKGG     33 SSNTT           45 YYYYY    117 GGGGG    125 MMVGG    198 GGGGG    213 YYKGE
21 DDREE     34 NNIGG           44 NNQQQ    118 YYYYY    124 PPHTT    199 LLLLM    212 SSTPE
20 EEMTT     35 SSSVV           43 PPPPP    119 FFFFF    123 HHRPP    200 KKQQQ    211 SSGKH
19 YY GG     36 TTSVV           42 GGGYA    120 VVVVV    122 GGGGG    201 TTCCC    210 GGVLP
18 SS NN     37 RRVRR           41 EEDEE        121 LLFFF            202 SSSSS    209 SSTTT
17 DD MM     38 GGGSS           40 FFWFF                              203 CCCCC    208 FFYYY
16 QQ           39 PPPPP                                              204 GGGGG    207 VVWYY
15 PP                                                                  205 PPPII 206 DDDDD
14 HH
13 RR              1  MM
12 GG              2  AA
11 AA              3  QQ
10 LL              4  QQ
9  RR              5  WW
8  QQ 7   LL 6     SS
```

Fig. 5.13. Amino acid sequences and schematic structures of human red, green, and blue iodopsins (cone pigments) and human and bovine rhodopsins (rod pigments) constructed using the single-letter code for amino acids. The distribution of the protein between the membrane bilayer and the aqueous interior of the rod disk (cone invagination) is indicated. The N-terminal residue of the opsin is in the interior of the disk (corresponding to the cell exterior of a cone), at the lower part of the figure.

Aligned sequences are shown using the homology numbering based on red-green iodopsins. To obtain the position in the rhodopsin sequence, subtract 16 (19 for numbering of the blue iodopsin). No skips are required for the alignment. Membrane bilayer segments of 24 amino acids (36 Å) are chosen on the basis of regions of hydrophobicity in the aligned sequences. Seven segments are found, a result that agrees with more formal methods of choosing hydrophobic segments. It is assumed that these segments occur as α-helices (numbered 1 to 7). The 312 lysine, bonded to the retinylidene moiety as an iminium ion, is located in the middle of helix 7 and is common to all known opsins. The second charged group common to four of the five opsins is asp 99, which must be the counter-ion to the iminium ion. The next most salient feature is the glu 138 found in the rhodopsins. At the same level in both rhodopsins is his 227, which thus is a good choice as a counter-ion (in protonated form) to glu 138. In red-green iodopsins, glu 102 is found at a location appropriate for pairing (weakly) with his 54. No counter-ion is found for the iminium ion in blue iodopsin. Amino acids with OH groups with an influence on the spectroscopic properties of the rhodopsins are marked with a black dot.

bilayer helices 3 and 5 and 1 and 2 with respect to one another. Asp 99 must be directed in the general direction of the chromophore since it is one helix turn above glu 102. By packing the helices around the 11-*cis*-retinylidene imine, I arrive at the central bilayer region structure depicted in Figure 5.14. The orientations of helices 1, 3, and 5 are different from those chosen on the basis of hydrophobicity (563). The arrangement in Figure 5.14 is only approximate, arbitrarily counterclockwise in helix order and schematic since the overall structure of rhodopsin is not known. Bacteriorhodopsin, for which both fine and overall structural details are established (699–702), and rhodopsin may be similar, but the similarity hypothesis cannot be applied without further information, in view of the rather different functions for the two proteins. Further model-building on rhodopsin will nevertheless be based on the arrangements found for bacteriorhodopsin, especially with neutron diffraction yielding information on the location of the deuterated chromophore (703, cf. model in 704).

Spectroscopic Parameters of Model

Among the questions that must be addressed in terms of the model are (1) the origin of the "opsin shift" and (2) the origin of the differences among iodopsins.

The model places groups and charges in specific locations, on the basis of which various types of interactions may be estimated. Two serious difficulties stand in the way of an exact calculation of electrostatic and other effects. First, the amino acid side chains can exist in numerous different local conformations, which may interconvert by "single group rotations" (705). The distances between charged groups cannot be then exactly specified. Second, the dielectric constant of the regions between the charges may vary in a complex way from 2 to 78 (706–710). The electrostatic interaction difference between the ground state and the excited state can thus be estimated only roughly. The calculations are carried out using the following parameters: distances measured between charged atoms (NH^+) and the centers of complex groups(his/COO^-); 0.5 as the excited state charge; either 2 or 4 for the dielectric constant (depending on proximity), and 10 in the case of his 54, for which ser 57 should influence the interaction. Charge-charge interactions were calculated by Coulomb's law for each pair of charges and the net stabilization (or destabilization) obtained for ground and excited states. Estimates for the model suggest that the stabilization of the excited state (positive charge appearing near the asp 99 and glu 138 in the rod) should be greater than the stabilization of the ground state, ca. 11–12 kcal/mole versus the 7.1 kcal/mole found experimentally. These results are consistent with the basic premises of the "point charge model," albeit with charge arrangements somewhat different from that used in the model (683, 684).

The sensitivity of the rhodopsin (iodopsin) maximum to the microenviron-

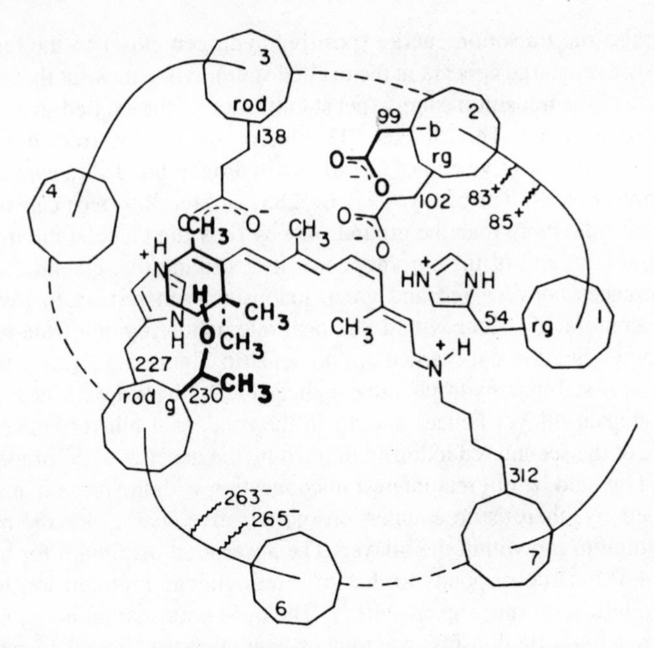

Fig. 5.14. Arrangement of critical groups in the central bilayer region of iodopsins and rhodopsins. Ion pairs in red-green iodopsins (his 54, glu 102) and rod rhodopsins (his 227, glu 138) are used to orient the helices, the net result being that the glu and asp carboxylate groups must be located very close to the retinylidene moiety. The OH group of thr 230 is close to the cyclohexene double bond. Other OH groups that differ in the red and green sequences are noted in the text and marked in Fig. 5.13. The positions of the 83^+, 85^+, 263^-, and 265^- groups are shown schematically.

ment is borne out by a striking property of retinylidene iminium systems: the shift of the absorption maximum to longer wavelengths in the presence of excess acid in solution. It is likely that the excess acid hydrogen-bonds to the anion associated with positive ions and diminishes ion-pairing. Weaker ion-pairing is equivalent to deshielding the positive charges (711). Other aspects of retinylidene iminium spectroscopic behavior are discussed later. The exact magnitude for the ''opsin shift'' may be smaller than that just cited if the 6,7-bond proves to be s-*trans* as in the case of bacteriorhodopsin (685, 712–715) but the 6-s-*cis* form is currently favored for rhodopsin (716, 717).

The spectroscopic shift between rod rhodopsins and red/green iodopsins is here accounted for by the ''point charge'' mechanism, but with a negative charge (glu 102) closer to the chromophore and a counter-ion which is farther away. A shift of 16 kcal/mole is estimated (5–6 kcal/mole more than that for the rod). The average of the measured red and green shifts is 13 kcal/mole. I ascribe the difference between red (565 nm) and green (535 nm) iodopsins to hydroxyl groups in the immediate vicinity (± one helix turn) of the retinylidene group. Hydroxyl groups at the cyclohexene end of the chromophoric moiety raise the transition energy (positive hydrogen closer to the region in which positive charge appears in the excited state), while those at the iminium should lower the transition energy (net stabilization of the excited state in α,β-conjugated systems). Thr 230, ser 233 (green), and ser 180 (red) destabilize the excited state (positive end of OH dipole hydrogen-bonded to cyclohexene double bond) (718). Thr 65, tyr 277, thr 285, and tyr 309 (red) can stabilize the excited state more than the ground state by providing a polar environment for the iminium end of the chromophore. It is striking that essentially all of the differences between red and green iodopsins with respect to hydroxyl-bearing amino acids occur within the local interaction region of the retinylidene group. (See the dot-marked amino acids in Figure 5.13.) Ser, thr, and cys may also stabilize hydrophobic α-helices (719); there are 26 such groups in the iodopsin bilayer helices and 15 in the rhodopsin bilayer helices. The sequence of the second red iodopsin implied by the genetics of color matching (720, 721) would be of great interest in connection with the present analysis.

Surprisingly, there is no counter-ion (no negative charge) for the blue iodopsin iminium ion within the bilayer. The absorption maximum for blue iodopsin (440 nm) corresponds to that of a retinylidene iminium ion without charge-stabilization (no "opsin shift"). The 6,7-conformation is not known. In the *Drosophila* rhodopsins, one may estimate that the charges (asp 99, lys 54, and lys172 [alignment number]) will cause an opsin shift to 480 nm, not far from the value of 470 nm reported for rho_{1-6} (722).

Linear dichroism measurements have suggested that an indole moiety of a tryptophan residue rotates as the retinylidene group reorients (723). Our model suggests that either trp 142 (helix 3), trp 177 (helix 4), or trp 281 (helix 6) should be near the cyclohexene ring of the retinylidene group.

Octopus Rhodopsin Sequence and Absorption

The amino acid sequence of octopus rhodopsin (*Paroctopus defleini*) has a glutamic acid at 305 (aligned set numbering, Figure 5.13) (724). The sequence is highly homologous to other rhodopsins and is aligned with great ease. A glu^- at position 305 in helix 7 can hydrogen-bond with the iminium ion proton of the retinylidene lysine at 312. The present model easily accounts for the blue shift in the absorption of the rhodopsin to ~470 nm (725), since such hydrogen-bonding would stabilize the ground state and destabilize the excited state. The orientation of the π-bonds in a fashion parallel to the plane of the bilayer is confirmed, and the geometric change of the NH shown in Figure 5.15 is made likely.

Meaning and Consequences of Rhodopsin Photoisomerization

The *cis-trans* isomerization in the rhodopsin S_1 state occurs within a restricted volume to yield bathorhodopsin, the close contact between the chiral

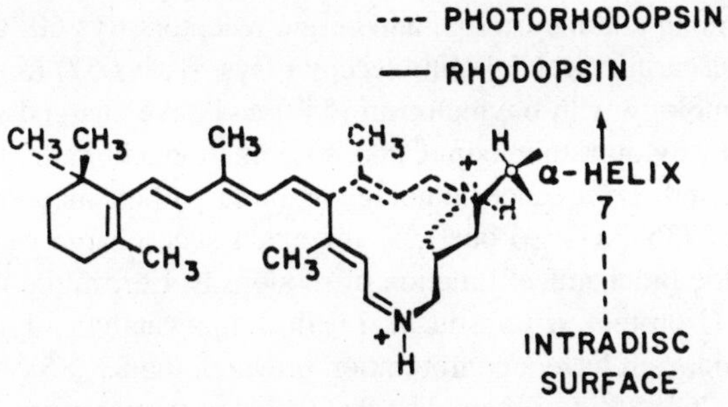

Fig. 5.15. Geometric change accompanying photoisomerization of rhodopsin to photorhodopsin. Photorhodopsin forms bathorhodopsin within 40 psec. Given the stereochemistry of the α-helix and an extended lysine side chain, the iminium ion will be found about one helix turn below the point at which the lysine side chain is attached to the helix. A concerted twist (bicycle-pedal) isomerization between C_{12} of the retinylidene chain and the penultimate carbon of the lysine in the S_1 state converts the *cis*-C_{11-12} double bond to a *trans* double bond. The ground-state photorhodopsin forms by radiationless internal conversion. The iminium ion is moved up ca. 4.5 Å to a position on the same level as the lys 312. The lower part of the structure is on intradiscal side (exterior of a cone cell).

protein and the retinylidene group being shown by circular dichroism (CD) (726). Taking into account the direction of the seventh α-helix, an extended conformation for the 312 lysine, the transition dipoles for light absorption of rhodopsin (ca. 18° from the bilayer plane) and bathorhodopsin (0° from the plane), and the normal photoisomerization exhibited by a C_{9-11} restricted rhodopsin (727), the geometry of the initial photoisomerization can be formulated (Fig. 5.15). The product of the initial isomerization is tentatively identified as photorhodopsin, already noted as an intermediate which decays rapidly to bathorhodopsin. The increase in the length of the conjugated system is compensated for by the decrease in lysine side chain extension. The motion required for the photoisomerization, as indicated in Figure 5.15, is related to "concerted twist" (728, 729), a modification of the "bicycle pedal" scheme (730, 731). In our model, the iminium nitrogen would be located at a position between helices 1 and 7 and at a level one helix turn below the position at which lys 312 joins helix 7. The iminium nitrogen would move about 4.5 Å toward the cytoplasmic surface of the bilayer to a position closer to helix 1 and at the level at which 312 lys joins helix 7 (cf. [732]). The occurrence of normal photoisomerization in the same C_{9-11} restricted rhodopsin alluded to above has been interpreted as suggesting a pathway different from that of natural rhodopsin (733). The π-bonds in the retinylidene group are parallel to the bilayer plane.

I now take note of the genetically homologous positive charges (arg 85 and lys 83) in the connecting link between helices 1 and 2 and the negative charges (glu 263 and glu 265) in the link between helices 5 and 6. Other G-protein receptors, human and hamster β_2-adrenergic receptors (lys 60, arg 63) (553, 734), and muscarinic acetylcholine receptor (lys 51, lys 57) (555) have considerable homology with bovine opsin (559) and have charged groups in the helix 1–helix 2 cytoplasmic loop (Chap. 8). Functional homology seems to be more important than sequence homology in the case of the nicotinic acetylcholine receptor (735–737), so one may infer that such charge centers may be involved in the biochemical function of rhodopsin, i.e., in the interaction of RX with the G-protein, α-transducin. I believe that catalytically inactive rhodopsin and iodopsin have a conformation in which the 83, 85 (+)-pair interacts with the 263, 265 (−)-pair. The fact that charge-preserving lysine modifications and 88% lysine acetylation do not interfere with G-protein activation (738) implies that (a) arginine is the key charged group and/or (b) an acetamido group can interact strongly enough to maintain the pathway. The amino acid side chains found in the 20 Å separating the iminium ion from the cytoplasmic surface (the upper side in Figure 5.13) are highly nonpolar (dielectric constant ca. 2). Almost 4 kcal/mole change in the interaction energy for the iminium ion with the 85–263 ion-pair at the bilayer "surface" could be expected, corresponding to a possible change of a factor of 1000 in stability of the initial conformation of rhodopsin. The C=N stretching frequency does not change between rhodopsin and bathorhodopsin, excluding electrostatic interactions of the iminium ion with a proximal negative charge as the origin of the energy storage (732). The photoisomerization can then be regarded as the operational mechanism for a "photoswitch."

Since the iminium ion is the only charge that moves within the bilayer in the blue iodopsin, its role in initiating the conformational changes that eventually produce catalytically active rhodopsin would seem to be essential. Although direct electrostatic interaction is an obvious consequence of the isomerization, other changes mediated through the helix may also contribute to the operation of the opsin "photoswitch." The fact that rhodopsin may be photoregenerated from intermediates like metarhodopsin (579) suggests that the conformational changes are limited in extent and closely linked to the geometric arrangement of the chromophore.

Formation of Metarhodopsin I

With a specific structure for rhodopsins (and iodopsins) in hand and a physical picture for the "photoswitch," one can address the question of how metarhodopsin II might catalyze the removal of GDP from its complex with transducin. One reasonable possibility is that metaR II associates in some way with the GDP in the T.GDP complex (which may interact with the metaR II in other ways) and promotes GDP replacement in T.GDP by GTP. On chemical grounds, an association site might consist of two positively charged groups on the same side of a β-strand. However, direct contact between the effector metaR II and the GDP binding site may not be necessary (558). The assumption of a common mechanism in all rhodopsins and iodopsins has been confirmed (739, 740) although cone transducin is different from that in the rod (741, 742). Thus, the positive pair should be present in all of the opsins. As already noted, such pairs can be found on the helix 1–helix 2 link on the cytoplasmic side in all opsins and other G-protein receptors. A pair of negatively charged groups (glu 263 and glu 265) within the connecting link between helices 5 and 6 could serve as counter-ions to 83 and 85. Their interaction would draw the connecting links together. In addition, the negatively charged pair might compete for the positive pair presumed to be present in T_α at the GDP binding site. A close association of the T_α unit of T with RX must exist on the basis of the inhibition of activity on blocking one of the T_α-thiol groups (743).

Absorption of a photon and operation of the photoswitch would lead to a change in the stability of the 85-265 pair, effectively freeing the 83, 85 positive pair for interaction with the negative charges of GDP in the T.GDP complex. The role of acetylated rhodopsin in G-protein activation (738) might be due to the partial positive charge on the nitrogen of the $NHCOCH_3$ group. A detailed scheme showing how metarhodopsin II might promote the dissociation of GDP from T.GDP and the formation of a T.GTP complex is given in Figure 5.16. The T_γ-subunits of bovine and frog transducins differ from one another and from the γ-subunits of the adenylate cyclase regulatory proteins, G_s and G_i. However, the β-subunits of all of the four proteins are similar (744). A plausible mechanism for the enhanced association of GTP versus GDP (745) in the complex involves the positively charged pair in T and one of the active positive charges in metaR II. Binding of GTP to T_α involves conformational change in the protein (745). Phosphorylation of metaR II by rhodopsin kinase (746, 747) apparently does not interfere with binding of metaR II to the G-protein complex but may enhance binding to the 48K protein (''arrestin'') in competition with that to transducin (748). Some G-protein effects on rhodopsin, such as stabilization of metaR II at the expense of metaR I, may be simulated by short peptides (311–329 and 340–350) from the T_α sequence (749).

Conclusions Based on Rhodopsin Model

Genetic homology, photophysics, and chemistry and chemical logic have been combined to yield a partial model of rhodopsin structure and dynamics. The occurrence of genes homologous to the opsin gene in a wide variety of species (archaebacteria, algae, invertebrates, vertebrates) (750) and the importance of G-protein receptors (559) (muscarinic acetylcholine receptor [555], β_2-adrenergic receptor [553, 734], olfactory receptor [443]) underlines the significance of rhodopsin structure and function to molecular neurobiology.

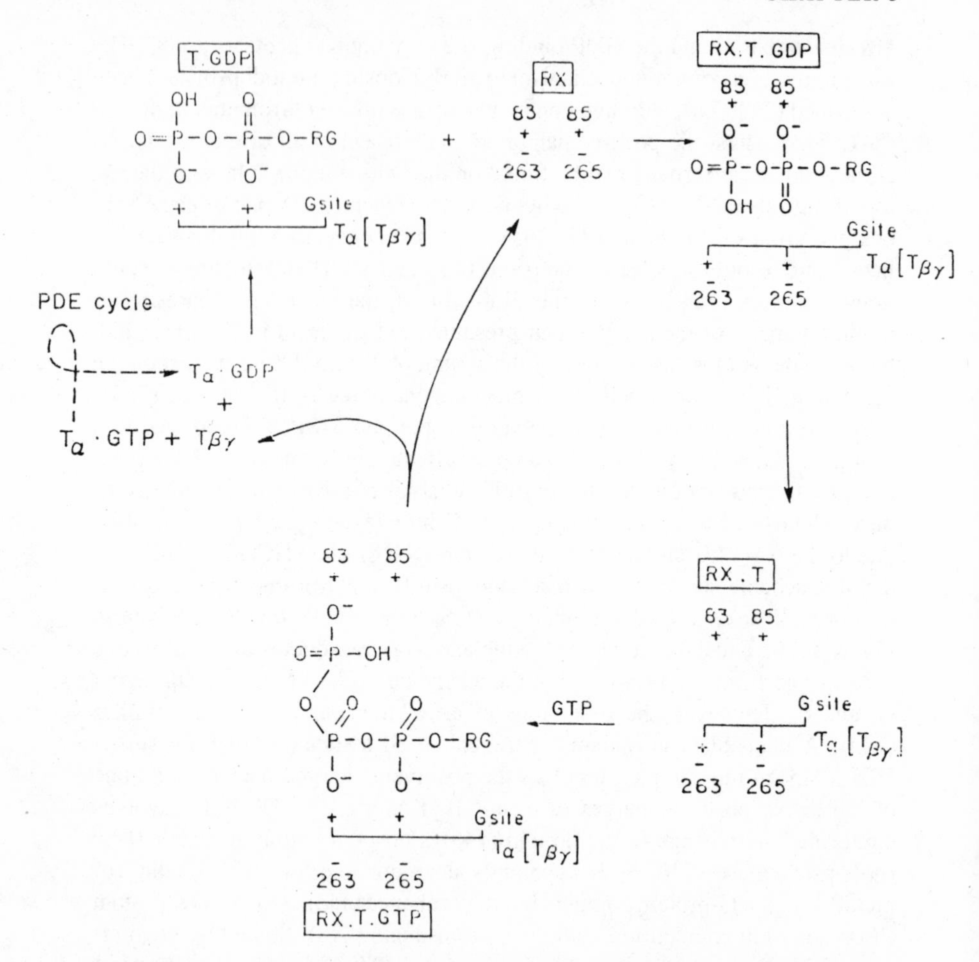

Fig. 5.16. A mechanism for the catalysis by metarhodopsin II (RX) of the exchange of GTP for GDP bound to T_α. Approach of T_α.GDP allows the 83, 85^+ pair of RX to exchange its links with the 263, 265^- pair for the GDP. The RX.GDP.T_α combination allows the GDP to dissociate and be replaced by GTP. There are now more + sites to stabilize the interaction of the GTP with the intermediate. Possibly the third − charge in GTP associates with a + charge in T_α in place of the + charge in RX, leading to the dissociation of RX and weakening the association between T_α and $T_{\beta\gamma}$. Hydrolysis of GTP to GDP within T_α.GTP, which occurs during the operation of the phosphodiesterase (PDE) cycle, leads to the dissociation of the complex and allows reassociation of T_α and $T_{\beta\gamma}$.

Genetics of Color Blindness

Color blindness affects about 8% of the male population. Less than half belong to the dichromat class in which either green or red sensitivity is affected (580). The heritability of color blindness is well known (751) and has been confirmed by blot hybridization of DNAs from color-blind individuals (675). A second type of color blindness is the anomalous trichromat, in which shifted absorption maxima (red ⇒ green, "protanomalous"; green ⇒ red, "deuteranomalous") are found for the pigments (752). The shifts in absorptions appear to depend on varying quantities of green pigment arising from a varying number of green pigment gene copies or on mistakes during crossing over between chromosome pairs rather than on point mutations (675). The red and green iodopsins (cone pigments) are 96% identical, and found on the X-chromosome (hence, sex-linked) but have only 43% similarity to the blue iodopsin. The iodopsins are 41% homologous to rhodopsin (674). Our model suggests that some variation in pigment absorption spectra might also arise from point mutations at serine or threonine sites in the middle of the bilayer. Color vision in the squirrel monkey can be explained by a polymorphism related to that of humans with color blindness (753).

Molecule: 11-*cis*-Retinylidene-ϵ-lysine

Rhodopsin light absorption is due to the chromophore, an 11-*cis*-retinylidene group bound as an imine to an ϵ-amino group of a lysine. Such imines ("Schiff bases") are usually protonated in neutral solution. The chromophore structure and the numbering of the carbons are shown in Figure 5.17. One might first consider how the range from 400 to 700 nm came to be "visible."

The solar spectrum at the earth's surface sets the parameters needed for

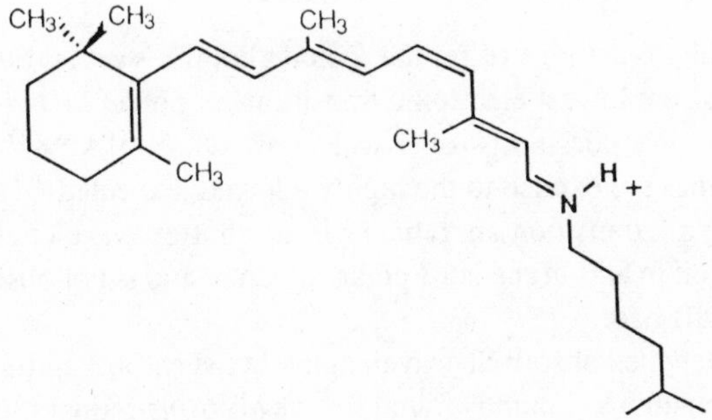

CH_3 CH_3 CH_3 CH_3 CH_3 H N +

Fig. 5.17. A protonated 6-s-*cis*-11-*cis*-12-s-*trans*-retinylidene-ϵ-lysine group, the chromophore of rhodopsin.

detection of light, since it is this radiation which living things must use. Some 43% of solar radiation occurs between 400 and 700 nm, and only 5% at shorter wavelengths (754). A light-detecting system should absorb past 400 nm, be derived from a group suitable for binding to a protein, and undergo physical changes which could be amplified through the protein. The retinylidene conjugated system fits these criteria and appeared quite early in evolution, being present in single cell algae, *Chlamydomonas* (62). However, the phototactic response is activated by retinal analogues incapable of *cis-trans* bond isomerization, suggesting that direct activation of *Chlamydomonas* rhodopsin is possible (755).

The lowest-energy electronic transitions ($S_0 \rightarrow S_1$) of conjugated systems decrease in energy and increase in intensity with increasing number of conjugated elements. These elements are normally π-systems, for which the excitation energies are considerably lower than those for σ-elements, i.e., C=C bonds absorb light at longer wavelengths than C-C bonds. A conjugated aldehyde, R-$(CH{=}CH)_n$CHO, is a good candidate for the precursor of the light-detecting protein. The aldehyde group is reactive enough to link to proteins and lowers the electronic excitation energy more than other plausible linking groups (COOH or CH_3CO). The variation in absorption maximum of conjugated aldehydes with the number of double bonds is illustrated in Table 5.3 (756). Additional examples more closely related to retinal have been reported in detail (757, 758).

A conjugated system long enough to absorb significantly in the range provided by the solar spectrum contains six or seven double bonds. The aldehydes have maxima at wavelengths shorter than the "visible," but the protonated imines present in the visual proteins, the rhodopsins and iodopsins, absorb light in the desired region. Visual pigments are derived from retinal, either retinal$_1$ (6 double bonds, with the C=O) or retinal$_2$ (7 double bonds, with the C=O).

Electronic Transitions in Retinylidene-ϵ-lysine

The electronic transitions of retinal include both $\pi \rightarrow \pi^*$ and $n \rightarrow \pi^*$ types. The ordering of the lowest electronic transitions in polyenes and polyenals is complicated by the possibility of a weak low-lying "$1A_g^{*-}$" state (654). Protonated imines are excited to the highly allowed, expected "$1B_u^{*+}$" state. The weak $n \rightarrow \pi^*$ transition in retinals is at shorter wavelengths than the $\pi \rightarrow \pi^*$ transition in all but the least polar solvents, and is not observed for the protonated Schiff base.

The retinaldehydes absorb at wavelengths between 360 and 370 nm. The longest wavelength maximum is found for the all-*trans*-retinal (368 nm), with small shifts to shorter wavelengths for 11-*cis*-retinal (365 nm), 9-*cis*-retinal (363 nm), and 13-*cis*-retinal (363 nm). The intensity of the maxima (absorp-

Table 5.3. Absorption Maxima and Intensities for Polyenals, CH_3-$(CH=CH)_n$CHO, in Dioxane

$(CH=CH)_n$ n	λ_{max} (nm)	ϵ_{max}
1	220	15,000
2	270	27,000
3	312	40,000
4	343	40,000
5	370	57,000
6	393	65,000
7	415	63,000

Source: These data and other related comparisons can be found in reference 756.

Table 5.4. Absorption Maxima of Retinals and Retinylidene N-Butylimines

Compound	3-MP, 77 K[a] λ_{max} (nm)	ϵ_{max}	EPA, 298 K[b] λ_{max} (nm)	ϵ_{max}
all-*trans*-Retinal	385	50,200	370	52,500
7-*cis*-Retinal			359	44,100[c]
9-*cis*-Retinal	384	27,800	368	30,600
11-*cis*-Retinal	386	37,300		
13-*cis*-Retinal	380	48,000		
all-*trans*-Retinylidene N-butylimine			364	49,600
9-*cis*-Retinylidene N-butylimine			364.5	47,100
11-*cis*-Retinylidene-N-butylimine (**11RI**)			367.5	42,800
13-*cis*-Retinylidene N-butylimine			361.5	35,400
11-*cis*-Retinylidene-N-butyliminium chloride (**11RIH**$^+$)			440[d]	
all-*trans*-Retinylidene N-butyliminium chloride			460	

[a] 3-MP; 3-Methylpentane, 10^{-4} M solutions.
[b] EPA; Ether-pentane-ethanol.
[c] In n-heptane.
[d] In methanol.

tion coefficients between 28,000 and 50,000) shows conclusively that the transitions are allowed $\pi \rightarrow \pi^*$ transitions.

The retinylidene imines absorb light intensely, but at shorter wavelengths than the corresponding retinaldehydes, with 11-*cis*-retinylidene-N-1-butylimine (**11RI**) having a λ_{max} at 350 nm and 11-*cis*-retinal a λ_{max} at 375 nm in methanol. The $n \rightarrow \pi^*$ transition for imines is also at shorter wavelengths than for carbonyl compounds. Absorption data for retinals and the corresponding N-butylimines are compared in Table 5.4.

The absorption maximum shifts to much longer wavelengths in protonated or quaternized imines (Table 5.4). The effect of the positive charge on the ground state is modest, but the excited state should be stabilized in light of the charge rearrangement expected on excitation (Eq. [5-4]).

(5-4)

$$C{=}C{-}C{=}N^{+} \xrightarrow{h\nu} C^{+}{-}C{=}C{-}N$$

Nevertheless the absorption maxima of protonated (or quaternized) model imines in methanol are at much shorter wavelengths than observed for the same chromophore in the protein, rhodopsin.

The "Opsin Shift"

The reason why the retinylidene chromophore combined with the protein (the "natural" iminium ion) absorbs at much longer wavelengths than apparently similar model iminium salts is not immediately obvious. The change in absorption maximum is called the "opsin shift" (683, 684, 759). Among the factors to be considered are geometry and environment. The *cis* or *trans* character of the double bonds does not have much influence on the maximum provided that the π-overlap is similar in both isomers (the 13-*cis*-iminium ion absorbs at shorter wavelength and with lower intensity than the other isomers listed in Table 5.4). Ground state single-bond conformations (s-*cis*- or s-*trans*-) affect π-overlap. Environmental effect can be divided between (a) solvent effects on the position of the maxima and (b) local effects, which include point charges. Since the rhodopsin model describes in some detail what might be expected for point charge interactions, I shall concentrate on other types of effects.

Conformations of Retinylidene Group

The conformational arrangements of the retinylidene group in the ground state must be identified in order to understand the opsin shift. Nmr spectra of 11-*cis*-retinal in acetone-d_3 show that the 12-s-*cis* and 12-s-*trans* conformations are in equilibrium (760), a conclusion supported by analysis of uv spectroscopic intensities of the aldehyde in methylcyclohexane. The s-*trans* conformer, although larger, is favored by increased external pressure (761). The conformation around this bond is 12-s-*trans* in rhodopsin according to resonance Raman spectroscopy (762–764). A distorted 12-s-*cis* conformation is present in crystalline 11-*cis*-retinal (765). I have adopted the 12-s-*trans* conformation for rhodopsin.

Solvent Effects: Retinylidene Iminium Ion Maxima

The opsin shift may be due to a solvent effect on the position of the maximum. The effect of solvent on the position of the longest wavelength maximum for the model compound, all-*trans*-retinylidene pyrrolidinium perchlorate (**PRI**$^+$), is illustrated by the data in Table 5.5 (766, 767).

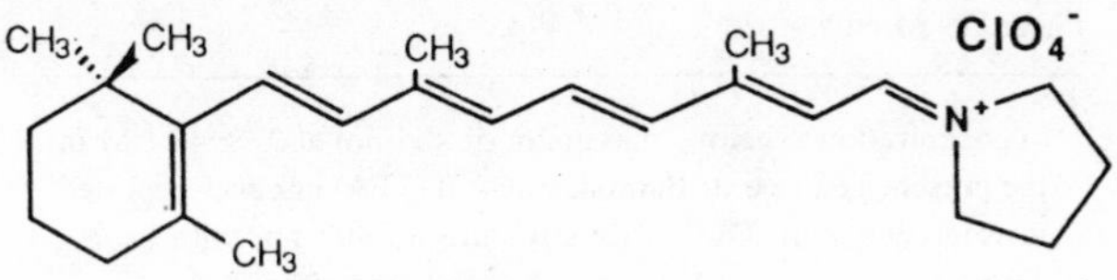

CH_3 CH_3 CH_3 CH_3 ClO_4^- N^+ CH_3

PRI$^+$

The $\pi \rightarrow \pi^*$ transition energies for α,β-unsaturated carbonyl compounds are normally solvent sensitive and usually correlated with **Z**-value. That the maxima are almost invariant for **PRI**$^+$ in dioxane, ethanol, and methanol may be due to the opposed effects of solvent on the ground and excited states at the iminium nitrogen. Polarizable solvents have a much larger effect on the position of the maximum. The ground state dipole moment of 9-*cis*-retinal is 4.69 D or 4.86 D, while that for all-*trans*-retinal is 5.26 D (768). The excited state of all-*trans*-retinal has a dipole moment 13.2 D greater than that of the ground state, or18.5 D. The polarizability of the excited state increases by 600 Å^3, even more than the change of 420 Å^3 for diphenyldecapentaene (769).

Solvent Polarity Parameters

A brief comment on the ways to measure solvent polarity is useful for understanding the absorption maxima listed in Table 5.5. Evaluation of solvent-sensitive properties requires well-defined reference parameters. A macroscopic parameter, dielectric constant, does not always give useful correlations of data. The first microscopic measure of solvent polarity, the **Y**-value, based on the solvolysis rate of *t*-butyl chloride, is particularly valuable for correlating solvolysis rates. **Y**-values are tedious to measure, somewhat complicated in physical basis, and characterizable for a limited number of solvents. The **Z**-value, based on the charge-transfer electronic transition of 1-

Table 5.5. Visible Absorption Maxima of all-*trans*-Retinylidene Pyrrolidinium Perchlorate in Various Solvents

Solvent	λ_{max} (nm)	$E_T(30)$ (kcal/mole)
Methanol	453	55.5
Ethanol	458	51.9
1,4-Dioxane	450	36.0
Benzene	464	34.5
Bromobenzene	480	37.5
1,2-Dichlorobenzene	503	38.1
Dichloromethane	496[a]	41.1
cis-Dichloroethene	508	—
trans-Dichloroethene	503	—
Phenol (9 M/ethanol)[b]	494	—

[a] A concentration-sensitive maximum of 484 nm at 2×10^{-4}M in the presence of 1 eq. trifluoroacetic acid (TFA) has been reported in reference 711. The maximum shifts to 497 nm with excess TFA.

[b] Retinylidene N-(2-hydroxy-1-propyl)iminium trichloracetate.

ethyl-4-carbomethoxypyridinium iodide, is easy to measure and has a readily understandable physical origin (Fig. 5.18). However, **Z**-values are difficult to obtain in a nonpolar solvent because of low salt solubility. The $\mathbf{E_T}$(30)-value is based on an intramolecular charge-transfer transition in a pyridinium phenol betaine, which dissolves in almost all solvents.

$\mathbf{E_T}$(30)-values and **Z**-values are useful as measures of solvent polarity. Solvent polarity is what is measured by a particular technique and may refer to different summations of molecular properties in different cases. For this reason, only simple reference processes should be used to derive solvent parameters. Descriptions of the origin and applications of solvent polarity parameters to spectroscopic and chemical problems may be found elsewhere (770–776).

Local Effects: Retinylidene Iminium Ion Maxima

A molecular electronic transition that results in charge rearrangement can be affected by the differential interaction of the ground and excited states with a charged group. The effect was suggested many years ago as an explanation for the opsin shift (777) and was demonstrated many years ago for the bicyclic α,β-unsaturated ketone systems **AB** and **AC**$^+$. In water, the λ_{max} are 245.5 nm and 233 nm, representing a difference of 2176 cm^{-1} in transition energies (778).

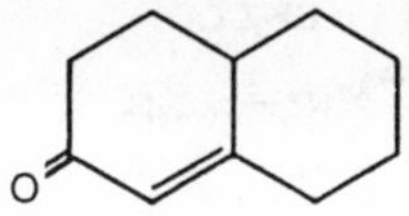

$N(CH_3)_2^+$ ClO_4^-

AB 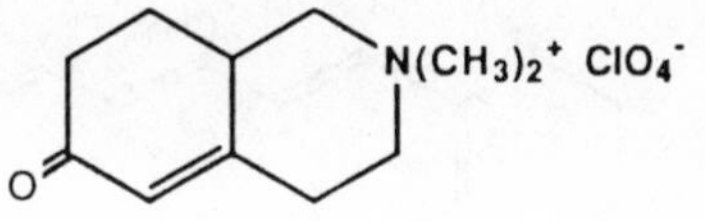 AC$^+$

An elegant solution to the problem of evaluating the effect of charges on retinylidene iminium ion absorption maxima has been to introduce either dimethylamino or carboxyl groups into positions that were not conjugated with the π-system. In the case of dimethylamino groups, the charge is created by adding a proton with acid. The charge may change both ground and excited state energy levels, and thereby alter the transition energy. According to the charge rearrangement shown for the $S_0 \rightarrow S_1$ transition (Eq. [5-4]), a positive charge near the iminium N^+ should destabilize the ground state and stabilize the excited state, leading to lower transition energies (779). A positive charge in the middle of the chain should have similar effects on both states, while a positive charge at the ionone ring should destabilize the excited state more than the ground state, leading to higher transition energies. These are exactly the effects reported (711).

Another striking property of iminium systems (**BI**$^+$, **9-ABI**$^+$, **4-ABI**$^+$, **PzI**$^+$), already mentioned in connection with the model, is the shift to longer wavelengths in the presence of excess acid. It is likely that the excess acid hydrogen-bonds to the anion associated with the positive ions and diminishes

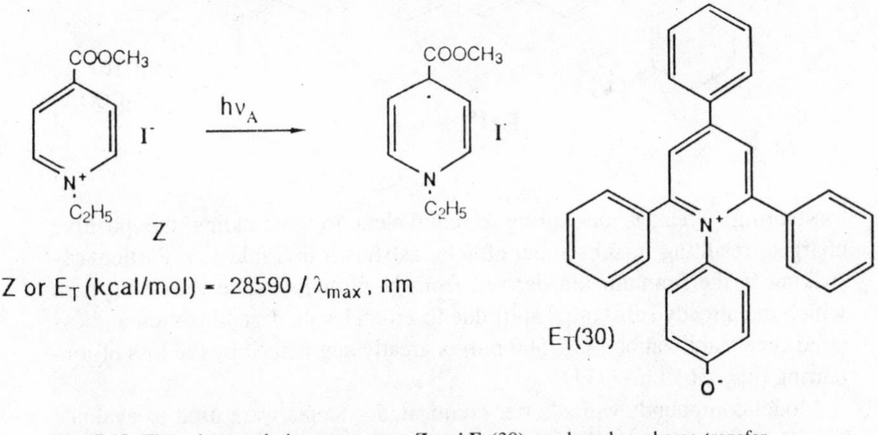

$COOCH_3$ I^- $h\nu_A$ $COOCH_3$ I^-

N^+ C_2H_5 Z N C_2H_5

Z or E_T(kcal/mol) = 28590 / λ_{max}, nm

$E_T(30)$ N^+ O^-

Fig. 5.18. The solvent polarity parameters, **Z** and $\mathbf{E_T}$(30), are based on charge-transfer transitions in pyridinium derivatives through the relationship **Z** or $\mathbf{E_T}$(30) (in kcal/mole) = 28,590/λ_{max} (in nm). **Z** is derived from the charge-transfer transition for 1-ethyl-4-carbomethoxypyridinium iodide. The charge-transfer transition of the pyridinium phenolate betaine is used to define the $\mathbf{E_T}$(30) value, the process involving electron transfer from the phenolate ion to the pyridinium ring.

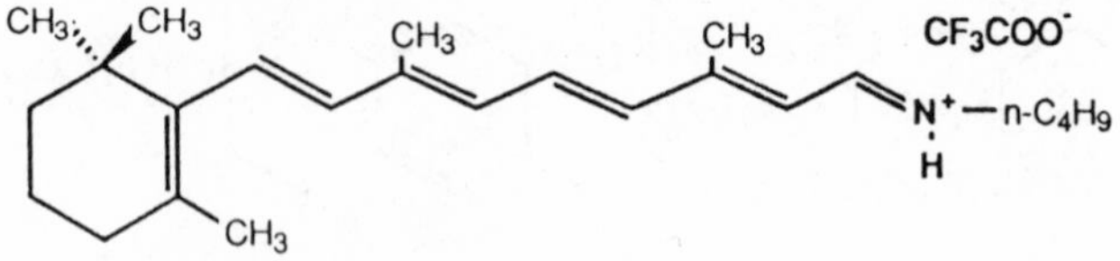

BI^+

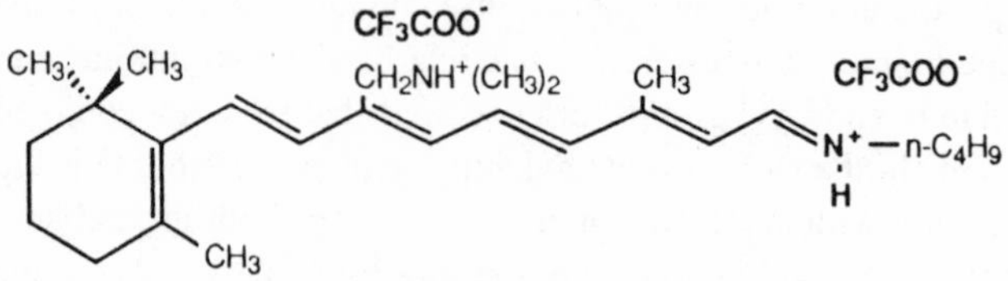

9-ABI^+

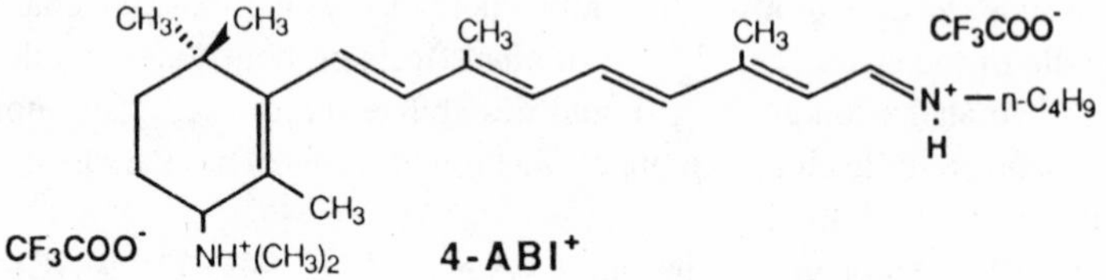

4-ABI^+

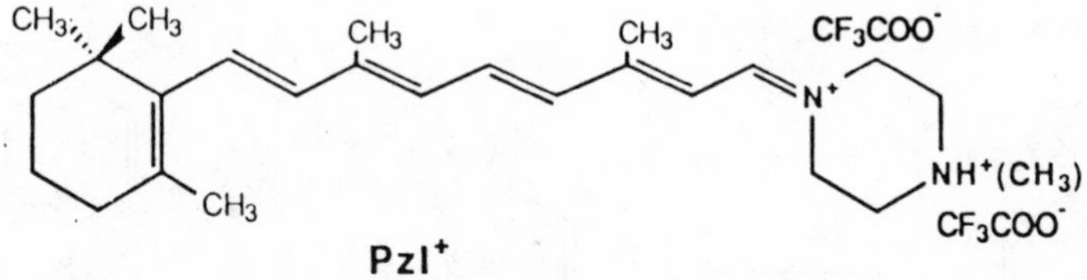

PzI^+

ion-pairing. Weaker ion-pairing is equivalent to deshielding the positive charges, resulting in substantial effects, as shown in Table 5.6. Particularly striking is the iminium ion derived from N-methylpyrazine (last entry), in which the already substantial shift due to ground state destabilization and excited state stabilization (λ_{max} 500 nm) is greatly augmented by the loss of ion-pairing (λ_{max} 561 nm) (711).

Model compounds with shorter conjugated systems were used to evaluate the effect of a negative charge attached to the π-system on the position of the absorption maximum. The acid form and the zwitterionic forms of iminium ions were compared (Fig. 5.19) to show that a negative charge has effects opposite to those described above for a positive charge (684).

Intimate interaction between the retinylidene group of rhodopsin and trp

Table 5.6. Absorption Maxima of Retinylidene Iminium Ions in Dichloromethane Containing Trifluoracetic Acid (TFA)

Ion	λ_{max} (nm) TFA, 1 eq.	TFA, excess	$\Delta\nu$ (cm^{-1})
BI$^+$	448	513	2828
9-ABI$^+$	423	455	1662
4-ABI$^+$	426	461	1782
PzI$^+$	500	561	2174

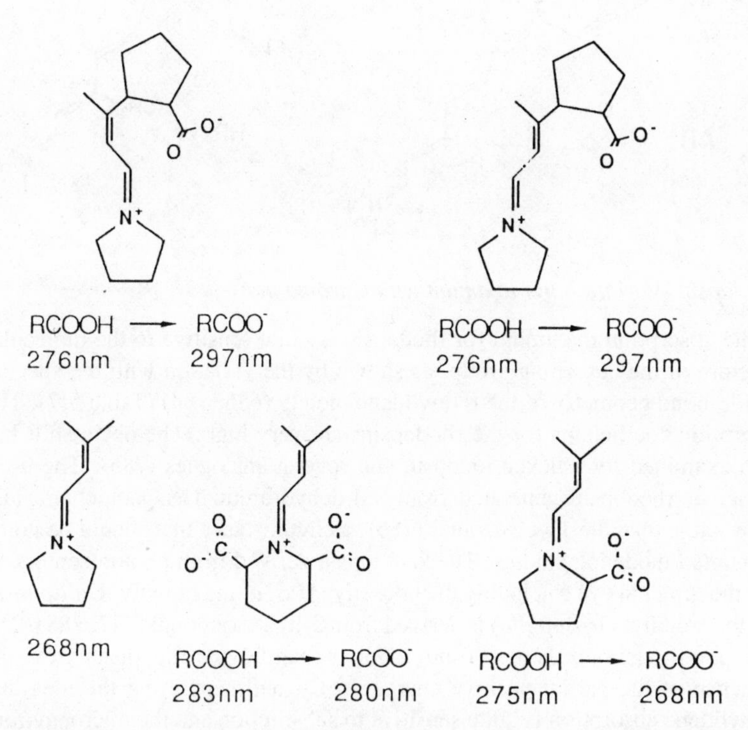

Fig. 5.19. Adjacent-charge effects on absorption maxima in various α,β-unsaturated iminium ions.

142 may be the origin of the loss of the visible CD spectrum in the rhodopsin derived from 5,6-dihydroretinal (780, 781). Nevertheless, some flexibility exists in the interaction of the trp 142 region of the opsin with the ionone ring. This is shown by the existence of photobleachable rhodopsins, in which the ring portion of the 9-*cis*-retinylidene moiety (with its *gem*-dimethyl group) has been replaced by a somewhat larger adamantylidene ring (rhodopsin unstable, **AR**) (782) or by two alkyl groups (**RR**) (783).

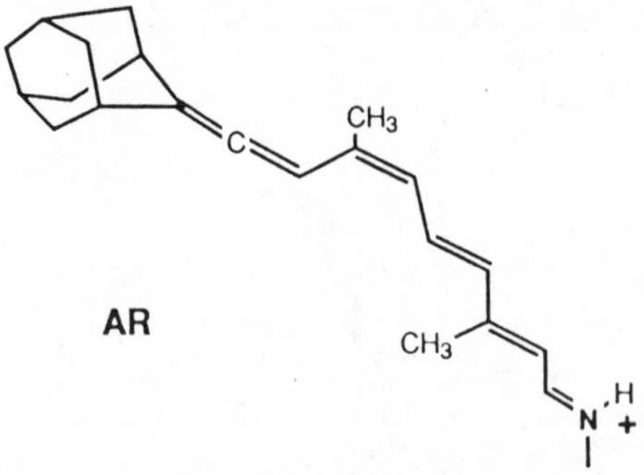

Rhodopsin: Variation in Maximum with Chromophore

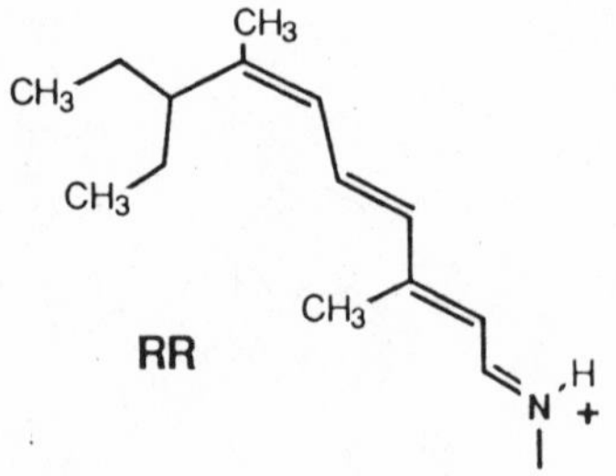

The absorption maximum for rhodopsin is quite sensitive to the molecular structure of the absorbing group, as shown by the variation with the specific double-bond geometry of the retinylidene moiety (685, 784) (Table 5.7). The absorption coefficients for the rhodopsins are very high. The opsin shift has been examined for chicken iodopsin and several analogues (785). The 9-*cis* isomer of rhodopsin generated from 3,4-dehydroretinal is photochemically more stable than the 11-*cis*-isomer (786), a circumstance that should be noted in detailed modeling studies. The xanthopsin derived from photoreceptors 1–6 of the dipteran eye (including the housefly, *Musca*, the blowfly, *Calliphora*, and the fruit fly, *Drosophila*) is derived from 3-hydroxyretinal (787, 788). The absorption maximum is at 490 nm, shorter than bovine rhodopsin from the influence of the partial positive charge on C_3, and supporting the idea that retinylidene absorption is quite sensitive to substitution and the microenvironment. The still shorter wavelength maximum (345 nm!) for the rhodopsin of the owl fly, *Ascalaphus macaronius*, requires further study, especially since 11-*cis*-retinylidene is claimed as the chromophore (789).

*Photoisomerization of 11-*cis*-Retinylidene* n*-butyliminium ion*

The photoisomerization of 11-*cis*-retinylidene *n*-butyliminium ion to the 11-*trans* compound occurs within 8 ps after a laser pulse with a quantum yield of 0.2. The quantum yield for photoisomerization of rhodopsin is 0.67. A rapid

Table 5.7. Absorption Maxima for Visual Pigments

Pigment	λ_{max} (nm)	ϵ_{max}	(Batho-)[a] λ_{max} (nm)	ϵ_{max}
Bovine rhodopsin (11-*cis*)	498	42,000	543	47,500
Bovine isorhodopsin I (9-*cis*)	483	44,000		
Bovine isorhodopsin II (9,13-di *cis*)	481	44,000		
Squid rhodopsin	498	46,600	543	65,000
Chicken iodopsin	575	46,500	640	70,000
Vertebrate porphyrosin (11-*cis*-3,4-dehydro)	522			
Vertebrate cyanopsin (11-*cis*-3,4-dehydro)	620			

Note: For pigments in 2% nonionic Ammonyx LO at 25 °C

[a] Batho form of pigment, formed by low-temperature irradiation.

and efficient photoisomerization process occurs without opsin. The protein, however, is obviously central to the process of translating a photon signal into a biochemical signal (790, 791).

Further Biochemical Reactions: Retinal Formation

The Schiff base is hydrolyzed after the rhodopsin has converted the light signal into a conformational change for transducin, producing apoprotein, opsin, and an aldehyde, 11-*cis*-retinal, for which there is chromatographic proof of identity (792). The course of the hydrolysis is given in Eq. (5-5) (793). An NAD-dependent dehydrogenase in retinal pigment epithelium forms 11-*cis*-retinal from 11-*cis*-retinol complexed to cellular retinal-binding protein (794). The *trans*-retinal produced in the ROS is transferred to the pigment epithelium, reduced, and isomerized to 11-*cis*-retinol by an eye-specific membrane-bound retinol isomerase (795).

(5-5)

$$RCH{=}NH^{+}R' \xrightarrow{HOH} RCH(OH)NHR' \longrightarrow RCH{=}O + R'NH_2$$

6 The Nicotinic Acetylcholine Receptor

Introduction to Neurotransmitter Receptors

Transfer of signals between neurons is achieved either by the direct transfer of potential change at an electrical synapse or by the transfer of signal molecules at a chemical synapse. Neurotransmitter molecules are usually packaged into organelles called synaptic vesicles. Calcium enters the release region under the influence of the depolarization on the presynaptic side and activates synapsin I, which moves the vesicle to the neuron membrane. Fusion of the vesicle membrane with the cell membrane leads to the release of the vesicle contents to the exterior. The main component, the neurotransmitter, diffuses across the synaptic cleft, the region between the releasing cell and the next neuron. At the recipient neuron membrane, the neurotransmitter combines with a receptor molecule. A number of accessory molecules are released together with the neurotransmitter. The accessory molecules can stimulate synapse formation and synthesis of mRNA for receptor protein. Nerve signal transfer, part of learning, may be accompanied in many cases by the stimuli which initiate the recording of memory after sufficient reiteration of the signal (Chap. 9). All aspects of neurotransmitter communication between neurons are vital to an understanding of the molecular basis of learning and memory. I will look in some detail at the nicotinic acetylcholine receptor, formulating a structural and dynamic model in order to understand the function and behavior of the receptor.

Neurotransmitter-Receptor Complexes

Cells in biological systems normally interact with external signals at the surface, with the exception of such stimuli as light, magnetic fields, and perhaps electric fields. The concept that particular elements of cells may act as receptors goes back to the days of Claude Bernard but was given specific focus by Ehrlich (immunoreceptors) and Langley (neurotransmitters) (796–798).

Biological receptors are proteins or protein complexes that interact with appropriate molecules (signals) to produce a combination capable of conformational change, leading ultimately to a biochemical or biological result. Among the results are ion flow into the cell or activation of an enzyme cascade, which then influences ion flow. Ion flow into the cell involves cations in the case of the acetylcholine receptor.acetylcholine (AChR.ACh) complex or chloride ions for the γ-aminobutyric acid receptor.γ-aminobutyric acid (GABAR.GABA) complex. The activation of an enzyme sequence leads to a change in the level of other molecules ("second messengers"). A stimulus-receptor combination that does not undergo the requisite change fails to produce the expected biological result (Eq. [6-1]); the inactive combining molecule is an "antagonist" to the effect of the active molecule ("agonist"). Acetylcholine is one of the most important neurotransmitters; the interaction of acetylcholine with the acetylcholine receptor leads to a flow of ions into the cell and the propagation of a wave of depolarization which may activate an axon signal. A clarification of the molecular basis for the action of acetylcholine must be a central goal for any general understanding of the nervous system.

(6-1)

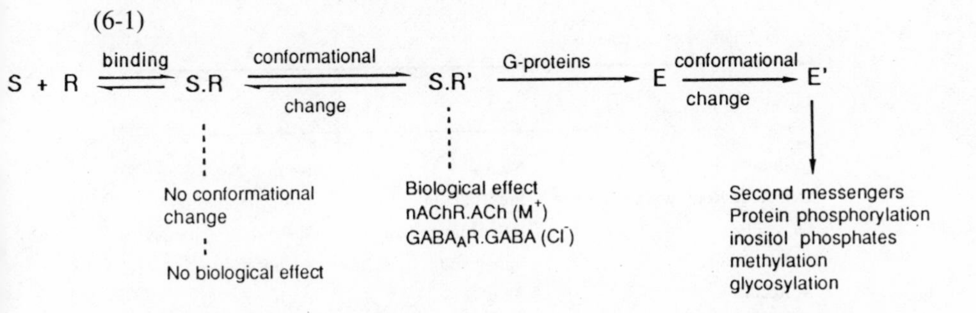

Acetylcholine: Historical Aspects

In 1921, release of a heart-stimulating substance, later proved to be acetylcholine, from the vagus nerve of frog and toad hearts was discovered (799). This confirmed previous suggestions of a chemical signal being transferred between cells, and implied the presence of specific molecular receptors. The inactivation of the signal substance, acetylcholine, in the heart was detected shortly thereafter (800). The fact that acetylcholine is the excitatory transmitter at particular vertebrate neuromuscular junctions was shown in 1936 (801), elucidating a biological role for a molecule first synthesized in 1867 (802).

Distribution of Acetylcholine Receptors

Acetylcholine is a major neurotransmitter in organisms over a good part of the evolutionary scale (803) including vertebrates (fish to mammals), arthropods (spiders), molluscs (shellfish), annelids (worms), nematodes (unsegmented threadlike worms, such as the hookworm), insects (locusts [804]), and echinoderms (starfish, sea urchins).

Cholinergic synapses are present in the mammalian peripheral nervous system (autonomic preganglionic fibers, parasympathetic postganglionic fibers, and motor fibers to skeletal muscles). In the mammalian central nervous system, there are cholinergic synapses on Renshaw cells and in the cerebellum, the cerebral cortex, the thalamus, the lateral and medial geniculate nuclei, and the caudate nucleus. Cholinergic synapses are probably widespread in both peripheral and central portions of invertebrate nervous systems (805).

Cholinergic synapses are similar in general structure to neuromuscular junctions, the connections at which nerves convey signals to muscles. A synapse is a structure composed of two complex organelles, a presynapton and a postsynapton (Fig. 1.5). Any specialized structures that facilitate the transfer of information across the synapse belong to these two organelles. The acetylcholine receptor is the key element of the postsynapton. Level analysis (Chap. 1) clarifies some of the organizational relationships among the elements of a synapse.

Level Analysis of a Synapse

Level	Synapse	
Cellular	Neuron 1	Neuron 2
Organelle Region	Presynapton	Postsynapton
Organelle	Releasing apparatus	Receptor apparatus
Polymolecules	Synapsin I	ACh Receptor
Molecule	ACh*	ACh

*Acetylcholine

Cholinergic synapses are divided into two major classes, "nicotinic" (nAChR) and "muscarinic" (mAChR), a distinction first made in 1914 according to activating agent (806). The nicotinic receptors are multisubunit complexes with molecular weights of ca. 290,000. Muscarinic receptors of several different subclasses are monomeric proteins with molecular weights of ca. 80,000 (807). The sequences of several mAChRs have been obtained by cDNA techniques. The mAChRs are homologous to rhodopsin and the β-adrenergic receptor, all being G-protein receptors (555, 556) (Chap. 8).

Immunological cross-reactivities, amino acid compositions and subunit compositions, and molecular weights all suggest that nicotinic receptors have a substantial similarity across a broad evolutionary spectrum. The embryological origin of electroactive tissues is muscle in a number of fishes (*Torpedo marmorata*, *Torpedo californica*) and eels (*Electrophorus electricus*).

Such tissues are a highly concentrated source of acetylcholine receptor (AChR). The innervated face of an electrocyte (electroplaque) of *Torpedo californica* is almost 100% covered with receptors (808, 809). One kilogram of wet *Torpedo* electric tissue contains 100 milligrams of AChR, and may yield

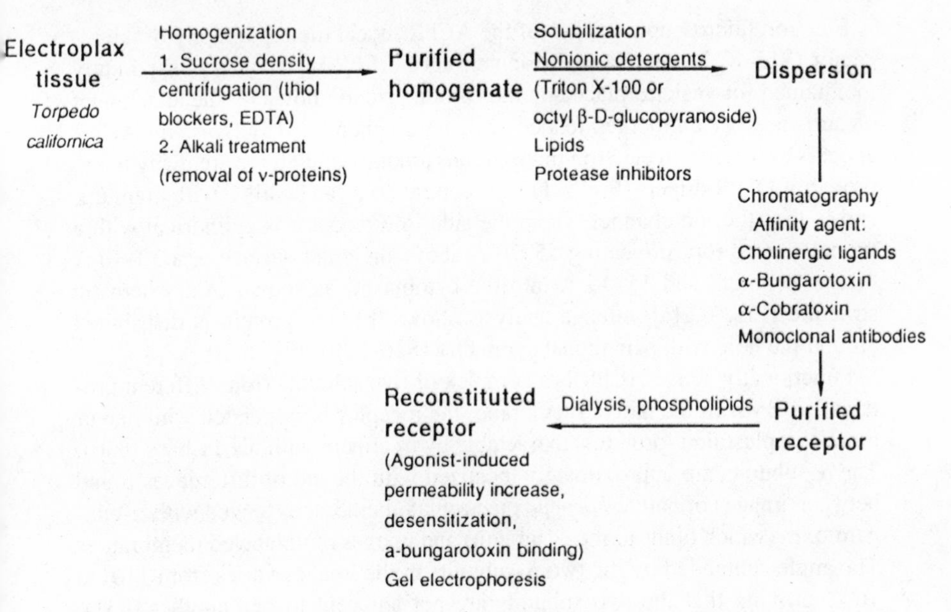

Scheme 6.1. Purification of acetylcholine receptor. Various aspects of the procedure are described in the references (808, 810–816).

20–40% of that in pure form. The procedures used involve dispersion with detergents (*and* phospholipids) in the presence of protease inhibitors, chromatographic columns carrying affinity agents such as snake toxins, and dialysis. The purified AChR can be reconstituted into a pharmacologically active form in vesicles (796) and is thus suitable for further analysis. Scheme 6.1 outlines the general procedure for the purification of acetylcholine receptors (810–816).

Acetylcholine Receptor: Structure

One wants to relate the agonist-binding and ion-gating functions of the receptor to the molecular structure. The primary structure, the amino acid sequence, is known for the AChR of *Torpedo californica*. Some information exists about secondary structures (α-helices, β-strands) in the receptors, and there are some notions about the tertiary structures (organization of the secondary structural segments) of the AChR.

The overall (quaternary) structure of the AChR is revealed by electron microscopy and image analysis, including a location for the ion channel in the center of the protein. Grosser aspects of receptor organization can be discerned.

Electron microscopic images of the AChR reveal the topography of the receptor (817–829). The natural orientation of AChR in bilayers is apparently maintained for vesicles placed on the carbon grid as shown by the attachment of anti-receptor antibodies loaded with gold spheres (818). *Torpedo* AChR appears as rosettes in electron micrographs among which there are many pairs, probably AChR dimers (Fig. 6.1). The central hole can be filled with uranium, and is thus the ion channel. From the side, the receptor is cylindrical with a funnel-shaped top, projecting 55–70 Å above the outer surface of a 30–40 Å bilayer segment and 15–45 Å into the cytoplasm, as shown in a schematic structure (Fig. 6.2[a]). Image analysis shows that the protein is distributed around the hole with pentagonal symmetry (826–829) (Fig. 6.2[b]).

Functionally active AChR is a complex of five subunits (four different proteins), $\alpha_2\beta\gamma\delta$. In the native membrane, the receptor is associated with one or more cytoplasmic ν-proteins, extractable by treatment with pH 11 base (830). The α-subunits are approximately localized with the aid of differences found between images of native *Torpedo* electroplax membranes tagged with α-bungarotoxin (which binds to the α-subunit) and images of unlabeled membranes. The angle subtended by the two α-subunits in the images varies from 110° to 160°, proving that the two subunits are not adjacent to one another (831). Differences in images of samples with or without the ν-proteins suggests that these are localized under the α-, γ-, and δ-subunits (823–825). However, the ν-proteins may be chemically linked to the β-subunit (832), showing that a preferred orientation for the ν-proteins is not easy to choose. Phosphorylation of the AChR is associated with a creatine kinase present in the ν-protein fraction (833–837).

The dimers present in some membranes are δ,δ-dimers linked through a disulfide bond; higher oligomers may also be present, possibly linked through β,β-disulfide bonds (838–840). The AChR dimers are converted to AChR monomers through reduction of the disulfide link with dithiothreitol. The AChR monomers within the dimer are arranged in a variety of ways, including side by side (841), with a symmetrical arrangement in "natural" membranes (823). There is little functional difference between monomers and dimers with respect to ion-gating properties. However, the dimers may be involved in the membrane organization of the AChR, since rows of receptors are formed in the presence of ν-proteins (808, 842). A wide range of lateral diffusion coefficients, from 5×10^{-11} cm^2/sec to 10^{-8} cm^2/sec, has been reported for nAChR in different environments (825). Receptor subunits are separately synthesized, inserted into the membrane, then organized into active receptors which are delivered to and trapped (843) at postsynapton sites. Monomeric AChR is thus mobile within the membrane (844).

The molecular weight of the AChR monomer (exclusive of 7% or less glycosylation [845], $<<1$% phosphorylation [846, 847], and 2–19 mol phospholipid/mole receptor [848, 849]) can be calculated as 267,757 from the subunit amino acid sequences (next section): α-, 50,116 (2 copies); β-, 53,681; γ-,

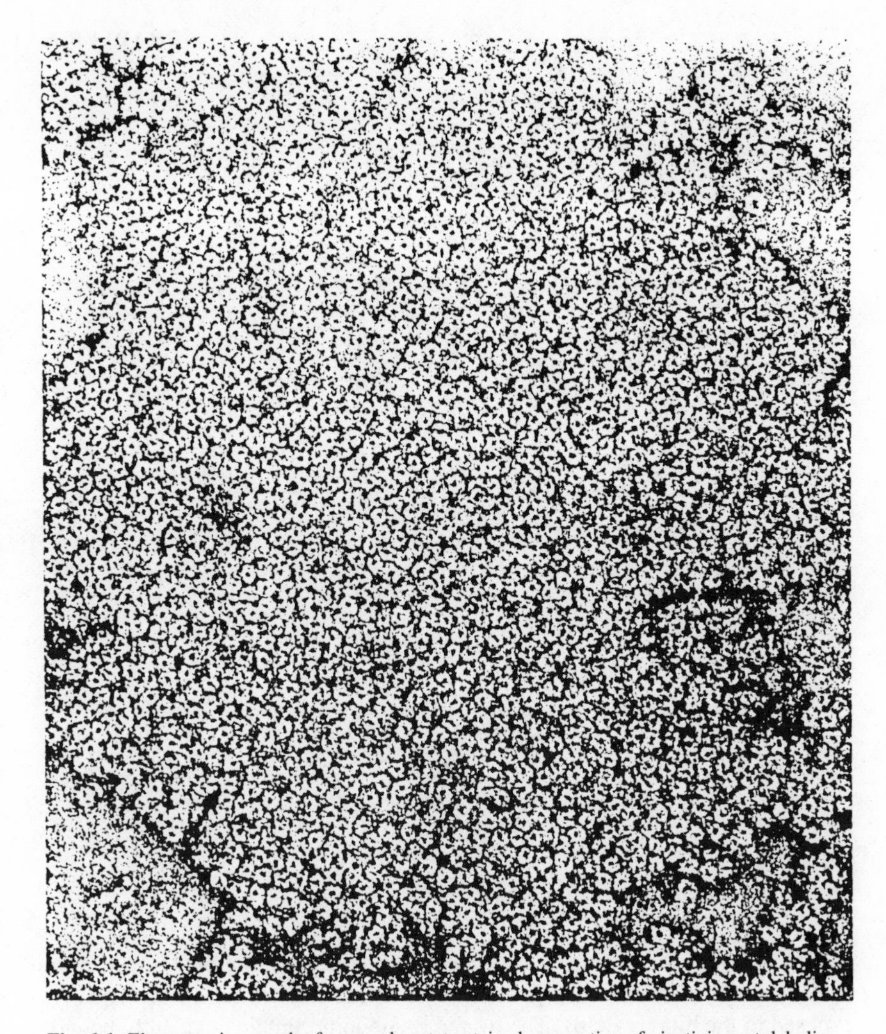

Fig. 6.1. Electron micrograph of a uranyl acetate-stained preparation of nicotinic acetylcholine receptor (nAChR) from *Torpedo californica* shows the dense packing of the native membrane. Some of the nAChR show up as pairs. The pairs are unique to *Torpedo* and are thought to be δ,δ-disulfide-linked dimers, since they are dissociated after treatment with dithiothreitol (DTT). The central ion channel in the middle of the pentameric structure is visible in the center of each of the monomers. (Photograph provided by Prof. R. M. Stroud and reproduced from the *Biophysical Journal* 37: 371–383 [1982], by copyright permission of the Biophysical Society [reference 821].)

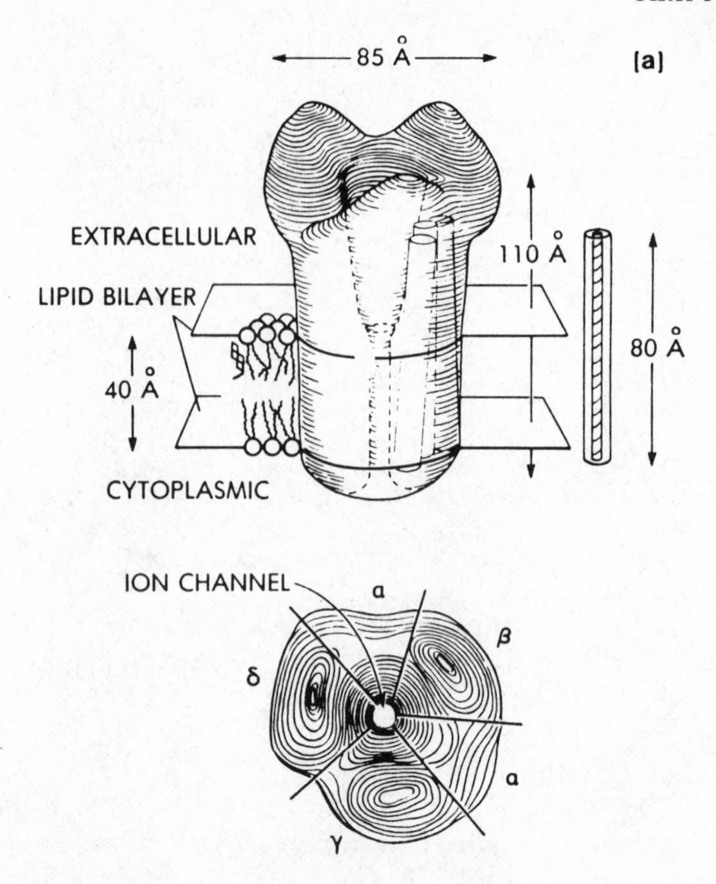

56,279; δ-, 57,565. A carbene generated within the phospholipid bilayer from [^{3}H]adamantane-diazirine labels all subunits, indicating that all are equally exposed to the bilayer (850). The α- and β-subunits contain covalently bound lipid, probably linked to an ϵ-amino group of a lysine (851). All subunits are partially hydrolyzed by trypsin from either side of the membrane, using AChR-loaded vesicles, showing that all subunits are transmembrane (852, 853).

Acetylcholine Receptor: Primary Structure

The first 54 amino acids of the AChR subunits were sequenced by the classical Edman technique. This important accomplishment (852, 854) was the starting point for the determination of the complete sequences of many receptors (855, 856) by DNA cloning techniques. A Japanese group (857) first reported results for the α-subunit, followed closely by a French group (858). The Japanese team then reported the sequences for the β- and δ-subunits (859)

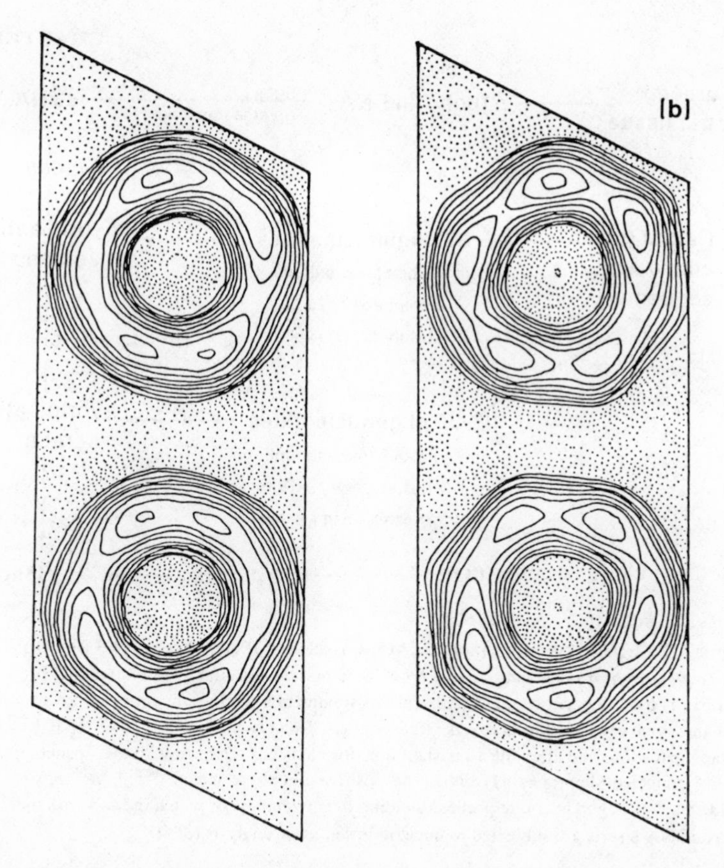

Fig. 6.2. Electron-micrographic image analyses of the nicotinic acetylcholine receptor. (a) A schematic representation of the topography of the acetylcholine receptor based on the first image analysis of electron micrographs. The receptor projects outside (55 Å) and inside (15 Å) the phospholipid bilayer (40 Å) of the cell. Although the pentameric structure is not very clear in the original work, more refined analyses show the pentagonal arrangement of the exobilayer, for which an example is shown in (b). The envelope of the receptor has changed somewhat, and is more in agreement with the model described in the text.

The extended structure suggested for the cytoplasmic portions of the nAChR in the most recent analyses might provide a basis for molecular arrangements if information from suitable cross-linking experiments were available. (Photographs provided by Prof. R. M. Stroud and reproduced from the *Biophysical Journal* 37; 371–383 [1982], by copyright permission of the Biophysical Society [reference 821].) Other excellent image analyses of the nAChR have appeared (826, 828, 924). The model arrangement for the exobilayer strands shown in Fig. 6.17 can be superimposed on the bulges shown in the images in (b). The conclusion, based on homology, that the exobilayer portions of the subunits should all be arranged in similar ways is strongly supported by the images derived from electron micrographs. The subunits around the central ion channel are in the order αβαγδ (838).

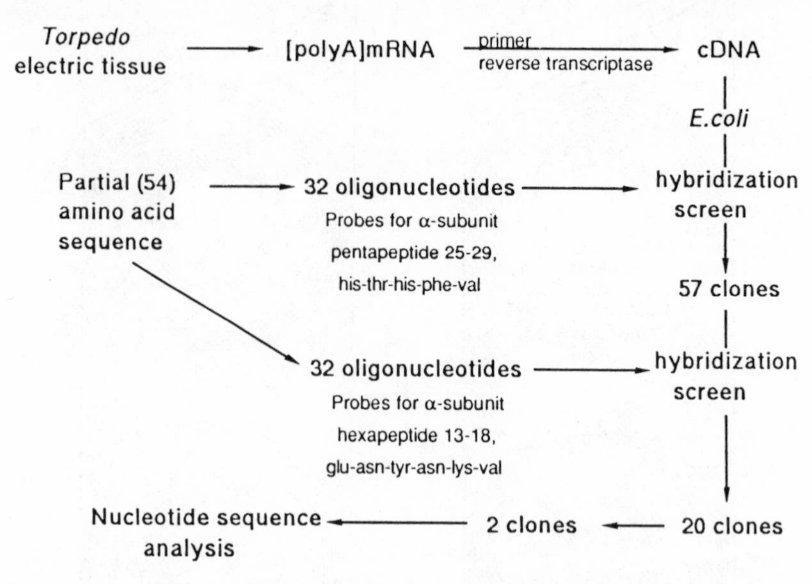

Scheme 6.2. Sequence determination of nAChR α-subunit. [PolyA]mRNA (2.3 μg), isolated from *Torpedo californica* electric tissue by chromatography of total RNA on oligo(dT)-cellulose, is used to generate a cDNA library by treatment with a vector-primer DNA (1.4 μg) (862) and avian myeloblastosis reverse transcriptase. *Escherichia coli* X1776 or HB101 is used for transformation, and ampicillin-resistant transformants are screened with the "pentapeptide" set of oligonucleotide probes. Hybridization-positive clones are screened with the "hexapeptide" set of oligonucleotide probes, yielding twenty clones, of which the two with the largest cDNA inserts are subjected to nucleotide sequence analysis (865).

but their sequence for the γ-subunit (860) was anticipated by an American group (861). The selection procedure for the clones is outlined in Scheme 6.2 (that for the α-subunit is shown; different oligonucleotide mixtures are used for the other subunits [862]). Nucleotide sequencing was carried out by the Maxam-Gilbert method (863–865). The genetic code is given in Table 6.1. A second genetic code may be based on the specificity with which transfer RNA (tRNA) is loaded with amino acid (866).

Acetylcholine Receptor: Secondary Structure

The levels of structural organization of proteins are primary structure (amino acid sequence), secondary structure (the arrangement of the polypeptide chain), tertiary structure (organization of the secondary structural segments), and quaternary structure (overall arrangement of the tertiary structures). The proteins may be components in higher levels of organization (Table 1.3). The acetylcholine receptor is a polymolecular system composed of five subunits. The secondary structure can be represented as a linear group or helix. The characteristics for important types of linear groups are listed in

Table 6.1. The Genetic Code

First Base: \ Second Base:	U	C	A	G	Third base:
U	phe	ser	tyr	cys	U
	phe	ser	tyr	cys	C
	leu	ser	END	END	A
	leu	ser	END	trp	G
C	leu	pro	his	arg	U
	leu	pro	his	arg	C
	leu	pro	gln	arg	A
	leu	pro	gln	arg	G
A	ile	thr	asn	ser	U
	ile	thr	asn	ser	C
	ile	thr	lys	arg	A
	met	thr	lys	arg	G
G	val	ala	asp	gly	U
	val	ala	asp	gly	C
	val	ala	glu	gly	A
	val	ala	glu	gly	G

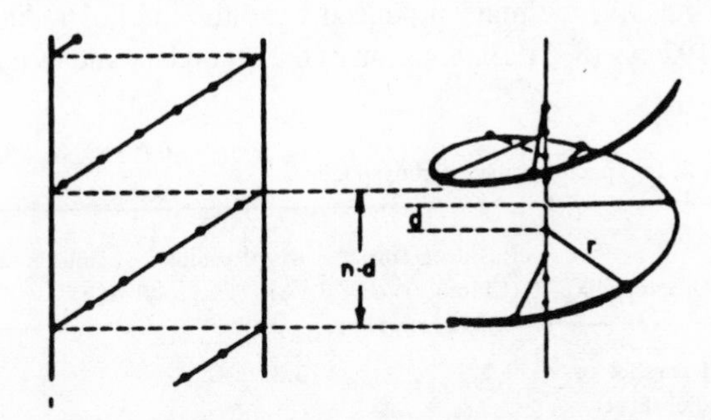

Fig. 6.3. A general helix representation along with its projection on a ''surface net'' or cylindrical plot. The ''net'' makes it easy to see the relationships between side chains within a given helix but is not suitable for revealing the interactions between groups on different helices. The parameters are n, residues per turn; d, axial shift per residue; $n{\cdot}d$, pitch of the helix; r, radius of the helix. (Adapted with permission from reference 867.)

Table 6.2, with the helical parameters illustrated in Figure 6.3. The structures of three different helices, including an α-helix and a β-pleated sheet, are shown in Fig.6.4. A complete discussion of all aspects of protein structure can be found elsewhere (867).

Can the secondary structure of the AChR subunit proteins be readily deduced from the primary sequences? The procedures that have been developed for this purpose are still quite approximate. In one approach, the propensity of particular residues to nucleate α-helices, β-sheets, or reverse turns (based on frequencies in a large set of proteins with known structures) is evaluated and then averaged over the nearest neighbors. An artificial intelligence approach begins with the choice of reverse turns to separate segments which can then be assigned to either α-helices or β-sheets. Information about the hydrophobicity or hydrophilicity can then be added to position secondary structural elements. The program ''learns'' by excluding certain predicted combinations, and yields a number (1–100) of plausible secondary structural assignments for many proteins (868).

Selection of hydrophobic and hydrophilic regions may be made on the basis of a plot of the side chain transfer energies (ethanol → water, for example) averaged over a number of residues against residue number (869). By using averaged transfer energies, the asymmetry of helices with respect to the distribution of polar and nonpolar groups (formalized as the ''helical hydrophobic moment'') may be estimated and used to choose amphiphilic helices (870–872). A related procedure has been applied to part of the AChR (873). The asymmetry of helices was also recognized with the ''helical wheel'' (874).

The hydrophobic portions of the AChR subunits are selected through identifying long sequences of hydrophobic amino acids. No other structural element is obvious. One binding site for acetylcholine should be close to a disulfide link; the reactive cysteine produced by dithiothreitol reduction of the disulfide is α 192 cys (875). Elaboration of the secondary and tertiary structure

Table 6.2. Polypeptide Linear Groups/Helices

Linear Group/Helix	Residues/Turn (Chirality)[a]	Rise/Residue (Å)	Helix Radius (Å)
Twisted parallel or antiparallel sheet	−2.3	3.3	1.0
α-Helix	+3.6	1.5	2.3
3_{10}-Helix	+3.0	2.0	1.9
Collagen helix	−3.3	2.9	1.6

Source: Adapted from reference 867.

[a] + indicates a right-handed helix; − indicates a left-handed helix.

of the AChR requires methods of identifying important features of the receptor, in order to refine or to add to the partial structures afforded by protein folding schemes. By using both theoretical and experimental results on the AChR, a fairly complete structural and dynamic model for the AChR may be constructed. The basis for several of the methods will be discussed before the model is presented.

Single Group Rotation (SGR) Theory

Proteins (receptors, enzymes, oxygen-binding agents, gene repressors, or fiber components) are large, complex molecules which may undergo a variety of chemical or conformational changes after association with their usual partner molecules. If the changes are too complex to understand easily, much may still be learned by focusing on a part, even a small part, of the overall process. For example, chemical (i.e., enzymatic) catalysis can be studied after identifying active-site groups, without necessarily knowing why the groups are active. Having a fairly detailed description of the motions of the atoms within myoglobin did not make clear how ligands entered the heme "pocket." The required additional insight was afforded by the crystal structure of the phenyl-iron derivative, which revealed the beautifully complex motions of the groups near the oxygen binding heme, and suggested which motions were involved in the passage of oxygen from the outside to its bound position (876).

In many cases, one knows the substrates for receptors, but does not know the nature of the groups at the binding site, or have any knowledge of what happens at a molecular level after the receptor-substrate complex is formed. Even when the active-site structure is known, as in the case of pyridine nucleotide dehydrogenases, one is unable to write a catalytic mechanism. The development of DNA cloning techniques and translation of the DNA sequences into protein amino acid sequences has led to a tremendous increase in the number of such linear amino acid sequences. The lack of an obvious way to construct a three-dimensional arrangement has made apparent the need for a variety of intermediate theoretical approaches, with which the problems of structure and function can be attacked and by means of which experiments can be designed (877–879).

Single group rotation theory involves a consideration of local, small conformational changes of restricted portions of proteins in the operation of biological effectors. Single side chains, or "single groups," are the logical first choice for active groups. I approach the study of stimulus-receptor complexes by (a) selecting binding groups for the substrate and (b) defining what might be the smallest biologically effective, conformational change for the combination of binding groups and substrate.

The rotation of one or more single groups ("SGR") is proposed as an identifiable conformational change involved in the transduction of certain molec-

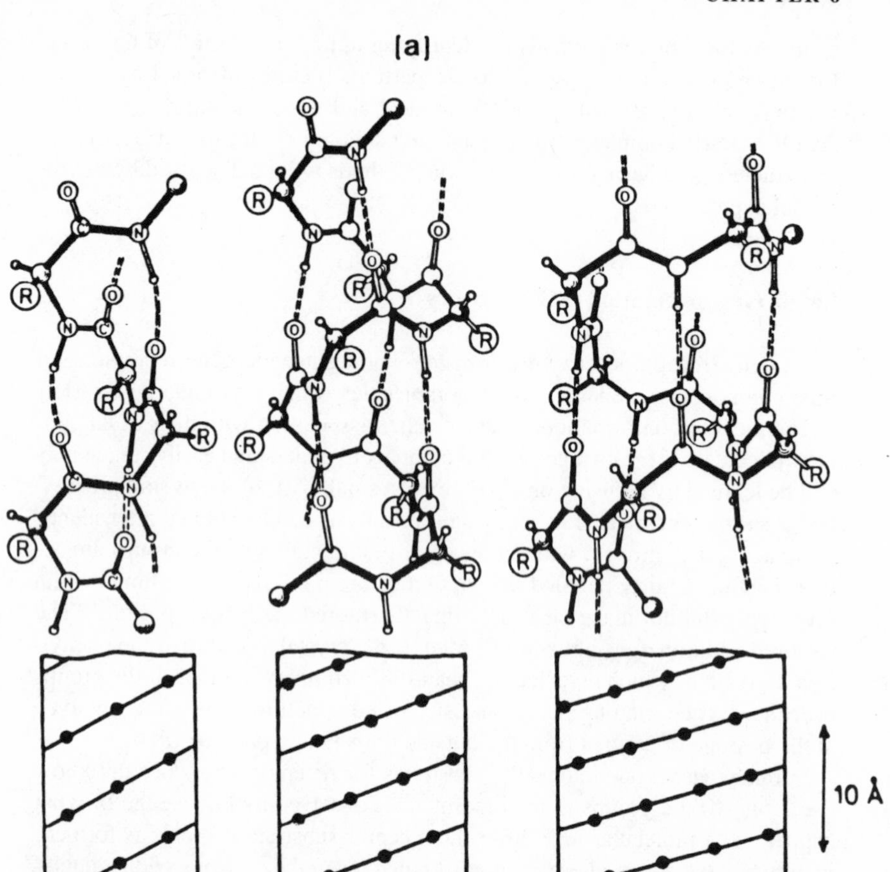

ular stimuli (signal molecules such as neurotransmitters), after formation of the signal molecule–receptor complex. SGR is also considered in developing a molecular mechanism of ion passage through membrane ion channels.

Background to the Development of SGR Theory

Conformational flexibility has been emphasized as a factor in receptor site–substrate combinations (880–882) without specific attention to the consequences of particular changes. In connection with the question of the molecular basis of neurotransmitter action, I tried to find an appropriate scale for the conformational changes needed for a plausible biological outcome. After designing a model receptor site for acetylcholine, I searched for the simplest conformational change which might have biological consequences. The study of initial local motions, as small as single group rotations (''SGR'') within the

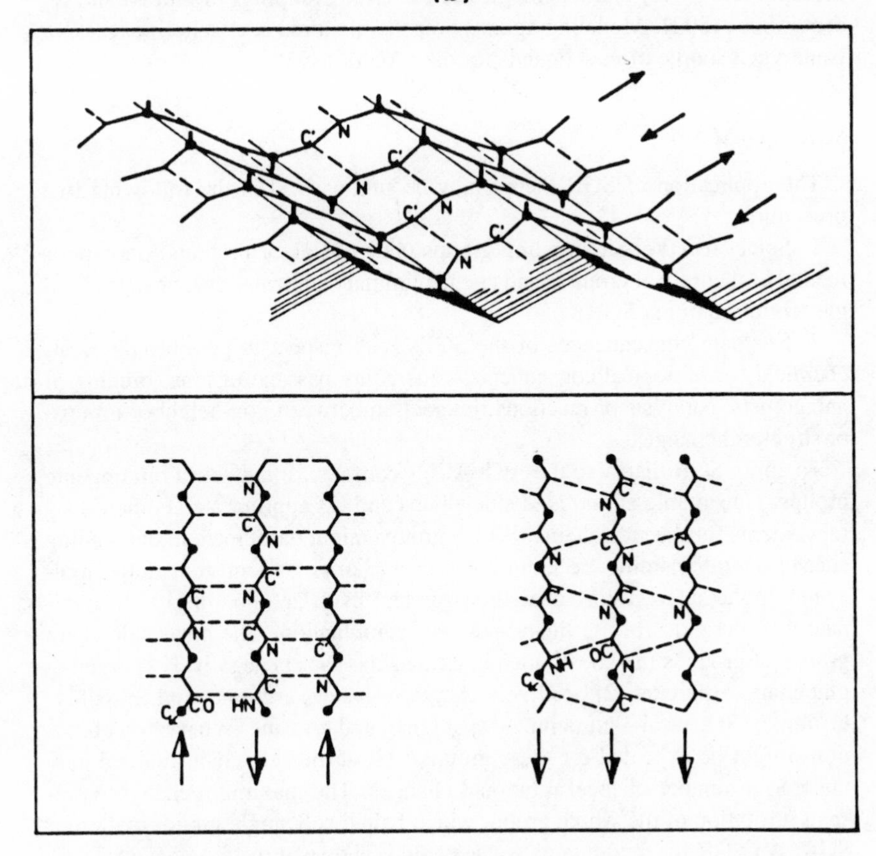

Fig. 6.4. Helical arrangements of polypeptide chains. (a) From left to right; 3_{10}-, α-, and π-helices. Cylindrical plots for the helices are shown below the models. (b) Antiparallel β-pleated sheet showing (upper drawing) side chain directions and (lower left) hydrogen-bond directions, which are also illustrated (lower right) for a parallel β-pleated sheet. Twisting of the strands is not shown. (Adapted with permission from (reference 867.)

receptor-ligand combination, turned out to be a useful approach to probing the behavior of receptor sites, and for developing useful model mechanisms for a number of important biological processes.

Previous analyses of receptor-ligand interactions focused on choosing the best static ligand conformation to combine with the receptor (883), an example being computer matching of small molecules in competition with thyroxine for the sites on the thyroxine-binding globulin (884–889).

The design of an inhibitor for the angiotensin-converting enzyme (D-3-mercapto-2-methylpropanoyl-L-proline, "captopril") was based on a static "map" of the receptor site made through analysis of chemical and kinetic properties of the enzyme (890, 891). Efforts have been made to map "pharmacophores'' (i.e., pharmacologically effective groupings of atoms) and receptor sites (892). Modeling agonists for the nicotinic acetylcholine receptor is not yet a source of new ligands for the nAChR (893).

SGR Theory

The application of SGR theory may be summarized in the following two procedures:

1. Select receptor-site binding groups (side chains of proteins) on experimental or theoretical grounds and create minimal conformational changes (single group rotations, SGR).

2. Evaluate consequences of the SGRs with respect to possible physical, chemical, or biological consequences, including passage of ions, binding of antagonists, catalysis of reactions, interaction between non-neighbor groups, or structural changes.

To apply SGR theory to the ACh:AChR complex, I first select binding-site groups: (1) suitable amino acid side chains and (2) appropriate geometric arrangements for the side chains. Which groups might participate in the binding sites of neurotransmitter-receptor complexes? For polar neurotransmitter molecules, polar interactions (charge-charge and hydrogen-bonding) will dominate the energetics of the ligand-receptor combination. The polar side chain groups of proteins may be divided into three classes: (1) negatively charged—glutamate, aspartate; (2) positively charged—lysine, arginine, and, possibly, histidine; (3) neutral—glutamine, asparagine, and tyrosine. What types of motions might be expected for these groups? All of the side chains named may undergo a number of local rotational changes. The maximum change represents a rotation of the whole group, which I shall call single group rotation or SGR. An SGR for the glutamate side chain is shown at the top of Figure 6.5. SGRs for other groups are summarized in Figure 6.5 using superimpositions of the initial and final conformations as compact representations of the SGR.

A moderate range of conformational changes of bound ligand groups can be accommodated by the SGRs of protein side chains. To examine the SGRs of groups, simple, two-dimensional drawings, made with a straight-edge and a geometric stencil, are adequate for many purposes.

Lysine and glutamate side chains are used to construct a model AChR binding site. (Aspartate is more or less equivalent to glutamate.) Lysine is the only reasonable choice for a full + (positive) charge at a well-defined position. At neutral pH, histidine is only partly protonated and the guanidino group of arginine has at least two (partial) + centers for binding to a nearby ligand. Conformationally mobile groups may be ''pinned'' by interaction with other polar groups to lower the rate of spontaneous conformational change. Glutamate (− [negative] charge) is used to ''pin'' lysine (+ charge) in the present model.

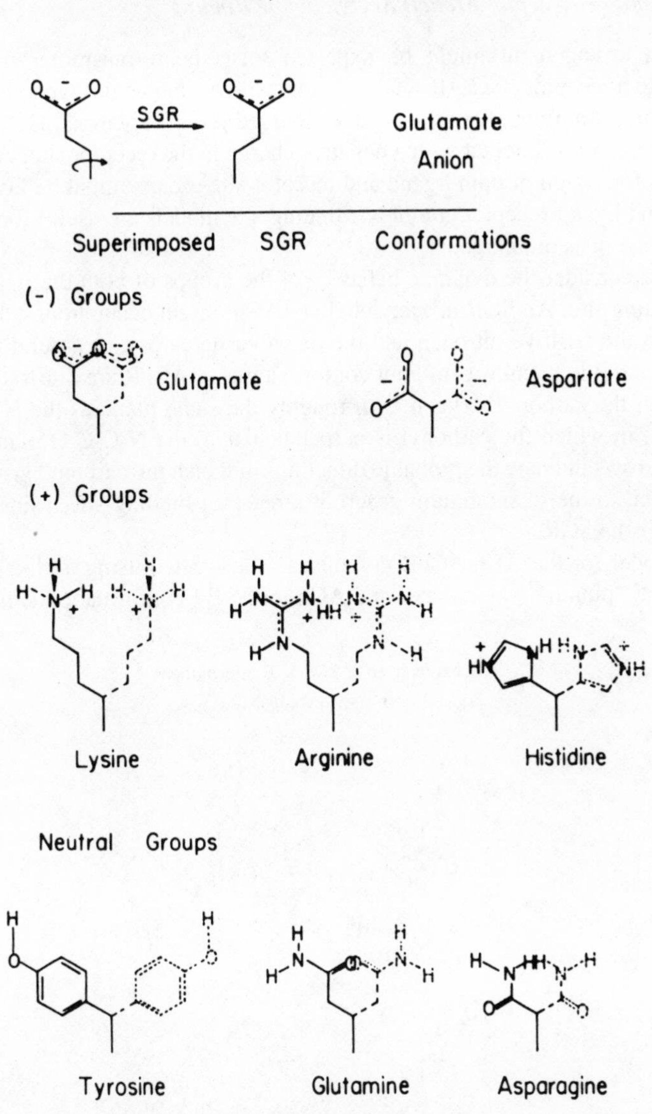

Fig. 6.5. Single group rotations (SGRs) for polar amino acid side chains of proteins. A single group rotation for glutamate ion is given as an example of the small conformational change embodied in an SGR. Superimposed extreme forms of SGRs are shown for all of the ordinary polar amino acid side chains found in proteins. The groups, according to charge, are (−) glu, asp; (+) lys, arg, his; and (0) tyr, glu, asn. Tyrosine is normally classified as a hydrophobic amino acid. Superimposition of the rotated form on top of the original conformation, using a dashed line for the group after single group rotation, offers a clear view of the actual changes in the location of side chain atoms. (Reprinted, with permission, from reference 737.)

Acetylcholine Receptor Model: Acetylcholine Binding

What arrangement might be expected for a neurotransmitter molecule bound to a receptor site? All small neurotransmitter molecules (acetylcholine, glutamate, γ-aminobutyrate) have a + charged group at neutral pH. The positive charge must interact with a negative charge in the receptor site. A particular conformation of both ligand and receptor site groups might be favored in the initial ligand-receptor complex. Binding-site models are useful for the design of agonists or antagonists.

I now consider the dynamic behavior of the groups of both the ligand and the binding site. An SGR in acetylcholine (ACh) might occur around the bond between the positive nitrogen and the neighboring carbon or around the carbons α and β to the nitrogen. Four conformations of AChR are illustrated, two in which the carbonyl oxygen is in roughly the same plane as the N-C-C-O and two in which the carbonyl is perpendicular to the N-C-C-O plane (Fig. 6.6). Arrows indicate the probable directions in which the carbonyl groups are H-bonded to the ϵ-ammonium group of a receptor binding-site lysine before and after the SGR.

A model for the ACh:AChR complex is constructed using a glutamate, a lysine, a "pinning" glutamate, and ACh in the B4 conformation (Fig. 6.7).

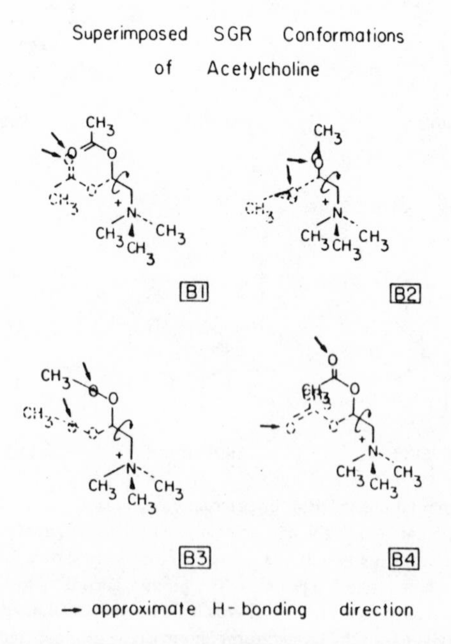

Fig. 6.6. Four conformations of the neurotransmitter molecule acetylcholine, formed by single group rotation (SGR) around the α,β-carbon-carbon bond. The arrows show the approximate direction of hydrogen-bonding between a receptor-site group and the carbonyl group.

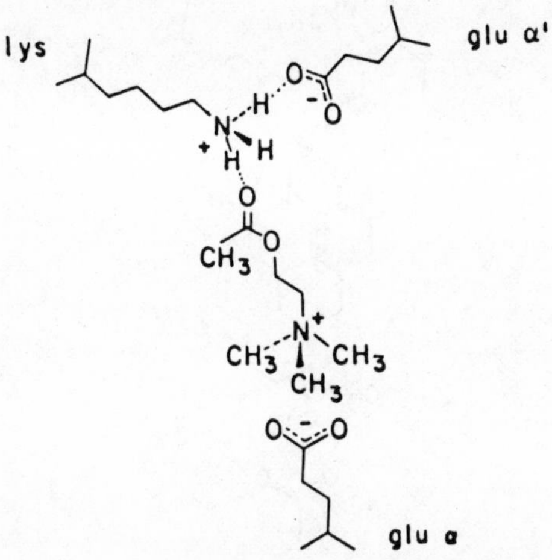

Acetylcholine: AChReceptor

Fig. 6.7. Model combining site for acetylcholine with groups able to interact with the agonist by chemically reasonable forces, such as electrostatic interactions and hydrogen-bonding. Three amino acid side chains are used to make up the binding site: a (−)-charged amino acid (electrostatic interaction with the (+)-trimethylammonio group); a (+)-ϵ-ammonium ion of a lysine (hydrogen-bonding to the acetyl carbonyl); and a second (−)-charged amino acid (electrostatic interaction and hydrogen-bonding with the lysine (+)-ϵ-ammonium ion). Two polypeptides (α and α′) are involved on geometric grounds. (Reprinted, with permission, from reference 737.)

An SGR in the ACh moves the carbonyl oxygen by 4.5 Å. Another SGR around the α,β-bond of the lysine chain moves the ϵ-ammonium ion by 4.5 Å. This pair of SGRs, one in the lysine chain and one in the ACh, results in a conformationally modified complex. A conformational change to the SGR rotated form of B1 seems to occur in desensitization of the ligand-AChR complex (894). The model shows that an SGR in the ACh can be correlated with an SGR in a receptor-site side chain. It is also striking that a known antagonist, *d*-tubocurarine, fits exactly into the space between the upper and lower glutamate ions as shown in Figure 6.8.

A Molecular Model for Ion Channels in Biomembranes

A membrane ion channel is a structure formed by one or more protein molecules and mediates the passage of ions across cell membranes. The problems of channel structure and mechanism are among the major unsolved problems of molecular neurobiology (895–902). Sequences have been obtained for a number of important ion channels, and a structural and dynamic model for the sodium channel (902) (Chap. 7) has been developed from the linear sequence

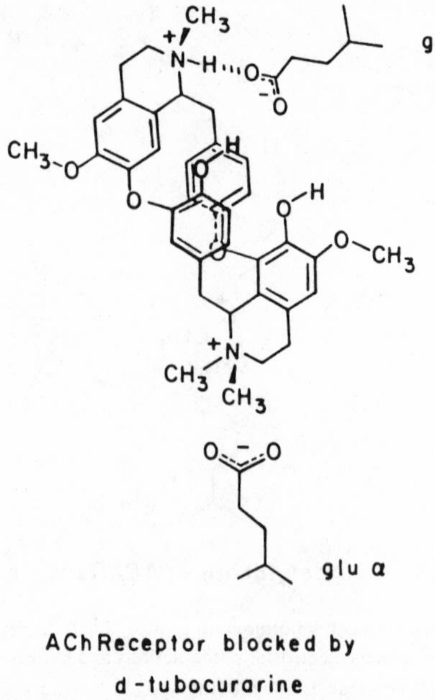

AChReceptor blocked by
d-tubocurarine

Fig. 6.8. Model nAChR binding site ''combined'' with *d*-tubocurarine. The group arrangement illustrated in Fig. 6.7 has been reproduced without alteration.

for the electric eel protein. An amphiphilic peptide assembles into a tetrameric channel structure (898) but crystal structures for the channel-forming peptide, gramicidin A, show that channel structure may vary with environment (899–901).

Ion Channel Model: Summary

There appear to be two basic solutions to the problem of allowing ions to pass from one side of a cell membrane to the other under control. A carrier encloses the ion in a hydrophobic shield; a channel allows the passage of ions under specified conditions. A carrier mechanism is too slow to account for the ion flows that occur in neurons. However, ion carriers (ionophores like valinomycin) can depolarize cells. A natural ion channel may be constructed in several ways. A common structural theme for an ion channel consists of a number of α-helical polypeptides carrying along one side a set of polar ''channel-active'' (CA) amino acids. One proposed helix sequence is glu(1), lys(5), glu(8), lys(12), . . . The charges may be alternating or on separate helices, as in the case of the sodium channel (Chap. 7). The channels permit the passage of ions by diffusion coupled to single group rotation (SGR) of the CA amino acids.

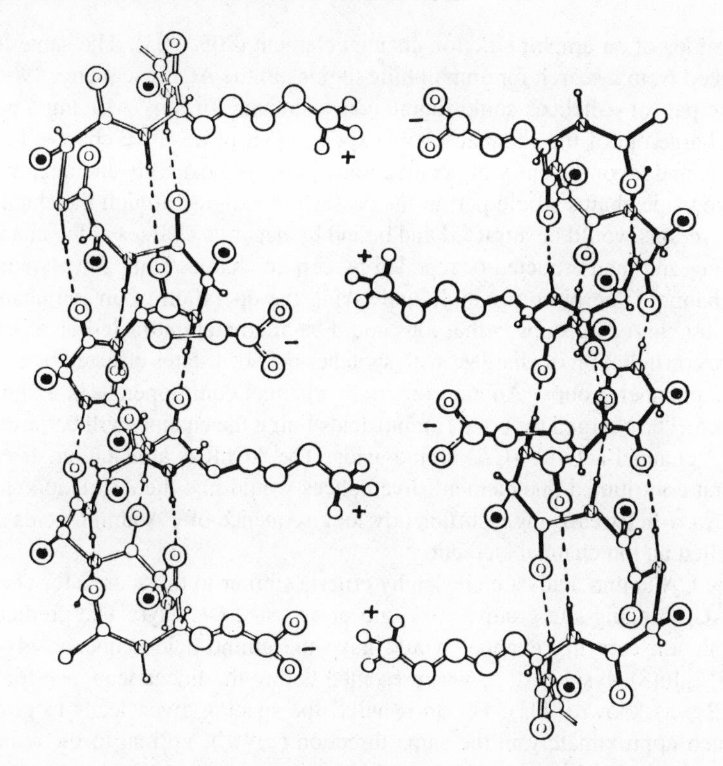

Fig. 6.9. Molecular model of a plausible ion channel based on polypeptides. The model suggested the formulation of the amphiphilic segment as part of the nicotinic acetylcholine receptor ion channel. The "channel-active" amino acid residues on the right-hand helix are 1 (glu), 5 (lys), 8 (glu), and 12 (lys). The apposed polypeptide carries lys, glu, lys, and glu. The charged groups are approximately 4.5 Å apart. The circles with black dots represent (partly) hydrophobic R groups that are the side chains of the other amino acids of the polypeptides. nAChR ion channel elements from the native protein sequence are given in Fig. 6.11; a dynamic model for the ion channel is shown in Fig. 6.27.

A Model for an Ion Channel

In the course of considering the consequences of an SGR within the ACh:AChR complex (Fig. 6.7), a second glutamate was added to "pin" the lysine in the position to which it had moved. Connecting the lysine to the two pinning glutamates by a minimum length polypeptide chain led to the sequence glu-x-x-x-lys-x-x-glu. Folding this sequence into an α-helix revealed an exciting new structural arrangement, an α-helix on which one side was occupied by polar groups—in this case, an alternating series of positively and negatively charged groups. The charges must be shielded in order to accommodate such sequences in the bilayer, a result achieved by apposing another helix with oppositely charged groups and interspersing hydrophobic amino acids (Fig. 6.9). This somewhat circuitous application of SGR theory thus led to the idea of an amphiphilic ion channel element (705, 903). The same idea emerged from a search for amphiphilic regions in the AChR sequence (904).

The pair of α-helices could function as a channel for ions. Moving a positive charge out of the channel would expose a pair of negative charges to the outside and favor the entry of positive ions. SGRs of positively and negatively charged side chains would permit the passage of cations through the channel. Positive ions would be attracted and bound by negative charges at the channel opening and then attracted or repelled by certain charge arrangements within the channel. The basic principle underlying the operation of an ion channel lined by charged groups is that ions could be alternately attracted or repelled by the constellation of charges, with switches between states effected by SGRs of the channel groups. An anion-carrying channel could operate in a similar fashion. The charged (or polar) amino acids lining the channel will be referred to as "channel-active" (CA) amino acids. The AChR is a pentamer. If each subunit contributed one element, five helices would line the AChR ion channel. An α-helix carrying a sufficiently long sequence of CA amino acids will be called an ion channel element.

The CA amino acids are chosen by criteria similar to those used for choosing ACh binding site groups, (−), glu or asp, and (+), lys. The prediction that an ion carrying channel would have the amino acid sequence gly(1), lys(5), glu(8), lys(12), . . . was presented before the linear sequence for the AChR was known (903). For an α-helix, the spacing given leads to groups directed approximately in the same direction (±40°), normal to the α-helix axis.

Nicotinic Acetylcholine Receptor: Summary of Model

Folding of the five polypeptide subunits ($\alpha_2\beta\gamma\delta$) of the nicotinic acetylcholine receptor (AChR) into a functional structural model is described in the following sections. The principles used to arrange the sequences into a structure include (1) Hydrophobicity → membrane crossing segments; (2) amphiphilic character → ion-carrying segments (ion channel with single group rotations), (3) molecular shape (elongated, pentagonal cylinder) → folding dimensions of exobilayer portion; (4) choice of acetylcholine binding sites → specific folding of exobilayer segments; (5) location of reducible disulfides (near agonist binding site) → additional specification of exobilayer arrangement; (6) genetic homology → consistency of functional group choices; (7) noncompetitive antagonist labeling → arrangement of bilayer helices. The AChR model is divided into three parts: (a) exobilayer: 11 antiparallel β-strands from each subunit; (b) bilayer: 4 hydrophobic α-helices and 1 amphiphilic α-helix from each subunit; and (c) cytoplasmic: one (folded) loop from each subunit.

The exobilayer strands form a closed "flower" (the "resting state") which is opened (''activated'') by agonists bound perpendicular to the strands in an agonist-extended conformation. Rearrangement of the agonists to a strand-parallel position in an extended or bent conformation, with partial closing of the ''flower,'' leads to a desensitized receptor. The actions of the agonists, acetylcholine and succinoyl and suberoyl bis-cholines, can be understood with the model. The opening and closing of the exobilayer ''flower'' controls access to an ion channel composed of the 5 amphiphilic bilayer helices. A molecular mechanism for ion flow in the channel can be written. Openings interrupted by short duration closings (50 μsec) can be related to channel group motions. The unusual photolabeling of intrabilayer serines in α-, β-, and δ-subunits but not in γ-subunits near the binding site for noncompetitive antagonists (NCAs) is explained along with a mechanism for the action of NCAs such as phencyclidine. The unusual α192cys–193cys disulfide may have a special peptide arrangement, such as a *cis*-peptide bond to a following proline (905). The position of phosphorylatable sites and proline-rich segments are noted for the cytoplasmic loops.

The dynamic behavior of the AChR channel and many different experimental results can be interpreted in terms of the model. An example is the lowering of ionic conductivity on substitution of bovine for *Torpedo* δ M2 segment. The model is useful for the design of experiments on the nAChR.

Models of this type can be used to understand the behavior of other members of the acetylcholine receptor ''superfamily,'' including the GABA receptor (855, 856).

Background for Model

Full sequence data have been published for the nicotinic acetylcholine receptor (AChR) of *Torpedo californica* (2,333 amino acids) and *Mus musculus* (mouse) (2,314 amino acids) as well as partial data for *Torpedo marmorata* (α-subunit), *Gallus domesticus* (chicken) (α-, γ-, and δ-subunits), *Bos taurus* (calf) (α-, β-, γ-, and ε-[fetal γ] subunits), and *Homo sapiens* (human) (α- and γ-subunits) (857–864, 906–913). Many of the sequences have been compared by Numa (914). The neural α-subunit from *Rattus rattus* (rat) AChR (915) is different from the muscle subunit in interesting ways. I derive a structural AChR model from the sequences of the subunits and a mechanism for the action of the AChR using the model. The model accounts for many of the properties of AChR as summarized by Maelicke (916, 917), Changeux (796, 808), Conti-Tronconi (918), Hucho (827), Karlin (919), and Barrantes (825). The present nAChR and sodium channel (902) (Chap. 7) models illustrate the complex relationship of molecular structure to physiological activity. Understanding the receptor is a necessary step on the way to defining all the processes that lead to storage of nerve signals, and therefore, learning and memory.

Construction of models is an important step in analyzing any complex process. Structural models can guide the experimenter into a better understanding of current information and to the formulation of new approaches to the problem. In the absence of a definitive structural determination, the structure and dynamic properties of a large molecular complex like the acetylcholine receptor (AChR) must be inferred from a great variety of facts aided by physical and chemical theory. Even if the elusive crystal structure for AChR were obtained from small crystals (40–100 μm), theories for the dynamic behavior of the protein molecule would still be necessary. I have tried to consider all possible information in building a model of the AChR, an approach which can be termed "holistic" and which provides explanations for many of the properties of the AChR.

The discussion is divided into sections, the titles giving an overview of the development of the model. Those who obtained the sequence data for AChR noted the presence of four hydrophobic segments within the sequence. Genetic homology between the subunits is substantial for these portions of the molecule, which are generally agreed to be the transmembrane portions of the receptor. An additional transmembrane segment was first inferred on the basis of single group rotation (SGR) theory and later on the basis of amphiphilic character. After the membrane portions of the AChR were identified, the molecular shape implied that much of the remainder of the protein had to be in the exobilayer region. Binding sites are selected for acetylcholine (ACh) (one near the reducible disulfide) and the molecule is folded within the dimensions of the AChR in order to define a structural model for the exobilayer region of the receptor. The dynamic behavior of the exobilayer and channel portions of the AChR is deduced from a consideration of the model.

Single Group Rotation (SGR) Theory: Brief Recapitulation

The idea of single group rotations (SGR) is developed at length in the first part of the chapter. A brief review will help to show how SGR is used in the modeling. At the time when only partial sequences for AChR (854) were known, I developed a general approach to modeling receptor structure in the absence of complete sequence and/or structural information (903). One considers (a) the nature of the amino acid groups which might interact with the molecule being bound and (b) the motions which the bound molecule or the amino acid side chains might undergo. A positional change is the result of a rotation of one conformation of a bound molecule or a side chain into another; such rotations are termed single group rotations (SGR). Various classes of amino acids and the corresponding SGR are illustrated in Figure 6.5.

A model combining site for acetylcholine is designed with groups able to interact with the agonist by chemically reasonable forces, such as electrostatic interactions and hydrogen-bonding. Three amino acid side chains are used to make up the binding site: (1) a (−)-charged amino acid [electrostatic interaction with the (+)-trimethyl-ammonio group], (2) a (+)-ϵ-ammonium ion of a lysine [hydrogen-bonding to the acetyl carbonyl] and (3) a second (−)-charged amino acid [electrostatic interaction and hydrogen-bonding with the lysine (+)-ϵ-ammonium ion] (Fig. 6.7). After association of the agonist with the binding site, a biological effect could arise through a minimum conformational change such as a single group rotation (SGR). Rotation in the ligand and/or the side chains in the protein leads to a rearranged ligand-receptor complex. SGRs may be involved in desensitization of the active ACh.AChR complex. SGRs are essential for explaining the dynamics of the sodium channel model (902).

Consider how a cation might traverse a membrane via a protein channel. The membrane portion of the protein is likely to be in the form of an α-helix to minimize peptide bond interactions with the hydrophobic portion of the bilayer. Initial binding of the cation occurs near negative charges (glu or asp). For the cation to progress farther into the channel after binding, one must weaken the attraction of the negative charge for the cation with another positive charge. This is easily done with a (+)-ϵ-ammonium ion of a lysine located one helix turn away. After an SGR of the lysine side chain, the positive charge is closer to the negative charge, and release of the cation is facilitated. Another SGR moves the charges apart. Extending this idea and arranging the negatively and positively charged amino acids along an α-helix could constitute a functional arrangement capable of serving as an ion channel. The AChR ion channel was predicted to have the sequence lys(1), glu(5), lys(8), glu(12) . . . (903).

AChR Models

The AChR is a complex of five subunits in the order αβαγδ (920), an arrangement favored by several groups (821, 824) even though biotin-labeling suggests αγαβδ (921–923). Image processing of electron micrographs of organized layers of AChR (821) shows the presence of five regions of increased density arranged in pentagonal symmetry (826–829) (Fig. 6.2[b]). Early structural sketches were drawn on the basis of readily identifiable hydrophobic segments (858, 860), but lacked the features appropriate for carrying ions through the membrane. The important genetic homologies between the subunits of the *Torpedo* AChR are shown in Figure 6.10. Three detailed structural models have been put forward by Guy (partial), Stroud (schematic), and Kosower.

Models of Bilayer Portion of AChR

Guy (873) utilizes the environmental polarity preferences of amino acid side chains in the AChR sequences to construct a model for the channel with five channel boundary helices (i.e., ion channel elements). The opening of the

AMINO ACID SEQUENCES, Acetylcholine Receptor Subunits

```
                  20                  40                  60                  80                  100
                   |                   |                   |                   |                   |
α SEHETRLVANLL-EN-YN-KVIRPVEHHTHFVDITVGLQLIQLISVDEVNQIVETNVRLRQQWIDVRLRWNPADYGGIKKIRLPSDDVWLPDLVLYNNAD
β SVMEDTLLSVLF-ET-YNPKV-RPAQTVGDKVTVRVGLTLTNLLILNEKIEEMTTNVFLNLAWTDYRLQWDPAAYEGIKDLRIPSSDVWQPDIVLMNNND
γ ENEEGRLIEKLL-GD-YDKRII-PAKTLDHIIDVTLKLTLTNLISLNEKEEALTTNVWIEIQWNDYRLSWNTSEYEGIDLVRIPSELLWLPDVVLENNVD
δ VNEEERLINDLLIVNKYNKHV-RPVKHNNEVVNIALSLTLSNLISLKETDETLTSNVWMDHAWYDHRLTWNASEYSDISILRLPPELVWIPDIVLQNNND

                  120                 140                 160                 180                 200
                   |                   |                   |                   |                   |
α GDFAIVHMTKLLLDYTGKIMWTPPAIFKSYCEIIVTHFPFDQQNCTMKLGIWTYDGTKVS--ISPES-----DRP------DLSTFMESGEWVMKDYRGW
β GSFEITLHVNVLVQHTGAVSWQPSAIYRSSCTIKVMYFPFDWQNCTMVFKSYTYDTSEVTLQ-HA--LDAKGEREVKEIVINKDAFTENGQWSIEHKPSR
γ GQFEVAYYANVLVYNDGSMYWLPPAIYRSTCPIAVTYFPFDWQNCSLVFRSQTYNAHEVNLQLSAE--EGE---AVEWIHIDPEDFTENGEWTIRHRPAK
δ GQYHVAYFCNVLVRPNGYVTWLPPAIFRSSCPINVLYFPFDWQNCSLKFTALNYDANEITMDLMTDTIDGK-DYPIEWIIIDPEAFTENGEWEIIHKPAK

                  220                 240                 260                 280                 300
                   |                   |                   |                   |                   |
α KH--WVYYTCCPD-TPYLDITYHFIMQRIPLYFVVNVIIPCLLFSFLTGLVFYLPTDSG-EKMTLSISVLLSLTVFLLVIVELIPSTSSAVPLIGKYMLF
β KN--W---RSD-DPS-YEDVTFYLIIQRKPLFYIVYTIIPCILISILAILVFYLPPDAG-EKMSLSISALLAVTVFLLLLADKVPETSLSVPIIIRYLMF
γ KNYNWQL-TKD-D-TDFQEIIFFLIIQRKPLFYIINIIAPCVLISSLVVLVYFLPAQAGGQKCTLSISVLLAQTIFLFLIAQKVPETSLNVPLIGKYLIF
δ KN---IYPDKFPNGTNYQDVTFYLIIRRKPLFYVINFITPCVLISFLASLAFYLPAESG-EKMSTAISVLLAQAVFLLLTSQRLPETALAVPLIGKYLMF
                            |----------M1----------|   |----------M2----------|       |-------

                  320                 340                 360                 380                 400
                   |                   |                   |                   |                   |
α TMIFFVISIIITVVVINTHHRSPSTHTMPQWVRKIFIDTIPNVMFFS----------------------------TMKRASKEKQENKIFADDI-DISD-
β IMILVAFSVILSVVVLNLHHRSPNTHTMPNWIRQIFIETLPPFLWIQ-----RPVTTPSPD----------SKPTIISRANDEYFIRK-PAGDFVCPVDN
γ VMFVSMLIVMNCVIVLNVSLRTPNTHSLSEKIKHLFLGFLPKYLGMQLEPSEETPEKPQP--------RRRSSFGIMIKAE-EYILKK-PRSELMFEEQK
δ IMSLVTGVIVNCGIVLNFHFRTPSTHVLSTRVKQIFLEKLPRILHMS-----RADESEQPDWQNDLKLRRSSSVGYISKAQ-EYFNIK-SRSELMFEKQS
  -----M3--------|
```

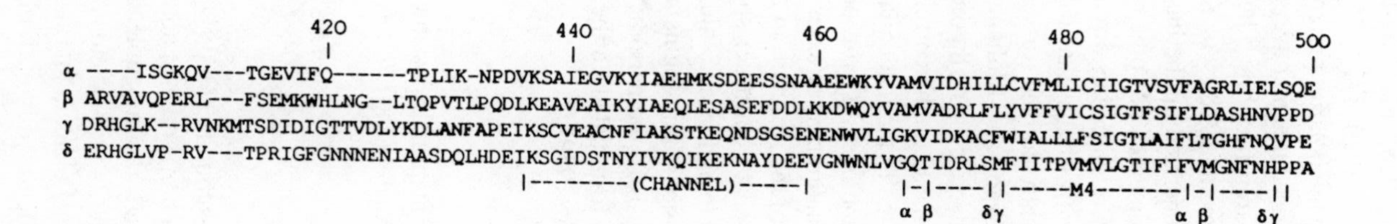

Fig. 6.10. Sequence homology between nicotinic AChR α-, β-, γ-, and δ-subunits (860, 873). The sequences (860) do not include the leader sequences. The numbering system does not correspond to the individual subunit sequence numbers, and is valid only for homology comparisons. Important segments in various parts of our model are identified in the figure, including the four hydrophobic transmembrane segments (M1, M2, M3, M4) and the ion channel elements (Channel). (Reprinted, with permission, from reference 737.)

channel via large-scale rearrangements is probably not consistent with observed rapid fluctuations in ion current (50 μsec) nor is a mechanism for ligand control of ion flow given. The Chou-Fasman approach (925), the Kyte-Doolittle scheme (869), the Eisenberg hydrophobic moment (870–872, 926), or the Guy analysis of partition energies (927) to evaluate the polarity preference of helices utilize rather uncertain transfer energies. Procedures that average a number of parameters for the purpose of identifying transmembrane helices may be convenient but would miss amphiphilic elements (928). As Kuntz (868, 929) has noted, there are many local minima generated on the basis of current methods for folding proteins. Chothia has commented on some of the difficulties involved in deriving such numbers (930). Creighton has discussed folding of proteins (931). Wallace (932) has specifically shown that folding schemes based on thermodynamic parameters empirically derived for soluble proteins are inappropriate for membrane proteins. Argos and Mohana Rao (933) have underlined the need for, as well as the inadequacies of, theoretical methods for folding proteins. The failure of methods based on the correlation of secondary structures and sequence has been attributed to the inadequacy of the crystallographic base (934, 935). A quantitative approach to hydrophobic interactions is based on the susceptibility to urea denaturation of a series of enzymes for which mutations are introduced to control the structure in a precise way (936). Other criteria must therefore be used to construct specific structures; protein unfolding schemes (937) may help to understand folding.

Finer-Moore and Stroud (904, 939) searched for amphiphilic segments belonging to α-helices or β-strands using Fourier analysis of the periodicity of hydrophobic amino acid sequences. The ion channel formulated on the basis of the search has been examined by molecular graphics (939).

A Total AChR Model

The holistic approach involves using as much information as possible from any and all sources to construct an AChR receptor model (735–737), as illustrated in Scheme 6.3.

AChR Model: Bilayer Portion

I initially identified a key channel element in the α-subunit (373–391) of the AChR (705) on the basis of SGR theory (channel function) and then noted that the amphiphilic character of the segment thus selected was compatible with its presence within the bilayer. The sequence α372–395 (24 amino acids needed to traverse the bilayer) was selected as the ion channel element and included 373–391. One additional ion channel element was identified in each of the β (401–424), γ (413–395), and δ (419–442) subunits (940, 941). The ion channel elements were assumed to be α-helices, with appropriately located charged side chains interacting via SGR. The amino acids ser, thr, and cys can stabilize

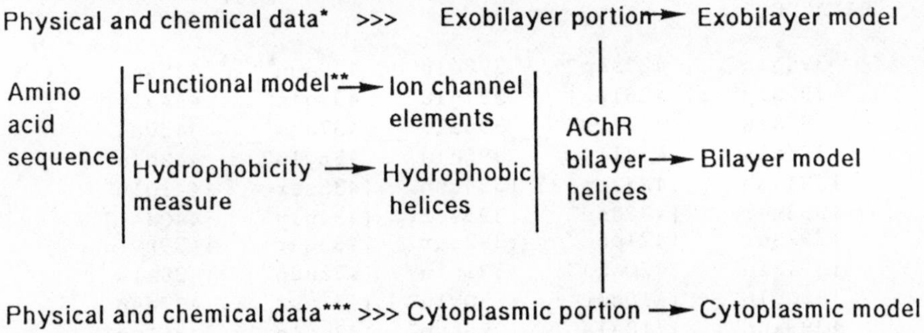

Physical and chemical data* >>> Exobilayer portion → Exobilayer model

Amino acid sequence | Functional model** → Ion channel elements | AChR bilayer helices → Bilayer model

Hydrophobicity measure → Hydrophobic helices

Physical and chemical data*** >>> Cytoplasmic portion → Cytoplasmic model

*Electron microscopy, x-ray scattering, functional model and single group rotation (SGR) theory, acetylcholine binding sites, thiol labeling, Raman data, genetic homology, deletion and mutation data

**Ion channel mechanism and structure based on SGR theory, genetic homology, deletion and mutation data, noncompetitive antagonist labeling data

***Electron microscopy, immunological data, phosphorylation sites

Scheme 6.3. Summary of holistic approach to constructing a model of the nicotinic acetylcholine receptor (nAChR).

the amphiphilic helices (719). The effects of site-directed mutagenesis (942) on function are consistent with the five-element ion channel. The aligned ion channel elements are shown together in Figure 6.11.

From the easily identified hydrophobic segments (four in each subunit), a 24–amino acid α-helical portion is placed within the bilayer, with prolines just outside the bilayer. The bilayer segments including the ion channel element for each subunit are shown in Figures 6.12–6.15. The bilayer helices account for 600/2333 (25.7%) of the AChR protein.

Model of Exobilayer Portion of AChR α-Subunit

The following criteria are used: (1) two binding sites must be present; (2) the ACh binding sites (certain +, − pairs) must be found; (3) a reducible disulfide bond is near an ACh binding site; (4) the exobilayer region must extend 55–70 Å from bilayer; (5) there must be a high antiparallel β-sheet content.

These points can be expanded as follows. (1) *Two binding sites*. AChR ion channel activation depends upon two molecules of ACh; at least two binding sites must exist. (2) *ACh binding site*. It is proposed on chemical grounds that half of a combining site consists of a lysine ϵ-ammonium ion and a negatively charged carboxylate group (glu/asp), chosen for electrostatic attraction and hydrogen-bonding to ACh. (3) *A reducible disulfide bond near an ACh binding site*. Carefully executed affinity alkylations with bromoacetylcholine re-

```
BILAYER____
           |
   398glu-  | 427asp-    398glu-   439asn     445asn
   397glu-  | 426lys+    397glu-   438glu-    444gly
__ 396ala __| 425lys+ __ 396ala __ 437asn ___ 443val __
   395ala     424leu     395ala    436glu-    442glu-
  |394asn    |423asp-   |394asn   |435ser    |441glu-
  |393ser    |422asp-   |393ser   |434gly    |440asp-
  |392ser    |421phe    |392ser   |433ser    |439tyr
  |391glu-   |420glu-   |391glu-  |432asp-   |438ala
  |390glu-   |419ser    |390glu-  |431asn    |437asn
  |389asp-   |418ala    |389asp-  |430gln    |436lys+
  |388ser    |417ser    |388ser   |429glu-   |435glu-
  |387lys+   |416glu-   |387lys+  |428lys+   |434lys+
  |386met    |415leu    |386met   |427thr    |433ile
  |385his    |414gln    |385his   |426ser    |432gln
  |384glu-   |413glu-   |384glu-  |425lys+   |431lys+
  |383ala    |412ala    |383ala   |424ala    |430val
  |382ile    |411ile    |382ile   |423ile    |429ile
  |381tyr    |410tyr    |381tyr   |422phe    |428tyr
  |380lys+   |409lys+   |380lys+  |421asn    |427asn
  |379val    |408ile    |379val   |420cys    |426thr
  |378gly    |407ala    |378gly   |419ala    |425ser
  |377glu-   |406glu-   |377glu-  |418glu-   |424asp-
  |376ile    |405val    |376ile   |417val    |423ile
  |375ala    |404ala    |375ala   |416cys    |422gly
  |374ser    |403glu-   |374ser   |415ser    |421ser
  |373lys+   |402lys+   |373lys+  |414lys+   |420lys+
   372val____ 401leu____ 372val____ 413ile____ 419ile____
   371asp-    400asp-    371asp-    412glu-    418glu-
   370pro     399gln     370pro     411pro     417asp-
   369asn     398pro     369asn     410ala     416his
     α          β          α          γ          δ

                  Channel Elements, AChR
```

Fig. 6.11. The ion channel elements, one from each subunit, of the nicotinic acetylcholine receptor. The elements are aligned in the way needed in the model for an active channel. (Reprinted, with permission, from reference 737.)

veal one reactive thiol per α-subunit, with a difference in the reactivities of the two α-subunits (943). The binding site near the disulfide may be blocked with or without elimination of ion channel function (944, 945). The labeled SH group is α192 (875).

The disulfide bond is unusual, being formed between the adjacent cysteines, α192–α193 (946), a disulfide link not observed until recently (947). Ovchinnikov (948) cited the disulfide link between cys 322 and cys 323 in bovine rhodopsin as the first between adjacent cys residues in a native protein and reported a similar link in the γ-subunit of the bovine retinal GTP-binding protein, transducin (949). The nmr spectrum (950) and crystal structure (951) of

```
                -                                 400lys+----401tyr
                -         268ser----269ser        399trp     402val
                -         267thr    270ala        398glu-    403ala
              210ile      266ser    271val        397glu-    404met
__________    211pro      265pro    272pro        396ala     405val
    |         212leu      264ile    273leu        395ala     406ile
    |         213tyr      263leu    274ile        394asn     407asp-
    |         214phe      262glu-   275gly        393ser     408his
    |         215val      261val    276lys+       392ser     409ile
 BILAYER      216val      260ile    277tyr       |391glu-    410leu
    |         217asn      259val    278met       |390glu-    411leu
    |         218val      258leu    279leu       |389asp-    412cys
    |         219ile      257leu    280phe       |388ser     413val
    |         220ile      256phe    281thr       |387lys+    414phe
    |         221pro      255val    282met       |386met     415met
    |         222cys      254thr    283ile       |385his     416leu
    |         223leu      253leu    284phe       |384glu-    417ile
    |         224leu      252ser    285val       |383ala     418cys
    |         225phe      251leu    286ile       |382ile     419ile
    |         226ser      250leu    287ser       |381tyr     420ile
    |         227phe      249val    288ser       |380lys+    421gly
    |         228leu      248ser    289ile       |379val     422thr
    |         229thr      247ile    290ile       |378gly     423val
    |         230gly      246ser    291ile       |377glu-    424ser
    |         231leu      245leu    292thr       |376ile     425val
    |         232val      244thr    293val       |375ala     426phe
    |         233phe      243met    294val       |374ser     427ala
    |         234tyr      242lys+   295val       |373lys+    428gly
__________    235leu      241glu-   296ile        372val     429arg+
              236pro      240gly    297asn        371asp-    430leu
              237thr      239ser    298thr        370pro     431ile
                   238asp-          299his        369asn     432glu-
          ______________________________________  _______    _______
Helices       αM1         αM2       αM3           αC         αM4
                             α-Subunit Bilayer Helices
```

Fig. 6.12. The amino acid sequences assigned as bilayer helices of the α-subunit of the nicotinic acetylcholine receptor. Helices 1–4 are hydrophobic (labeled α*Mn*, $n = 1$–4); the ion channel element is amphiphilic (labeled αC). The ion channel element (marked with vertical lines) is assigned on functional grounds, using single group rotation (SGR) theory to recognize side chains communicating along one side of an α-helix. The amphiphilic ion channel element includes a substantial number of hydrophobic amino acids. (Reprinted, with permission, from reference 737.)

the cyclic disulfide of L-cysteinyl-L-cysteine are known. Malformin A is a fungal cyclic pentapeptide with a disulfide link between adjacent D-cysteinyl residues (952). Computer graphics studies on the model receptor suggest that there is a *cis*-peptide link between the cys-cys disulfide fragment and the successor amino acid, proline (905). Six of the known α-subunit sequences have the CCP tripeptide moiety in the same position, but the rat neural AChR has a glu in place of pro (915).

	β M1	β M2	β M3	β C	β M4
	–				
	208thr				
	209phe				
	210tyr				
	211leu				
	212ile				
	213ile			429gln	
	214gln	274ser---	275leu	428trp	430tyr
	215arg$^+$	273thr	276ser	427asp$^-$	431val
	216lys$^+$	272glu$^-$	277val	426lys$^+$	432ala
	217pro	271pro	278pro	425lys$^+$	433met
	218leu	270val	279ile	424leu	434val
	219phe	269lys$^+$	280ile	423asp$^-$	435ala
	220tyr	268asp$^-$	281ile	422asp$^-$	436asp$^-$
	221ile	267ala	282arg$^+$	421phe	437arg$^+$
	222val	266leu	283tyr	420glu$^-$	438leu
B	223tyr	265leu	284leu	419ser	439phe
I	224thr	264leu	285met	418ala	440leu
L	225ile	263leu	286phe	417ser	441tyr
A	226ile	262phe	287ile	416glu$^-$	442val
Y	227pro	261val	288met	415leu	443phe
E	228cys	260thr	289ile	414gln	444phe
R	229ile	259val	290leu	413glu$^-$	445val
	230leu	258ala	291val	412ala	446ile
	231ile	257leu	292ala	411ile	447cys
	232ser	256leu	293phe	410tyr	448ser
	233ile	255ala	294ser	409lys$^+$	449ile
	234leu	254ser	295val	408ile	450gly
	235ala	253ile	296ile	407ala	451thr
	236ile	252ser	297leu	406glu$^-$	452phe
	237leu	251leu	298ser	405val	453ser
	238val	250ser	299val	404ala	454ile
	239phe	249met	300val	403glu$^-$	455phe
	240tyr	248lys$^+$	301val	402lys$^+$	456leu
	241leu	247glu$^-$	302leu	401leu	457asp$^-$
	242pro	246gly	303asn	400asp$^-$	458ala
	243pro	245ala	304leu	399gln	459ser
		244asp$^-$	305his	398pro	460his

Helices

β–Subunit Bilayer Helices

Fig. 6.13. The amino acid sequences assigned as bilayer helices of the β-subunit of the nicotinic acetylcholine receptor. Helices 1–4 are hydrophobic (labeled βM*n*, $n = 1$–4); the ion channel element is amphiphilic (labeled βC). Further details are given in the caption for Fig. 6.12. (Reprinted, with permission, from reference 737.)

```
        214ile
        215ile                                         441val
        216gln     277ser---278leu        440trp       442leu
        217arg+    276thr   279asn        439asn       443ile
        218lys+    275glu-  280val        438glu-      444gly
        219pro     274pro   281pro        437asn       445lys+
  _____
      | 220leu     273val   282leu        436glu-      446val
      | 221phe     272lys+  283ile       |435ser       447ile
      | 222tyr     271gln   284gly       |434gly       448asp-
      | 223ile     270ala   285lys+      |433ser       449lys+
      | 224ile     269ile   286tyr       |432asp-      450ala
      B 225asn     268leu   287leu       |431asn       451cys
      I 226ile     267phe   288ile       |430gln       452phe
      L 227ile     266leu   289phe       |429glu-      453trp
      A 228ala     265phe   290val       |428lys+      454ile
      Y 229pro     264ile   291met       |427thr       455ala
      E 230cys     263thr   292phe       |426ser       456leu
      R 231val     262gln   293val       |425lys+      457leu
      | 232leu     261ala   294ser       |424ala       458leu
      | 233ile     260leu   295met       |423ile       459phe
      | 234ser     259leu   296leu       |422phe       460ser
      | 235ser     258val   297ile       |421asn       461ile
      | 236leu     257ser   298val       |420cys       462gly
      | 237val     256ile   299met       |419ala       463thr
      | 238val     255ser   300asn       |418glu-      464leu
      | 239leu     254leu   301cys       |417val       465ala
      | 240val     253thr   302val       |416cys       466ile
      | 241tyr     252cys   303ile       |415ser       467phe
      | 242phe     251lys+  304val       |414lys+      468leu
     _| 243leu     250gln   305leu        413ile       469thr
        244pro     249gly   306asn        412glu-      470gly
        245ala     248gly   307val        411pro       471his
        246gln----247ala    308ser        410ala       472phe
        ______________________________    ______     ________
         γ M1      γ M2      γ M3          γ C         γ M4       Helices
```

γ-Subunit Bilayer Helices

Fig. 6.14. The amino acid sequences assigned as bilayer helices of the γ-subunit of the nicotinic acetylcholine receptor. Helices 1–4 are hydrophobic (labeled γM*n*, $n = 1$–4); the ion channel element is amphiphilic (labeled γC). Further details are given in the caption for Fig. 6.12. (Reprinted, with permission, from reference 737.)

	δ M1	δ M2	δ M3	δ C	δ M4
	218tyr				
	219leu				
	220ile				
	221ile			447asn	
	222arg$^+$	282ala----	283leu	446trp	448leu
	223arg$^+$	281thr	284ala	445asn	449val
	224lys$^+$	280glu$^-$	285val	444gly	450gly
	225pro	279pro	286pro	443val	451gln
	226leu	278leu	287leu	442glu$^-$	452thr
	227phe	277arg$^+$	288ile	441glu$^-$	453ile
	228tyr	276gln	289gly	440asp$^-$	454asp$^-$
	229val	275ser	290lys$^+$	439tyr	455arg$^+$
	230ile	274thr	291tyr	438ala	456leu
B	231asn	273leu	292leu	437asn	457ser
I	232phe	272leu	293met	436lys$^+$	458met
L	233ile	271leu	294phe	435glu$^-$	459phe
A	234thr	270phe	295ile	434lys$^+$	460ile
Y	235pro	269val	296met	433ile	461ile
E	236cys	268ala	297ser	432gln	462thr
R	237val	267gln	298leu	431lys$^+$	463pro
	238leu	266ala	299val	430val	464val
	239ile	265leu	300thr	429ile	465met
	240ser	264leu	301gly	428tyr	466val
	241phe	263val	302val	427asn	467leu
	242leu	262ser	303ile	426thr	468gly
	243ala	261ile	304val	425ser	469thr
	244ser	260ala	305asn	424asp$^-$	470ile
	245leu	259thr	306cys	423ile	471phe
	246ala	258ser	307gly	422gly	472ile
	247phe	257met	308ile	421ser	473phe
	248tyr	256lys$^+$	309val	420lys$^+$	474val
	249leu	255glu$^-$	310leu	419ile	475met
	250pro	254gly	311asn	418glu$^-$	476gly
	251ala	253ser	312phe	417asp$^-$	477asn
	252glu$^-$			416his	478phe
	δ M1	δ M2	δ M3	δ C	δ M4 Helices

δ-Subunit Bilayer Helices

Fig. 6.15. The amino acid sequences assigned as bilayer helices of the δ-subunit of the nicotinic acetylcholine receptor. Helices 1–4 are hydrophobic (labeled δMn, $n = 1$–4); the ion channel element is amphiphilic (labeled δC). Further details are given in the caption for Fig. 6.12. (Reprinted, with permission, from reference 737.)

(4) *55–70 Å Extension from bilayer*. Electron microscopy indicates that the structural elements of the exobilayer portion of the acetylcholine receptor extend approximately 55–70 Å from the bilayer (821, 826–829, 924). (5) *High antiparallel β-sheet content*. Raman spectroscopy indicates that antiparallel β-sheet is the most abundant form of polypeptide (34%) present in the AChR (953, 954). The α-helical content (25%) (953, 955) is accounted for as bilayer α-helices.

I therefore arranged strands of 17–18 amino acids (56–60 Å) (not including turns) as folded antiparallel β-sheets (X1-X11, Fig. 6.16). I searched for apposed lys$^+$ and glu$^-$ (or asp$^-$). Since labeling of the thiol of the reduced AChR is effected by a thiol-reactive acetylcholine analogue, one pair of half binding sites were placed near the disulfide. Twists in the strands, although likely (867, 956), have not been incorporated into the model.

Surprisingly, there were not too many alternative arrangements for the strand structure shown. Four binding sites were found, two near the bilayer [(1) 172glu$^-$, 179lys$^+$ and (2) 107lys$^+$, 97asp$^-$] and two near the 192–193 disulfide [(3) 166asp$^-$, 185lys$^+$ and (4) 115lys$^+$, 89asp$^-$].

The spatial organization of the exobilayer strands is decided on the basis of several criteria: (1) the binding site groups (strands X5, X6, X9, X10) and the 192–193 disulfide (X11) are on the side of the ion channel; (2) the X6 strand is close to the X7 strand; (3) strand X8 is closely linked to X7 by a disulfide; (4) strands X1–X4 and X7–X8 are away from the side of the ion channel; (5) X4 is close to X5. To satisfy criterion (2) about the proximity of X6 and X7, a third rank of strands with X1 and X2 must exist. Remarkably, the arrangement resembles that found in the most refined images generated from electron micrographs of the AChR (826–829, 924). A schematic diagram shows the general relationship of the upper binding site groups to the strands for the groups near the disulfide (Fig. 6.17).

Wild-type mRNA (in the correct proportions) for AChR when expressed in a *Xenopus laevis* öocyte yields protein that behaves normally with respect to electrophysiological measurements and α-bungarotoxin binding (942, 957). Substitution of ser for any of the α-subunit exobilayer cys (128cys, 142cys, 192cys, or 193cys) led to AChR with no electrophysiological activity, but α-BTX binding capacity was partly retained after replacement of either 192cys or 193cys. These results signify the importance of the 128cys-142cys disulfide link to the conformation of the AChR.

Models of Exobilayer Portions of AChR β-, γ-, and δ-Subunits

The excellent genetic homology between the subunits in the exobilayer sequence (796) (Fig. 6.10) suggests that exobilayer portions of the β-, γ, and δ-subunits are similar in structure to that of the α-subunit. The conformationally

```
  157ser 156val 155lys+
 158ile          154thr                                       82ser    81pro
159ser            153gly                                      83asp-   80leu   41ile   40leu   1ser
160pro            152asp-                                     84asp-   79arg+  42ser   39gln   2glu-
161glu-            151tyr                   119thr             85val    78ile   43val   38ile   3his
162ser   189tyr--150thr--190tyr  120pro          118trp       86trp    77lys+  44asp-  37leu   4glu-
163asp-  188val    149trp  191thr   121pro       117met       87leu    76lys+  45glu-  36gln   5thr
164arg+  187trp    148ile  192cys|  122ala       116ile       88pro    75ile   46val   35leu   6arg+
165pro   186his    147gly  193cys|  123ile       115lys+      89asp-   74gly   47asn   34gly   7leu
166asp-  185lys+   146leu  194pro   124phe       114gly       90leu    73gly   48gln   33val   8val
167leu   184trp    145lys+ 195asp-  125lys+      113thr       91val    72tyr   49ile   32thr   9ala
168ser   183gly    144met  196thr   126ser       112tyr       92leu    71asp-  50val   31ile   10asn
169thr   182arg+   143thr  197pro   127tyr       111asp-      93tyr    70ala   51glu-  30asp-  11leu
170phe   181tyr    142cys---198tyr--128cys       110leu       94asn    69pro   52thr   29val   12leu
171met   180asp-   141asn  199leu   129glu-      109leu       95asn    68asn   53asn   28phe   13glu-
172glu-  179lys+   140gln  200asp-  130ile       108leu       96ala    67trp   54val   27his   14asn
173ser   178met    139gln  201ile   131ile       107lys+      97asp-   66arg+  55arg+  26thr   15tyr
174gly   177val    138asp- 202thr   132val       106thr       98gly    65leu   56leu   25his   16asn
175glu-  176trp    137phe  203tyr   133thr       105met       99asp-   64arg+  57arg+  24his   17lys+
                   136pro  204his   134his       104his       100phe   63val   58gln   23glu-  18val
                     135phe-205phe --|           103val       101ala   62asp-  59gln   22val   19ile
                           206ile                      102ile           61ile  60trp   21pro   20arg+
                           207met
                           208gln
                           209arg+

X9       X10       X8      X11      X7           X6           X5       X4      X3      X2      X1

               Exobilayer Sequences α-Subunit, AChR
```

Fig. 6.16. The exobilayer arrangement of the nAChR α-subunit according to the criteria explained in the text. (Reprinted, with permission, from reference 737.)

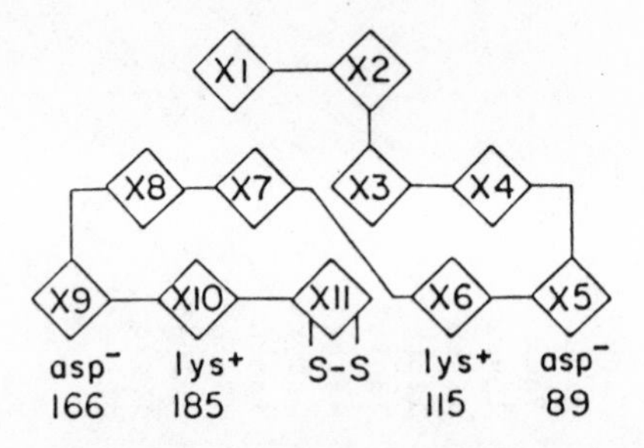

Fig. 6.17. Schematic exobilayer strand arrangement with the upper binding-site groups of the nAChR α-subunit noted. (Reprinted, with permission, from reference 737.) The overall geometry corresponds to the shape of the exobilayer portion of a subunit, as suggested by image analysis of electron micrographs of the receptor.

important disulfides, except for the unusual α-cys-cys disulfide, are present: β-, γ-: 128cys–142cys; δ-: 130cys–144cys. Structures are formulated in analogy to that of the exobilayer α-subunit (Figures 6.18–6.20).

Models of the Cytoplasmic Portions of the AChR

Electron micrographs of the *Torpedo californica* AChR suggest that the cytoplasmic portion does not extend very far (ca. 15 Å) into the cell (826, 924, 958, 959). Highly schematic models of the cytoplasmic regions are formulated by supposing that the protein is restricted to a plane immediately below the bilayer (Figures 6.21–6.24).

These models show the following:

1. The area of the cytoplasmic structures for the β-, γ-, and δ-subunits might be greater than the area of the corresponding α-subunit, given the 30–40% difference in lengths of the cytoplasmic portions in the model. The extended cytoplasmic structures of the β-, γ-, and δ-subunits might interact with the shorter cytoplasmic portions of the α-subunit.

2. Excess positive charge is found in the cytoplasmic portions of the AChR. The sites subject to phosphorylation are tyr and ser (847) and appear to be quite far from the ion channel. Ignoring the histidines, complete phosphorylation neutralizes the excess positive charge already present. The positive and negative charges in each cytoplasmic segment are summarized as follows [P (n^-) = number of phosphorylated sites (additional number of negative charges due to -$OPO_3H_2^-$ or -$OPO_3H^=$)]: α: 10^+, 9^-, 1P(1 to 2^-); β: 10^+, 9^-, 1P(1 to 2^-); γ: 20^+, 17^-, 2P(2 to 4^-); δ: 18^+, 16^-, 3P(3 to 6^-). The rate of desensitization of *Torpedo californica* AChR is increased 7–8-fold by phosphorylation on γ354ser and δ361ser although there is little effect on channel conductance or mean channel open time (960–964). The relevant phosphorylation may be mediated by cAMP, as shown by the correlation of

```
        155glu- 156val                                               82ser 81pro
        154ser   157thr                                             83ser   80ile   41leu   40leu   1ser
        153thr   158leu                                             84asp-  79arg+  42ile   39asn   2val
  ----  152asp- --159gln -------------------------------119gln      85val   78leu   43leu   38thr   3met
   |    151tyr   160his   191lys+  192pro   118trp                  86trp   77asp-  44asn   37leu   4glu-
120pro  150thr   161ala   190his   193ser   117ser                  87gln   76lys+  45glu-  36thr   5asp-
121ser  149tyr   162leu   189glu-  194arg+  116thr                  88pro   75ile   46lys+  35leu   6thr
122ala  148ser   163asp-  188ile   195lys+  115ala                  89asp-  74gly   47ile   34gly   7leu
123ile  147lys+  164ala   187ser   196asn   114gly                  90ile   73glu-  48glu-  33val   8leu
124phe  146phe   165lys+  186trp   197trp   113thr                  91val   72tyr   49glu-  32arg+  9ser
125tyr  145val   166gly   185gln   198arg+  112his                  92leu   71ala   50met   31val   10val
126arg+ 144met   167glu-  184gly   199ser   111gln                  93met   70ala   51thr   30thr   11leu
127ser  143thr   168arg+  183asn   200asp-  110val                  94asn   69pro   52thr   29val   12phe
128cys--142cys   169glu-  182glu-  201asp-  109leu                  95asn   68asp-  53asn   28lys+  13glu-
129thr  141asn   170val   181thr   202pro   108val                  96asn   67trp   54val   27asp-  14thr
130ile  140gln   171lys+  180phe   203ser   107asn                  97asp-  66gln   55phe   26gly   15tyr
131lys+ 139trp   172glu-  179ala   204tyr   106val                  98gly   65leu   56leu   25val   16asn
132val  138asp-  173ile   178asp-  205glu-  105his                  99ser   64arg+  57asn   24thr   17pro
133met  137phe   174val   177lys+  206asp-  104leu                  100phe  63tyr   58leu   23gln   18lys+
134tyr  136pro    175ile 176asn    207val   103thr                  101glu- 62asp-  59ala   22ala   19val
   135phe                                             102ile                 61thr  60trp   21pro   20arg+


X7      X8       X9       X10      X11      X6       X5             X4      X3      X2      X1

              Exobilayer Sequences β-Subunit, AChR
```

Fig. 6.18. The exobilayer arrangement of the nAChR β-subunit according to the criteria noted in the text, including a 128cys–142cys disulfide bond. (Reprinted, with permission, from reference 737.)

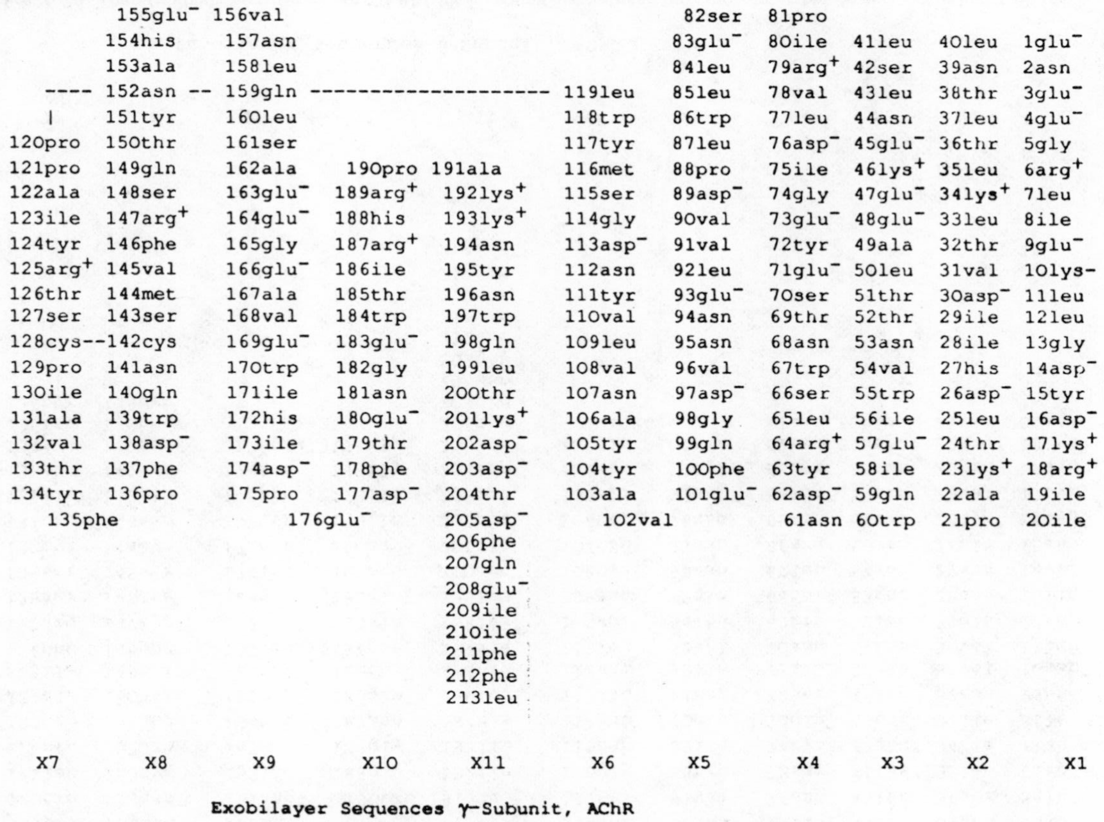

```
           155glu-  156val                                                    82ser   81pro
           154his   157asn                                                    83glu-  80ile   41leu   40leu   1glu-
           153ala   158leu                                                    84leu   79arg+  42ser   39asn   2asn
    ----   152asn -- 159gln --------------------------------------  119leu   85leu   78val   43leu   38thr   3glu-
      |    151tyr   160leu                                         118trp   86trp   77leu   44asn   37leu   4glu-
120pro     150thr   161ser                                         117tyr   87leu   76asp-  45glu-  36thr   5gly
121pro     149gln   162ala      190pro  191ala                     116met   88pro   75ile   46lys+  35leu   6arg+
122ala     148ser   163glu-    189arg+  192lys+                    115ser   89asp-  74gly   47glu-  34lys+  7leu
123ile     147arg+  164glu-    188his   193lys+                    114gly   90val   73glu-  48glu-  33leu   8ile
124tyr     146phe   165gly     187arg+  194asn                     113asp-  91val   72tyr   49ala   32thr   9glu-
125arg+    145val   166glu-    186ile   195tyr                     112asn   92leu   71glu-  50leu   31val   10lys-
126thr      144met   167ala     185thr   196asn                     111tyr   93glu-  70ser   51thr   30asp-  11leu
127ser      143ser   168val     184trp   197trp                     110val   94asn   69thr   52thr   29ile   12leu
128cys--142cys      169glu-    183glu-  198gln                     109leu   95asn   68asn   53asn   28ile   13gly
129pro      141asn   170trp     182gly   199leu                     108val   96val   67trp   54val   27his   14asp-
130ile      140gln   171ile     181asn   200thr                     107asn   97asp-  66ser   55trp   26asp-  15tyr
131ala      139trp   172his     180glu-  201lys+                    106ala   98gly   65leu   56ile   25leu   16asp-
132val      138asp-  173ile     179thr   202asp-                    105tyr   99gln   64arg+  57glu-  24thr   17lys+
133thr      137phe   174asp-    178phe   203asp-                    104tyr   100phe  63tyr   58ile   23lys+  18arg+
134tyr      136pro   175pro     177asp-  204thr                     103ala   101glu- 62asp-  59gln   22ala   19ile
   135phe                  176glu-      205asp-               102val              61asn   60trp   21pro   20ile
                                        206phe
                                        207gln
                                        208glu-
                                        209ile
                                        210ile
                                        211phe
                                        212phe
                                        213leu

X7          X8       X9         X10      X11                  X6       X5      X4      X3      X2      X1
```

Exobilayer Sequences γ-Subunit, AChR

Fig. 6.19. The exobilayer arrangement of the nAChR γ-subunit according to the criteria noted in the text, including a 128cys–142cys disulfide bond. (Reprinted, with permission, from reference 737.)

```
                                                                  82leu  81arg+
                                                                 83pro   80leu  41asn  40ser  1val
        157glu-  158ile                                          84pro   79ile  42leu  39leu  2asn
       156asn    159thr                                          85glu-  78ser  43ile  38thr  3glu-
119thr- 155ala -- 160met -------------------------- 118val       86leu   77ile  44ser  37leu  4glu-
120trp  154asp-  161asp-  190trp  191glu-           117tyr       87val   76asp- 45leu  36ser  5glu-
121leu  153tyr   162leu   189glu-  192ile           116gly       88trp   75ser  46lys+ 35leu  6arg+
122pro  152asn   163met   188gly   193ile           115asn       89ile   74tyr  47glu- 34ala  7leu
123pro  151leu   164thr   187asn   194his           114pro       90pro   73glu- 48thr  33ile  8ile
124ala  150ala   165asp-  186glu-  195lys+          113arg+      91asp-  72ser  49asp- 32asn  9asn
125ile  149thr   166thr   185thr   196pro           112val       92ile   71ala  50glu- 31val  10asp-
126phe  148phe   167ile   184phe   197ala           111leu       93val   70asn  51thr  30val  11leu
127arg+ 147lys+  168asp-  183ala   198lys+          110val       94leu   69trp  52leu  29glu- 12leu
128ser  146leu   169gly   182glu-  199lys+          109asn       95gln   68thr  53thr  28asn  13ile
129ser  145ser   170lys+  181pro   200asn           108cys       96asn   67leu  54ser  27asn  14val
130cys--144cys   171asp-  180asp-  201ile           107phe       97asn   66arg+ 55asn  26his  15asn
131pro  143asn   172tyr   179ile   202tyr           106tyr       98asn   65his  56val  25lys+ 16lys+
132ile  142gln   173pro   178ile   203pro           105ala       99asp-  64asp- 57trp  24val  17tyr
133asn  141trp   174ile   177ile   204asp-          104val       100gly  63tyr  58met  23pro  18asn
134val  140asp-   175glu- 176trp   205lys+          103his   101gln      62trp 59asp- 22arg+ 19lys+
135leu  139phe                     206phe               102tyr           61ala 60his  21val 20his
136tyr  138pro                      207pro
   137phe                          208asn
                                   209gly
                                   210thr
                                   211asn
                                   212tyr
                                   213gln
                                   214asp-

 X7      X8       X9      X10      X11       X6      X5      X4     X3     X2     X1
```

Exobilayer Sequences δ-Subunit, AChR

Fig. 6.20. The exobilayer arrangement of the nAChR δ-subunit according to the criteria noted in the text, including a 130cys–

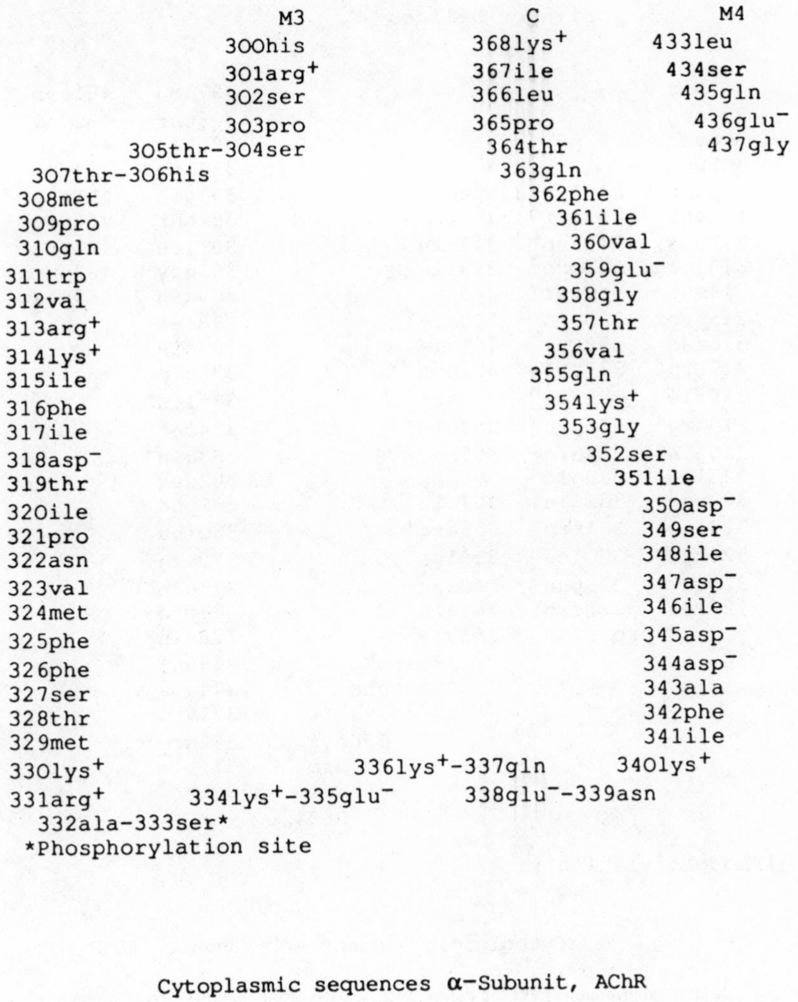

M3 C M4

300his 368lys$^+$ 433leu
301arg$^+$ 367ile 434ser
302ser 366leu 435gln
303pro 365pro 436glu$^-$
305thr-304ser 364thr 437gly
307thr-306his 363gln
308met 362phe
309pro 361ile
310gln 360val
311trp 359glu$^-$
312val 358gly
313arg$^+$ 357thr
314lys$^+$ 356val
315ile 355gln
316phe 354lys$^+$
317ile 353gly
318asp$^-$ 352ser
319thr 351ile
320ile 350asp$^-$
321pro 349ser
322asn 348ile
323val 347asp$^-$
324met 346ile
325phe 345asp$^-$
326phe 344asp$^-$
327ser 343ala
328thr 342phe
329met 341ile
330lys$^+$ 336lys$^+$-337gln 340lys$^+$
331arg$^+$ 334lys$^+$-335glu$^-$ 338glu$^-$-339asn
332ala-333ser*
*Phosphorylation site

Cytoplasmic sequences α–Subunit, AChR

Fig. 6.21. Schematic arrangement of the cytoplasmic section of the nicotinic acetylcholine receptor α-subunit. A phosphorylation site is starred. The long segment must be folded so as to be almost parallel to the bilayer, in view of the 15–20 Å length of the cytoplasmic portion of the receptor (826). (Reprinted, with permission, from reference 737.)

```
                 bilayer helices
     M3                                            C         M4
     |                                             |         |
   306his                                        397leu    461asn
   307arg+                                       396thr    462val
   308ser                                        395val    463pro
   309pro                                        394pro    464pro
   310asn    344lys+345pro                       393gln    465asp-
   311thr    343ser    346thr                    392thr    466asn
   312his    342asp-   347ile                    391leu    467pro
   313thr    341pro    348ile                    390gly    468phe
   314met    340ser    349ser                    389asn    469ala
   315pro    339pro    350arg+                   388leu
   316asn    338thr    351ala                    387his
   317trp    337thr    352asn                    386trp
   318ile    336val    353asp-                   385lys+
   319arg+   335pro    354glu-                   384met
   320gln    334arg+   355tyr*                   383glu-
   321ile    333gln    356phe                    382ser
   322phe    332ile    357ile                    381phe
   323ile    331trp    358arg+                   380leu
   324glu-   330leu    359lys+                   379arg+
   325thr    329phe    360pro                    378glu-
   326leu    328pro    361ala                    377pro
       327pro          362gly                    376gln
                           363asp-               375val
                              364phe             374ala
                                365val           373val
                                   366cys        372arg+
                                   367pro        371ala
                                   368val     370asn
                                        369asp-

*Phosphorylation site

                       Cytoplasmic sequences β-Subunit, AChR
```

Fig. 6.22. General arrangement of the cytoplasmic section of the nicotinic acetylcholine receptor β-subunit. Phosphorylation sites are starred. Note the presence of a large number of prolines (3/9) in the C-terminal ''tail'' of the subunit. The long segment must be folded so as to be almost parallel to the bilayer, in view of the 15–20 Å length of the cytoplasmic portion of the receptor (826). (Reprinted, with permission, from reference 737.)

M3 C M4

309leu
310arg$^+$
311thr
312pro
313asn
314thr
315his
316ser
317leu
318ser
319glu$^-$
320lys$^+$
321ile
322lys$^+$
323his
324leu
325phe
326leu
327gly
328phe
329leu
330pro
331lys$^+$
332tyr
333leu
334gly
335met
336gln
337leu
338glu$^-$
339pro-340ser
341glu$^-$
342glu$^-$
343thr
344pro
345glu$^-$
346lys$^+$
347pro
348gln
349pro
350arg$^+$-
351arg$^+$
352arg$^+$
353ser
354ser*
355phe
356gly
357ile
358met
359ile
360lys$^+$
361ala
362glu$^-$
363glu$^-$
364tyr*
365ile
366leu
367lys$^+$
368lys$^+$-369pro
370arg$^+$
371ser
372glu$^-$
373leu
374met
375phe
376glu$^-$
377glu$^-$
378gln
379lys$^+$
380asp$^-$-381arg$^+$-382his
383gly
384leu
385lys$^+$
386arg$^+$
387val
388asn
389lys$^+$
390met
391thr
392ser
393asp$^-$
394ile
395asp$^-$
396ile
397gly
398thr
399thr
400val
401asp$^-$
402leu
403tyr
404lys$^+$
405asp$^-$
406leu
407ala
408asn
409phe

473asn
474gln
475val
476pro
477glu$^-$
478phe
479pro
480phe
481pro
482gly
483asp$^-$
484pro
485arg$^+$
486lys$^+$
487tyr
488val
489pro

*Phosphorylation sites

Cytoplasmic sequences γ-Subunit, AChR

Fig. 6.23. General arrangement of the cytoplasmic section of the nicotinic acetylcholine receptor γ-subunit. Phosphorylation sites are starred. Note the presence of a large number of prolines (5/17) in the C-terminal ''tail'' of the subunit. The C-terminal sequence of the γ-subunit (5/17 prolines) of the calf nAChR is completely analogous (with substantial homology) to that of *T.californica* and the chicken. The latter has an ''extra'' proline at the end. The long segment must be folded so as to be almost parallel to the bilayer, in view of the 15–20 Å length of the cytoplasmic portion of the receptor (826). (Reprinted, with permission, from reference 737.)

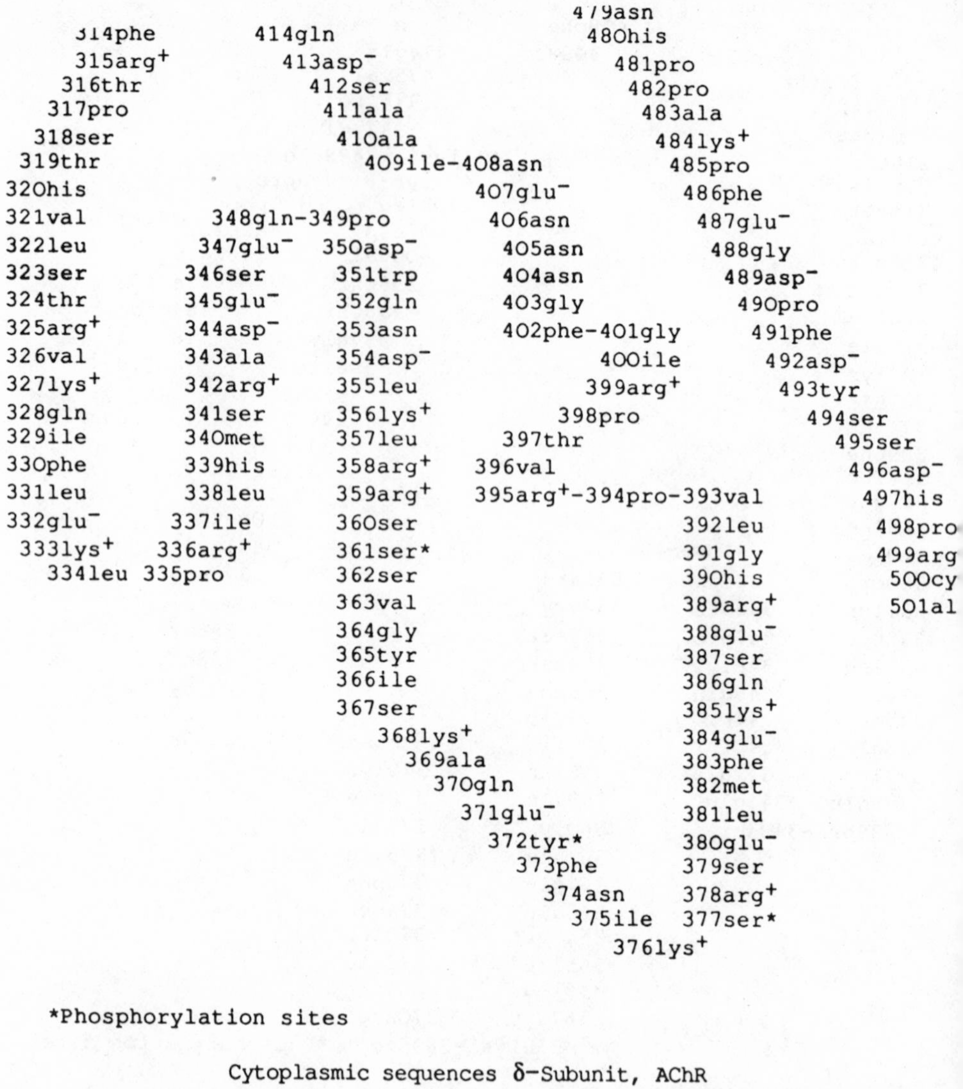

479asn
314phe 414gln 480his
315arg$^+$ 413asp$^-$ 481pro
316thr 412ser 482pro
317pro 411ala 483ala
318ser 410ala 484lys$^+$
319thr 409ile-408asn 485pro
320his 407glu$^-$ 486phe
321val 348gln-349pro 406asn 487glu$^-$
322leu 347glu$^-$ 350asp$^-$ 405asn 488gly
323ser 346ser 351trp 404asn 489asp$^-$
324thr 345glu$^-$ 352gln 403gly 490pro
325arg$^+$ 344asp$^-$ 353asn 402phe-401gly 491phe
326val 343ala 354asp$^-$ 400ile 492asp$^-$
327lys$^+$ 342arg$^+$ 355leu 399arg$^+$ 493tyr
328gln 341ser 356lys$^+$ 398pro 494ser
329ile 340met 357leu 397thr 495ser
330phe 339his 358arg$^+$ 396val 496asp$^-$
331leu 338leu 359arg$^+$ 395arg$^+$-394pro-393val 497his
332glu$^-$ 337ile 360ser 392leu 498pro
333lys$^+$ 336arg$^+$ 361ser* 391gly 499arg
334leu 335pro 362ser 390his 500cy
363val 389arg$^+$ 501al
364gly 388glu$^-$
365tyr 387ser
366ile 386gln
367ser 385lys$^+$
368lys$^+$ 384glu$^-$
369ala 383phe
370gln 382met
371glu$^-$ 381leu
372tyr* 380glu$^-$
373phe 379ser
374asn 378arg$^+$
375ile 377ser*
376lys$^+$

*Phosphorylation sites

Cytoplasmic sequences δ-Subunit, AChR

Fig. 6.24. General arrangement of the cytoplasmic section of the nicotinic acetylcholine receptor δ-subunit. Phosphorylation sites are starred. Note the presence of a large number of prolines (5/23) in the C-terminal "tail" of the subunit. The C-terminal sequence of the chicken is very similar to that of the *Torpedo* with respect to the fraction of proline (5/21). One proline is in a different position and one group of three consecutive prolines is present. The long segment must be folded so as to be almost parallel to the bilayer, in view of the 15–20 Å length of the cytoplasmic portion of the receptor (826). (Reprinted, with permission, from reference 737.)

desensitization with the concentration of forskolin (965–967), which activates hormone-mediated adenylate cyclase (968). Phosphorylation on other sites also increases the rate of desensitization (969). It is interesting that the degree of phosphorylation of the AChR of *Torpedo marmorata* increases on maturation (970). Phosphorylation of the δ-subunit may decrease on assembly (971). The major postsynaptic density protein is a component of a Ca^{++}/calmodulin dependent kinase (972).

3. A remarkable number of prolines are found in C-terminal sequences of the β- (3/10), γ- (5/14), and δ- (5/21) subunits. The proline arrangement of the C-terminal sequence of the γ-subunit (5/17 prolines) of the calf AChR is analogous to that of *T.californica* (906). There are no prolines among the 15 amino acids of the C-terminus in six α-subunits and one proline in this portion of rat neural α-subunit. The sequences which include the prolines in the C-termini resemble those with polyproline II structures (867, 930) and might anchor the AChR to subsynaptic filaments or coated vesicles (973). Collagen (I and III) appears to be involved in AChR aggregation (974). The role of pro-pro interactions in collagen-like packing has been examined (975). The putative AChR binding protein (43K, ν-protein) (976–979) associates with actin (976) and is found in isolated acetylcholine receptor clusters (980). However, the ratio of 43K protein to nicotinic acetylcholine receptor protein is 1:1 whether or not the receptors are clustered (981, 982). Facile cross-linking of the 43K protein via disulfide links produced through oxidation (983) may reflect a mechanism for connecting the AChR to the postsynaptic cytoskeleton (984) with the S-S bond resisting intracellular reduction because of the lipophilic character of the ν-protein (983). At the neuromuscular junction, an extracellular protein, "agrin" (M_r 50,000–100,000), causes aggregation of both AChR and acetylcholinesterase (AChE) (58, 59, 985, 986). The distribution of AChR depends upon biosynthesis, membrane insertion, and degradation (987).

Dynamics of the AChR According to the Model

The model should exhibit (1) a resting (closed) state which requires two molecules of agonist ligand for opening; (2) competition of antagonists and agonists for binding sites; (3) binding sites for noncompetitive blocking agents (anesthetics); (4) partial agonists; (5) ion channel function (916, 917).

The exobilayer portion of the AChR model consists of a pentagonal arrangement of strands carrying a substantial number of charged groups within the pentagon ("cup" or "flower"). The oppositely charged sides ("petals") should attract one another, except for the repulsion of groups of like charge and the high dielectric constant of the aqueous medium. A bending angle of approximately 7° with respect to the vertical would be sufficient for exclusion of an agonist. I propose that the resting state of AChR is "poised" as a closed ("resting") form, into which cations cannot easily enter. The upper side of Figure 6.25 shows how the petals of the flower can close. Agonists (e.g., two ACh) can enter and interact with both sides of the cup, opening the cup and allowing ions to diffuse to the ion channel at the bilayer. The charged groups in the exobilayer cup are shown in Figures 6.26(a) (upper binding region) and 6.26(b) (lower binding region). Using known bond distances, one can easily calculate that the combination of binding site groups with an acetylcholine molecule, associated through electrostatic and hydrogen-bonds, can easily span a distance of 26–30 Å. Significant changes in the orientation of the γ- and δ-subunit exobilayer strands have been detected in the transformation of the resting to the desensitized state by image analysis of electron micrographs of the nAChR (826).

Ion Passage through Channel

The channel elements shown in Figure 6.11 are represented schematically as a cylinder in Figure 6.27. The charged groups are indicated by positive or negative signs, and are divided into seven levels. I consider only the entry of sodium ion, the most common external ion. Many of the positively charged groups in the channel will associate with or dissociate from the negatively charged groups via single group rotations (SGRs) of the side chains. A likely arrangement for the resting state is shown in cylinder B of Figure 6.27, showing group interactions between levels 2–3, 4–5, and 6–7. Sodium is attracted into the negatively charged upper portion of the channel (cylinder C of Fig. 6.27). With an interhelix distance of ca.10 Å, the interior of the ion channel would accommodate a hydrated sodium ion (d = 5.2 Å). The sodium ion should easily continue to levels 4 (cylinder D, Fig. 6.27) and 6 (cylinder E, Fig. 6.27) and then into the interior of the cell. The model of the ion channel thus leads to an attractive molecular mechanism for ion flow.

Binding Site for Noncompetitive Antagonists (NCA)

An important group of antagonist molecules, including histrionicotoxin, phencyclidine, and chlorpromazine, block the ion channel without competing with ACh. Remarkably, irradiation of ^{3}H-chlorpromazine (988, 989) or ^{3}H-methyltriphenylphosphonium ion (919, 990) and *Torpedo marmorata* AChR labels δ262ser. The photolabel was found in homologous positions on the α- and β-, but not on the γ-subunit (991–993). In view of the highly likely homology of the *marmorata* "M2" helix with that of *Torpedo californica*, an explanation for the labeling and the antagonist action of the NCAs can be developed.

Both of the agents used for labeling as well as the NCAs are positively charged. A negative charge was therefore sought on the same helix (helix-

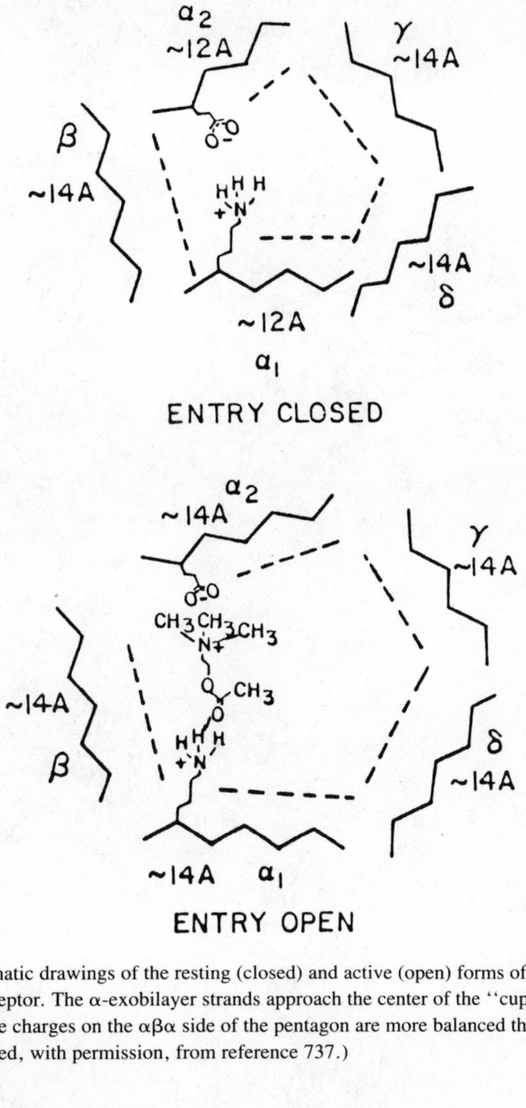

Fig. 6.25. Schematic drawings of the resting (closed) and active (open) forms of the nicotinic acetylcholine receptor. The α-exobilayer strands approach the center of the ''cup'' more than the other strands. The charges on the αβα side of the pentagon are more balanced than those on the γδ side. (Reprinted, with permission, from reference 737.)

specific because the M2 helices in only three of the four subunits are labeled). For each helix that is labeled, a negatively charged residue (glu^-) is found two turns closer to the cytoplasm (see Figs. 6.12–6.15). The labeled positions and the associated glu^- for each subunit are α [254ser, 241glu], β [254ser, 247glu] (989), and δ [262ser, 255glu]. In the γ-subunit, which is not labeled, the residue at the position two turns nearer the cytoplasm is uncharged glutamine (γ250gln).

Why do the NCAs combine with this particular position? Inspection of the

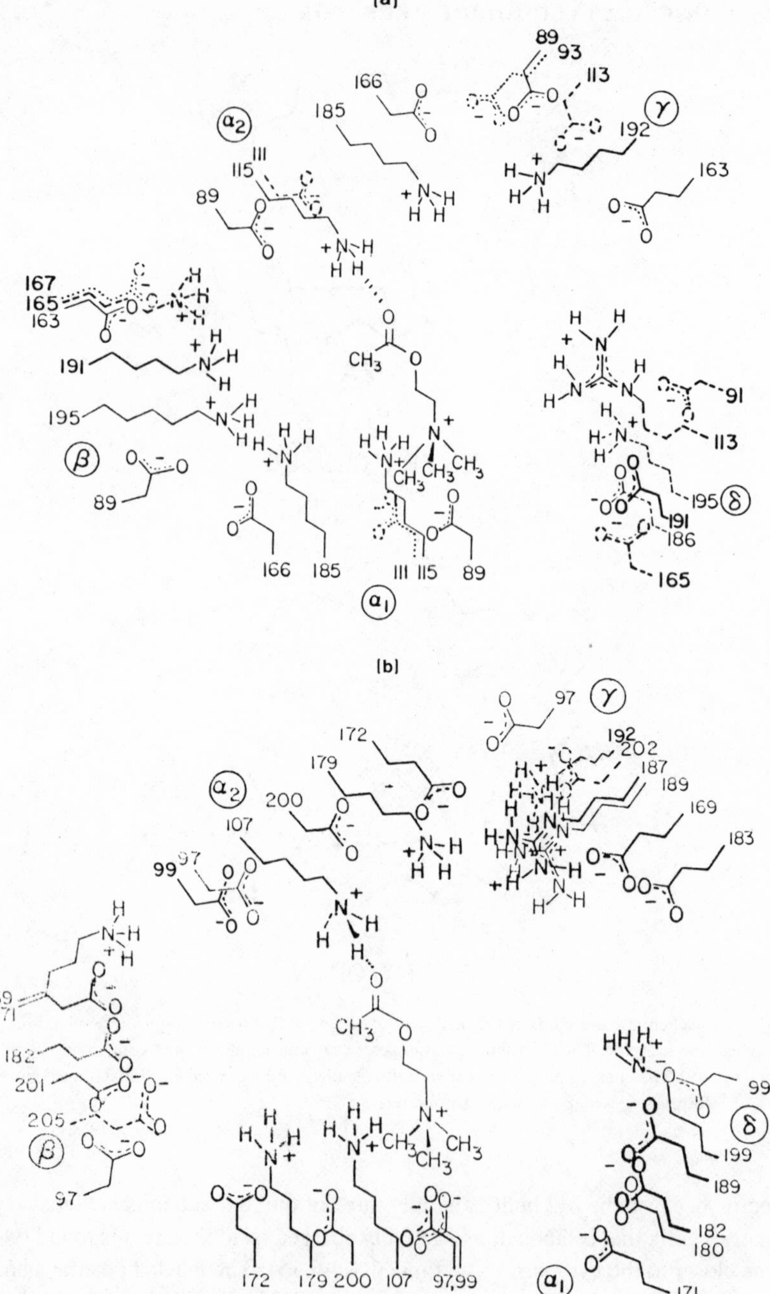

Fig. 6.26. The exobilayer region of the nicotinic acetylcholine receptor as viewed from the outside of the cell. For clarity, the region has been divided into upper (a) and lower (b) portions. The γ- and δ-sections are displaced outward to avoid confusion in the drawing. The nature of the line used for the groups indicates the relative distance from the outside of each portion, the order being 1,———; 2, = = = =; 3,----; 4,—; 5,..... (Reprinted, with permission, from reference 737.)

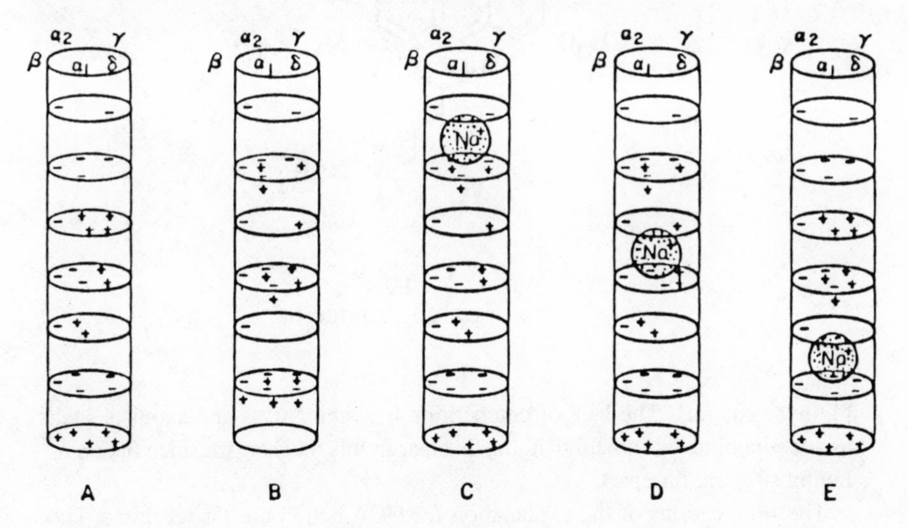

Fig. 6.27. A schematic model for the nicotinic acetylcholine receptor ion channel. A: Charge distribution based on Fig. 6.11. B: Charge distribution after allowing for Coulombic attraction and single group rotation. For simplicity, only + charges have been moved. C: Entry of a hydrated sodium ion into the channel. D: Further movement of the sodium ion. E: A particularly favorable location for the sodium ion. Not all stages of the sodium ion motion are illustrated. (Reprinted, with permission, from reference 737.)

NCA molecular structures shows that these are very hydrophobic molecules carrying a positive charge. Interaction energies would be maximized in a hydrophobic region with an appropriately located negative charge, exactly as found. Models suggest that the binding site fits the NCAs.

An NCA with a longer hydrophobic group, ^{3}H-azidoquinacrine (formula **AQ**) photolabels the α-subunit between 208glu and 243leu in the presence of ACh after rapid mixing and irradiation within 20 msec to avoid desensitization (994). The labeling is blocked by other NCAs, suggesting that 241 glu$^-$ is the binding site. Our model suggests that labeling will occur 3 helix turns above the glu$^-$ within the bilayer, probably at 225phe (Fig. 6.12).

How can the antagonist effect on the receptor be explained? I suppose that the glu$^-$ occupied by the NCA should normally be associated with a mechanistically important positive charge. From the location of the glu$^-$ in the model, one choice might be the nearby lys$^+$ of the ion channel element (amphiphilic helix). The interaction of the glu$^-$ with the channel lys$^+$ might help in allowing the Na^+ ion to pass easily. One lys$^+$ left in the channel might block the passage of cations (see Fig. 6.27, cylinder E). The groups which should thus be involved are α [373lys, 241glu], β [402lys, 247glu], and δ

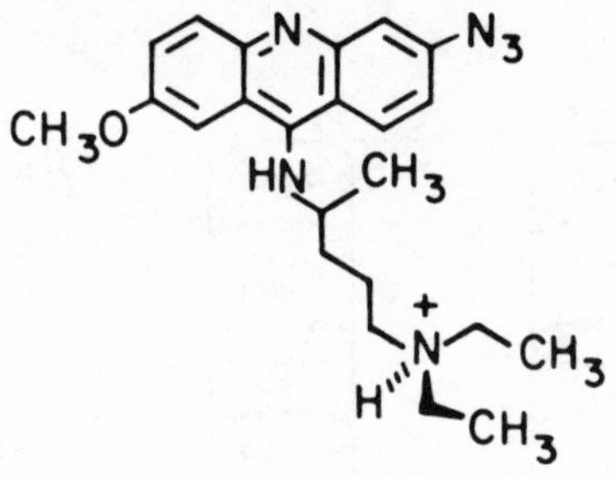

AQ
azidoquinacrine

[420lys, 255glu]. The lack of competition between NCAs and agonists such as acetylcholine or carbamylcholine is clear in this mechanism since the combining sites are far apart.

The implications of the explanation for NCA action are far reaching. The proximity of the M2 helix and the ion channel element are established. The control mechanisms for the ion channel are seen to be more complicated than previously supposed. One might ask whether or not there is a natural occupant of the NCA binding site, knowing that there are both natural (histrionicotoxin) and synthetic (phencyclidine) occupants of the site. The ease with which the mechanisms for the labeling and NCA action can be formulated using the bilayer portion of the AChR model emphasizes the value of the model without requiring that it be completely correct.

Ethidium (ED) binds at the NCA site (995). Exciting the fluorescent bischoline (BCNI), a SubCh model (996) bound to AChR containing ED results in energy transfer and ED fluorescence (997). The efficiency of the process implies a separation of 48 Å (998, 999). A separation of this magnitude (ca. 50 Å) is implied by the binding locations assigned to agonists (exobilayer region) and noncompetitive antagonists (lower part of bilayer) in the present model.

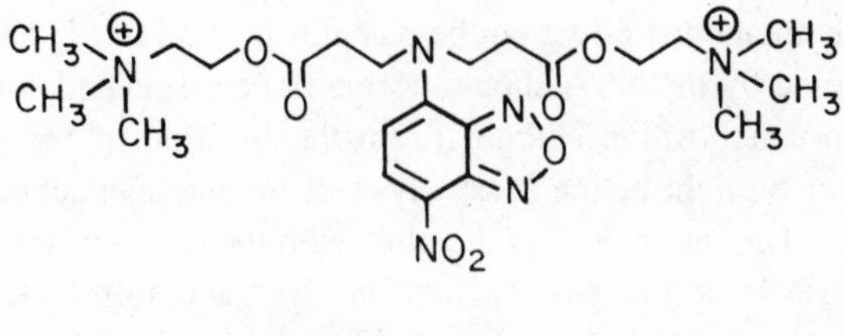

BCNI

H_2N NH_2 $N^{\oplus}$ C_2H_5

ED, Ethidium

Access to Noncompetitive Antagonist (NCA) Binding Site

The rate of NCA association with the NCA binding site is increased greatly in the presence of the agonist, carbamoylcholine (1000). The opening of the exobilayer flower in our mechanism (to the ''open channel state'') would allow NCAs greater access to the bilayer region of the AChR. At the level of the bilayer, the hydrophobic portion of the NCA could interact with groups in the region between the ion channel element and M2. The NCA would then bind to an appropriate negative charge on M2.

Arrangement of Bilayer Helices

The differential labeling of cys-424 in *Torpedo marmorata* as compared with *Torpedo californica* by a hydrophobic photoactivated probe (1000) suggested that M4 (the last hydrophobic helix) is exposed to the bilayer. Given the lengths of the peptides connecting the bilayer helices (M1–M2 short, M2–M3 short, M3–ICE long, and ICE–M4 short, ICE = ion channel element), and the proximity of M2 to ICE (see above), an arrangement for the α-subunit bilayer helices can be formulated as shown in Scheme 6.4.

Arrangement of α-subunit bilayer helices

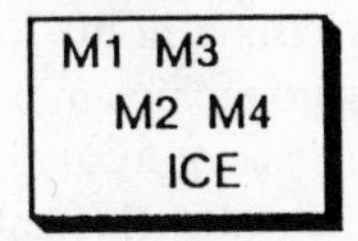

Scheme 6.4. Geometric relationships between bilayer helices.

Given the considerable sequence homology among the different subunits, it is reasonable to suppose that the other subunits are arranged in the same fashion. An overall structure for the bilayer portion of the AChR (Fig. 6.28) reveals that the M2 and M4 helices of adjacent subunits should be close to one

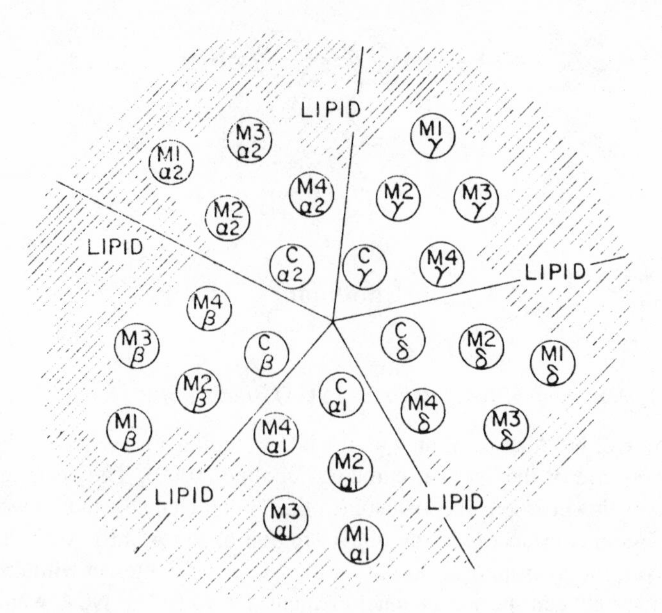

Fig. 6.28. Arrangement of bilayer helices in nicotinic acetylcholine receptors. The placement of the helices is based on the central position of the ion channel elements (''C'') and the pentagonal symmetry of the nAChR. The ion channel is in the center. The M2 helix in all subunits (except γ) is labeled by appropriate noncompetitive antagonists (NCA). The effect of the NCA on blocking the channel suggests that the M2 helix is adjacent to the ion channel element. The short connections between M1 and M2 and between M2 and M3 dictates the placement of these helices. One NCA appears to label M1 in the α-subunit. A long polypeptide, postulated to be the cytoplasmic segment of each subunit, connects M3 and C. A short peptide connects C and M4, defining its location. Fitting together the helices gave a beautiful symmetry to the bilayer helices, with M2 and M4 in different subunits postulated to be near one another. (Reprinted, with permission, from reference 737.)

another. Examination of the assigned sequences suggests that there are many polar interactions favoring the arrangement shown, with a pattern that is repeated over the five subunits in considerable detail (Fig. 6.29). The ion-pair interaction should be reasonably strong in the low dielectric medium established by the large number of hydrophobic amino acid side chains in this region, although a more careful analysis of local dielectric properties should be carried out (708, 709, 1001). All channel elements have a negatively charged group which can form an ion-pair with a positively charged group on M4. In each subunit there may be two pairs of polar interactions between M3 and M2. The negatively charged groups in the M2 helix of the α-, β-, and δ-subunits are the proposed binding sites for the NCAs. The effect of the NCAs on the channel may therefore be indirect. The binding of N-phenylmaleimide to the cysteines in the γ-ion channel element blocks ion flux without affecting ligand

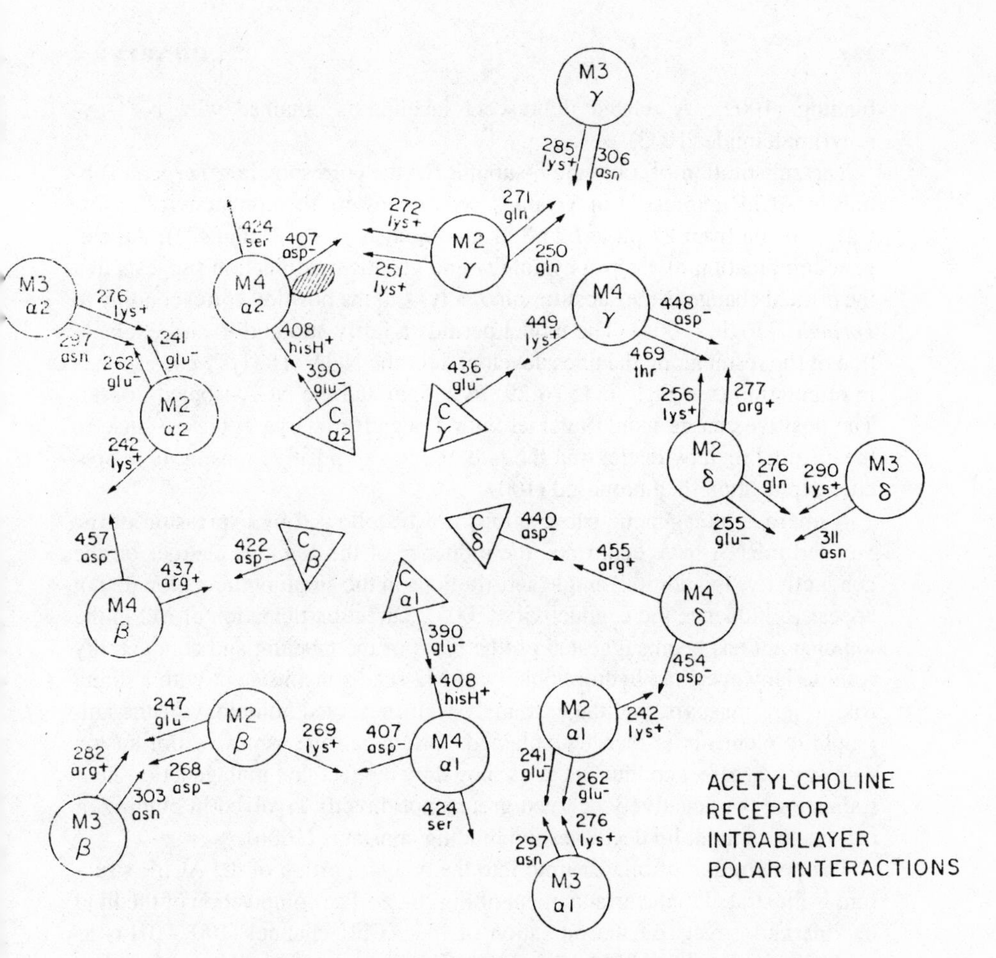

Fig. 6.29. Intrabilayer polar interactions in nicotinic acetylcholine receptors. The presence of charged groups in nominally hydrophobic and nonpolar portions of the structure leads to the idea that ion-pairs help to link together the various subunits. Considerable homology is apparent, with both ion-pairs and hydrogen-bonds participating in the stabilization of the structure. Each channel element forms an ion-pair with the M4 of the same subunit. The placement of the groups between C and M4 is schematic, since the exact angle of projection from the α-helices cannot be specified. The direction shown for ser 424 is consistent with the labeling of the homologous cys 424 in *Torpedo marmorata* by an agent delivered from the lipid region of the bilayer. (Reprinted, with permission, from reference 737.)

binding (1002). A related fluorescent labeling is obtained with N-(1-pyrenyl)maleimide (1003).

The substitution of a bovine δ-subunit for the corresponding *Torpedo* subunit in AChR expressed in *Xenopus* öocytes lowers the conductivity in low Ca^{++} media from 87 pS to 62 pS (1 picoSiemen = 10^{-12} $ohms^{-1}$). An elegant combination of electrophysiology and genetic modification suggests that the critical change is the substitution of a lys^{+} at the position corresponding to *Torpedo* 276gln (1004). The model permits a fairly straightforward explanation of the result along the lines developed for the NCA. The lys^{+} corresponds in orientation (see Figs. 6.15, 6.29) to 255glu and the NCA-labeled 262ser. The positive charge would interact with a negative charge at the entrance to the channel thereby decreasing the rate of entry of positive ions, a more specific explanation than proposed (1004).

Numerous mutagenetic substitutions in M2 followed by expression of the mutated mRNA have confirmed the influence of the negative charges on the conductivity. In addition, single substitutions in the amphiphilic region do not appear to influence the conductivity (1005). Direct participation of M2 in the ion channel has been suggested on the basis of the labeling and conductivity results. However, the hydrophobicity of M2 seems inconsistent with a direct role in ion transport, and the extremely well-preserved homology of the amphiphilic elements is then unexplained. Groups at one remove from an ion pathway can affect conductance, as shown by the fact that mutagenetic ''neutralization'' of negatively charged groups not directly involved in binding of cations to calmodulin decreases ion binding constants (1006).

The penetration of bilayer lipid into the bilayer portion of the AChR structure is illustrated in the arrangement of Fig. 6.28. The composition of the lipid has marked effects on the operation of the AChR channel (1007–1014) as might be expected for this model. The AChR contains strongly bound cholesterol as well as closely associated phospholipid.

A more dramatic image of the acetylcholine receptor is presented in the schematic drawing of Fig. 6.30. The relationship of the exobilayer strands with binding-site groups and 192–193 disulfide link to the α-subunit bilayer helices and the associated cytoplasmic sequence are shown.

Ion Pathway through Exobilayer Cup

There are locally excess exobilayer carboxylate groups in our model (upper portion, γ76, 89, 93, 113, and lower portion, δ171, 180, 182, 189). These may be the sites of Ca^{++} binding in the closed form (1015). Terbium(III) distributions may parallel that of Ca^{++}; 18 Tb^{+++} near entrance, 11 Tb^{+++} near exit, and 4–5 Tb^{+++} in the channel (1016) must be associated with many of the available—rather than the net—negative charges. On activation, the cup

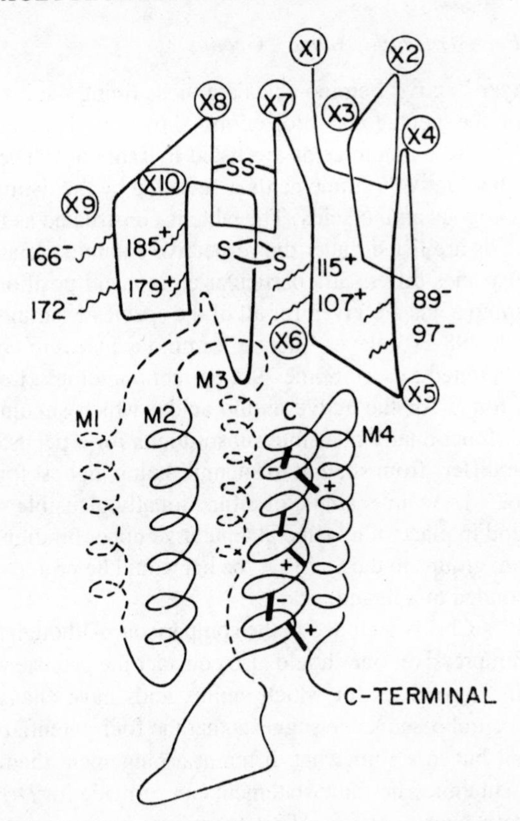

Fig. 6.30. Relationship of exobilayer strands (X1–X10, X11 with cys-cys-SS bond not marked) to the bilayer helices (M1–M3, C = channel, M4) and the cytoplasmic segments in an α-subunit. Binding-site groups are named, and the charges on the ion channel element indicated. (Reprinted, with permission, from reference 737.)

opens and the Ca^{++} is released. The participation of the exobilayer portions of the β- and γ-subunits in controlling ion flow is suggested by the inhibition of single channel activity by antibodies specific to those subunits (1017).

Tests of Model

The model may be probed through (1) measures of evolutionary persistence of exobilayer and ion channel "active" groups; (2) effects of genetically engineered deletions and mutations; (3) response to agonists and antagonists; (4) behavior of reduced receptors; (5) possibility of producing rapid current fluctuations.

Evolutionary Persistence of "Active" Groups

The exobilayer "active" amino acids are those lining the inside of the cup as specified in the pentagonal projections shown in Figures 6.26(a) and 6.26(b). The "active" amino acids are listed in Table 6.3. The evolutionary persistence of the "active" amino acids is revealed by the listing in Table 6.3 of the nonhomologous amino acids. The table is constructed as follows. "Active" amino acids are listed under the respective subunits. If any subunit sequence of one species differs at a particular amino acid position from that of another, the amino acids are given for all of the species examined in the order specified. In all, 298 exobilayer "active" amino acids were considered (the α-subunit was counted twice). Some 79% perfect homology (i.e., completely identical) was found for the active amino acids, which should be taken together with 7% functionally plausible substitutions for a persistence of 86%. The percentage differs from subunit to subunit, being highest for α (94%) and lowest for β (67%). As an example of a functionally plausible substitution, a histidine is found in place of a lysine at a putative upper binding site. At pH 7 with carboxylate groups in the vicinity, the his would be protonated and could be hydrogen-bonded to a ligand.

Rat "neural" AChR is included in the comparison. Although the homology noted above is impressive, one should also consider the persistence of putative binding sites in those cases for which amino acids have changed. A closer look at the rat neural α sequence suggests that the four potential ACh binding sites are present but in a somewhat different arrangement than proposed for the *Torpedo* α-subunit. The same statement can be made for *Drosophila* α on the basis of the sequence (1018). That functional homology does not equal sequence homology has been established by electrophysiological measurements on hybrid AChR receptors derived from subunits of more than one species (1019, 1020) or the fetal and adult forms of AChR of one species (1021). The binding of subunit specific monoclonal antibodies did not change during development over the time of receptor clustering and a change in ion channel properties; the subunit composition of mammalian muscle AChR apparently does not change over this period (1022). Although γ-subunit-specific antibodies might not crossreact with the ϵ-subunit-specific antibodies, the öocyte measurements (1021) show a change in ion channel behavior. There are differences between the γ- and ϵ-M2 and ion channel element composition (910) which might account for a difference in channel conductance.

The channel active amino acids (total 182) are compared for all subunits in Table 6.4. The homologies with the channel amino acids of Torpedo are extensive and almost complete for the seven α-subunits (92% identity, 96% including functionally plausible substitutions [f.p.s.]) and three δ-subunits (76% identity, 100% including f.p.s.), but less so for the β- (71% including f.p.s.) and γ- (81% including f.p.s.) subunits.

The evolutionary persistence of the 480 exobilayer and channel "active"

Table 6.3. Homologies among Exobilayer "Active" Amino Acids

				Homology	Inexact Species Homologies[b]			
				Sequence	α	β	γ[c]	δ
α	β	γ	δ	Number[a]	1 2 3 4 5 6 7	1 4 5	1 3 4 5 6 8	1 3 4
89	89	89	91	92			DDDDDE	DEE
		93		96				
97	97	97	99	100				
99				102				
	101	101		104		EDD	EEEEEG	
107				110				
111			113	114	DDEDDQK			RYY
		113		116			DDDDDG	
115				118	KKKHHHE			
		163	165	166				DEE
	163			169		DDS		
	165			171		KEE		
	167		171	173		EEQ		DNS
	169			175		ERE		
	171	169		177		KEQ	EEEEES	
166			180	182	DDDDDDN			
			182	184				
172	182		186	188		ENE		
		183		191				
			191	193				
179				194	KKKKKKI			
		187		195			RKQRQD	
	191	189	195	197		KPK	RRRRRC	KRR
		192		200			KRKKKV	
185	195		199	201		KLL		
		202		211			DDEEQG	
	201			213		DQD		
	205			218		EKE		
200				219				

Source: This table is taken from reference 737.

Note: The "active" amino acids (those present on the inner side of the exobilayer cup [see text] and carrying a charge in the side chain) are listed for each subunit using the subunit numbering. Except as noted, the homologies established by those who reported the sequence are used.

[a] Numbering is given according to the homology alignment in reference 860.

[b] Species homologies were exact unless given in the table. The comparisons were made using the alignments of references 864 and 906–909. The species compared are α (*Torpedo californica* [1], *Torpedo marmorata* [2], *Gallus domesticus* [chicken, 3], *Mus musculus* [mouse, 4], *Bos taurus* [calf, 5] *Homo sapiens* [human, 6], and *Rattus rattus* [rat, 7] [neural]); β (*Torpedo californica*, calf and mouse); γ (*Torpedo californica*, chicken, calf, human, fetal calf [8], and mouse); and δ (*Torpedo californica*, chicken, and mouse). Single-letter codes are used for the amino acids. In the case of inexact homology, the amino acid present at the given position is given for all species in the order noted for each subunit.

[c] The calf ϵ-subunit (fetal γ-subunit) is reported in reference 910.

amino acids is 88% (including f.p.s.), higher than might have been expected from the overall homology. The persistence would be even higher if the persistence of functional homology could have been included in the estimates for functionally plausible substitutions. If homology among "active" amino acids is greater than expected, it follows that other amino acids will be less homologous. Additional data on other species for β- and δ-subunits as well as further analysis of functional arrangements may help in the search for "active" groups. The genetic homologies are the more striking in view of the different chromosomal locations found for the various subunits of the AChR in *Mus musculus domesticus* and *Mus spretus* (α, chromosome 17, β, chromosome 11, γ,δ, chromosome 1) (1023).

Immunological methods including electron microscopy have shown that the β- and δ-subunits terminate in the cytoplasm (1024, 1025). The γ360–377 sequence is located in the cytoplasm (1026–1028). A cytoplasmic location for the αC-terminus is suggested by antibody binding experiments using weak detergents (saponin or lithium 3,5-diiodosalicylate) to permeabilize the bilayer of the cell membrane (1026–1029).

Deletions and Mutations

A thorough and wide-ranging study of the AChR using site-directed mutagenesis of the α-subunit (1029) shows that all or most of the sequence between

Table 6.4. Homologies among Ion Channel "Active" Amino Acids

				Homology	Inexact Species Homologies[b]			
				Sequence	α	β[c]	γ	δ
α	β	γ	δ	Number[a]	1 2 3 4 5 6 7	1 4 5	1 3 4 5 6 8	1 3 4
373	402	414	420	437K		KRR	KRQQQR	
377	406	418	424D	441E	Q	EDS	EEDEED	
380	409	421N	427N	444K		KRS		
384	413	425K	431K	448E		EGR	KNRRCS	KKN
387	416E	428	434	451K	S	EEQ	KRRHNR	KRR
391	420	432D	438A	455E	V	EED	DDHHHA	AS S
394N	432D	435S	441	458E	E	DS D	S/ SNED	

Note: See Table 6.3, but this applies to channel "active" amino acids. Table is taken from reference 737.

[a] Numbering is given according to the homology alignment in reference 860.

[b] See footnote b, Table 6.3. Here, the slash (/) indicates that no match was made in the homology alignment.

[c] Careful examination of the β, γ, and δ mouse sequences suggested that the alignments given in reference 913 are not the most appropriate. Our selections are as follows (channel "active" amino acids): β, 395arg$^+$, 398asp$^-$, 402arg$^+$, 405gly, 409glu$^-$, 412glu$^-$, 416ser; γ, 411gln, 415asp$^-$, 418asn, 422arg$^+$, 425arg$^+$, 429his, 432ser; δ, 415lys$^+$, 419asp$^-$, 422asn, 426asn, 429arg$^+$, 433ser, 436glu$^-$.

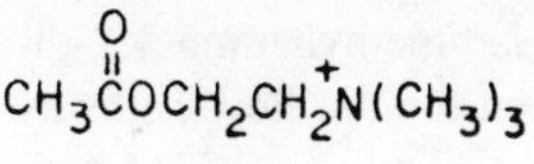

$CH_3\overset{O}{\overset{\|}{C}}OCH_2CH_2\overset{+}{N}(CH_3)_3$

1, ACh

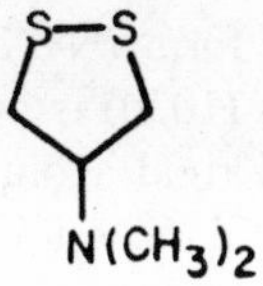

$N(CH_3)_2$

2, NTX

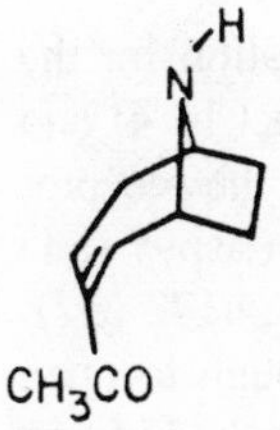

CH_3CO

3, Ana

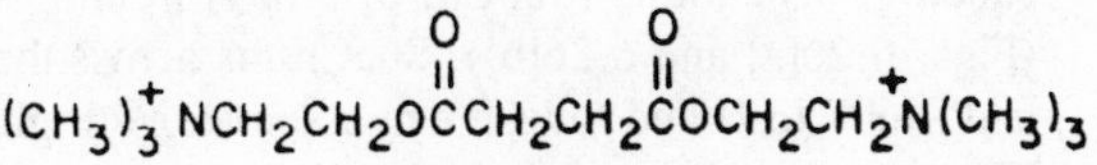

$(CH_3)_3\overset{+}{N}CH_2CH_2O\overset{O}{\overset{\|}{C}}CH_2CH_2\overset{O}{\overset{\|}{C}}OCH_2CH_2\overset{+}{N}(CH_3)_3$

4, SucCh

$(CH_3)_3\overset{+}{N}CH_2CH_2O\overset{O}{\overset{\|}{C}}(CH_2)_6\overset{O}{\overset{\|}{C}}OCH_2CH_2\overset{+}{N}(CH_3)_3$

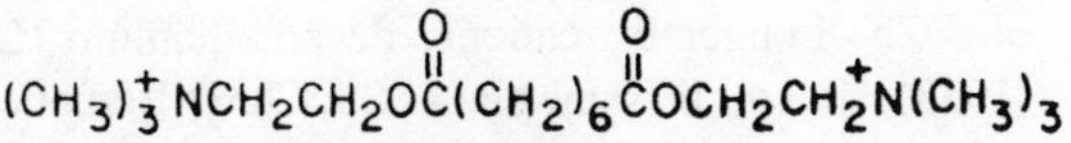

5, SubCh

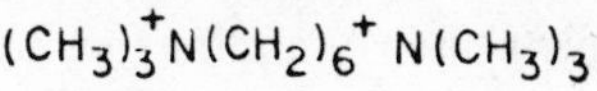

$(CH_3)_3\overset{+}{N}(CH_2)_6\overset{+}{N}(CH_3)_3$

6, Hex

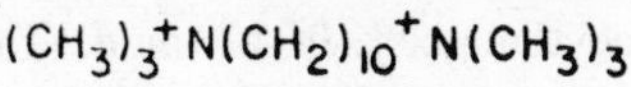

$(CH_3)_3\overset{+}{N}(CH_2)_{10}\overset{+}{N}(CH_3)_3$

7, Dec

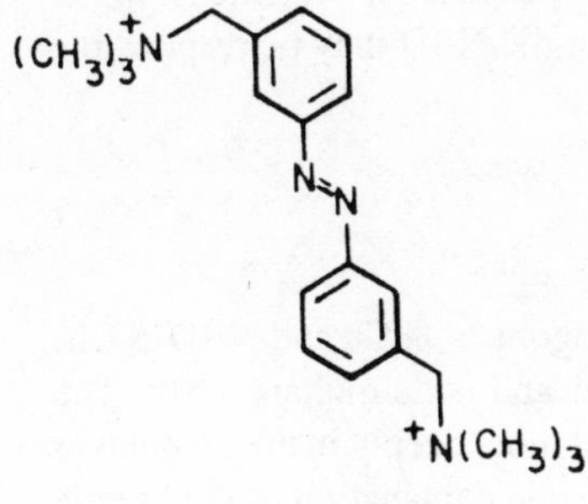

$(CH_3)_3\overset{+}{N}$

$\overset{+}{N}(CH_3)_3$

8, trans-bis-Q

NO_2

HN

$(CH_2)_5$

$\overset{+}{N}(CH_3)_3$

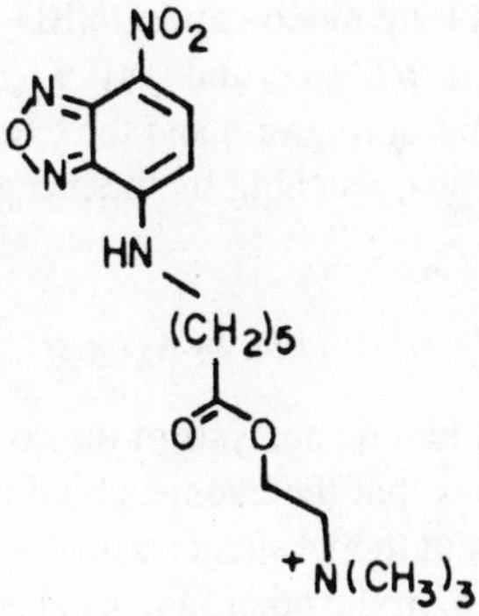

9, NBD-5-Ch

α376–389 is central to the response of the receptor to acetylcholine, but has only a small effect on α-bungarotoxin (α-BTX) binding. Thus, the ion channel element identified on functional grounds (705) is confirmed in location and physiological activity.

Agonists and Antagonists

Formulas for many of the compounds referred to in this section are illustrated, identified by number and abbreviation. Activation of the receptor requires two molecules of ACh (**1**). Our model has four binding sites but only two can be occupied simultaneously. Binding of ACh in the fully opened form is shown in Figures 6.26(a) and 6.26(b).

The protonated form ($NTXH^+$) of nereistoxin (2-dimethylamino-1,2-dithiolane (NTX, **2**) (1030) is not large enough to open the ion flow pathway completely. Small rigid agonists, such as anatoxin-a (Ana, **3**) (1031) and, most likely, the rigid nicotinoid, pyrido[3,4-b]norhomotropane (1032, 1033), fit into the site almost as well as acetylcholine.

Perhaps the most striking success of the model is an explanation for the activity of long bis-cationic agonists. Succinoyl-bis-choline (SucCh, **4**) and suberoyl-bis-choline (SubCh, **5**) (1034) fit almost perfectly into the receptor (Figs. 6.26[a] and 6.26[b]). SucCh fits across the space between asp97 (α1) and asp97 (α2) and SubCh can associate with glu172 (α1) and glu172 (α2). The short bis-cation, hexamethonium (Hex, **6**), is not long enough to open the cup but binds well enough to interfere (as an antagonist) with the binding of ACh. Longer bis-cations, decamethonium (Dec, **7**) and *trans*-bis-Q (**8**) (1035), fit into the same site as SucCh. Different cells may respond differently to long bis-cationic molecules. SubCh and SucCh are agonists for human medulloblastoma cells, TE671 (model for CNS neurons), but antagonists for rat pheochromocytoma cell, PC12 (model for autonomic neurons), a difference which may be due to different AChR molecules or a differently modified AChR (1036).

A long mono-cation, NBD-5-choline (NBD-5-Ch, **9**) (1037), fits a receptor site, asp97 (α1) and lys179 (α2), quite well. A complex between the lysine ε-ammonium group and the excited-state anion of the NBD may be responsible for the quenching by the receptor.

Ordered Binding of Agonist

A kinetic analysis of the combination of the agonists, ACh and NBD-5-Ch, shows that the agonists bind one after the other and not simultaneously. The present model suggests that one agonist is bound more deeply in the exobilayer cup than the other, and has been given as a possible explanation of the kinetic results (1038, 1039). The ''trapping'' of an open-channel blocker, the bis-quaternary ion chlorisondamine (CID, **10**) (1040), may occur through strong binding at the ''deep'' site, with release being facilitated through binding of ACh at the ''upper'' site. It should be noted, however, that the bis-amine precursor of CID is an NCA (805), and that the slow irreversible inactivation observed (1040) might be related to demethylation, or even alkylation of the NCA site.

Behavior of Reduced Receptor

The dithiothreitol (DTT) reduced AChR has some, although diminished, activity toward agonists. The increased flexibility of strand X11 might make it more difficult for agonist ligands to activate the receptor and open the pas-

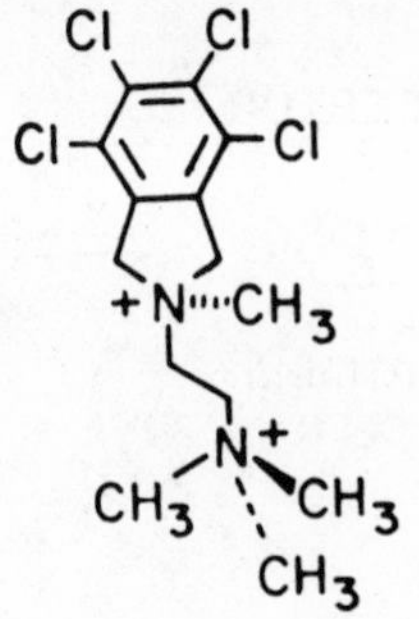

CID

10, Chlorisondamine

sage to cations. One antagonist (Hex) toward the normal AChR becomes an agonist with the reduced receptor.

Thiol-reactive agonists differ in effect according to length (919), bromoacetylcholine reacting and depolarizing the receptor, whereas the somewhat longer N-(4-trimethylammoniomethylphenyl)maleimide reacts as an affinity label (1000 times as rapidly as N-ethylmaleimide) and blocks the receptor. Reduced receptor alkylated with bromoacetylcholine exhibits longer channel open times, but exhibits desensitization (1041). The longer thiol-bound agonist must quickly assume a parallel conformation in which the cup is closed, whereas the shorter agonist opens the ion pathway to the channel in a perpendicular arrangement and then subsequently isomerizes to a parallel or bent (desensitized) conformation.

Current Fluctuations

The discovery of current fluctuations in the microsecond and millisecond time range (1042–1046) can be understood in terms of the conformational isomerizations of the many groups in the channel.

Desensitization and Resensitization

One of the striking properties of the AChR is the conversion to an inactive (''desensitized'') form after activation by agonist (916, 917) even for pure receptor in vesicles (1047–1050). In our model, desensitization occurs when agonist stabilizes the closed form of the cup. A summary of these steps is outlined in the scheme in Figure 6.31. A three-dimensional representation of the three states (resting, active, and desensitized) is shown in Figure 6.32. With respect to the exobilayer strand direction, parallel rather than perpendic-

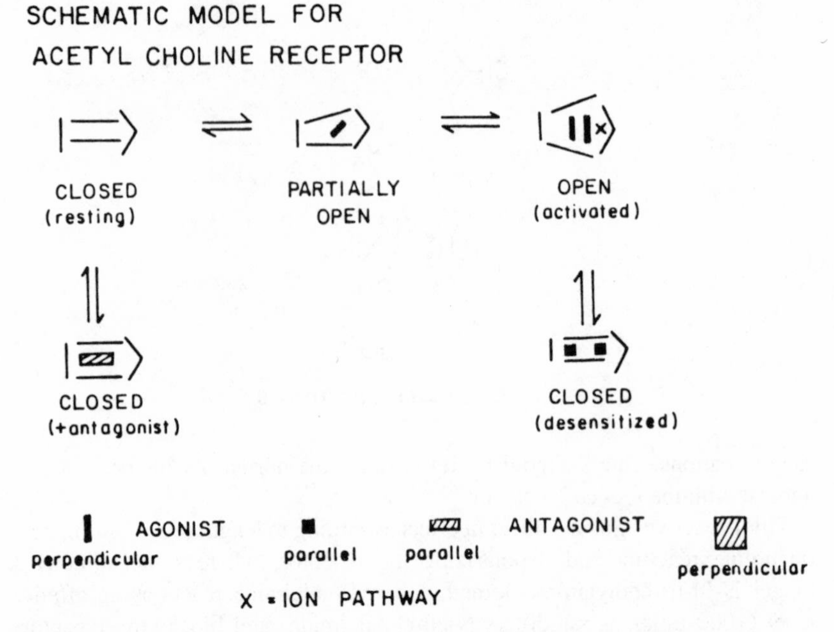

Fig. 6.31. Schematic summary of nicotinic acetylcholine receptor states. *Parallel* means parallel to the strands and perpendicular to the bilayer. The conformation of acetylcholine in the desensitized receptor is *bent*, with the methyl groups of the trimethylammonio group about 3 Å from the methyl of the acetate group (894), an arrangement close to that of the SGR rotated B1 conformation in Fig. 6.6. *Perpendicular* refers to a direction perpendicular to the strands and parallel to the bilayer. Agonists open the closed exobilayer cup of the receptor by binding in a *perpendicular* manner, then rearrange to a somewhat more stable state in which the molecules are now *parallel* to the strands and able to interact with an upper or lower binding site and an intermediate group such as 111asp. The exact geometry depends upon the length of the agonist. A *bent* conformation of the acetylcholine might also account for desensitization, the binding-site groups on one α-subunit replacing oppositely charged groups on two α-subunits. Both *parallel* and *bent* arrangements would allow the strands to come closer together, closing the path to the ion channel. Antagonists appear to be "thicker" molecules than agonists, and contain hydrophobic regions. Antagonists prefer the *parallel* direction but might also have some agonist activity if bound in the *perpendicular* direction. (Reprinted, with permission, from reference 737.)

ular binding should result in desensitization. Nuclear Overhauser effects in the nmr spectrum of acetylcholine have shown that the acetylcholine molecule in the desensitized receptor has a different conformation than in aqueous solution or in the crystal. The conformation of acetylcholine in the desensitized receptor is *bent*, with the methyl groups of the trimethylammonio group about 3 Å from the methyl of the acetate group (894), close to the SGR rotated B1 conformation in Figure 6.6. Thus, a bent conformation of the acetylcholine might also account for desensitization, the binding-site groups on one α-subunit replacing oppositely charged groups on two α-subunits. Although a molecular mechanics calculation suggests that the conformation is higher in energy by

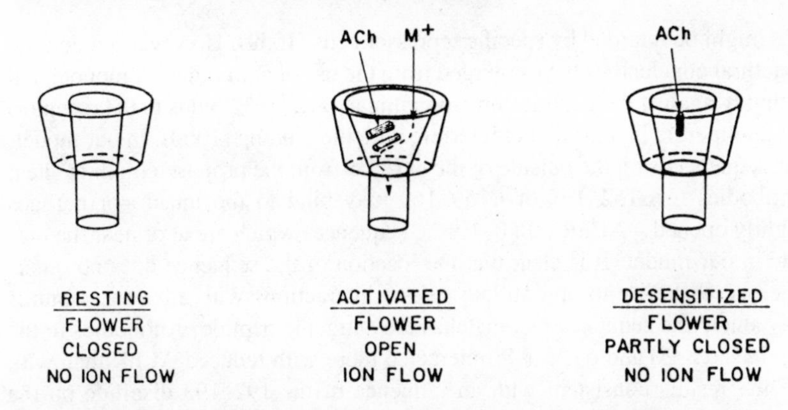

Fig. 6.32. Three-dimensional model (''flower''—usually a blossom, but can be read as something that flows) representing in the simplest possible way the operation of the nicotinic acetylcholine receptor. The resting state has an inner set of strands (inner portion of the cup) close enough to block ion influx. Entry of two acetylcholine molecules in an arrangement *perpendicular* to the strands forces open the inner cup, creating the activated state and revealing the ion pathway. After some time,the agonist (ACh) molecules rearrange to a *parallel* (or *bent*) position, and the cup partially closes, forming the desensitized state, with the ion pathway blocked. (Reprinted, with permission, from reference 737.)

ca. 7 kcal/mol than the extended equilibrium form, hydrogen-bonding and other interactions in the binding site could compensate for the energy disadvantage (894). The bent or parallel binding must be somewhat stronger than perpendicular binding (1051). Computer graphics indicate that 111asp might be a suitable group for binding to a positive charge in the ''parallel'' agonist, even for anatoxin-a (905, 1052, 1053). Resensitization could occur without activation by partial opening of the cup and dissociation of the agonist. A powerful result derived from the model is a straightforward explanation for the desensitization which succeeds activation by a wide variety of agonists. The model accounts for the lower rate of reduction of native AChR by dithiothreitol in the presence of desensitizing quantities of agonists. The α192–193 disulfide of the closed form of the unreduced AChR would be less accessible to the reducing agent. ''Thicker'' antagonists, while blocking access to ligands and failing to open the ion pathway through the cup, would not diminish the disulfide reactivity to the same degree, as observed (1054).

Antibodies

Extensive libraries of antibodies to the AChR have been prepared; the disease *myasthenia gravis* is an autoimmune disease due, in part, to the action of anti-AChR antibodies which either affect AChR response (916, 917, 1055) or cause accelerated internalization and degradation of AChR (1056–1058). The loss might be retarded by specific repressor cells (1059). However, no definite structural conclusions have emerged from the use of antibodies. A monoclonal antibody against the peptide corresponding to α127–132 binds to the receptor, but cholinergic ligands do not interfere with the binding (1060). In our model, this sequence is on the outside of the cup, far from the proposed binding sites. Antibodies to α152–167 or α159–169 may bind to the intact—or perhaps slightly opened—AChR (1060–1062), sequences which are at or near the outside in our model. It is clear that the location of the sequence does not guarantee accessibility to an antibody, since interactions with adjacent subunits may shield the sequence that matches the antigenic peptide. Antibodies to the peptides α1–20 and α126–143 interacted more with reduced AChR than with AChR, results consistent with an influence of the 192–193 disulfide on the conformation of the receptor (1063).

Myasthenic serum selectively blocks the ''slow'' (embryonic) AChR receptors which normally are transformed into the ''fast'' (adult) AChR in rat muscle (1064). Elucidation of the molecular basis for the change will be valuable in interpreting the dynamics of the AChR. Monoclonal antibodies against β- and γ-cytoplasmic segments interfere with channel function (1017). The β-antibody binds between β368–406 (1065), most likely between β368–397, using the cytoplasmic sequence assignment made in our model. Single-channel measurements on AChR in reconstituted planar lipid bilayers (1066, 1067) represent an experimental paradigm which allows antibodies to be added to either side of the receptor without encountering the problems, noted later, of binding in the presence of weak detergents.

Snake Neurotoxins

Peptides with sequences α173–204 (1068), α160–216 (1069), and α185–196 (1070, 1071) bind α-bungarotoxin (α-BTX); the corresponding sequences in AChR are found on the outside in our model. A snake neurotoxin, α-dendrotoxin (from *Dendroaspis viridis* [1072, 1073]), binds to purified *Torpedo californica* AChR with a stoichiometry of 4:1 and displaces α-BTX (1074). Some nicotinic AChRs do not bind α-BTX (1075–1077). Thus, the binding sites of snake neurotoxins may or may not overlap with ligand binding sites of AChR. The rate constants for the binding of snake neurotoxins to AChR vary considerably with the size and nature of the toxin, from 6.7×10^3–20×10^6 $M^{-1}sec^{-1}$ (1078–1080). A thorough review of structure-function relationships for postsynaptic neurotoxins action has appeared (1081).

Potential Artifacts in Structural Analysis

Maelicke and coworkers (1082) have found that particular antibodies to the *T.californica* α-subunit, including part of the putative amphiphilic sequence, are sparsely attached on the cytoplasmic side (ca. 16% or less) to AChR reconstituted into vesicles, as shown by gold labeling. Analysis of this finding (1083) suggested that the antibody might have combined with a minority form of the AChR or had even promoted the formation of such a form in the presence of the mild detergent. The latter is used to permeabilize the vesicles for the introduction of the antibody into the vesicles. The minority form would be a conformation of the AChR in which the ion channel element is not completely within the bilayer. It seems reasonable to suppose that the amphiphilic channel element is the last to be inserted into the bilayer during AChR assembly and the one most likely to be displaced from its position by detergent treatments which destabilize the bilayer or membrane proteins. Monoclonal antibodies are known which induce conformational changes in AChR vis-à-vis agonist binding sites (1084–1086), underlining the idea that an antibody may form a complex with a form of the receptor different from the native form.

Binding of antibodies to native proteins in the presence of detergents can lead to artifacts. The conclusion that certain sequences are cytoplasmic rather than extracellular as proposed in most models, including the one described in this book, is unwarranted. The antisera (i.e., polyclonal antibodies) to an α389–408 + tyr peptide bind only to the extent of 6–10% with detergent-solubilized receptor (1087). Additional treatments that permeabilize vesicles bearing the receptor will necessarily reveal portions of the receptor that were previously inaccessible. The cytoplasmic assignment of the 389–408 sequence leads to mass distributions (extracellular:bilayer:cytoplasmic ratio) at variance with electron micrographic analyses, and must be regarded with considerable skepticism, even by the author (1088).

The whole question of the internal motions of the receptor, including oscillations or librations of the membrane helices, deserves consideration for its relationship to receptor action, to the mechanism of insertion of the receptor into and through the bilayer, and to receptor degradation. The analysis of protein dynamics and function has been attempted for "small" proteins like myoglobin (1089).

Conclusions

A useful model for the nicotinic acetylcholine receptor has been devised, based on many different facts and theories. The most important sources for data and ideas were the sequence; the overall structure (electron microscopy); thiol labeling; hydrophobicity of segments; binding-site model and ion channel selection (single group rotation theory); genetic homology and evolutionary stability of "active" amino acids; ligand structures; plausible physical ideas about the operation of the receptor; and labeling by noncompetitive antagonists.

Experimental approaches to testing the proposed model include cross-linking putative proximate groups; preparation of antibodies to peptides; molecular graphics studies of the exobilayer region (905); structural determination by X-ray crystallography; and identification of labeled sites, especially those that have been cross-linked. Definitive identification of the specific cysteine (tentatively δ108cys [1090]) involved in AChR dimer formation in *Torpedo* could be useful in refining the model. Additional nicotinic AChRs may become available as a result of the successful expression of *T.californica* α-subunit in yeast (1091), although the correct orientation is not achieved until all subunits are present (1092). The cross-linking of proximate carboxylic acid and amino groups followed by enzymatic degradation to small peptides could be carried out in the manner described for lysozyme (1093, 1094). Chromatographic separation (HPLC) and mass spectrometric determination of peptide composition (1095) could provide enough information to evaluate the three-dimensional structure of the exobilayer portion of the receptor proposed in the present model. The present model is a useful basis for design of experiments, for theoretical analysis of biological channels, and for the design of agents designed to affect the operation of the acetylcholine receptor. The principles developed in the analysis of the nAChR should be applicable to other receptors and channels in the nervous system.

7 Molecular Models for Sodium Channels

Introduction to Channels

Ion channels are proteins that allow ions to pass through biological membranes, either under ligand control (nicotinic acetylcholine receptor, nAChR) or under voltage control, like the sodium channel. Voltage-gated channels are the primary active element in the transfer of depolarization within neurons to the axon terminal, where either direct depolarization of a successor cell occurs (electrical coupling) or neurotransmitters are released (chemical coupling). The channels are most prominent in the axon. Channels are named for the ions that are most easily transferred, and include sodium channels, potassium channels, and calcium channels. Chloride channels, either ligand-gated—like the $GABA_A$ receptor—or voltage gated, will be taken up in the next chapter.

Channels play a role in learning since modulation of channel behavior changes the characteristics of the signal being transferred (Chap. 9). A structural and dynamic model of the sodium channel can help in the analysis of how ionic currents influence the operation of neurons.

Overview of the Sodium Channel Model

Chemical logic (choice of hydrophobic and amphiphilic domains; electrostatic interactions), genetic homologies, and single group rotation (SGR) theory for channel operation are used to construct molecular models from the primary structures of eel and rat brain sodium channels. Both structural and dynamic aspects of the channels can be discussed in terms of (1) a detailed model for the bilayer portion and (2) amino acid sequences for an outer control element (OCE) (tentatively, the Hodgkin-Huxley m-gate) and an inner control element (ICE) (tentatively, the Hodgkin-Huxley h-gate). The factors treated include gating current, sensitivity to changes in membrane potential (threshold for action potential), channel opening, a binding site for sodium, selectivity for sodium over potassium, capacity for rapid sodium flow, sensitivity to batrachotoxin (BTX) and other toxins, and inactivation. Genetic homologies between ''active'' portions of the eel and the three rat brain sodium channels are largely consistent with the models. Similar mechanisms for the eel and rat channels are accompanied by differences in charge distributions. Identification of potentially acylatable intrabilayer thiol groups is made, since many membrane proteins appear to possess such groups. Mechanisms for both local anesthetic (LA) and general anesthetic action on the sodium channel are put forward. BTX binds to the LA site. Portions of the ICE are identified as possible analogues to local anesthetics. The model is consistent with the neurophysiological results obtained with "anti-C_1^+ peptide" antibody. Molecular graphics studies on the central portion of the eel sodium channel model are described.

Brief Background

In many cases, the sodium channel of the axon is the central element in the transfer of integrated information from the dendrites and cell body of neurons to the axon terminal. The transfer signal is an action potential propagated through the axon and arises from the operation of the sodium and potassium channels of the axon. A deep understanding of how channels produce action potentials (897) or the functioning of other components of the neuron (receptors and other biopolymers) cannot be achieved without knowledge about their molecular structure (1096, 1097).

The sodium channel is a glycoprotein, initially in a resting state, through which sodium enters a neuron down its concentration gradient after activation by a moderate depolarization of the normal membrane potential has been achieved. The flow of sodium reaches a maximum in less than a millisecond and is usually turned off (inactivated) within several milliseconds. The local depolarization is usually lost in a somewhat longer time by the operation of an outward flow of potassium ions through a potassium channel. The minor imbalance in the usual intracellular high K^+/low Na^+ ratio is corrected after a time by the operation of the metabolically driven "Na^+/K^+ pump" (Na^+/K^+ ATPase). Structural models for the ATPase might be designed on the basis of the sequence (1098,1099).

Sodium Channel Sequence

Numa and his coworkers (1100) have elucidated the primary structures of the sodium channel of the electric eel, *Electrophorus electricus*, and three sodium channels from rat brain (1101, 1102). The eel channel is an 1820-amino-acid sequence, calculated molecular weight 208,321. The α-subunits of the rat channels have molecular weights of 228,758 (I, 2009 amino acids), 227,840 (II, 2005 amino acids) and 221,375 (III, 1951 amino acids). The channel isolated from rat brain is accompanied by two subunits, noncovalently-bound β_1 and disulfide-bound β_2 (1103–1105). Removal of the β_1-subunit and reconstitution of the $\alpha\beta_2$ combination into phospholipid vesicles is associated with the loss of STX (saxitoxin) and LqTX (*Leiurus quinquestriatus*) binding as well as veratridine-mediated Na^+ influx (1106). However, a functional rat brain sodium channel can be expressed in *Xenopus* öocytes using only α-subunit messenger RNA (1107–1109). The β_1-subunit appears to stabilize the sodium channel in a vesicle but might not be necessary in all membranes. The inactivation rates of the channels generated in öocytes from high molecular weight RNA (α-channel only) are apparently lower than those from unfractionated RNA (1110).

The eel channel protein was isolated in purified form (1111, 1112), degraded with trypsin, and the resulting peptides separated and sequenced. Blot hybridization using cDNA from immunologically positive transformants with oligodesoxynucleotides corresponding to all possible variants of one of the tryptic peptides led to the isolation of a positive clone and eventually to cDNA containing the information for the channel protein. A cDNA library from rat brain was screened using probes from the eel cDNA (1101).

The DNA sequences were obtained by the method of Maxam and Gilbert (865). The Kyoto group has analyzed the eel sequences, finding genetic homologies between four sections of the protein. A minimum of 12 hydrophobic helices were identified and the four unusual positively charged sequences (and the possibility that these might be 3_{10}-helices) were noted along with the presence of unusual long sequences of negatively charged amino acids. A specific channel was not identified.

Acetylcholine Receptor Channel

Reiteration of some points made in Chapter 6 provides a context for the discussion of the sodium channel model. A functional group arrangement lys(1) . . . glu(4) . . . lys(8) . . . was suggested by single group rotation (SGR) theory as a basis for a plausible ion channel in the acetylcholine receptor (nAChR) (903). The predicted sequence was found (705) in the amino acid sequences elucidated by Numa and others (857–861) and was utilized together with other facts and chemical logic to construct a model for the nAChR (735–737). The ion channel sequences found by using SGR theory were amphiphilic and essentially the same as those identified on the basis of amphiphilic character (873, 904).

Current Approach

Beginning with the principle that a charged amphiphilic segment was needed to form an ion channel, a model for the sodium channel was constructed. Although the details of the sodium ion channel structure are not identical to those of the nAChR ion channel model, the underlying principle is similar (902). I will try to account for both structural and dynamic aspects of the channels, including gating current, sensitivity to changes in membrane potential (threshold potential), channel opening, a binding site for sodium, selectivity for sodium over potassium, capacity for rapid sodium flow, sensitivity to batrachotoxin (or other toxins), and inactivation. Genetic homologies between "active" portions of the eel and the three rat brain sodium channels support the models. Identification of potentially acylatable intrabilayer thiol groups is made, since many membrane proteins appear to possess such groups (1113, 1114). Mechanisms for both local anesthetic (LA) and general anesthetic action on the sodium channel are put forward, with a new and important role being assigned to the LA binding site.

Neurophysiological responses to antibodies against peptides selected on the basis of their relationship to the "active" portions of the channel are consistent with our model (1115, 1116). A satisfactory model provides a basis for the design of neuroimmunology, mutagenesis, and labeling experiments, as well as explanations for the behavior of the channel. Models by Catterall (1096), Greenblatt *et al.* (1117), and Guy (1118, 1119) for the sodium channel differ substantially from mine. Models should be approached with appropriate skepticism, and tested in all possible ways. Eventually, we require a molecular structure precise enough to allow us to design drugs and to understand all significant aspects of channel behavior.

Constructing the Model: Folding Proteins

General methods for constructing three-dimensional arrangements in the absence of X-ray crystallographic structures for proteins on the basis of a linear sequence are still unavailable. Analogies and homologies in sequence may be useful in predicting structural elements of proteins if there is a reference X-ray crystallographic structure (934, 935, 1097). In the absence of known structures, especially for membrane proteins, folding "rules" (what structures might be expected for a given sequence) must be used.

Since the currently available rules are problematic (Chap. 6) (737, 930, 932, 1120), a variety of information on the shape and folding of the target protein must be used. The hydrophobic portions of the sequence are usually easy to recognize, either by scanning the sequence or with the aid of a program that assigns "hydrophobicity" indices to individual side chains. Charged groups may be present in some hydrophobic helices, serving to stabilize the overall arrangement, as in the nAChR structural model (737).

Search for Ion Channel

The eel sodium channel amino acid sequence does not have an ion channel like that of the nAChR model. The positively charged sequences noted by Numa are amphiphilic, having many hydrophobic groups interspersed with the charged groups, and thus are candidates for ion channel elements. I placed the four sequences of this type in the bilayer and accepted the suggestion by Numa (1100) that the repeat period of three amino acids for these sequences could correspond to a 3_{10}-helix. The 3_{10}-helix occurs with less frequency than α-helices in proteins (867) but is common enough to be a reasonable choice (1121). Stretches of 3_{10}-helix with 20 a.a. have been found in a 30 kDa "natural insecticide'' from *Bacillus thuringiensis* (1122). The 3_{10}-helix has been characterized in the crystal structure of an octapeptide derived from α-aminoisobutyric acid (1123). After placing all the 3_{10}-helices in the bilayer, it is then necessary to balance the locally large number of positive charges (20^+) now within the bilayer. The only compact sources of a large number of negatively charged groups are the highly unusual neighboring sequences present in the middle of the sodium channel sequence. I took the logical but surprising step of inserting two hydrophilic, negatively charged segments (net 22^-) into the bilayer, arranged as 3_{10}-helices on the basis of length (18 amino acids) and as a match for the positively charged helices. The functional part of the channel thus contains six 3_{10}-helices. The ion channel elements, C_1^+, C_2^+, C_3^-, C_4^-, C_5^+, and C_6^+, are listed in Figure 7.1. An amphiphilic peptide with the sequence RVIRLARIARVLRLIRAAKGIR forms voltage-gated channels in lipid bilayers which are selective for Na^+ over Cl^- by 6:1, suggesting that the channel contains anions (1123). Competition between the channel (e.g., C_3^- and C_4^-) and medium anions for association with the positive charges in amphiphilic sequences might occur during sodium channel operation, but will not be further considered here.

The homology between the eel and rat (I/II/III) (+)-sequences is extraordinarily good (94.4–100%), counting both identity and functionally plausible substitutions [FPS, $(+) \rightarrow (+)$, $(-) \rightarrow (+)$, hydrophobic → hydrophobic] (1100). The eel and rat (I/II/III) (−)-sequences are more difficult to compare, but the extent of local homologies is at least equal to that of the overall channel homology. The overall mechanism (later) indicates that segment-by-segment homology may be less significant than overall functional homology, as has been noted for one type of nAChR (737).

Hydrophobic Bilayer Helices

A sufficiently long sequence of hydrophobic amino acids will be more stable within the hydrophobic phospholipid bilayer rather than the polar environments of the cytoplasm or cell exterior. The sequence will fold in a way that will minimize the interaction of the polar peptide links with the apolar milieu, with an α-helix a highly preferred choice. Such hydrophobic sequences are therefore presumed to form bilayer α-helices. Helix-end criteria have not been tested (878).

The genetic homology discovered by Numa between four sections of the protein, and confirmed by Greenblatt *et al.* (1117), is taken into account, some polar groups perhaps stabilizing hydrophobic helices (719). Twenty hydrophobic helices are proposed, a number greater than that originally suggested (902), with the helix length limited to 24 amino acids (ca. 36 Å) to fit the bilayer. A role for salt and/or polar bridges within the bilayer has been suggested for the nAChR; some charged groups are found in the bilayer region of rhodopsin (564). The helices are similar to, but not the same as, those of Numa (1100). Between zero and five of the somewhat hydrophilic amino acids ser,

Ion channel elements

210 arg$^+$	657 arg$^+$	1093 ala	1417 arg$^+$	910 asp$^-$	942 glu$^-$
211 thr	658 ser	1094 ile	1418 val	909 glu$^-$	943 glu$^-$
212 phe	659 leu	1095 lys$^+$	1419 ile	908 glu$^-$	944 glu$^-$
213 arg$^+$	660 arg$^+$	1096 asn	1420 arg$^+$	907 asp$^-$	945 glu$^-$
214 val	661 leu	1097 leu	1421 leu	906 ser	946 glu$^-$
215 leu	662 leu	1098 arg$^+$	1422 ala	905 ser	947 glu$^-$
216 arg$^+$	663 arg$^+$	1099 thr	1423 arg$^+$	904 asp$^-$	948 glu$^-$
217 ala	664 ile	1100 ile	1424 ile	903 val	949 glu$^-$
218 leu	665 phe	1101 arg$^+$	1425 ala	902 leu	950 glu$^-$
219 lys$^+$	666 lys$^+$	1102 ala	1426 arg$^+$	901 gly	951 pro
220 thr	667 leu	1103 leu	1427 val	900 glu$^-$	952 glu$^-$
221 ile	668 ala	1104 arg$^+$	1428 leu	899 glu$^-$	953 glu$^-$
222 thr	669 lys$^+$	1105 pro	1429 arg$^+$	898 glu$^-$	954 leu
223 ile	670 ser	1106 leu	1430 leu	897 ile	955 glu$^-$
224 phe	671 trp	1107 arg$^+$	1431 ile	896 glu$^-$	956 ser
225 pro	672 pro	1108 ala	1432 arg$^+$	895 ser	957 lys$^+$
226 gly	673 thr	1109 leu	1433 ala	894 glu$^-$	958 asp$^-$
227 leu	674 leu	1110 ser	1434 ala	893 gly	959 pro
$^1C^+$	$^2C^+$	$^5C^+$	$^6C^+$	$^3C^-$	$^4C^-$

Fig. 7.1. Eel electroplax sodium ion channel elements.

thr, and cys are present in the bilayer helices (average 2.3/helix). The sequences for the hydrophobic helices are shown in Figures 7.2(a) and 7.2(b). The total number of bilayer helices (ion channel elements plus hydrophobic helices) in our model is 26. In the nAChR model, 25 bilayer helices have been proposed for 2333 amino acids.

There is substantial genetic homology between the eel and rat (I/II/III) hydrophobic segments, with the sum of identity (I) and FPS amino acids being for most helices between 83 and 100%. The average I-homology is ca. 62% (1101), rising to 87% between the rat I/II channels. A rough count of FPS in the amino acid sequences of the rat with respect to the eel suggests that the I + FPS homology for the eel/rat comparison might be 74%.

Arrangement of Ion Channel Elements

The constraints on the local arrangements are clear once the ion channel elements have been chosen. The (+)-elements are amphiphilic and therefore must be in contact with the more hydrophobic bilayer helices of the protein. The geometry of the (+)-elements is defined by the spacing of the (+)-charges and the proposition that all should be pointed in the same directions. A 3_{10}-helix satisfies this criterion. These criteria imply that the hydrophilic (−)-elements must lie within a region surrounded by (+)-elements. To match the charge spacing of the (+)-helices, it is presumed that the (−)-elements are also 3_{10}-helices.

A square arrangement of the amphiphilic (+)-helices minimizes (+)-(+) repulsion and yields an internal space just large enough to admit the completely polar (−)-helices. The interaction of the the (+)-helices with the (−)-helices is maximized if the charged (mostly arginine) groups are placed between the pairs of (−) groups. The 3_{10}-helices are oriented (''outer'' portion) so that the least charge faces in the direction of the hydrophobic helices. The amino acids located on the outer and inner sides (A and B) of the (−)-helices as well as those on the inner side of the (+)-helices (the side oriented toward the center) are shown in Figure 7.3.

To minimize the number of charges in the region facing the hydrophobic portion of the channel protein, the inner sets of (−) groups are apposed to one another. The charge distributions for this arrangement are shown level by level on pentagons in Figure 7.3, a presentation that requires a word of explanation. The 3_{10}-helices (18 amino acids to span the bilayer) have been divided into 6 ''layers.'' The charges in each ''layer'' have been projected onto a plane. Since there are 10 charge-carrying points per ''layer,'' two apposed pentagons are a natural way to present the charge distribution in each layer. The average net charge on any side of the combination through the bilayer ranges from −2 to +1. A significant feature of the arrangement is that the net charge at a given level of the channel changes from negative (level 1) at the top of the bilayer to neutral at all other levels, the net charges being −2, 0, 0, 0, 0, 0.

[a]

H1	H2	H3	H4	H5	H6	H7	H8	H9	H10
141 ile	146 ala	200 phe	267 leu	379 met	584 ser	595 gln	646 gly	710 phe	767 leu
140 thr	147 trp	199 glu⁻	266 asn	380 val	583 met	596 ser	645 leu	709 leu	768 ala
139 met	148 ser	198 thr	265 gly	381 phe	582 phe	597 leu	644 glu⁻	708 gln	769 val
138 phe	149 lys⁺	197 ile	264 met	382 phe	581 leu	598 leu	643 leu	707 phe	770 tyr
137 ile	150 ile	196 tyr	263 phe	383 ile	580 thr	599 ser	642 leu	706 gly	771 met
136 cys	151 val	195 thr	262 leu	384 met	579 asn	600 ala	641 ser	705 val	772 met
135 asn	152 glu⁻	194 met	261 gln	385 val	578 leu	601 gly	640 leu	704 leu	773 val
134 ser	153 tyr	193 thr	260 met	386 ile	577 ile	602 asn	639 ser	703 ala	774 ile
133 phe	154 thr	192 val	259 gly	387 phe	576 ile	603 leu	638 val	702 phe	775 ile
132 ile	155 phe	191 val	258 ala	388 leu	575 cys	604 val	637 ile	701 ile	776 ile
131 thr	156 thr	190 ser	257 leu	389 gly	574 leu	605 phe	636 ile	700 phe	777 gly
130 phe	157 gly	189 phe	256 thr	390 ser	573 thr	606 thr	635 ser	699 val	778 asn
129 met	158 ile	188 asp⁻	255 phe	391 phe	572 ile	607 thr	634 asp⁻	698 ile	779 leu
128 ile	159 tyr	187 leu	254 val	392 tyr	571 phe	608 ile	633 phe	697 ile	780 val
127 phe	160 thr	186 trp	253 ala	393 leu	570 leu	609 phe	632 ile	696 ala	781 met
126 phe	161 phe	185 asn	252 leu	394 ile	569 asp⁻	610 ala	631 asn	695 leu	782 leu
125 asn	162 glu⁻	184 trp	251 ser	395 asn	568 thr	611 ala	630 trp	694 val	783 asn
124 phe	163 val	183 pro	250 phe	396 leu	567 phe	612 glu⁻	629 thr	693 ile	784 leu
123 ala	164 ile	182 asp⁻	249 val	397 ile	566 pro	613 met	628 gln	692 thr	785 phe
122 ser	165 val	181 arg⁺	248 thr	398 leu	565 asp⁻	614 val	627 gln	691 leu	786 leu
121 asn	166 lys⁺	180 leu	247 leu	399 ala	564 met	615 leu	626 phe	690 asn	787 ala
120 val	167 val	179 phe	246 ile	400 val	563 met	616 lys⁺	625 tyr	689 gly	788 leu
119 phe	168 leu	178 thr	245 val	401 val	562 val	617 ile	624 tyr	688 leu	789 leu
118 val	169 ser	177 phe	244 val	402 ala	561 phe	618 ile	623 tyr	687 ala	790 leu

[b]

H11	H12	H13	H14	H15	H16	H17	H18	H19	H20
1024 ile	1033 ile	1088 tyr	1151 phe	1238 val	1341 val	1353 ile	1410 phe	1463 phe	1544 ile
1023 tyr	1034 leu	1087 gly	1150 leu	1239 tyr	1340 met	1354 leu	1409 tyr	1462 met	1545 thr
1022 ile	1035 glu$^-$	1086 leu	1149 asn	1240 met	1339 met	1355 ser	1408 lys$^+$	1461 ile	1546 phe
1021 asp$^-$	1036 tyr	1085 leu	1148 val	1241 tyr	1338 ala	1356 gln	1407 glu$^-$	1460 leu	1547 phe
1020 glu$^-$	1037 ala	1084 ser	1147 gly	1242 leu	1337 val	1357 ile	1406 ile	1459 phe	1548 cys
1019 phe	1038 asp$^-$	1083 ser	1146 met	1243 tyr	1336 met	1358 asn	1405 ile	1458 leu	1549 ser
1018 ala	1039 lys$^+$	1082 thr	1145 ile	1244 phe	1335 asn	1359 val	1404 asp$^-$	1457 leu	1550 tyr
1017 leu	1040 val	1081 ile	1144 ser	1245 val	1334 ile	1360 ile	1403 ser	1456 leu	1551 ile
1016 val	1041 phe	1080 gly	1143 phe	1246 ile	1333 cys	1361 phe	1402 leu	1455 gly	1552 ile
1015 gly	1042 thr	1079 met	1142 ile	1247 phe	1332 ile	1362 val	1401 leu	1454 ile	1553 leu
1014 ser	1043 tyr	1078 ile	1141 leu	1248 ile	1331 leu	1363 ile	1400 leu	1453 asn	1554 ser
1013 ser	1044 val	1077 ser	1140 trp	1249 val	1330 ala	1364 ile	1399 gly	1452 phe	1555 phe
1012 leu	1045 phe	1076 ala	1139 phe	1250 phe	1329 met	1365 phe	1398 ile	1451 leu	1556 leu
1011 leu	1046 ile	1075 gly	1138 met	1251 gly	1328 ile	1366 thr	1397 ile	1450 ala	1557 val
1010 ile	1047 val	1074 val	1137 leu	1252 ala	1327 phe	1367 val	1396 ser	1449 pro	1558 val
1009 met	1048 glu$^-$	1073 ile	1136 cys	1253 phe	1326 ile	1368 glu$^-$	1395 ile	1448 leu	1559 val
1008 phe	1049 met	1072 val	1135 val	1254 phe	1325 asp$^-$	1369 cys	1394 val	1447 ser	1560 asn
1007 ile	1050 leu	1071 phe	1134 leu	1255 thr	1324 thr	1370 leu	1393 val	1446 met	1561 met
1006 ile	1051 leu	1070 asp$^-$	1133 leu	1256 leu	1323 phe	1371 leu	1392 val	1445 met	1562 tyr
1005 phe	1052 lys$^+$	1069 leu	1132 val	1257 asn	1322 pro	1372 lys$^+$	1391 ala	1444 leu	1563 ile
1004 thr	1053 trp	1068 trp	1131 asn	1258 leu	1321 gln	1373 leu	1390 phe	1443 ala	1564 ala
1003 glu$^-$	1054 val	1067 cys	1130 met	1259 phe	1320 thr	1374 leu	1389 asp	1442 phe	1565 ile
1002 phe	1055 ala	1066 trp	1129 ile	1260 ile	1319 val	1375 ala	1388 phe	1441 leu	1566 ile
1001 tyr	1056 tyr	1065 ala	1128 ser	1261 gly	1318 ile	1376 leu	1387 val	1440 leu	1567 leu

Fig. 7.2. (a) Hydrophobic helices H1–H10 selected from the eel sodium ion channel sequence. (b) Hydrophobic helices H11–H20 selected from the eel sodium ion channel sequence.

$^{1}C^{+}$	$^{4}C^{-}$ (outer)	$^{5}C^{+}$
210 arg^{+}	942 glu^{-}	1095 lys^{+}
213 arg^{+}	945 glu^{-}	1098 arg^{+}
216 arg^{+}	948 glu^{-}	1101 arg^{+}
219 lys^{+}	951 pro	1104 arg^{+}
222 thr	954 leu	1107 arg^{+}
225 pro	957 lys^{+}	1110 ser

944 glu^{-}	943 glu^{-}
947 glu^{-}	946 glu^{-}
950 glu^{-}	949 glu^{-}
953 glu^{-}	952 glu^{-}
956 ser	955 glu^{-}
959 pro	958 asp^{-}
$^{4}C^{-}$ (innerA)	$^{4}C^{-}$ (innerB)

$^{3}C^{-}$ (innerA)	$^{3}C^{-}$ (innerB)
908 glu^{-}	910 asp^{-}
905 ser	907 asp^{-}
902 leu	904 asp^{-}
899 glu^{-}	901 gly
896 glu^{-}	898 glu^{-}
893 gly	895 ser

657 arg^{+}	909 glu^{-}	1417 arg^{+}
660 arg^{+}	906 ser	1420 arg^{+}
663 arg^{+}	903 val	1423 arg^{+}
666 lys^{+}	900 glu^{-}	1426 arg^{+}
669 lys^{+}	897 ile	1429 arg^{+}
672 pro	894 glu^{-}	1432 arg^{+}
$^{2}C^{+}$	$^{3}C^{-}$ (outer)	$^{6}C^{+}$

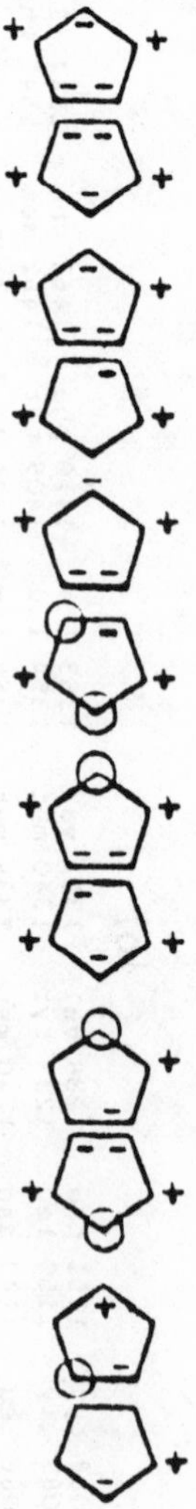

1 2 3 4 5 6

Fig. 7.3. Ion channel active amino acids and charge distributions. The inner amino acids of the channel line the pathway that the sodium ion traverses through the bilayer. (*A* and *B* are arbitrary designations of the apposed sodium ligand groups.) The charges at each level (one turn of the 3_{10}-helix) have been projected onto a plane, shown in pentagons at the right of the amino acid lists (see text).

The charge distributions in rat channel I are 0, +4, +1, +1, +2, 0, and −1, and in rat channel II, 0, +3, 0, +3, −1 and 0. Clearly, the net positive charges on the channel require that charges in the regions around the channel helices must contribute to the operation of the channel. A search was therefore made for mutations within the eel channel protein that replaced neutral amino acids with charged amino acids. There are 6 such mutations involving both rat I and rat II within or right next to segments assigned as bilayer helices. Rat III mutations correspond. [Eel numbering 121 N → HHH (H1), 146 A → DDD (H2), 628 Q → EEE (H8), 1075 G → DDD (H13), and 1321 Q → RKR.] Taking into account the probable disposition of the helices, and the change 1025 W → DEE which occurs immediately under H11, there is an increase of between 2 and 4 negative charges in the rat I, II and III channels within the bilayer near the + charges of the channel. The overall charge distribution within the protein is curious regardless of model. According to the model, the extracellular, bilayer, and intracellular charges are, respectively, in the eel channel, −19, −19, −12, in the rat I channel, −17, −7, +18, and, in the rat II channel, −22, −10, and +30.

Sodium Binding Site

I applied single group rotation (SGR) to the topmost set of groups on the channel elements and found that, after SGR of the (−) groups of the inner portion, a plausible binding site for sodium was created (Fig. 7.4). The site might discriminate (13/1) in favor of the larger ion (Na^+, hydrated radius 2.76 Å) over the smaller ion (K^+, hydrated radius 2.32 Å). The bilayer helices may be arranged around the central channel elements, with the four (+) helices around the (−) helices according to (a) connectivity, with helices separated by short peptide links being near one another, and (b) charge, with helices bearing negative charge being placed near the (+) helices. A schematic overall arrangement of the bilayer helices of the sodium channel is shown in Figure 7.5. A compact representation showing the sequence numbers for the beginning and end of each bilayer helix is shown in Figure 7.6. Seven of the potential N-glycosylation sites [asn-X-ser(thr)] (X not pro) at 205, 278, 288, 317, 591, 1160, and 1174 are outside the bilayer. The molecular weight of the glycosyl portion of the channel is ca. 60,800 (1100, 1124, 1125).

Overall Folding Pattern

The N- and C-termini of the sodium channel are almost certainly intracellular, based on (a) intracellular side localization of the C-terminus by gold-labeled antibody to the eel 1783–1794 region (1126); (b) no leader peptide in cDNA sequence (1101); and (c) probable fourfold symmetry of protein (1096, 1100, 1117).

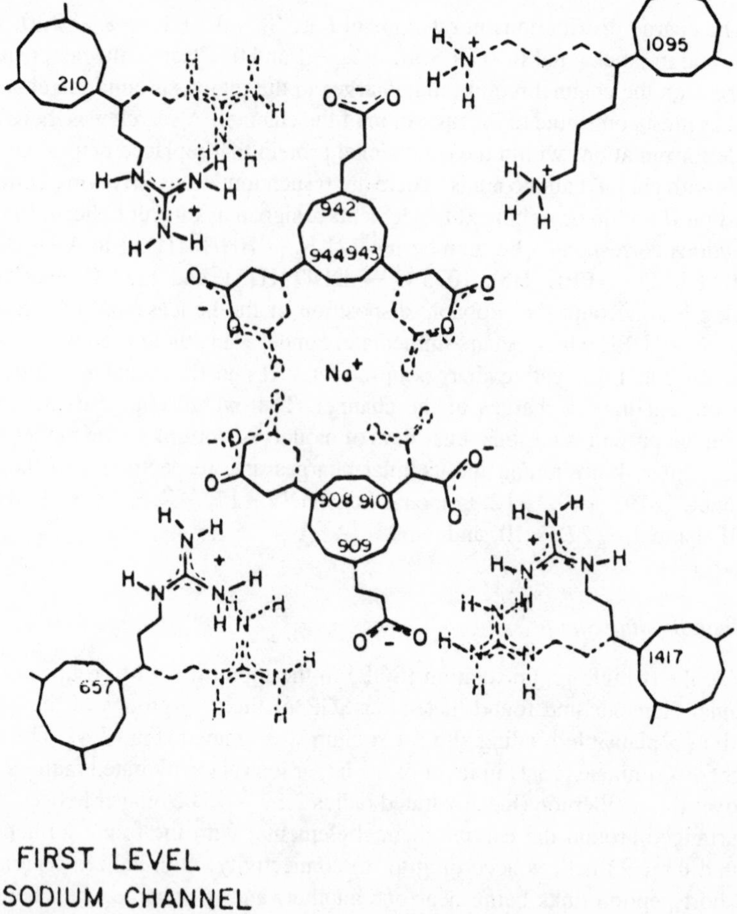

FIRST LEVEL
SODIUM CHANNEL

Fig. 7.4. The level-1 arrangement of the sodium channel. The two extreme single group rotation (SGR) conformations for arginine and glutamate are shown with dashed lines superimposed on solid lines. Two remarkable features are readily discerned. First, SGR of the glutamate side chains away from positions near the guanidino groups of arginine leads to a plausible arrangement for binding a cation. Second, the guanidino groups of the arginine may interact with either the inner or outer (Fig. 7.3) carboxylate groups according to the SGR conformation.

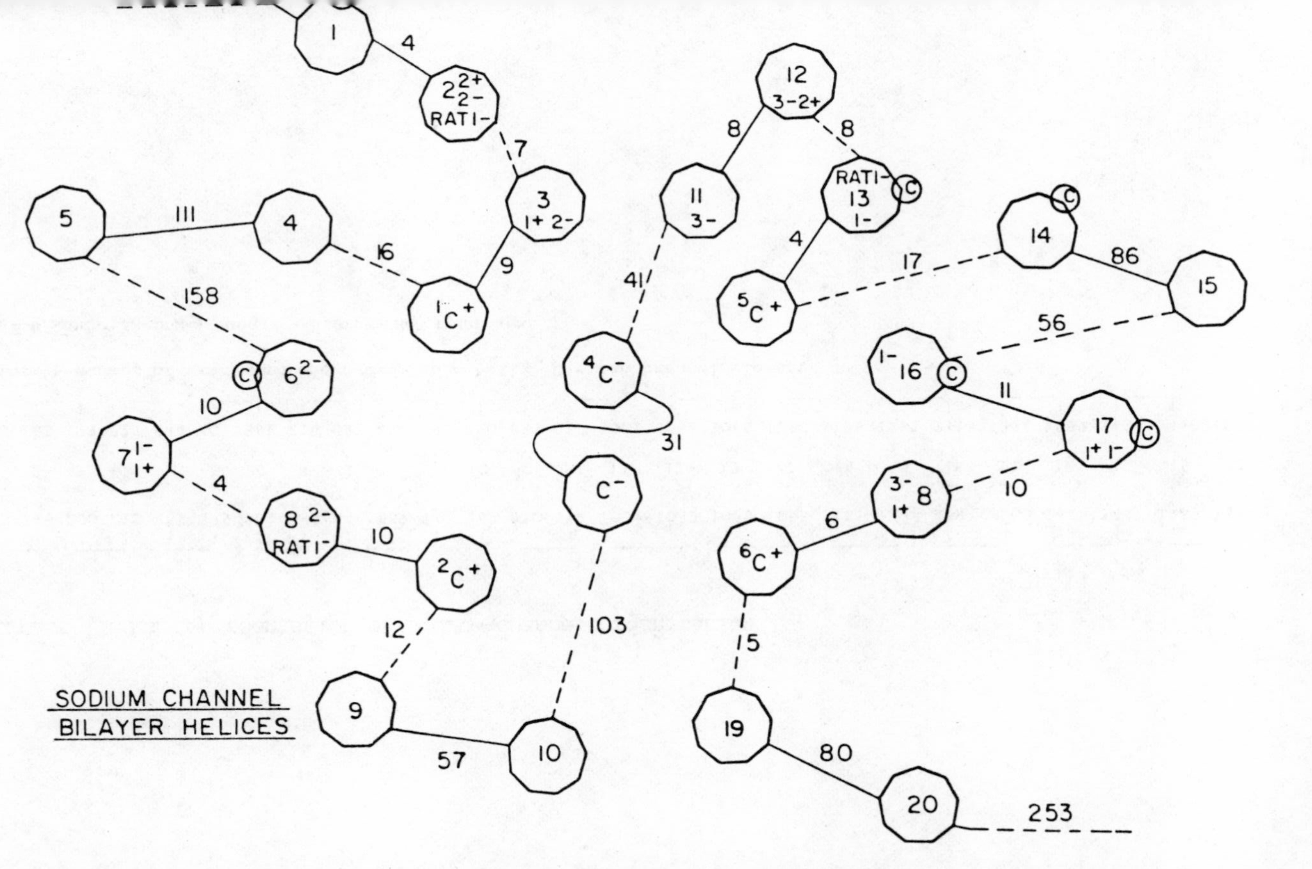

Fig. 7.5. Arrangement of eel ion channel elements and hydrophobic helices. The solid lines are connections between elements or helices outside the bilayer; dashed lines represent connections on the cytoplasmic side. The connecting links contain the indicated number of amino acids. The signs on the channel elements indicate the predominant charge on a given 3_{10}-helix. The top of the third channel element is amino acid 910, exactly halfway through the 1820-amino-acid sequence of the channel protein. © = cysteine.

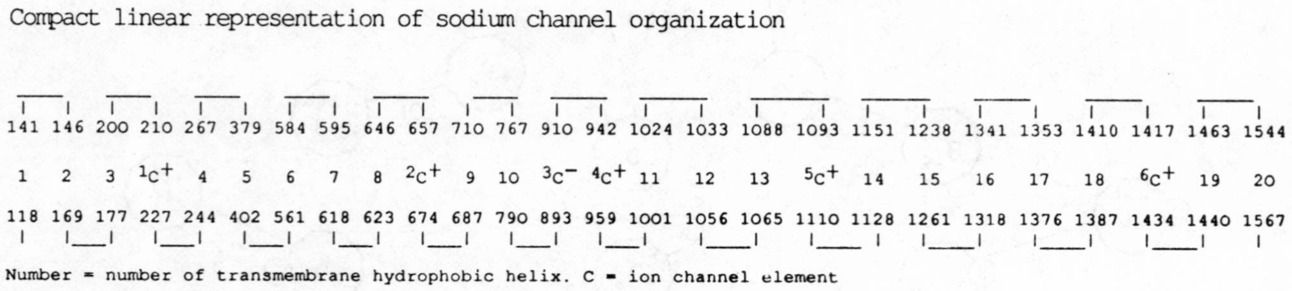

Compact linear representation of sodium channel organization

```
141 146 200 210 267 379 584 595 646 657 710 767 910 942 1024 1033 1088 1093 1151 1238 1341 1353 1410 1417 1463 1544

 1   2   3  1C+  4   5   6   7   8  2C+  9  10  3C- 4C+ 11   12   13  5C+  14   15   16   17   18  6C+  19   20

118 169 177 227 244 402 561 618 623 674 687 790 893 959 1001 1056 1065 1110 1128 1261 1318 1376 1387 1434 1440 1567
```

Number = number of transmembrane hydrophobic helix. C = ion channel element

Fig. 7.6. Compact representation of sodium channel organization.

Channel Dynamics

The conductance of the sodium channel (3–9 pS,1 picoSiemen = 10^{-12} $ohms^{-1}$) is 1000 times greater than expected for ion carrier transport (1096). The transport rate for Na^+ through the sodium channel, ca. 10^7 ions/ sec, is within an order of magnitude of that in water. At least two additional protein segments are required to control the ion flow, one outside the bilayer (outer control element or OCE), a second inside the cell (inner control element or ICE).

Outer Control Element (OCE)

The two elements, OCE and ICE, are needed to control the operation of the sodium-carrying portion of the channel. The OCE closes an otherwise open binding site and channel; the ICE closes an active channel. A logical choice for the OCE is the 31-amino-acid sequence connecting the (−) ion channel elements (Fig. 7.7). This sequence contains groups of a charge type and position (941 lys^+ and/or 913 lys^+, 915 lys^+) which can block the opening to the channel and maintain a closed state. It is possible that the OCE corresponds to the m-gate of Hodgkin and Huxley (1127).

With a normal membrane potential, (+) groups (probably 2) from the OCE are associated with the level 1 groups of the sodium channel. Comparison of the eel and rat channels suggests that 913 and 915 lys^+ in the eel OCE correspond to (eel numbering) 915 and 917 lys^+ in the rat OCE. The rat OCE is somewhat shorter and carries much more negative charge than the eel OCE (Δ −5), possibly to counterbalance the decrease in negative charge present in the (−) channel elements (Δ +3).

Inner Control Element (ICE)

Once the channel is "activated," some closure mechanism is required. Inactivation is affected by intracellular proteolysis, as described later, suggesting that a portion of the channel is directly involved. The protein segment can be called the inner control element (ICE). The ICE may correspond to the h-gate of Hodgkin and Huxley (1127). The ICE is almost certainly the sequence between 1261 gly and 1318 ile. An antibody prepared against the rat channel II peptide from 1491–1508 interferes from the intracellular side with inactivation of the sodium channel in rat "myoballs" (colchicine-treated myotubes formed by fusion of rat skeletal muscle cells) (1128). The proposed ICE is very highly conserved in sequence from eel to rat sodium channels. The sequence includes a substantial number of (+) groups (11^+ versus 6^-) (Fig. 7.7). Inactivation could arise from (+) charges interacting with the (−) charges near the intracellular exit of the channel. The ICE is thus similar to cationic blocking agents, the local anesthetics (see below).

Outer control element (OCE)

934 ser ---- 933 pro ----- 932 ser ---- 931 tyr ----- 930 asp^{-}

935 glu^{-}	917 ala --------	918 leu	929 val
936 gln	916 his	919 asn	928 thr
937 asp^{-}	915 lys^{+}	920 asp^{-}	927 ser
938 pro	914 lys^{+}	921 glu^{-}	926 cys
939 leu	913 lys^{+}	922 asp^{-}	925 val
940 ala	912 asn	923 ser ---	924 ser
941 lys^{+}	911 thr		

Inner control element (ICE)

1261 gly	1281 met	1301 ala
1262 val	1282 thr	1302 lys^{+}
1263 ile	1283 glu^{-}	1303 cys
1264 ile	1284 glu^{-}	1304 ile
1265 asp^{-}	1285 gln	1305 pro
1266 asn	1286 lys^{+}	1306 arg^{+}
1267 phe	1287 lys^{+}	1307 pro
1268 asn	1288 tyr	1308 ser
1269 arg^{+}	1289 tyr	1309 asn
1270 gln	1290 asn	1310 val
1271 lys^{+}	1291 ala	1311 val
1272 gln	1292 met	1312 gln
1273 lys^{+}	1293 lys^{+}	1313 gly
1274 leu	1294 lys^{+}	1314 val
1275 gly	1295 leu	1315 val
1276 gly	1296 gly	1316 tyr
1277 glu^{-}	1297 ser	1317 asp^{-}
1278 asp^{-}	1298 lys^{+}	1318 ile
1279 leu	1299 lys^{+}	
1280 phe	1300 ala	

Fig. 7.7. The sequences of the outer and inner control elements (OCE, ICE) of the sodium channel. The OCE is the segment between C_3^- and C_4^-. The ICE is contained in the segment between H15 and H16.

Channel Activation by Depolarization

At a membrane potential near the ''threshold'' in depolarization, the OCE moves away and opens up the binding site, allowing Na^+ to flow into the cell. The change in local ionic concentration arising from the flow of Na^+ within the cell (either from a neighboring sodium channel or from a ligand-gated channel) diminishes the attraction of the interior for the OCE. The attraction is mediated via the (+) groups moving up and the (−) groups moving down, with the outer channel region more strongly affected because of a lower dielectric constant. A schematic view of the process is shown in Figure 7.8. The inactivation process includes the approach of the inner control element (ICE) to the exit region of the channel.

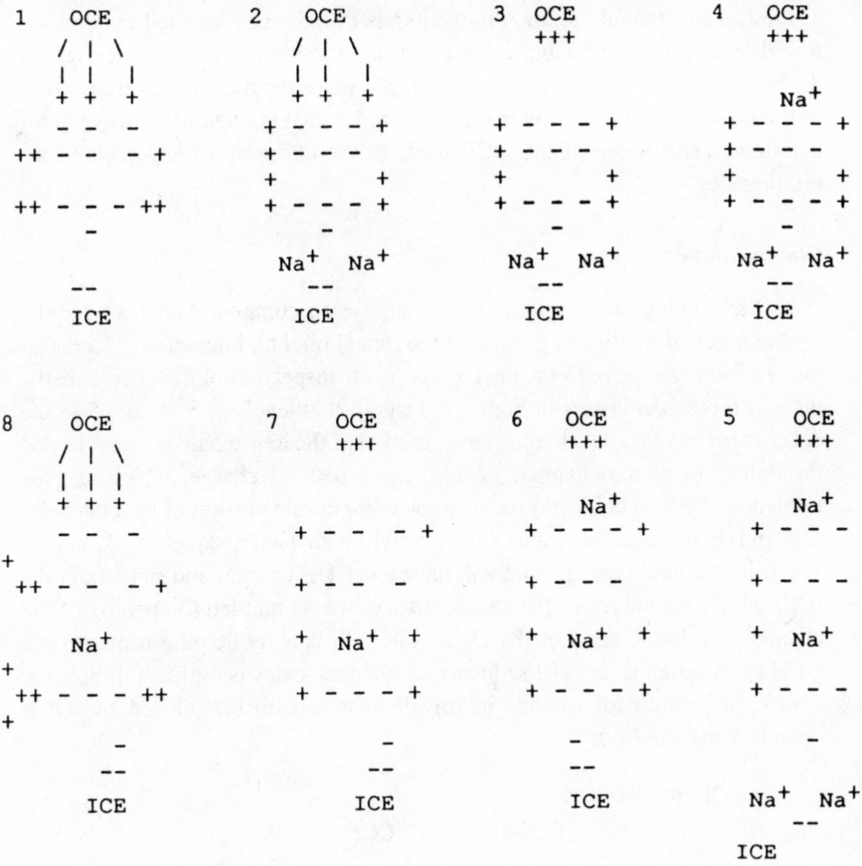

```
1   OCE          2   OCE          3  OCE          4   OCE
   / | \            / | \            +++              +++
   |  |  |          |  |  |
   +  +  +          +  +  +                           Na+
   -  -  -        + -  -  - +     + - - - +        + - - - +
++ -  -  - +      + -  -  -       + - - -          + - - -
                  +         +     +       +        +       +
++ -  -  - ++     + -  -  - +     + - - - +        + - - - +
   -                 -                -                -
                   Na+  Na+       Na+  Na+         Na+  Na+
   --                --               --               --
   ICE               ICE              ICE              ICE


8   OCE              7    OCE        6   OCE          5    OCE
   / | \                  +++            +++               +++
   |  |  |                                Na+               Na+
   +  +  +                             + - - - +         + - - -
   -  -  -            + - -  - +
+
++ -  -  - +          + - -  -         + - - -           + - - -

   Na+                +  Na+  +        +  Na+  +         +  Na+
+
++ -  -  - ++         + - - - +        + - - - +         + - - -
+
       -                   -               -                 -
       --                  --              --
      ICE                 ICE             ICE            Na+  Na+
                                                             --
                                                            ICE
```

Schematic for Channel Operation

Fig. 7.8. Dynamics of sodium channel operation. OCE, outer control element; ICE, inner control element.

Gating Current

The motion of the channel group charges can account for the gating current (estimated as equivalent to the transmembrane motion of at least 4 charges, or 144 D for a 36 Å bilayer) (1127–1130), which precedes channel opening (1131). Charge motion in the eel model channel could involve SGRs of 20^+ over 4.5 Å (arg^+ or lys^+) and 22^- over 2.5 Å (glu^- or asp^-), corresponding to 150 D, in agreement with the estimate from gating current. The corresponding figures for the rat I and rat II model channels are 125 D and 128 D respectively, whereas 300 D ($8.5\ e^-$, 36 Å) is derived indirectly from rat brain channel II measurements on protein expressed in *Xenopus laevis* öocytes (1132).

There is a substantial variation in the number of charges reported to be transferred across the membrane. In addition, the motion of other charges within the bilayer has not been included in the estimates for the charges transferred in the models. Some caution is thus required in making and using these comparisons. The release of the OCE leads to the exposure of a potential ion-binding site.

Channel States

At the binding site, arginine (+) charges can compete with Na^+ for the (−) charges of the ligand groups. If the ligand binding interactions decrease, the Na^+ ion can move to the next lower level. Inspection of the charge distributions on the pentagons in Figure 7.3 shows that levels 2, 3, 4, and 5 in the eel channel have no net charge. Thus, motion of the ions would not be retarded through most of the channel. Level 6 has a net +1 charge, which may be influenced by (−) charged groups in the cytoplasmic portion of the channel.

A maximum of 3 Na^+ ions can occupy the channel at levels 1, 3, and 5. The fully occupied eel channel will have a net 1+ charge, and can attract the ICE, which then reverses the charge displacements that led to opening of the channel via dissociation of the OCE. The ICE acts by helping to return the OCE to its initial state. The sequence of channel states is outlined in Scheme 7.1. A plausible name for the intermediate state with ICE closed and OCE open is ''anesthetized.''

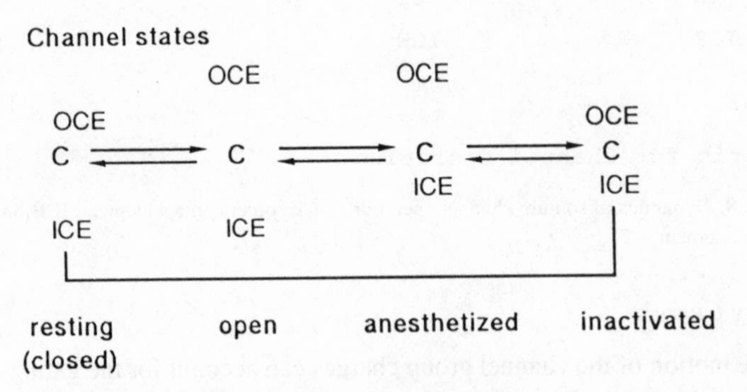

Scheme 7.1. Schematic summary of sodium channel states and the positions of the outer and inner control elements (OCE and ICE) in the various states. Depolarization of the membrane leads to opening of the channel. Motion of the ICE produces an internally blocked or anesthetized channel, followed by return of the OCE and ICE to the positions occupied in the resting state.

Toxin Effects and Toxin Classes

There are five types of sodium channel toxins, channel blockers [tetrodotoxin (TTX), saxitoxin (STX), μ-conotoxin], channel activators [batrachotoxin (BTX), grayanotoxin (GTX), veratridine, aconitine], channel inactivation blockers [α-scorpion toxin (North African *Leiurus quinquestriatus*), sea anemone toxins], activation enhancer [β-scorpion toxin, American *Centuroides sculpturatus*] (1134, 1135), and persistent activators [ciguatoxins]. One makes the tacit assumption that the effects obtained with toxins acting on sodium channels from organisms other than the eel may be used for interpretation of the *Electrophorus* channel structure (1136). Further, I assume that the conclusions apply to the rat channels I, II, and III. Although BTX seems without effect on the sodium channels of the frog which produces it, the same channels are affected by veratridine and GTX (1137). Both TTX and veratridine act on eel channels (1138). Long-term experiments with toxins should be interpreted with caution, since a reversible decrease in the number of sodium channels on chick muscle has been observed after activation with BTX (1139).

Channel Blockers: Tetrodotoxin and Saxitoxin

A chemical analogy exists between the guanidino groups of tetrodotoxin (TTX) and saxitoxin (STX) and the arginine groups of the channel. A plausible blocking site for STX and TTX is the binding site for Na^+. More information is needed on the other parts of the TTX binding site of the channel protein. The active form of the sodium channel with a high affinity for saxitoxin can be stabilized by intramolecular cross-linking with carbodiimides (1140).

Channel Activators: Batrachotoxin

I have proposed that the compounds of the BTX-channel activator group have a constellation of oxygen atoms (the "oxygen triad") suitable for complexation with the ε-ammonium ion of a lysine side chain (1141). The oxygen-oxygen distances in the triad vary within the group from 2.26 to 4.97 Å, as shown in Figure 7.9. Externally added BTX does not activate the sodium channel immediately, implying that the BTX must penetrate the bilayer, and act on the interior of the bilayer (897). Binding to the local anesthetic site would prevent the binding of the ICE and thus prevent inactivation. Association of the oxygen triad with a (+) group at the lower end of a (+)-charged helix of the channel (eel 1432 arg^+ becomes lys^+ in both rat channels I and II) could enhance the binding of the BTX, while the proximity of the BTX–binding site to the channel might lower the conductivity by partial occlusion.

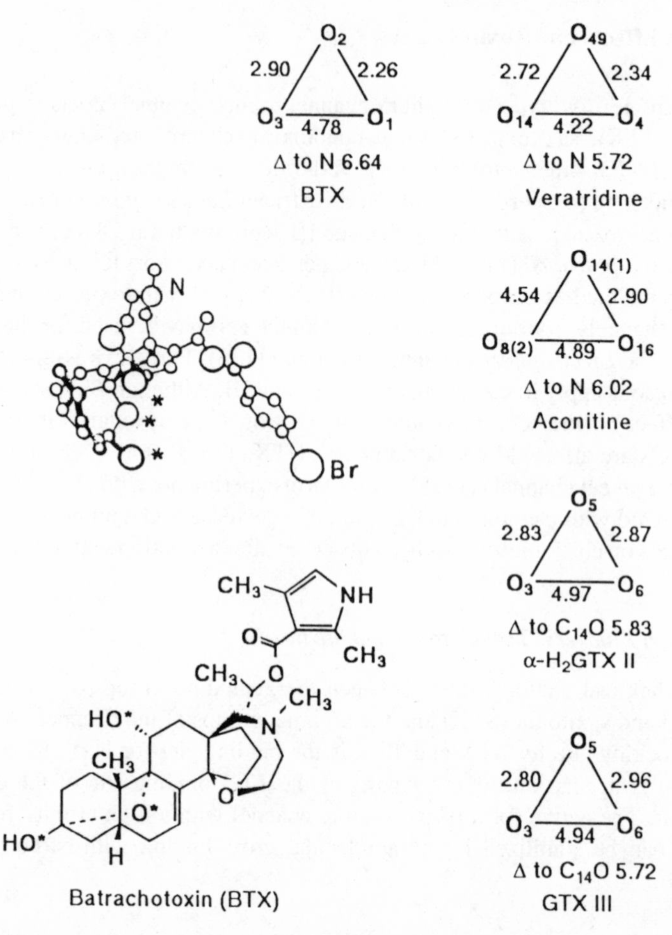

Fig. 7.9. "Oxygen triads" of various toxins that prolong sodium channel opening. The triads may complex with an ϵ-ammonium ion of a lysine side chain in the ICE or the lower part of the channel. The triad of batrachotoxin is starred in the formula and the crystal structure.

Keeping a lysine away from the channel could lower the rate of inactivation and might account for the shift of activation to more negative voltages with the BTX group of compounds (1142, 1143) since one (+)-charge is already "neutralized" by the BTX. The LA site is occupied by the ICE or by BTX, so that gating charge is still immobilized. If the LA site is not occupied as happens as a result of pronase treatment, gating charge is no longer immobilized (1144). The cooperative effect of BTX-class compounds (BTX and veratridine), which complex with the LA site, and polypeptide toxins, which complex with the OCE (1145, 1146), is readily explained since different sites are involved. The use of BTX in the study of sodium channels has been reviewed (1147, 1148). Another "oxygen triad" toxin, α-dihydrograyanotoxin

II, opens sodium channels from the inside of the squid axonal membranes (1149).

STX blocks BTX-activated channels in a voltage-dependent manner (1150). STX complexation with the Na^+-binding site, or in place of an arginine from one of the C^+ channel elements, should occur with voltage dependence (Fig. 7.7). The binding site is opened up by dissociation of the OCE above the threshold voltage; the positive charges in the C^+ channel elements should change in average position with depolarization.

Other Toxins

Complexation of the OCE with the other toxins can account for their activity. The three-dimensional structure of one scorpion toxin has a triangular arrangement of charged amino acids (2 glu$^-$, 1 lys$^+$, 52 glu$^-$) (1151) which might interact with a corresponding OCE combination, the toxin-OCE complex inhibiting closure of the channel. Photoattachment of the ScTx occurs at the segment between (eel numbering) 275 and 379, confirming the extracellular placement of the segment in most models (1096).

Pyrethroids

The pyrethroid deltamethrin (10 μM) prolongs the open time of sodium channels in neuroblastoma cells without affecting single channel conductance. Hydrophobic deltamethrin may enter the sodium channel molecule in the region between a (−)- and two (+)-channel elements, preventing the (+) charges from moving away from the (−) charges, and thereby increasing the open time of the channel (1152). The slow activation (as well as slow inactivation) of squid axon sodium channel by deltamethrin (1153) can be explained by a delay in the time for a channel element (+)-charge to move up toward the OCE. Different pyrethrin isomers (*cis*- and *trans*-tetramethrin, for example) give somewhat different limiting effects, and yet compete for the same site (1154, 1155).

Toxin Effects on Channel Selectivity

Both equilibrium and kinetics play a role in determining the selectivity of the channel. The permeabilities to various ions in neuroblastoma cells and skeletal myotubes ($Na^+ : K^+ : Rb^+ : (NH_2)_2C{=}NH_2^+ : CH_3NH_3^+$, 1 : 0.086 : 0.012 : 0.13 : 0.007) change in the presence of channel-opening toxins (1 : 0.4 : 0.15 : 0.65 : 0.14) (1156). Using a simple kinetic model for the passage of ions, the amount of cation delivered via the channel to the interior of a cell would be determined by the ratio of the rate at which ions pass through the channel and the rate at which inactivation occurs times an equilibrium constant for ion binding. The more slowly the ion moves through the channel, the greater is the inactivation rate. In the presence of toxins, the OCE-toxin complex forms, inactivation does not occur, and the "intrinsic rates" of cation diffusion through the channel would control the amount of cation going through the membrane.

Effects of Other Agents

Methylation of carboxylate groups of the sodium channel at frog nodes of Ranvier with trimethyloxonium ion (TMO) leads to the loss of TTX sensitivity and a lower channel conductance (1157). Extracellular TMO abolishes TTX- and STX-block of single BTX-activated rat brain sodium channels and greatly diminishes the voltage-dependent block by Ca^{++}, these effects being prevented by STX and Ca^{++}, respectively. A methoxycarbonyl group at the Na^+ binding site in the present model would be a somewhat weaker ligand for Na^+, but much weaker with respect to interaction with a guanidino group of TTX or STX (Eq. [7-1]) (1158). Blockage by a carbodiimide, 1-ethyl-3-(3′-dimethylaminopropyl)carbodiimide, which may link a carboxylic acid group to an amino group, diminished the effectiveness of STX in blocking the sodium current (1159, compare with 1140).

$$(7\text{-}1) \qquad (CH_3)_3O^+ + RCOO^- \rightarrow RCOOCH_3$$

The removal of sodium channel inactivation by the internal perfusion of squid axons with N-bromoacetamide has been ascribed to cleavage of a peptide link next to tyrosine (1160). Cleavage of the ICE at 1288 tyr or 1289 tyr would interfere with the conformational change needed to initiate inactivation. Internal perfusion of squid axons with pronase removes sodium channel inactivation (1161), a result consistent with the loss of an effective ICE. Pronase is an enzyme mixture in which the main activity is due to alkaline proteinase b, an endopeptidase similar in activity to trypsin, with cleavage expected at the peptide bond of arginine or lysine. The effect of pronase would be due to the removal of a cationic peptide, and was shown to be replaced by another cation (1162). The reappearance of inactivation (lost after pronase treatment) through introduction of polyglycylarginine (1163) or octylguanidinium ion (1164) suggests that a polypeptidic or hydrophobic guanidinium ion can simulate the action of the ICE. In the model, there are three pairs of lys^+ in the ICE sequence, one or more of which could serve as a candidate for the proposed positively charged amino acid. It is interesting that the 1286 lys^+ and 1287 lys^+ are next to the 1288 tyr or 1289 tyr mentioned above. The effects of 1-propylguanidinium ion in decreasing peak Na^+ conductance have been adduced in support of a role for an arg^+ group in inactivation (1165, 1166). There are two arg^+, 1269 and 1306, in the ICE. Some interaction of Ca^{++} with the Na^+ site can account for the calcium-induced decrease in sodium conductance (1167).

Antibody Effects: Antipeptide Antibodies

Antibodies against solubilized sodium channels are species-specific and not physiologically active. However, Meiri *et al.* (1168, 1169) have succeeded in generating antibodies against native eel electroplax membranes which block mammalian nerve conduction.

Peptides selected from the eel channel sequence have been used to generate antibodies that are neurophysiologically effective against dissociated rat ganglion (DRG) cells. In particular, an anti-C_1^+ antibody generated against the eel channel 210–223 sequence, RTFRVLRALKTITI, shifted the steady-state sodium channel inactivation parameter h_∞ to more negative potentials and changed the kinetics of activation (1115).

That an anti-C_1^+ antibody based on the eel channel 210–223 sequence was active against mammalian sodium channel was not surprising in view of the almost complete homology of these sequences between eel and rat. However, epitopes (antigenic determinants) may not even have sequence similarity with the peptide used as antigen (1170, 1171), requiring caution in interpretation. I looked for contiguous triplets of amino acids as the epitope, with the results shown in Scheme 7.2, using eel sodium channel numbering and the amino acid single letter code. Triplets are listed under the sequence so that identical triplets are in the same column.

In addition to the FRV (1416-1417-1418) triplet noted in (1115), and the triplets of the C_1^+ itself, two additional contiguous triplets might be epitopes to the polyclonal antibodies generated against C_1^+-peptide. One is VLR, 655-656-657. The two first amino acids are outside the assigned region for C_2^+. The second is LRA, 371-372-373. In the model, this connects bilayer helices 2 and 3. Several other triplets were found but are considered to be too buried in the bilayer for strong consideration.

Anesthetics

Anesthetics (Greek, *anaesthetos*, insensible) clearly decrease perception, either generally or locally, but their actions are difficult to define in physical terms (1172). The major target for local anesthetics appears to be the sodium channel (1173, 1174), but the structural diversity of the compounds that can induce general anesthesia certainly precludes a single receptor as the target (1175, 1176). Anesthesia is a reversible, drug-induced perturbation of neuronal behavior, arising from a complex combination of many effects (1176).

Local Anesthetics

Local anesthetics inhibit the flow of ions through sodium channels more effectively in the activated than in the resting state (1177). The quaternary

TRIPLET EPITOPES RELATED TO C_1^+

E = EEL, I = RAT I, II = RAT II

C_1^+ :RTFRVLRALKTITI

210-211-212-213-214-215-216-217-218-219-220-221-222-223

Triplet		Epitope
210-211-212		(C_1^+) E/I/II
211-212-213		(C_1^+) E/I/II
212-213-214		(C_1^+) E/I/II
1416-1417-1418	1out+2in	(C_6^+) E/I/II
213-214-215		(C_1^+) E/I/II
1426-1427-1428		(C_6^+) Eonly
214-215-216		(C_1^+) E/I/II
655-656-657	2out+1in	(C_2^+) E/I/II
1427-1428-1429		(C_6^+) Eonly
215-216-217		(C_1^+) E/I/II
371-372-373		(Out) E/I/II
1106-1107-1108		(C_5^+) E/I/II
216-217-218		(C_1^+) E/I/II
232-233-234		(in) Eonly
744-745-746		(Out) Eonly
1101-1102-1103		(C_5^+) E/I/II
217-218-219		(C_1^+) E/I/II
218-219-220		(C_1^+) E/I/II
227-228-229		(in) E/I/II
219-220-221		(C_1^+) E/I/II
228-229-230		(in) E/I/II
220-221-222		(C_1^+) Eonly
221-222-223		(C_1^+) Eonly

Scheme 7.2. Summary of possible epitopes (antigenic amino acid combinations) present in the eel sodium channel. Only contiguous combinations are considered. Epitopes apart from those present in the C_1^+ peptide could be responsible for the reactivity of the channel toward anti-peptide antibodies.

derivative (QX-314) of the local anesthetic lidocaine acts only on the inside of the squid axon and not when added on the outside. By inference, the site for the action of local anesthetics is at the intracellular exit of the sodium channel, a location that contains a reasonable binding place for local anesthetics. Cocaine is a local anesthetic introduced by Freud (1178) but abandoned because of addiction, presumably due to action at another receptor (1179). The coordinates from the crystal structure for cocaine (1180, 1181) can be used to

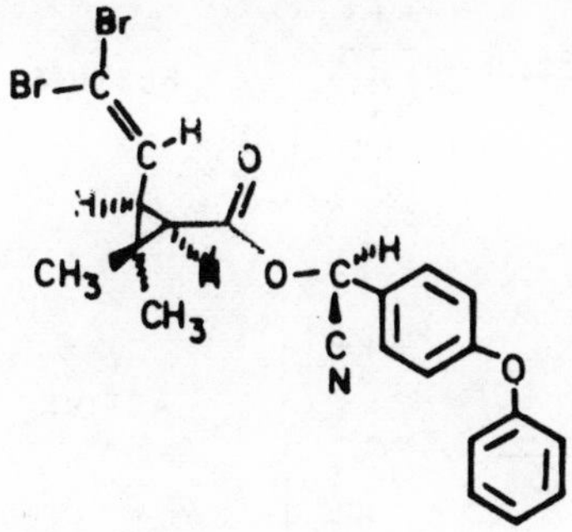

DELTAMETHRIN

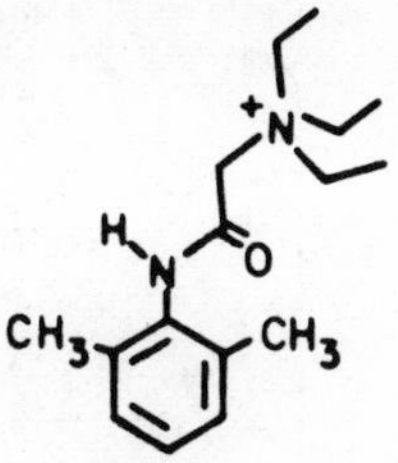

QX-314

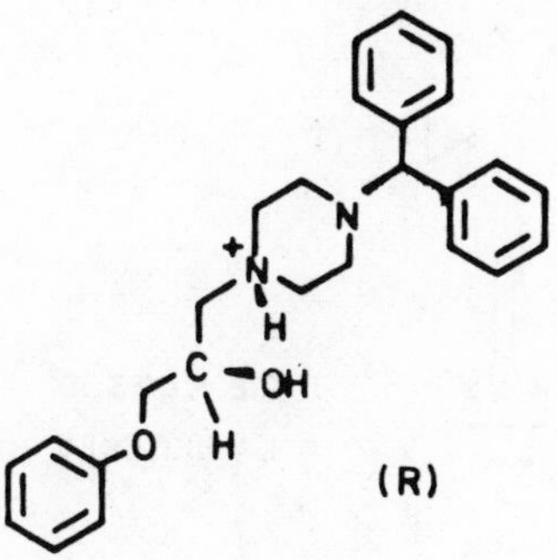

DPI 205-429 (R)

i_{Na} {R down, S up}

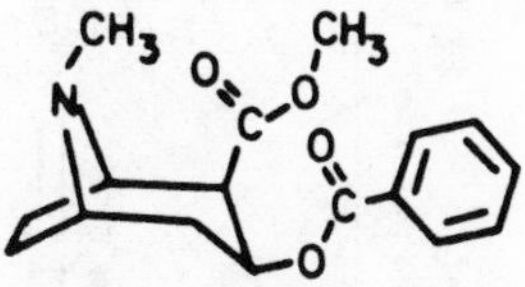

COCAINE

model cocaine for placement in a partial sodium channel model using molecular graphics.

Molecular graphics studies suggest that a binding site on the central part of the sodium channel of moderate specificity binds local anesthetics through a combination of ionic and hydrophobic interactions. An aryl group on the bound molecule physically blocks the main pathway for sodium ions to exit from the channel and pass into the cell. Proximity of the BTX and LA binding sites would lead to anesthetic-promoted dissociation of bound BTX (1182, 1183). A schematic diagram of the equilibria involved in the effect of local anesthetics (LA) on the sodium channel is shown in Figure 7.10. The Na^+ channel is activated by BTX, but combination with an LA promotes the dissociation of bound BTX. The ability of LA to compete with BTX for binding

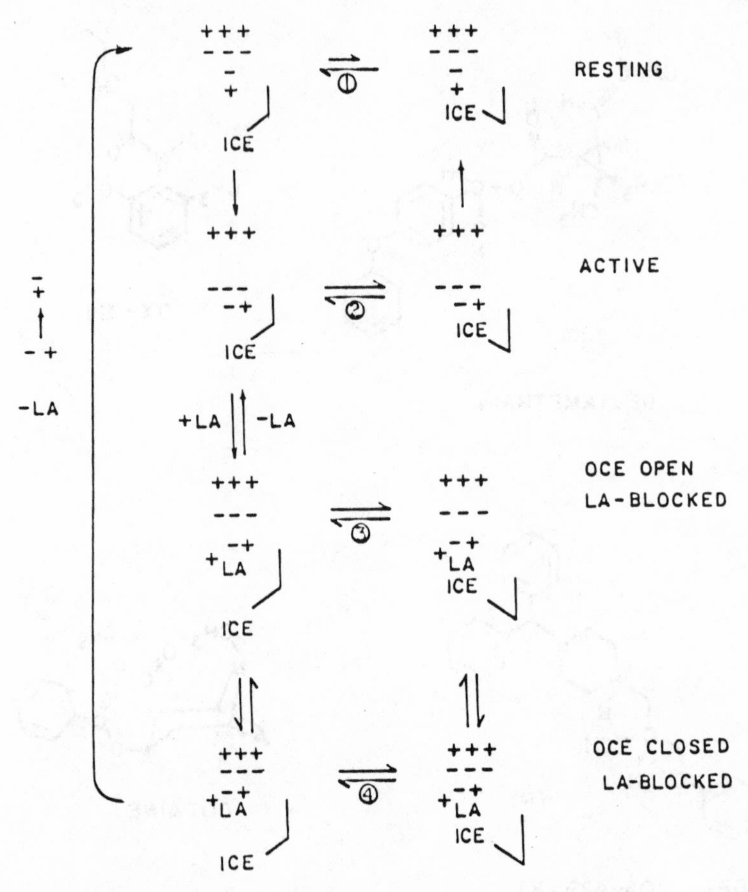

Fig. 7.10. Equilibria involving local anesthetics (LA) and the sodium channel at various stages of sodium channel operation.

to sodium channel is an index of the effectiveness of the LA (1184). The specific location implied for the LA is in accord with a variation in the effectiveness of use-dependent sodium channel block with LA structure (1185, 1186). The competition of BTX with LA is consistent with the idea of a common binding site.

The cardiotonic agent DPI acts on sodium channels. The two enantiomers have quite opposite effects, although they appear to be bound to the same sites with about the same binding constants. The S-DPI (*205–429*) increases peak Na^+ current and slows down the kinetics of inactivation. The R-DPI (*205–430*) diminishes peak Na^+ current and blocks Na^+ channel activity. Local anesthetics and both R-DPI and S-DPI are similar in that they promote the dissociation of BTX benzoate, suggesting that all bind at the same site (1187). Indeed, comparison of the structures of R-DPI and QX-314 show that these are quite similar in a number of ways, such as the location of polar binding groups, positive charge, and hydrophobic groups. A molecular graphics study of the interaction of the enantiomeric forms of DPI would be quite interesting.

The ICE as the Natural Anesthetic

There is some resemblance between local anesthetics (most of which would be protonated at neutral pH) and the cationic molecules which replace the proteolyzed intracellular portions of the sodium channel responsible for inactivation. In analogy to the opiate peptides being the natural analog of morphine, the ICE might be the natural analog of local anesthetics. The appropriate combination seems to be a lys or arg followed or preceded by a hydrophobic amino acid. These combinations are preserved from eel to rat for 1273 lys^+–1274 ile (rat, phe), 1287 lys^+–1288 tyr, 1292 met–1293 lys^+, 1294 lys^+–1295 leu, and 1306 arg^+–1307 pro (1305 pro). A relationship between blocking by quaternary LA and inactivation has been pointed out (1188).

General Anesthetics

The strong correlation between hydrophobicity and anesthetic potency implies that the bilayer is involved in the mechanism of anesthetic action (1175). Addition of anesthetic molecules increases the "fluidity" of bilayer molecules as measured by spin probes. Some of the anesthetic molecules may be associated with the sodium channel protein as implied by the action of anesthetics in studies on bacterial luciferase. Specific association with membrane binding sites, presumably on proteins, is suggested by saturation of such sites by halothane (1189, 1190). There are many possible binding sites in the sodium channel protein. The thickness of the membrane may be influenced by general anesthetics, and affect the operation of the sodium channel via the strength of the field across the membrane. Thinner bilayers presumably result from the addition of "short-chain" phospholipids like diheptanoyl phosphatidylcholine to squid axons, shifting activation and inactivation of the sodium channels to more positive voltages (1191).

Specific Groups

Thiol Groups

The model permits the identification of potentially acylatable intrabilayer thiol groups, since many membrane proteins appear to possess such groups (1113). Almost all of the bilayer cysteines are found in helices on the "outside" (H1, H6, H13, H14, H16, H17) of the sodium channel molecule. One plausible idea is that these thiol groups are acylated by long-chain alkane carboxylic acids yielding thioesters compatible with the hydrophobic bilayer phospholipid acyl groups. Two additional bilayer groups are found in the rat I and rat II channel models. About 25 palmitoyl (C_{16}) and stearoyl (C_{18}) groups are covalently bound to the sodium channel in a ratio of 3:1 (1192). If only the external ring of bilayer helices is involved, all of the thiol groups (cys) and most of the hydroxy groups (ser, thr, and tyr) in these helices are acylated. A thiol group in the nAChR was identified by photolabeling as being in contact with the phospholipid bilayer (1193).

Phosphate Groups: Phosphorylation

Modification of the sodium channel by phosphorylation with cAMP-dependent protein kinase has been demonstrated in synaptosomes (1194) and cultured brain neurons (1195). Tryptic degradation of channel protein localizes the sites of phosphorylation between 448–630 of the rat I or 450–639 of rat II proteins, both results being consistent with current models for the channel. However, no physiological effects have yet been found for this type of modification of the sodium channel.

Prolines: Distribution

Unusual concentrations of proline at the ends of the nAChR β-, γ-, and δ-subunits have been noted and may serve as an intracellular anchor (737). In this connection, I note that several intracellular regions of the sodium channel have unusually high proline contents: in the eel (54–63, 4 pro/10 a.a.), rat I/II (56–64, 3 pro/9 a.a.), rat (1300–1307, 4 pro/8 a.a.), eel (1619–1625, 3 pro/7 a.a.), rat I (1618–1625, 4 pro/8 a.a.), rat II (1618–1625, 3 pro/8 a.a.), and eel (1772–1788, 5 pro/17 a.a.). Each of the channels has a total of 74 prolines (less than 5% of the total number of amino acids). The identified proline-rich regions have less than 1.5% of the total a.a., but have 9/27 (eel), 11/25 (rat I), and 10/25 (rat II) prolines.

Shape and Biological Stability

In the proposed model, the mass of the sodium channel is more or less evenly distributed between the outside (23.4%) and inside (44.3%) of the cell, with a major fraction (32.3%) within the bilayer. Given the 65% α-helix content determined for purified α-polypeptide of the rat brain sodium channel (1196), substantial stretches of α-helical protein must be present outside of the bilayer.

Turnover

Inhibition of N-linked protein glycosylation in cultured cells with tunicamycin diminishes the number of STX-binding sites to ca. 20% of the initial number in 48 hours. Turnover of sodium channels is indicated with a $t_{1/2}$ of ~20 hours (1197, 1198). Newly synthesized intracellular α-subunit can be detected within a few minutes after a pulse of ^{35}S-methionine and matures over several hours through glycosylation (1199).

Other Models

A number of other models, some more specific than others, have attempted to relate certain aspects of the sodium channel sequence to a structure. Catterall (1096, 1200) suggested a "sliding helix model" in which the positively charged α-helixes spiral upwards on depolarization by 5 Å and 60°. The (+) charges are matched by (−) charges in other unspecified α-helixes, the latter remaining more or less in place. Each (+)-helix would move more or less independently. Although the proposal formally accounts for the gating current, it leaves unclear how and where the Na^+ ions would traverse the channel. In addition, the large-scale motions involved might not be consistent with the speed of sodium channel response to depolarization (20–50 microsecond range, with peak currents being reached within 1 msec).

The Greenblatt *et al.* model (1117) does not have a clear mechanism for channel operation and control. The SGRs of the charged groups are ignored in considering the geometry of their interaction, and the charge motion required to explain the gating current seems too large unless some substantial motions of the helices occur. The original or modified Guy models (1118) lack suitable control mechanisms and require a large-scale "screwlike" motion of an α-helix carrying positive charges. Both models rely very much on protein folding predictions for their details.

Antibody experiments (1115, 1116) favor the present model as the best working hypothesis for further experiments. Antibody to eel 930–941 (corresponding to the OCE) produced labeling primarily on the cytoplasmic face in cryostat sections of eel electroplax (1201). Such results must be regarded with extreme caution, as has been pointed out in connection with antibody labeling experiments on the nAChR (737, 1170, 1171), because hydrophilic peptides within the membrane can leave the membrane easily. In the case of our model, the LIFO ("last in, first out") principle would be particularly important. Release of natural constraints would expose the channel to the intrusion of water or anions, allowing the exit of the negatively charged segments into the cytoplasm together with the dissociation of the OCE. The reverse process would occur during assembly of the channel. The contrast between the neurophysiological results on intact nerve (in which a similar but monoclonal antibody against the eel 930–942 peptide was active against frog sciatic nerve when applied from the external surface of the membrane) (1116) and the antibody experiments (1201) is striking and explicable. In any case, models of any type must be regarded with caution and skepticism. Additional constraints on folding patterns of the sodium channel are needed. These might be provided by cross-linking and other labeling (α-scorpion toxin or phosphorylation [1096]) experiments.

Conclusions

The sodium channel model proposed here is a remarkably balanced and beautiful structure, which deserves the comment, "*Si non e vero, e ben trovato*" ("If it is not true, it might as well be"). The structural features discovered with the model may well deserve incorporation into a synthetic model. In addition, these features are a chemical expression of a new physical model for channel gating (1202), and define a molecular basis for local anesthetic blocking of the open channel (1203).

A Molecular Graphical Model of the Eel Sodium Channel

The manipulation of models is a required step in examining and developing the structural, spatial, and dynamic relationships of molecules. Biologically important molecules like the sodium channel are so large that a physical model would be prohibitively large and expensive, and even if one were built, it would be too unwieldy for detailed analysis. In addition, looking at internal segments of large molecules would be difficult; moving the segments or small ligand molecules around, next to impossible. The development of computer graphics for molecules, now commonly referred to as molecular graphics, has brought to fruition the treatment of large molecules like proteins (1204).

The eel sodium channel model described in the earlier part of the chapter requires careful scrutiny on the molecular level. In the present study, I have examined only the central—ion-binding and ion-transporting—portion of the molecule. I have checked the relationships between the channel side chain groups, the adequacy of the proposed sodium binding site, and have searched for a possible local anesthetic binding site (1205).

The amino acids constituting the channel elements C_1^+, C_2^+, C_3^-, C_4^-, C_5^+, and C_6^+ were incorporated into 3_{10}-helices and arranged with the hydrophobic (+)-helices outside and the (−)-helices inside. A projection through the array in a direction parallel to the bilayer helices is shown in Figure 7.11, with the central "+" representing a sodium at the binding site. To test the transient neutrality of the channel groups, each positively charged side chain

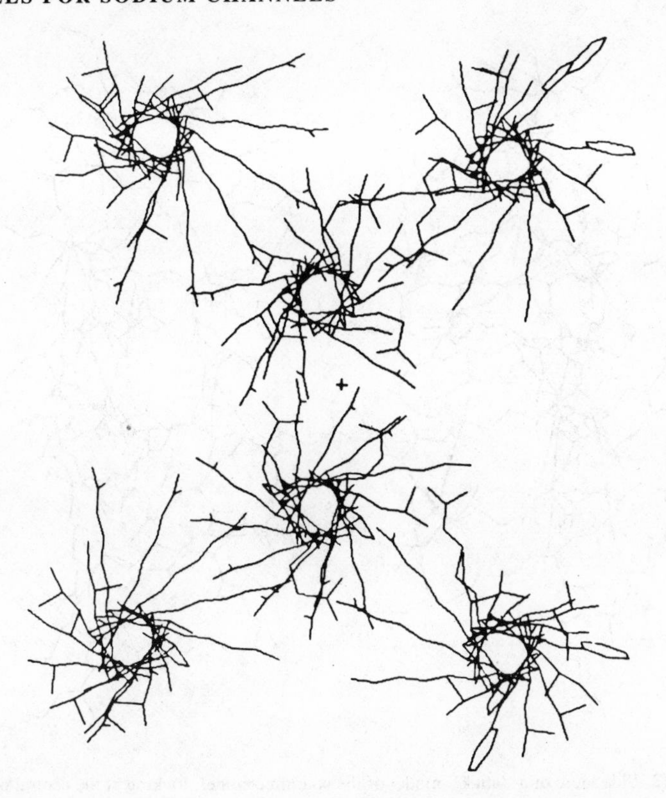

Fig. 7.11. A "stick" model of the sodium channel, looking down through the central part of the sodium channel from the outside. Almost every side chain on the two central helices carries a carboxylic acid group (glutamate), while most of the chains that project toward the center are guanidino (arginine). The (+) sign between the two central helices represents a Na^+ ion. The "stick" molecular graphics were generated with an Evans and Sutherland PS340, using a VAX 750 host at M.I.T. The FRODO program (version 6.0) was utilized to generate and manipulate the model. The color-filled stereo models were executed from the coordinates generated by the PS-VAX combination on the Apollo-Iris network at the N.I.H. (Collaboration with G. A. Petsko and R. J. Feldmann.)

was rotated around its bond to the polypeptide so as to bring the charged end close to the oppositely charged end of a negatively charged group. The local conformational change is exactly what was referred to elsewhere as single group rotation or SGR (Chap. 6). The distance over which the guanidino end of an arginine can move with an SGR is ca. 4.5 Å; the corresponding distance for a glutamate side chain is ca 2.5Å. It was found that every charged group in the channel could "communicate" with a counter-charged group, consistent with the idea of the model.

A side view of a projection through the central portion of the eel sodium channel is shown in Figure 7.12. The carboxylate groups in the sodium binding site are rotated to positions that face the positive charge (the sodium cat-

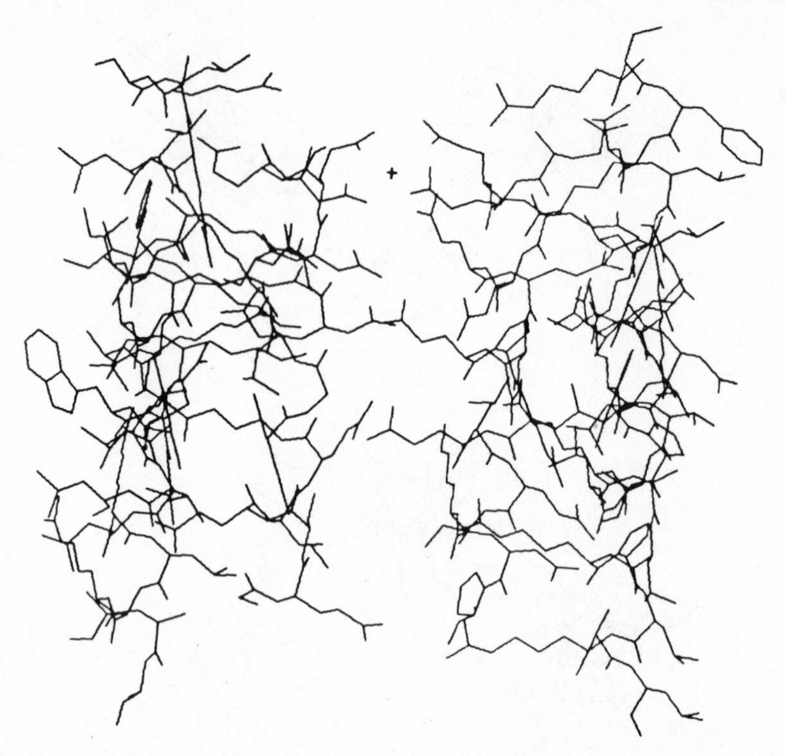

Fig. 7.12. Side view of a "stick" model of the sodium channel, looking at the central part of the sodium channel. The (+) sign at the top between the two central helices represents a Na^+ ion within the sodium binding site.

ion). A good chelating binding site for a positive ion is then visible. The oxygen-oxygen distance in the model binding site is ca. 5.5 Å, quite suitable for a hydrated sodium cation.

I wished to study the interaction of a local anesthetic with the sodium channel (NaC). The only local anesthetic for which a crystal structure was available was cocaine, a classical local anesthetic (LA) with strong addicting tendencies, the last presumably as a consequence from binding to other sites. Knowing that the LA binding site was approached from the cytoplasmic side, the positively charged LA molecule was brought to a readily available negatively charged group on the cytoplasmic side of the channel. With relatively little adjustment, the structure of a NaC.LA complex produced a surprise; the phenyl group of the LA blocked the ion pathway. The mechanism would obviously apply to all of the common LA, each of which has a phenyl group in approximately the same position. There may be additional binding due to participation of a portion of the ICE in interactions with a hydrophobic part of the LA. Further study of more complex interactions would be of interest. The complex of cocaine and NaC is shown in Figure 7.13, the phenyl (orange)

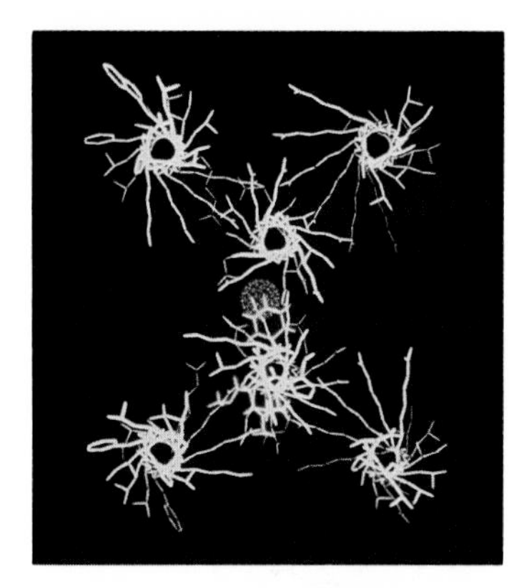

Fig. 7.13. (left) A color view of a ''stick'' model of the eel sodium channel blocked by cocaine from the cytoplasmic side. The blue dotted sphere represents a sodium ion. The orange phenyl group of the cocaine blocks the pathway of the sodium ion through the cytoplasmic outlet of the channel.

Fig. 7.14. (middle) Color stereo views of a space-filling model of the eel sodium channel from the side, with a cocaine molecule (turquoise) bound to the cytoplasmic side of the channel. Positively charged amino acids are dark blue and negatively charged amino acids are red. The uncharged amino acids are yellow.

Fig. 7.15. (bottom) Color stereo views of a space-filling model of the eel sodium channel from the cytoplasmic side, with a cocaine molecule (turquoise) blocking the channel. Positively charged amino acids are dark blue and negatively charged amino acids are red. The uncharged amino acids are yellow.

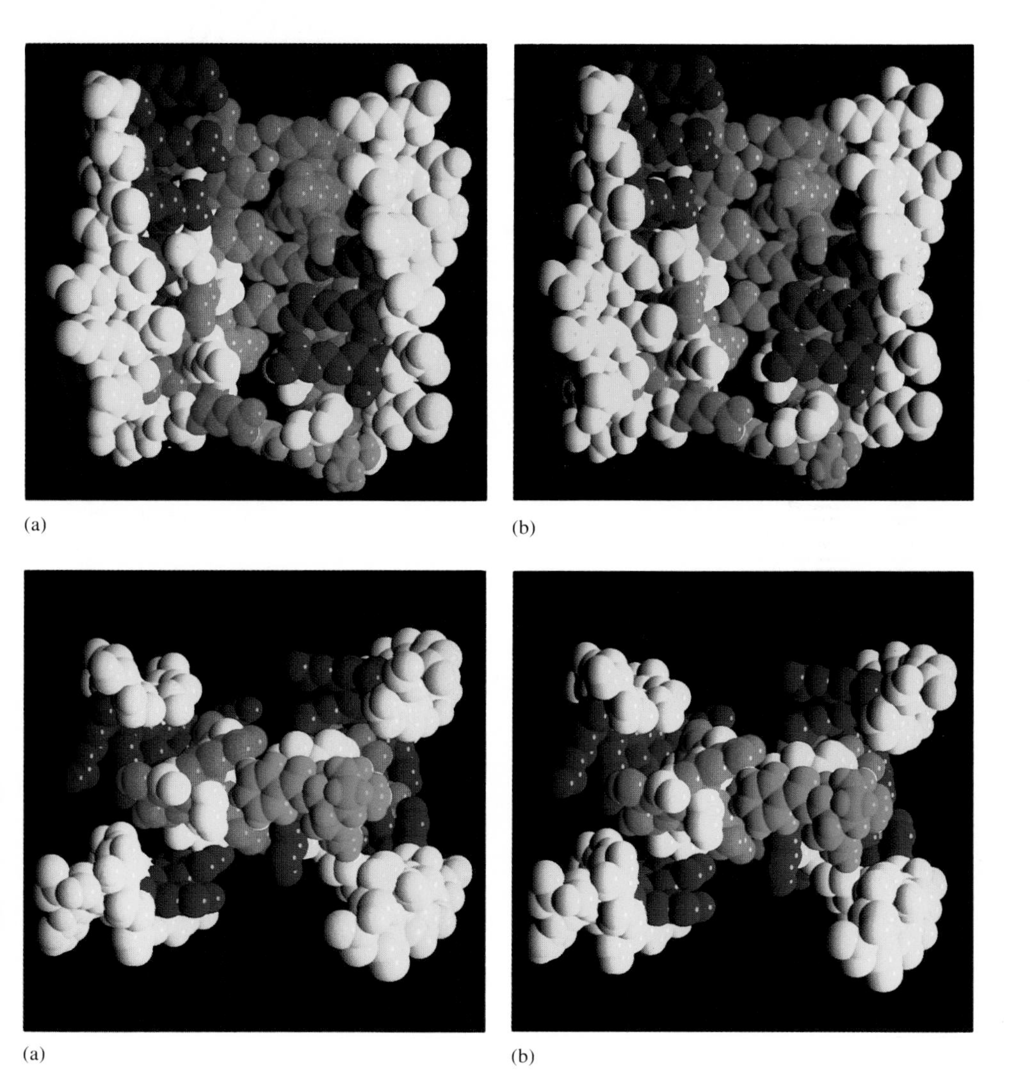

(a) (b)

(a) (b)

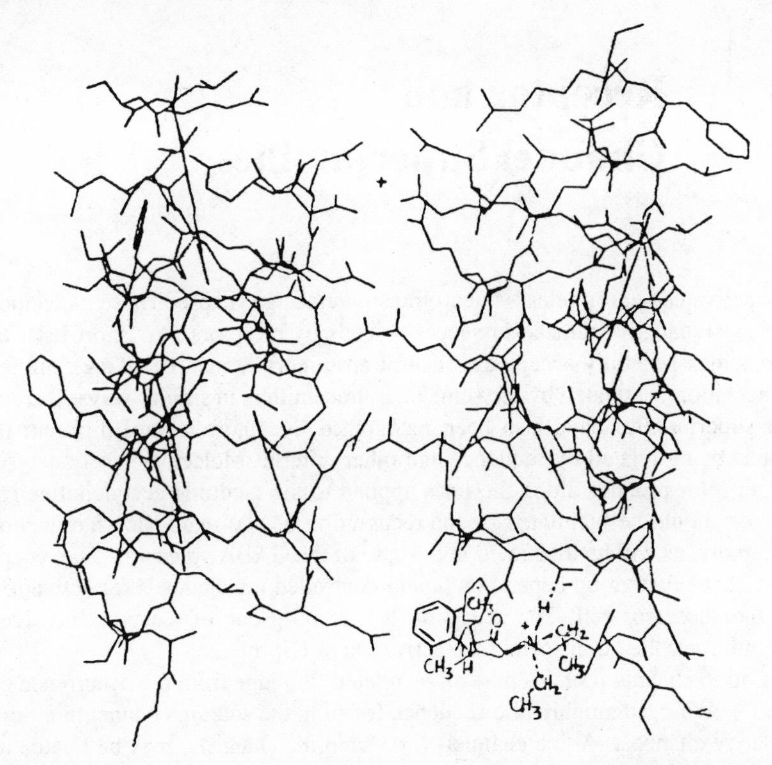

Fig. 7.16. Side view of a "stick" model of the eel sodium channel with a molecule of lidocaine, a local anesthetic.

blocking the (blue) sodium ion's entry into the cell. Stereo views show how the phenyl group of the cocaine blocks the channel, with a side view in Figures 7.14(a) and 7.14(b) and a view from the cytoplasm in Figures 7.15(a) and 7.15(b).

The insertion of lidocaine, a common LA, into a side view of the central part of the eel sodium channel illustrates in a general way the position of the phenyl group in the ion pathway, showing how an LA might block sodium conduction (Fig. 7.16).

8 Receptor and Channel Superfamilies

An activated neuron releases neurotransmitters at a synapse. These molecules act as signals, recognized in successor cells by receptors. A "short list" of molecules generally accepted as neurotransmitters exists. There are families of receptors, activated by the same neurotransmitters in similar ways. Receptor superfamilies have also been recognized, receptors toward different ligands being related by sequence and other criteria. Molecular-level analysis of receptor proteins along the lines applied to the nicotinic acetylcholine receptor should be facilitated by the recognition of two main superfamilies of receptors, as will be illustrated in the case of the $GABA_A$ receptor. The receptors of the first group depend on ligand-controlled ion channels for activation of the successor cell. The action of the second group of receptors involves ligand-controlled transmembrane activation of G-proteins.

Cation channel proteins may all be related, to judge from the occurrence of the (+)-charged amphiphilic sequence found in the sodium channel in all the putative channels. Anion channels (i.e., chloride channels) may be related to ligand-controlled ion channels.

Receptor Superfamilies I

The ion channels of the first superfamily receptors are activated by ligands. The search for receptors is based on the knowledge that particular ligand molecules are neurotransmitters. Well-defined criteria underlie the search for neurotransmitters, providing a solid basis for the search for the receptor proteins.

Criteria for Neurotransmitter Assignments

At least eight criteria may be utilized for deciding whether or not a given compound is a neurotransmitter (1206).

Anatomical. 1. Presence: The candidate transmitter must be present in presynaptic portions of the neuron.

Biochemical. 2. Precursors, Converting Enzymes: Precursors and enzymes which catalyze the synthesis of the neurotransmitter molecule must be present at or near the site of presumed action. 3. Specific Inactivation: Removal of the neurotransmitter candidate should lead to the loss of action within a physiologically reasonable time.

Physiological. 4. Release: Stimulation of specified neurons should cause the release of physiologically significant amounts of the putative neurotransmitter. 5. Stimulates Response: Direct application of the candidate transmitter at or near the target synapse should produce responses equivalent to those obtained by other means of stimulating release of the substance. 6. Potential Change: Action of the substance should lead to either hyperpolarization or depolarization of the postsynaptic membrane. 7. Specific Receptors: Receptors that respond to or bind the neurotransmitter should be present at or near the presumed site of action.

Pharmacological. 8. Antagonists: Certain compounds called antagonists should interfere with the response to the candidate substance through actions on synthesis, storage, release, action, inactivation, and/or reuptake.

Neurotransmitters that are widely accepted as having fulfilled all or most of the foregoing criteria are listed in Table 8.1 (1207).

Ligand-gated Ion Channel Receptor Superfamily

The γ-aminobutyric acid (GABA) and glycine (gly) receptors have some evolutionary connection to the nicotinic acetylcholine receptor (nAChR), a relationship adduced on the basis of amino acid sequence homology and domain distribution, particularly of hydrophobic segments (855, 1208, 1209). The GABA and gly receptors are activated by GABA or gly to allow the entry of chloride ion, whereas activation of the nAChR by acetylcholine leads to the

Table 8.1. Neurotransmitters

Chemical Name	Short Name	Formula
Acetylcholine	ACh	$(CH_3)_3N^+CH_2CH_2OC(=O)CH_3$
Glutamic acid	Glu	$NH_3{}^+CH(COO^-)CH_2CH_2COOH$
γ-Aminobutyric acid	GABA	$NH_3{}^+CH_2CH_2CH_2COO^-$
Glycine	gly	$NH_3{}^+CH_2COO^-$
Dopamine	dopa	$3,4\text{-}(HO)_2C_6H_3CH_2CH_2NH_3{}^+$
Noradrenaline[a]	NA	$3,4\text{-}(HO)_2C_6H_3CH(OH)CH_2NH_3{}^+$
Adrenaline[b]	A	$3,4\text{-}(HO)_2C_6H_3CH(OH)CH_2NH_2{}^+CH_3$
Serotonin	5HT	5-Hydroxytryptamine
Histamine	H	β-(5-Imidazolyl)ethylamine

[a] Norepinephrine.
[b] Epinephrine.

entry of small cations. The common denominator for the three receptors is that ligand-induced activation of the receptor opens an ion channel. The sequence homologies together with the common activation behavior implies strongly that the three receptors belong to a superfamily of ligand-gated ion channel receptors. Another candidate member of the superfamily is glutamate receptor, for which there is much pharmacological but little structural information. It is generally accepted that the strong homologies among the receptors of the nicotinic acetylcholine receptor family reflect common structural features. There is thus a good chance that structural conclusions for one receptor in a superfamily can be utilized in deducing certain aspects of the structure of another member of the superfamily. This principle will be applied in constructing a partial model for the $GABA_A$ receptor.

Nicotinic Acetylcholine Receptor Family

The nicotinic acetylcholine receptor (nAChR) consists of five subunits, $\alpha_2\beta\gamma\delta$. The nAChR from the endplate of the neuromuscular junction in the calf has an ϵ-subunit in place of the γ-subunit present in the fetal nAChR. Stimulating synthesis of nAChR by denervation produces extrajunctional nAChR, with electrophysiological properties resembling those of the fetal nAChR, $\alpha_2\beta\gamma\delta$. The adult form is the junctional form, $\alpha_2\beta\epsilon\delta$ (1210). In the rat, a family of genes, $\alpha 2$, $\alpha 3$, and $\alpha 4$, has been found on the basis of homology to the $\alpha 1$ of the muscle nAChR (1211, 1212). An additional gene, $\beta 2$, encodes a protein that lacks the cys-cys which is characteristic of the α-subunit of the nAChR. The $\beta 2$ protein combined with $\alpha 2$, $\alpha 3$, or $\alpha 4$ proteins forms functional nAChR in *Xenopus* öocytes (1211). Such receptors are not blocked by α-bungarotoxin (α-BTX) and are thus pharmacologically similar to ganglionic type nAChR. However, bungarotoxin 3.1 (Bgt 3.1), although indistinguishable from α-BTX (a 66-amino-acid MW 7300 polypeptide) by electrophoretic or amino acid analyses (1213), blocks the nAChR derived from $\alpha 3$-$\beta 2$ and $\alpha 4$-$\beta 2$ combinations. Bgt 3.1 also blocks nAChR from chick ciliary ganglions (1214). Another α-subunit, $\alpha 5$, occurs in the hippocampus (1212) and may be part of the nAChR detected by single channel recordings in hippocampal neuron cultures by applying either acetylcholine or the potent agonist, anatoxin-a (see Chapter 6) (1215). It is likely that neuronal nAChR is pentameric, like muscle nAChR, on the basis of the 5-subunit stoichiometry for the glycine receptor (1216).

The nAChR family includes fairly similar receptors that participate in different stages of development and those that differ somewhat more in sequence, subunit composition, and possibly subunit stoichiometry. Elucidating the structural and dynamic relationships within the family will become easier to carry out after the structure and dynamic properties of a particular nAChR are understood.

Gamma-aminobutyric Acid Receptor

Gamma-aminobutyric acid (GABA) is the major inhibitory neurotransmitter in the vertebrate brain. A receptor activated by GABA (the $GABA_A$ receptor) controls the entry of chloride ion via an ion channel. A number of important drugs act on the $GABA_A$ receptor, including "anxiolytics" (benzodiazepines) and "hypnotics" (barbiturates). A structural and dynamic model for the $GABA_A$ receptor is thus important in analyzing the action of some important pharmacological agents, and for the design of new agents. In addition, the balance of depolarization and inhibition (of depolarization) determines the nature of the signal within a neuron, and therefore the detailed structure of the representation within a polyneuronal system (Chaps. 1 and 9).

Affinity-purified GABA receptor from bovine cortex probably consists of five subunits, $\alpha_3\beta_2$ or $\alpha_2\beta_3$, the stoichiometry being inferred on the basis of the glycine receptor (1216). Cyanogen bromide cleavage of the receptor or the purified α-subunit led to peptides that could be separated by high-performance liquid chromatography. Synthetic oligodeoxyribonucleotide probes were designed on the basis of the sequences of several of the peptides and used to screen cDNA libraries from bovine and calf cerebral cortex. Apart from the signal peptides, sequences for the α- (429 amino acids) and β-subunits (449 amino acids) have been identified (855).

The cooperative character of the α,β-combination has been confirmed by demonstrating GABA responsiveness of receptor expressed in *Xenopus* öocytes after injection of the appropriate mRNAs. However, single α- or β-subunits can form chloride channels like those of the native $GABA_A$ receptor (1217).

Comparison of the α- and β-subunit amino acid sequences showed that 35% are identical, and that the homology is 57% when conservative substitutions (functionally plausible substitutions, FPS) are included. The hydrophobic character of segments is evaluated by running average analysis (869) of the sequence. A sufficiently long hydrophobic segment is most stable within the bilayer; four such segments (M1–M4) are found in each of the subunits. The two subunits probably arise from a common precursor in evolution (1218). The four segments, M1–M4, are distributed within the overall sequences of both subunits in a way that is strikingly similar to the distribution of M1–M4 within the nAChR. In addition, there is considerable sequence homology, especially in certain regions, between the $GABA_A$ receptor and the nAChR (855). The exocyclic disulfide link at α128–142 in nAChR occurs at α139–153 in the $GABA_A$ receptor. The close relationship in segment distribution, in amino acid sequence homologies, and in domain distribution (exobilayer versus bilayer) establishes that the $GABA_A$ receptor and the nAChR family of receptors belong to a receptor superfamily. However, the unusual disulfide between adjacent cysteines, present in almost all nAChR α-subunits, is not found in either subunit of the GABA receptor.

Modeling the GABA-Receptor

The classification of the $GABA_A$ receptor and nAChR in the same superfamily (855) suggests another idea, that structural inferences about the latter receptor might be useful for the former. Like the nAChR model, I divide the GABA receptor into three parts; exobilayer, bilayer, and cytoplasmic. Without experimental evidence on the shape of the GABA receptor, the folding is taken as analogous to that proposed for the subunits of the nAChR. The bilayer portion of the protein should contain the channel for chloride ions. I develop the analogy between nAChR and $GABA_A$ in the next section and construct a partial model for the $GABA_A$ receptor exobilayer portion. The model predicts successfully the excess positive charge on the interior strands that constitute the ligand-responsive portion of the receptor. Separate binding sites for anxiolytics (benzodiazepines) and hypnotics (barbiturates) are suggested.

Partial Exobilayer Model for γ-Aminobutyric Acid ($GABA_A$) Receptor

The superfamily relationship between the $GABA_A$ and nACh receptors implies that a structural similarity might exist. Cartoons for receptors are attractive but do not have sufficient physical and chemical detail to explain receptor behavior on the molecular level. Using the scheme developed for the nicotinic acetylcholine receptor (737), the exobilayer portion of the $GABA_A$ receptor is divided into 11 strands of 20 amino acids each. An overall view of the strand arrangement is shown in Figure 8.1. The lower strands—5, 6, 9, 10, 11—are on the inner side of the exobilayer portion of the receptor. Charged amino

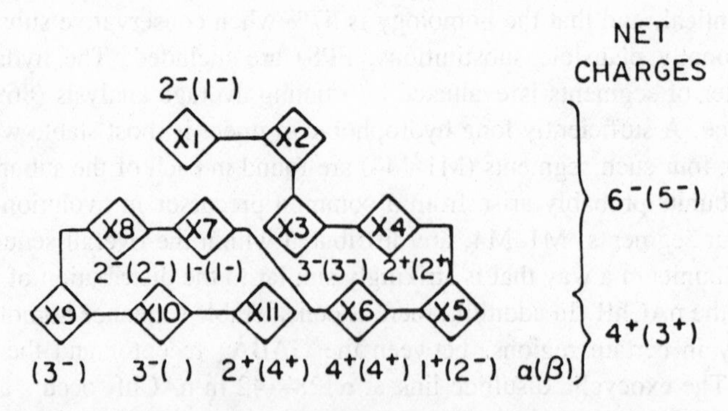

Fig. 8.1. Exobilayer strand arrangements for the α- and β-subunits of the γ-aminobutyric acid ($GABA_A$) receptor. Net charges for each strand of the α-subunit are shown (in parentheses for the β-subunit strands). The charges on the strands are unsymmetrically distributed in two ways. First, the net charge of the outer strands is negative, while that of the inner strands is positive. Second, the distribution of charges on the inner strands is quite unsymmetrical, the right-hand and center strands being very positive while the strands on the left are negative. The strand arrangements are the same as those for the nAChR (737.)

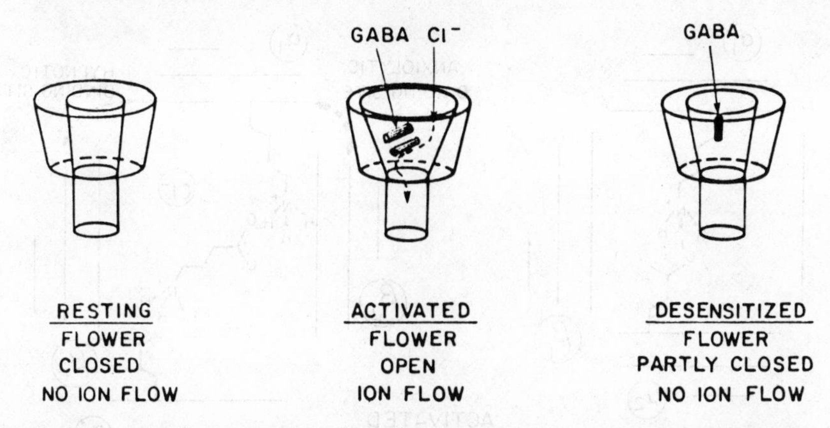

Fig. 8.2. Activation and desensitization mechanism for the $GABA_A$ receptor. Two GABA molecules bind to the interior of the exobilayer portion of the receptor, opening a pathway for chloride ions. A reorientation of the GABA molecule or molecules leads to a partial closing of the pathway and blocks the flow of chloride ions. (Adapted from reference 737.)

acids (lys^+, arg^+, glu^-, and asp^-) are counted and net charges shown next to each strand. No corrections are made for the possibility that adjacent amino acids might point in opposite directions on a β-strand. Surprisingly, the net charge is asymmetrically distributed, with the inner strands positive and the outer strands negative (856). Barnard *et al.* (855, 1218) pointed out that there were many positive charges near the bilayer, but could not define their relationships.

The "flower" model for nAChR is a reasonable choice as a model for the $GABA_A$ receptor (Fig. 8.2), since the phenomena of activation and desensitization are parallel in the two cases (1219) (Scheme 8.1). The composition of the receptor, $\alpha_3\beta_2$ or $\alpha_2\beta_3$, is presumably like that of nAChR. Activation of the receptor opens a pathway to chloride ion, which should involve the positively charged regions of the strands. Activation is achieved by two molecules of γ-aminobutyric acid binding between two subunits of the receptor, either two α- or two β-subunits. Desensitization involves rotation of the $GABA_A$

$$\mathrm{GR} \underset{}{\overset{\mathrm{GABA}}{\rightleftharpoons}} \mathrm{GABA.GR} \underset{}{\overset{\mathrm{GABA}}{\rightleftharpoons}} (\mathrm{GABA})_2\mathrm{GR} \longrightarrow (\mathrm{GABA})_n\mathrm{GR}$$

resting — activated — desensitized

GR = GABA receptor
GABA = γ-aminobutyric acid
n = 1 or 2

Scheme 8.1. Summary of the state sequence for the combination of the γ-aminobutyric acid receptor (GR) with γ-aminobutyric acid (GABA). The activated receptor requires two GABA molecules.

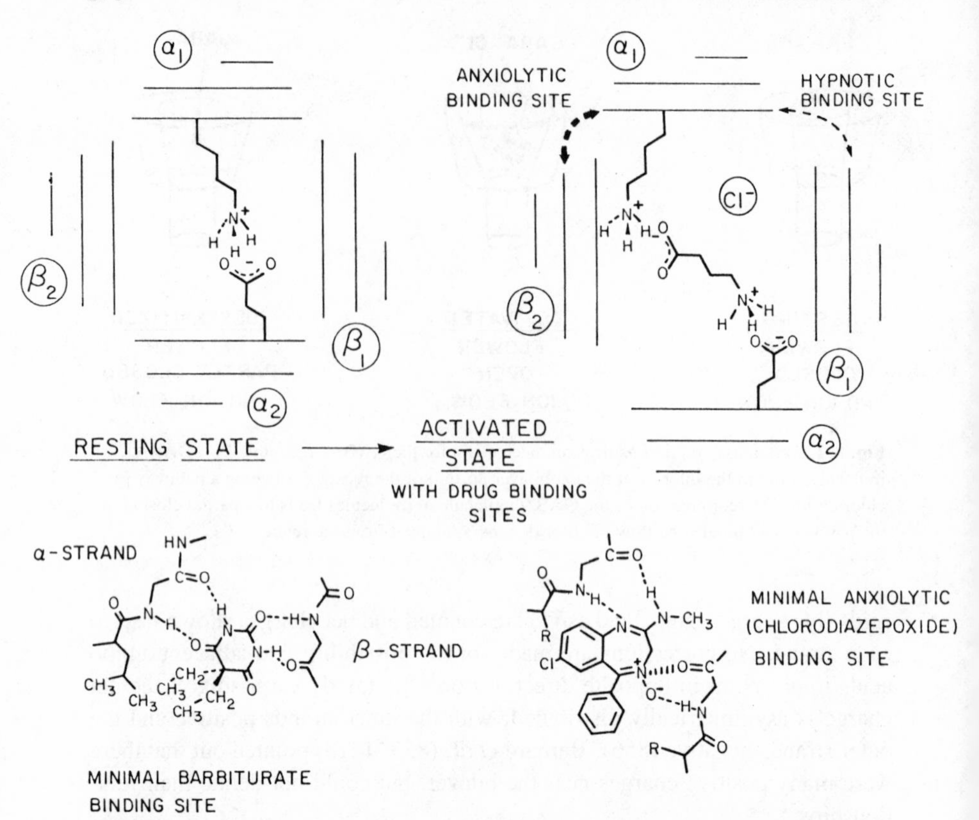

Fig. 8.3. Mechanism of activation of the $GABA_A$ receptor by GABA, in which the resting (closed) form of the receptor is converted into the expanded (opened) form of the receptor. Lines represent the strands shown individually in Fig. 8.1. The closed form is estimated to have a separation of 16 Å between opposed subunits, whereas the opened form has an α-α separation of ca. 21 Å. The binding-site groups for GABA are shown in a particular initial conformation in the resting form, but they change orientation by single group rotation in the activated form. The binding sites for anxiolytics and hypnotics between two antiparallel peptide strands are illustrated with detailed hydrogen-bonding schemes for a barbiturate and a benzodiazepine. The drug binding sites must include hydrophobic groups, shown as the side chain of leucine for barbiturate and as R for the benzodiazepine. A pentameric receptor, in analogy with the glycine receptor (1216), could be interpreted in the same way, using somewhat different separation distances. The proposed binding sites for receptor modulators would be the same.

molecule(s) to a position parallel to the strands, allowing the flower to close somewhat and cutting off the flow of chloride ion (856).

Anxiolytic and hypnotic drugs inhibit nervous activity by increasing the open time for the $GABA_A$ chloride channel. A schematic mechanism for the conversion of the resting (closed channel) to the activated state (open channel) is shown in Figure 8.3. The scheme was formulated in terms of a four-subunit stoichiometry for the receptor (855). The principles apply to a pentameric receptor. Single group rotation (737) changes the orientation of the binding groups, and combination of GABA with the receptor (an α-α site is arbitrarily chosen) expands the receptor from ca. 16 Å to 21 Å for the α-α dimension. The expansion opens a pathway to the bilayer channel, which is unspecified at this time (855, 1218). Prolonging the channel open time might be achieved through binding of drugs to the α5/β8 and β4/α9 strand combinations. The drug complex stabilizes the activated, expanded state of the receptor, retarding closure of the chloride pathway. The benzodiazepine binding site must be accessible to the milieu, since a benzodiazepine affinity column is used to purify the receptor. The model accounts for two different types of prolongations for channel opening, one due to anxiolytics and one due to depressants. Since the chloride pathway is close to the α5/β8 corner, I suppose this to be the depressant binding site. Channel blocking inhibitors like picrotoxinin and *t*-butylbicyclophosphorothionate (1220, 1221) bind on the interior to several positively charged groups. A detailed hydrogen-bonding scheme for the binding of barbiturate to two peptide strands shows quite clearly that the strands must be antiparallel (856) (Fig. 8.3).

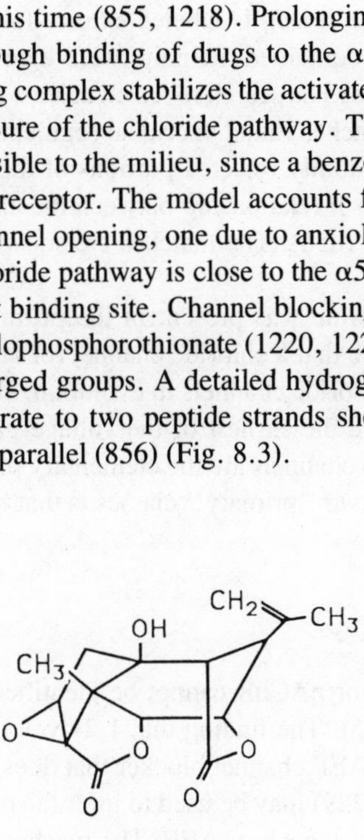

Picrotoxinin

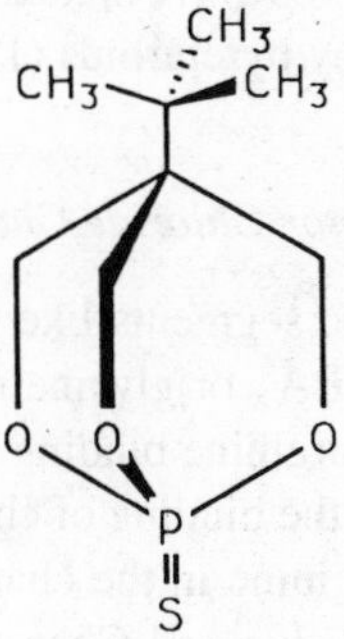

t-butylbicyclophosphorothionate

$GABA_A$ Receptor and Anion Exchange Protein

The anion exchange protein (AEP) of membranes (band 3 of red-blood-cell membranes) is similar in some respects to the γ-aminobutyric acid ($GABA_A$) receptor. Both proteins are inhibited and labeled by diisocyanatostilbenedisulfonate (DIDS), both transport chloride and bicarbonate ions (1222, 1223), and both are membrane proteins. The finding that bicarbonate ion could exit from cells through the chloride channel of the $GABA_A$ receptor suggested that a search be made for a possible connection between the receptor and anion exchange proteins (AEP). Amino acid sequences were chosen for examination on the basis of their importance to the activity of AEP. Starting with the lysines known to be labeled in band 3 protein, searches of the amino acid sequences of the $GABA_A$ receptor α- and β-subunits were carried out with critical pairs. Pairs made of the labeled lysines plus a neighboring amino acid were selected, all contained within the amino acid sequences compared in the analysis of AEP proteins (Fig. 5, 1224), the "membrane domains" of HKB3, human non-erythroid band 3 protein (1224), MEB3, mouse erythrocyte band 3 (1225), and a 72-peptide amino acid sequence of HEB3, human erythrocyte band 3 (1226). The tyrosine included in the search is labeled extracellularly by radioiodination (1227). Sequences that contained the pairs were then examined for matches in neighboring segments. The search revealed at least four reasonably homologous sequences. The matches found at critical segments of the $GABA_A$ receptor are not present in randomly selected portions of the receptor amino acid sequences. Thus, there is a relationship between the anion exchange protein and the $GABA_A$ receptor (856). The matches are shown in Fig. 8.4.

Given the high probability that the chloride was present in the primitive environment, it seems reasonable to propose that a chloride channel (or anion exchange protein) preceded ligand-gated chloride channels in evolution. Hille (897) suggested that calcium channels were the earliest of the voltage-gated channels. Apparently, chloride and calcium channels are the elementary primitive channels (856). A link between these two "primary" classes is that both are inhibited by pyrethroids (1228).

$GABA_A$ Receptor Chloride Channel in Bilayer

Amphiphilic segments like those noted for nAChR cannot be identified in either the $GABA_A$ or glycine receptors (855). The finding that 1,2-cyclohexanedione, an arginine binding agent, is an AEP channel blocker that does not interfere with the binding of chloride ion (1229) may be used to infer the presence of an arginine in the channel region of the r.b.c. AEP. The mechanism proposed by Falke and Chan (1230) for translocation of ions in AEP seems quite different from that envisaged here for the $GABA_A$ receptor. Nevertheless, it is similar in the sense that a physical barrier opens or closes the pathway for chloride ions. The question of the presence of another transmembrane segment between M3 and M4, as well as the related question of the orientation of M4, is best left for the future, pending further experimental data. As pointed out elsewhere (737), there is no really persuasive evidence concerning the orientation of M4 in nAChR and whether or not it ends extra- or intracellularly. Anions with a diameter of 5.6 Å or less can be transported through the mouse $GABA_A$ channel. As in the case of the AEP, binding sites for chloride ion exist within the channel. The existence of binding sites is indicated by the saturation of conductance with increasing chloride ion concentration (1231). A channel composed of five α-helices, each 11 Å in diameter, would be sufficient to account for a pore of 5.6 Å diameter. The nicotinic acetylcholine receptor channel is larger (7.4 Å) (1232) and is also composed of five α-helices.

HOMOLOGY SEARCH

The MEB numbering system (1225) is used to denote the amino acid sequences (1224).

Sequences containing pairs derived from MEB3 proteins:

MEB 448	EKTRNLMGVSELLISTA	Lys 449 (1225)
MEB 556	FSKLIKIFQDYPLQQTY	Lys 558, 561 (1225)
MEB 606	LRKFKNSTYFPGKLRRV	Lys 608, 610, 618 (1224, 1226)
MEB 645	TYTQKLSVPDGLKVSNS	Tyr 646 (1224, 1227)

FPS = functionally plausible substitutions

Matches:

		Identical	FPS	Similar	TOTAL
117					
KLLRITEDGTL	$GABA_A$-α				
KLIKIFQDYPL	MEB3	4/11	3/11	1/11	8/11
KLVKIFQEHPL	HKB3	4/11	3/11	1/11	8/11
557					
380					
RKPMSSREGYGR	$GABA_A$-β				
RKFKNSTYFPGK	MEB3	4/12	1/12		5/12
RKFKNSRFFPGR	HKB3	6/12	–		6/12
RKFKNSSYFPGK	HEB3	4/12	1/12		5/12
607					
196					
KKVEFTTGAYPRL	$GABA_A$-β				
RKFKNSTYFPGKL	MEB3	3/13	4/13	–	7/13
RKFKNSRFFPGRI	HKB3	2/13	4/13	–	6/13
RKFKNSSYFPGKL	HEB3	2/13	5/13	–	7/13
607					
347					
SYTPNLARGD	$GABA_A$-α				
TYTQKLSVPD	MEB3	4/10	1/10	–	5/10
TYTQKLSVPS	HKB3	3/10	1/10	–	4/10
TYTQ...	HEB3	2/4	1/4	–	3/4
646					

Fig. 8.4. Homologies between membrane anion exchange proteins (AEP) and $GABA_A$ receptor. Pairs based on labeled lysines (or a labeled tyrosine) from the AEP were used to search $GABA_A$ sequences. Sequences that included matched pairs were then examined further. Two sequences in each of the $GABA_A$ subunits were found to show substantial homology to anion exchange protein sequences. Further homologies may be present. The abbreviations are defined as follows: MEB3 = mouse erythrocyte band 3 protein; HEB3 = human erythrocyte band 3 protein, HKB3 = human non-erythroid band 3 protein.

Glycine Receptor

Glycine (gly) is the major neurotransmitter for inhibitory neurons in the spinal cord and brainstem. A receptor activated by GABA (the $GABA_A$ receptor) controls the entry of chloride ion via an ion channel. Taking advantage of the specific affinity of the alkaloid strychnine for the glycine receptor, the glycine receptor has been purified and shown to contain three polypeptides with 48K, 58K, and 93K molecular masses. One of the subunits (48K) can be photolabeled with strychnine. The 48K and 58K polypeptides are glycosylated integral membrane proteins and, thus, the components of the glycine receptor. The 58K subunit shows some immunological cross-reaction and therefore homology to the strychnine-binding subunit (1233). The 93K polypeptide seems to be a cytoplasmic peripheral membrane protein (1208).

The purified 48K subunit was degraded by V8 protease to peptides that are sequenced with the aid of HPLC detection of phenylthiohydantoins produced by the Edman degradation. 35-Mer oligonucleotides were synthesized using all possible codons to base 19 and a frequency-selected variety for the remainder. The oligonucleotides were used to probe a cDNA library, revealing a number of hybridizable clones. One clone was used to develop the entire coding sequence for the 48K glycine receptor subunit (1208).

The sequence of the rat subunit exhibits a number of significant features. First, there is considerable homology with either subunit of the bovine $GABA_A$ receptor (44%, including functionally plausible substitutions) (1218). Second, the distribution of hydrophobic helices is very similar to that of the $GABA_A$ receptor. Third, the receptor consists of five subunits (three of 48K and two of 58K) as shown by cross-linking with cleavable disulfide agents, molecular weight estimation by gel electrophoresis, and reduction of the separated pentamer into subunits (1216). These criteria allow the placement of the glycine receptor in the nAChR superfamily.

Using the nAChR exobilayer model, the exobilayer portion of the strychnine-binding subunit may be folded into 11 strands. Careful attention to the orientation of the charged groups leads to a model with a charge distribution on the interior (the part facing the chloride ion pathway) that is not too different from that found for the $GABA_A$ receptor. A molecular graphics analysis has implicated a portion of the strychnine molecule as being very similar to glycine (1234). Possibly, strychnine binds to the receptor in the desensitized form as proposed for antagonists to the nAChR.

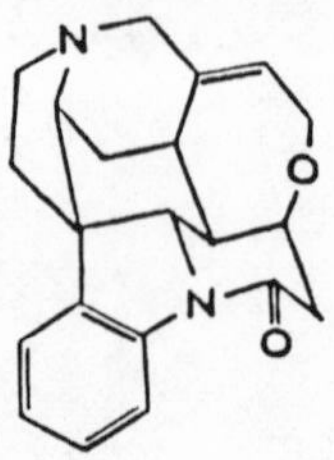

Receptor Superfamilies II: Ligand-activated G-protein Receptors

Receptors that transfer ligand activation across cell membranes to a family of "G-proteins," or guanine nucleotide regulatory proteins, constitute the second receptor superfamily. The ligands for such receptors are diverse, and include hormones (the α- and β-adrenergic receptors, α-AR, β-AR), neurotransmitters (muscarinic acetylcholine M_1 and M_2 receptors, mAChR, serotonin [5-hydroxytryptamine] receptors, 5HTR1c), photon-activatable retinylidene groups (rhodopsins and iodopsins), and neuropeptides (substance K receptor, SKR).

The G-proteins consist of three subunits (α, β, and γ) with extensive structural similarities between like subunits (1235). G-proteins are activated by ligands to replace guanosine diphosphate (GDP) by guanosine triphosphate (GTP), followed by dissociation of the GTP-loaded α-subunit. The GTP-carrying α-subunit causes different physiological effects depending on the subunit type, which varies with the G-protein source (1236). The latter include the adenylate cyclase (AdCase) stimulatory G_s-protein (α_s), the AdCase inhibitory G-protein (α_i), or a major functionally undefined G-protein (G_o, o = other) which may be channel-activating (α_o) (1237). Transducin is the G-protein that is activated by metarhodopsin-II; the parallels between light-activation of a G-protein and hormone activation have already been pointed out (Fig. 5.8, Chap. 5). The action of GTP-loaded α-subunits is terminated by hydrolysis of GTP to GDP.

The family resemblance among the G-protein receptors is illustrated by sequence comparisons of SKR, substance K receptor (bovine) (1238), ham2βAR, β_2-adrenergic receptor (hamster) (553), h2βAR, β_2-adrenergic receptor (human) (734), h1βAR, β_1-adrenergic receptor (human) (1239), h2αAR, α_2-adrenergic receptor (human) (1240), $mAChR_1$, muscarinic receptor (pig cerebral) (555), $mAChR_2$, muscarinic receptor (pig cardiac) (556), and Ops, rhodopsin (bovine rod) (564, 672) in Figure 8.5. The serotonin (5-hydroxytryptamine) receptor, HTRlc, belongs to the ligand-gated G-protein receptor family on the basis of sequence and probable structure (1241).

Beta-adrenergic Receptors

Adrenergic receptors respond to noradrenaline (norepinephrine) and belong to two main classes, α and β, each group being divided into at least two subclasses, α_1, α_2 and β_1, β_2. The α_1-adrenergic receptors stimulate breakdown of polyphosphoinositides; α_2-adrenergic receptors inhibit adenylate cyclase. The cyclase is stimulated by β_1- and β_2-adrenergic receptors. The inhibition of cAMP formation via G_i may occur either by direct inhibition of the cyclase (1236) or by diminishing the concentration of the GTP-$G_{\alpha s}$ complex through the release of βγ-subunits arising from the formation of a GTP-$G_{\alpha i}$ complex (1242).

Comparison of G-Protein Receptor Sequences

```
SKR       MGACVVMTDINISSGLDSNATGITAFSMPGWQLALWTAAYLALVLVAVMGNATVIWIILAHQRMRTVTNYFIVNLALADLCMAAFNAAFNFVYASHNIWYFGRAFCY  107
hamβAR   MGPPGNDSDFLLTTNGSHVPDHDVTEERDEAWVVG-MAILMSVIVLAIVFGNVLVITAIAKFERLQTVTNYFITSLACADLVMGLAVVPFGASHILMKMWNFGNFWCE  107
hβ2AR              MGQP----G%AFLLA-PNRSHAPDHDVT@IVMSLIVLAIVFGNVLVITAIAKFERLQTVTNYFITSLACADLVMGLAVVPFGAANILMKHWTFGMFWCE  107
hβ1AR           MGAGVLVLGA#GNLSSA-APLPDGAAT*LS"LLMALIVLLIVAGNVLVIVAIAKTPRLQTLTNLFIMSLASADLVMGLLVVPFGATIVVWGRWEYGSFFCE  132
hα2AR              MGSLQPDA&SWNGTE-APGGGARATPYS!CLAGLLMLLTVFGNVLVIIAVFTSRALKAPQNLFLVSLASADILVATLVIPFNLANEVMGYWYFGKTWCE  107
mAChR1             MNTSAPPAVSPNITVLAPGKGPWQVAFIGITTGLLSLATVTGNLLVLISFKVNTELKTVNNYFLLSLACADLIIGTFSMNLYTTYLLMGHWALGTLACD   99
mAChR2                MNNSTNSSNSGLALTSPYKTFEVVFIVLVAGSLSLVTIIGNILVMVSIKVNRHLQTVNNYFLFSLACADLIIGVFSMNLYTLYTVIGYWPLGPVVCD   97
Ops     MNGTEGPNFYVPFSNKTGVVRSPFEAPQYYLAEPWQFSMLAAYMFLLIMLGFPINFLTLYVTVQHKKLRTPLNYILLNLAVADLFMVFGGFTTTLYTSLHGYFVFGPTGCN  111
                                            |---M1-----------------|            |---M2-----------------|

SKR       FQNLFPITAMFVSIYSMTAIAADRYMAIVHPFQPRLSAPGTRA--VIAGIWLVALAL-AFPQCFYSTITTDEG---ATKCVVAWPEDSGGKMLLLYHLIVIALIYF-LP  209
hamβAR   FWTSIDVLCVTASIETLCVIAVDRYIAITSPFKYQSLLTKNKARMVILMVWIVSGLTSFLPIQMHWYRATHQK---AIDCYHKETCCDFF-TNQAYA-IASSIVSFYVP  211
hβ2AR    FWTSIDVLCVTASIETLCVIAVDRYFAITSPFKYQSLLTKNKARVIILMVWIVSGLTSFLPIQMHWYRATHQE---AINCYANETCCDFF-TNQAYA-IASSIVSFYVP  211
hβ1AR    LWTSVDVLCVTASIETLCVIALDRYLAITSPFRYQSLLTRARARGLVCTVWAISALVSFLPILMHWWRAESDE---ARRCYNDPKCCDFV-TNRAYA-IASSVVSFYVP  236
hα2AR    IYLALDVLFCTSSIVHLCAISLDRYWSITQAIEYNLKRTPRRIKAIIITVWVISAVISFPPLISIEKKGGGGGPQPA-----EPRC-EI--NDQKWYVISSCIGSFFAP  208
mAChR1   LWLALDYVASNASVMNLLLISFDRYFSVTRPLSYRAKRTPRRAALMIGLAWLVSFVLWA-PAILFWQYLVGERTVLAGQCYIQ-----FL--SQPIITFGTAMAAFYLP  200
mAChR2   LWLALDYVVSNASVMNLLIISFDRYFCVTKPLTYPVKRTTKMAGMMIAAAWVLSFILWA-PAILFWQFIVGVRTVEDGECYIQ-----FF--SNAAVTFGTAIAAFYLP  198
Ops      LEGFFATLGGEIALWSLVVLAIERYVVVCKPMS-NFRFGENHAIMGVAFTWVMALACAA-PPLVGWSRYIPEG--MQCSCGIDYYTPHE-ETNNESFVIYMFVVHFIIP  215
          |---M3-----------------|                 |---M4------------------|                            |---M5-------

SKR      LVVMFVAYSVIGLTLWRRSV                PGHQAHGANLRHLQAK--KKFVKTMVLVVVTFAICWLPYHLYFI-LGTFQEDIYCHKFI-QQVYLALFWL-AMSSTMYN  303
hamβAR   LVVMVFVYSRVFQVAKRQLQ--(20)----GRSGHGLRRSSKFCLK-EHKALKTLGIIMGTFTLCWLPFFIVNI-VHVIQDN-----LIPKEVYILLNWLGYVNSAF-N  322
hβ2AR    LVIMVFVYSRVFQEAKRQLQ--(20)----GRTGHGLRRSSKFCLK-EHKALKTLGIIMGTFTLCWLPFFIVNI-VHVIQDN-----LIRKEVYILLNWIGYVNSGF-N  322
hβ1AR    LCIMAFVYLRVFREAQKQVK--(45)----ANGRAGKRRPSRLVALREQKALKTLGIIMGVFTLCWLPFFLANV-VKAFHRE-----LVPDRLFVFFNWLGYANSAF-N  373
hα2AR    CLIMILVYVRIYQIAKRRTR--(123)---SASGLPRRRAGAGGQNREKRFTFVLAVVIGVFVVCWFPFFFTYT-LTAVGC------SVFRTLFKFFFWFGYCNSSL-N  432
mAChR1   VTVMCTLYWRIYRETENRAR--(124)---GKEQLAKRKTF-SLVK-EKKAARTLSAILLAFIVTWTPYNIMVL-VSTFCKD-----CVPETLWELGYWLCYVNSTI-N  414
mAChR2   VIIMTVLYWHISRASKSRIK--(149)---MTKQPAKKKP--PPSR-EKKVTRTILAILLAFIITWAPYNVMVL-INTFCAP-----CIPNTVWTIGYWLCYINSTI-N  436
Ops      LIVIFFCYGQLVFTVKEAAA                ---QQQESA---TTQKAEKEVTRMVIIMVIAFLICWLPYAGVAFYIFTHQGSD----FGPIFMTIPAFFAK--TSAVYN  302
          ----------|                                     |---M6-----------------|                 |---M7-------------
```

```
SKR      PIIY-CCLNHRFRSGFRLAFRCCPWVTPTEEDKMELTYTPSLSTRVNRCHTKEIFFMSGDVAPSEAVNGQAESPQAGVSTEP                384
hamβAR   PLIY--CRSPDFRIAFQELL--CLRRSSSKAYGNGYSSNSNGKTDYMGEASGCQLGQEKESERLCEDPPGTESFVNCQGTVPSLSLDSQGRNCSTNDSPL  418
hβ2AR    PLIY--CRSPDFRIAFQELL--CLRRSSLKAYGNGYSSNGNTGEQSGYHV------QEKENKLLCEDLPGTEDFVGHQGTVPSDNIDSQGRNCSTNDSIL  413
hβ1AR    PIIY--CRSPDFRKAFQCLL-CCARRAARRRHATHGDRPRASGCLARPGPPPSPGAASDDDDDDVVGATPPARLLEPWAGCNGGAAADSDSSLDEPCRPGFASESKV  477
hα2AR    PVIY-TIFNHDFRRAFKKIL-CRGDRKRIV                                                                       450
mAChR1   PMCYALC-NKAFRDTFRLLLLCRWDKRRWRKIPKRPGSVHRTPSRQC                                                      460
mAChR2   PACYALC-NATFKKTFKHLLMCHYKNIGATR                                                                      466
Ops      PVIY-IMMNKQFRNCMVTTL-CCGKNPLGDDEASTTVSKTETSQVAPA                                                     348
         -------|

SKR      Substance K receptor (bovine)              (1238)      hβ2AR  % NGS  @ QQRDEVWVVGHG
hamβAR   β2-Adrenergic receptor (hamster)           (553)       hβ1AR  # SEP
hβ2AR    β2-Adrenergic receptor (human)             (734)              * -(24)- AARLLVPASPPASLLPPASESPEP
hβ1AR    β1-Adrenergic receptor (human)             (1239)             " QQWT----AGHG
hα2AR    α2-Adrenergic receptor (human)             (1240)      hα2AR  & GNA  !LQVT----LTLV
mAChR1   muscarinic receptor, M1 (pig cerebral)     (555)
mAChR2   muscarinic receptor, M2 (pig cardiac)      (556)
Ops      opsin (bovine rod)                         (564, 672)
```

Fig. 8.5. Comparison of sequences for substance K receptor, α_1-, α_2-, β_1- and β_2-adrenergic receptors, muscarinic M_1 and M_2 receptors, and rod opsin, all G-protein receptors. The alignment of Masu *et al.* (1238) was used as the basis, with some modifications after including the alignments of Kobilka *et al.* (1240) and small obvious adjustments to improve the matches. To simplify the presentation, some nonhomologous sequences have been omitted, these being either noted below or denoted by the number of amino acids in the omitted insert. The homologies in the segments assigned to hydrophobic helices are very substantial, especially for some of the amino acids regarded as essential for ligand binding. The membrane helices, marked M1–M7, are selected following the choices for opsin (564).

The discovery that the sequence of the hamster β_2-receptor had considerable homology to that of rod opsin (553) opened the way to recognizing the G-protein superfamily of receptors. The family relationship was further substantiated by comparison with β_2-receptor sequences from humans and turkeys, the latter receptor possibly having β_1-receptor character (734, 1243, 1244). The homologies between the opsins and various adrenergic receptors have been extensively documented (559, 1243) (Fig. 8.5).

There are a number of important molecular questions about the G-protein receptors. What is the overall structure? Where is the ligand binding site? What is the mechanism by which the conformational change is induced by binding of the ligand? On the basis of the strong homologies with opsins and a careful search for putative hydrophobic transmembrane sequences, it is inferred that there are seven transmembrane helices, probably α-helices. By analogy with the intrabilayer location for the retinylidene moiety of rhodopsins and iodopsins and on the basis of its presence in all G-protein receptors, asp 83 has been suggested as the part of the binding site for catecholamines (559). The numbering, from rod rhodopsin, corresponds to asp 99 in iodopsin, Figure 5.14. A location well within the bilayer is supported by the results with mutant β_2-receptor, from which it may be concluded that the beginning, the end, and a number of the hydrophilic loops are not important for binding of the specific antagonist ^{125}iodocyanopindolol (1245, 1246).

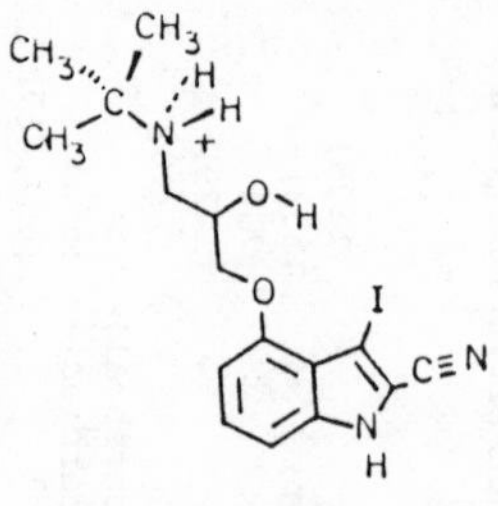

IODOCYANOPINDOLOL

A slightly different assignment for the location of M3 places asp 113 (β_2-AR) near the outside of the bilayer. The numbering is given in Figure 8.5 for β_2-AR; the same amino acid is present at the homologous position in all AR. It seems possible that the intrabilayer aspartic acid residues on M2 (β-AR asp 79, present at homologous positions in all G-protein receptors) and M3 can form a carboxylic acid dimer. Substitution of asn for asp 130 (β-AR) gave a receptor with unusually high agonist binding affinity (1247). Binding of a ligand, either noradrenaline to β-AR or acetylcholine to mAChR, would (a) alter the local conformations and (b) introduce an ion-pair at an initially neutral location. Other interactions near the presumed binding site can be recognized. Arginines are located just within the inner side of the bilayer on M3

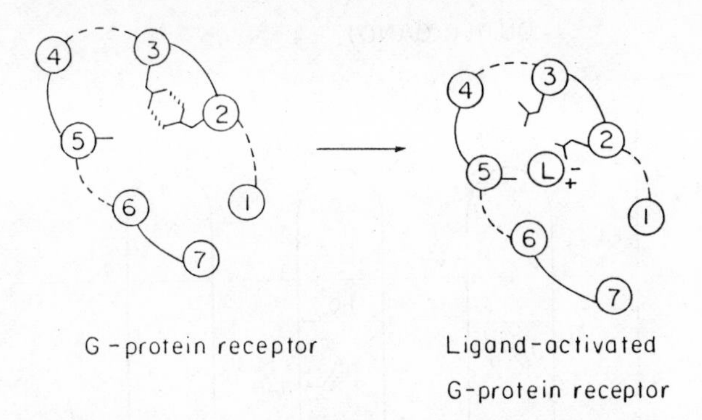

Fig. 8.6. A general model for a G-protein receptor, looking down into the membrane from the exterior of the cell. The arrangement of the helices is based on the arrangement of the seven helices in bacteriorhodopsin. The aspartates in M2 and M3 are shown as hydrogen-bonded to one another (similar to a carboxylic acid dimer in a low-polarity solvent), with some of the groups on M5 indicated as a line. Binding of ligand changes the conformation, and leads to changes in the interaction of the receptor with a G-protein. An antagonist presumably stabilizes the original conformation.

and M5. Ser 207 (β_2-AR, homologous positions in other AR) is strategically placed at the middle of M5 only in AR, including the D_2 dopamine receptor (1247). It seems likely that the conformational changes can account for the effect on G-proteins. The ligand binding is shown from the outside in Figure 8.6, with a schematic side view in Figure 8.7. The bilayer thickness is about 36 Å; adrenaline can link groups over 12–13 Å, not counting the lengths of the binding groups themselves (Fig. 8.8). A molecular graphics study of the G-protein receptors could produce a more definitive analysis of the sketchy mechanism outlined here.

The overall behavior of the adrenergic receptor has a resemblance to that of the nAChR (Scheme 8.2), for which the process has been explained as a conformational change in both ligand and receptor. Phosphorylation appears to be the major pathway for desensitization. The molecular details are unknown, but it has been suggested that coupling of the receptor to the G-protein is decreased by phosphorylation (1248). A kinase catalyzes the phosphorylation of β-adrenergic receptor and rhodopsin (1249), and is thus similar to opsin kinase or arrestin (748). The latter has been isolated (747) and characterized as a member of a superfamily of such kinases (1250). It is perhaps significant that

$$\text{Receptor}^* \xrightarrow{\text{Ligand}} \underset{\text{Activated}}{\text{Receptor.ligand}} \xrightarrow{\text{Phosphorylation}} \underset{\text{Desensitized}}{\text{P-receptor.ligand}}$$

*adrenergic receptor

Scheme 8.2. Activation and desensitization through phosphorylation of adrenergic receptors.

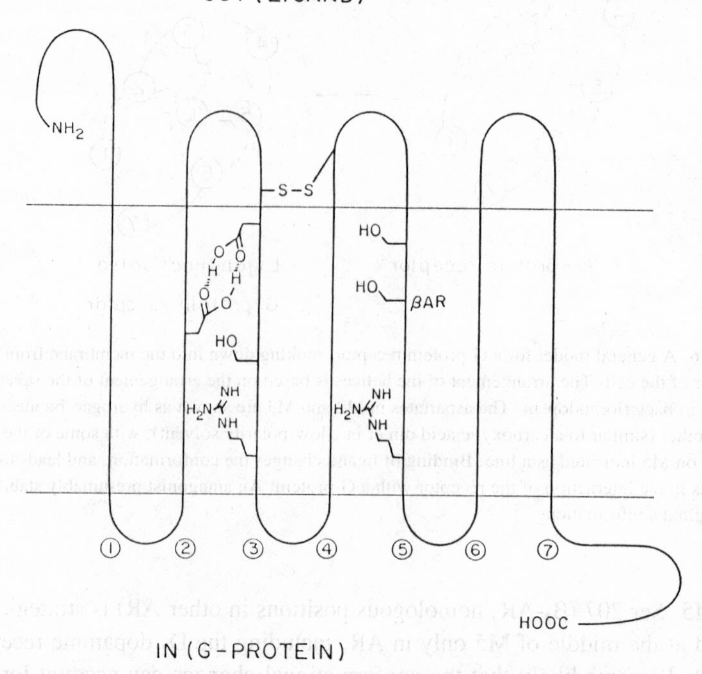

Fig. 8.7. Seven transmembrane helices of a G-protein receptor. Some of the probable ligand binding groups are shown. The OH group in the middle of M5 is present only in adrenergic receptors, including the dopamine receptor (1247), and may be used for interaction with the 3-OH group of a catecholamine.

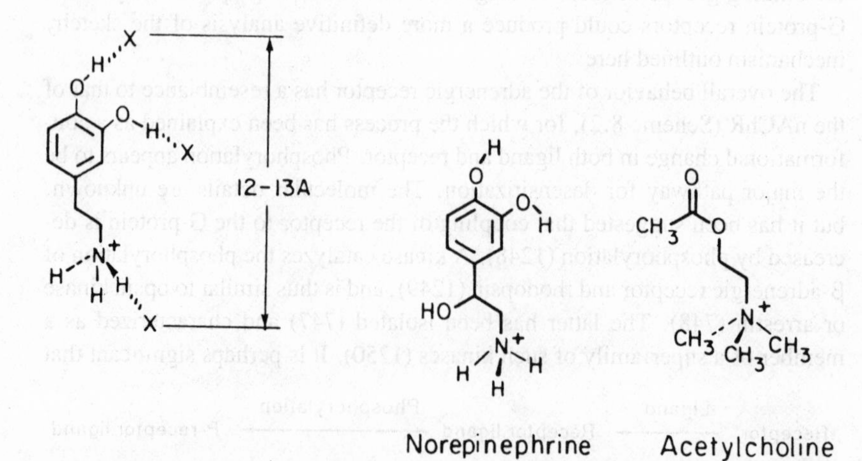

Fig. 8.8. Hydrogen-bonding arrangements for the binding of a catecholamine derivative to an adrenergic receptor. The formulas of acetylcholine and noradrenaline (norepinephrine) are given to show the similarity in length and shape.

phosphorylation substantially increases the rate at which the nAChR is desensitized (960, 1251).

Muscarinic Acetylcholine Receptors

The active agent, muscarine, from the beautiful but deadly mushroom *Amanita muscaria* stimulates acetylcholine receptors entirely different from those classified as "nicotinic" (nAChR) (Chap. 6). The nAChRs respond to acetylcholine with rapid, millisecond-time-scale opening of a cation channel. Muscarinic AChR (mAChR) responds slowly over seconds by a second messenger mechanism, typically involving G-proteins. MAChRs are located mainly in the central nervous system and on smooth muscles, and have been classified pharmacologically into M_1 and M_2 subtypes on the basis of affinity for the antagonist, pirenzepine. A related compound, telenzepine, is somewhat more specific in its binding to M_1 receptors. *Second messengers* are explicitly defined as substances that are produced in response to a receptor stimulus, the stimulus being regarded as a messenger. Second messengers include G-proteins—or more specifically, the Gα.GTP complex—and inositol 1,4,5-trisphosphate (IP_3).

The mAChRs induce various physiological responses, such as increases in cGMP, decreases in cAMP, changes in ion permeability, and formation of

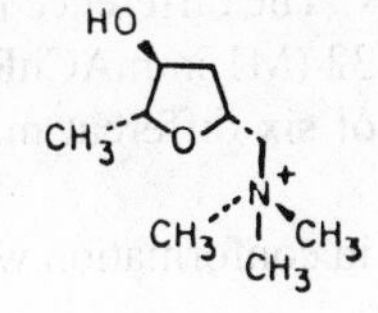

Muscarine

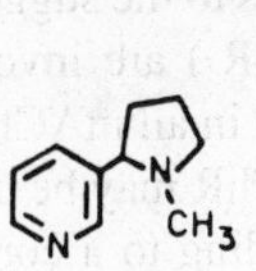

Nicotine

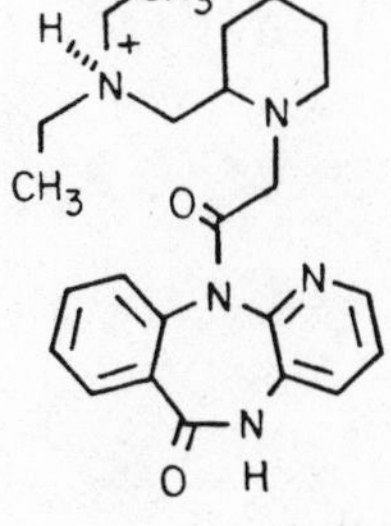

AF-DX-116 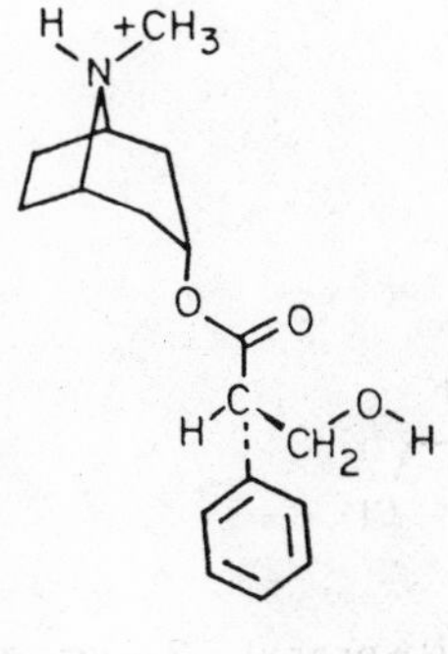 Atropine 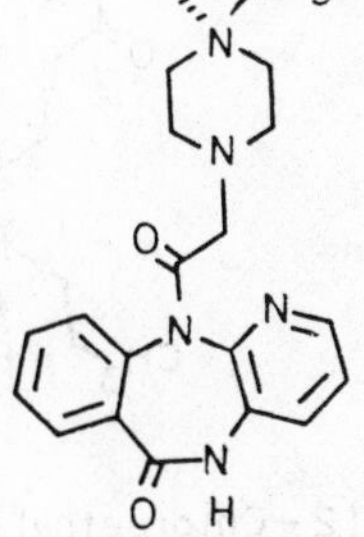 Pirenzepine 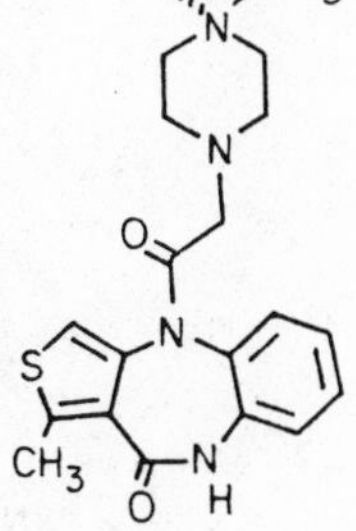 Telenzepine

inositol polyphosphate second messengers. Reconstituted purified mAChR from porcine brain binds acetylcholine more effectively in the presence of two different G-proteins, G_i or G_o (1252). Are there multiple mAChRs for various functions, or is there only one mAChR acting in different ways?

A partial amino acid sequence, met-pro-met-val-asp, obtained by sequencing a peptide from tryptic degradation of purified mAChR, defined a set of oligodeoxyribonucleotide probes used to screen a cDNA library from porcine cerebrum. The selected clones were used in further screening, finally resulting in a DNA fragment used to determine the DNA and thus the amino acid sequence for porcine $mAChR_1$ (555). Investigation of cDNA derived from porcine cardiac muscle yielded the sequence of porcine $mAChR_2$ (556, 1253). Both are shown in Figure 8.5. A third porcine muscarinic receptor, similar to M_3, has been sequenced (1254). Expression of porcine $mAChR_1$ and $mAChR_2$ receptors in *Xenopus* öocytes revealed that pirenzepine was bound 50 times more strongly to cerebral $mAChR_1$, AF-DX 116 (1255) some 5 times more strongly to cardiac $mAChR_2$, and atropine more or less equally to both subtypes (1256, 1257). At least four additional rat and human mAChR have been sequenced, expressed, and examined for pirenzepine binding (1258, 1259). The multiple sequences are of importance in the process of identifying sequences and amino acids that might be related to the binding of ligands.

Irreversible labeling by CPD of acidic amino acids in hydrophobic regions of the mAChR leads to the suggestion that aspartates 71 (M2 in $mAChR_1$) and 105 (M3 in $mAChR_1$) are involved in ligand binding (1260, 1261). These residues are present in all mAChR and AR. The difference in antagonist binding to various mAChR may be due to thr 32 (M1 in $mAChR_1$) and thr 76 (M2 in $mAChR_1$) according to a comparison of six different mAChR (1258), including the two shown in Figure 8.5.

The general mechanism for the change in conformation with ligand binding

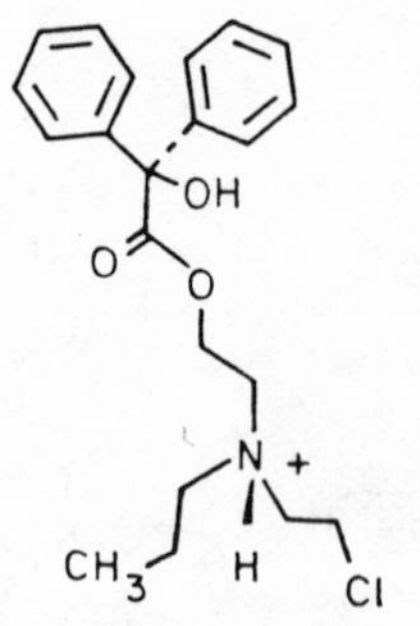

N-(2-Chloroethyl)-N-propyl-2-aminoethyl
2,2-diphenyl-2-hydroxyacetate

for adrenaline (Figs. 8.6 and 8.7) should apply to acetylcholine binding. The conformationally altered mAChR can promote the replacement of GDP by GTP in more than one type of G-protein. A porcine cardiac $mAChR_2$ expressed in ovary cells can either inhibit adenylate cyclase or promote phosphoinositol turnover (1262). Interaction with two different G-proteins is detected with mAChR in reconstituted porcine cerebral $mAChR_1$ in vesicles (1252).

A G-protein.GTP complex (GG) induced by the action of acetylcholine on cardiac mAChR can directly activate K^+ channels. Islet-activating protein (IAP) specifically ADP-ribosylates G_i or G_o and interferes with K^+ channel activation, which is not affected by concentration changes of second messengers like cAMP or cGMP, indicating a direct involvement of GG (1263, 1264). Noradrenaline and γ-aminobutyric acid inhibit voltage-dependent Ca^{++} channels in a similar way, acting through AR and $GABA_B$ receptors (1265). Purified G_α.GTP stimulated K^+ (1266) and Ca^{++} channels in guinea pig and bovine membranes (1267, 1268). The K^+ channel and a slow, inward Ca^{++} current are both inhibited by a GTP analogue, GppNHp (1269). Scheme 8.3 summarizes the consequences of combining ACh with mAChR.

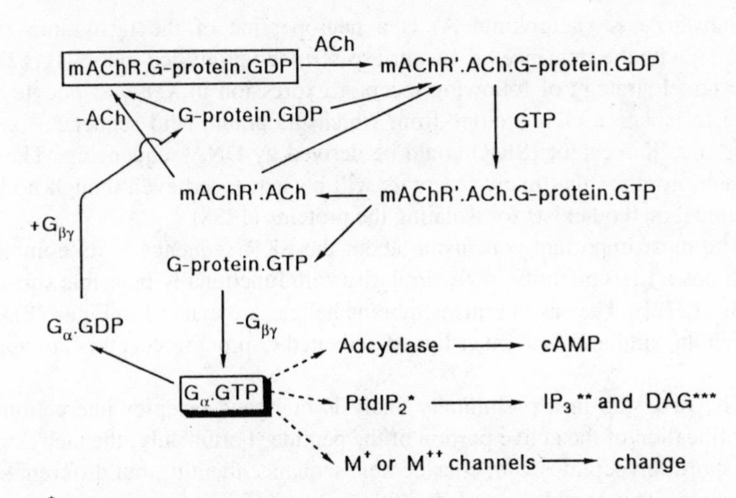

Scheme 8.3. Combination of acetylcholine (ACh) with muscarinic acetylcholine receptor (mAChR). The activated receptor exchanges guanosine triphosphate (GTP) for guanosine diphosphate (GDP), which then gives rise to G_α.GTP, an activator for Adcyclase (cyclic adenosine monophosphate, cAMP), for the enzyme that produces inositol triphosphate (IP_3), or for certain cation channels. The overall pattern of reactions resembles that found for other G-proteins, such as the transducin of the retina.

Rhodopsin (Light Receptor)

Rhodopsin is the paradigm for ligand-activated G-protein receptors, and has been described in great detail in Chapter 5. Opsin, the protein of rhodopsin and iodopsin that lacks the light-responsive ligand, has been sequenced for a variety of species and color receptors. Bovine rod cell opsin is compared to adrenergic and muscarinic receptors in Figure 8.5. The presence of seven transmembrane segments presumed to be α-helices is clear in every G-protein receptor. Based on the intrabilayer location of the retinylidene "ligand" of rhodopsin, it has been suspected that other G-protein receptors will bind ligands in the bilayer region. This opinion is supported by the demonstration that most of the hydrophilic regions of an adrenergic receptor are not important for the binding of ligand, and the indication that labeling of a muscarinic receptor takes place within a hydrophobic region. The involvement of a G-protein in the pathway from light activation to nerve signal was worked out for rhodopsin with the G-protein, G_t or transducin.

Substance K Receptor

Substance K (neurokinin A) is a neuropeptide of the tachykinin class (1270), with the sequence his-lys-thr-asp-ser-phe-val-gly-leu-met-NH_2 (1271). The novel strategy of following receptor expression in *Xenopus* öocyte was used to isolate a cDNA clone from which the amino acid sequence for the substance K receptor (SKR) could be derived by DNA sequencing. The approach promises that many receptors will be sequenced even though no biochemical method exists for isolating the proteins (1238).

The most important conclusion about the SKR sequence is its homology with opsin (21% identity, 46% similarity with functionally plausible substitutions [1272]). The seven transmembrane helices are marked in Figure 8.5, in which the similarities to other ligand-activated G-protein receptors are apparent.

The first step in a preliminary analysis of ligand-receptor interactions is identification of the active portion of the peptide. Fortunately, the tachykinins are short polypeptides with considerable sequence identity, but different sites of action. The peptides listed in Table 8.2 have four common amino acids, with slight modifications in biologically active synthetic peptide analogues (1273). The "ligand-active" portion of neurokinin A (substance K) is likely to be the first four amino acids, his-lys-thr-asp. The active portion of the truncated substance P and neurokinin B analogues must be the first one or two amino acid residues.

The second step in the analysis is a search for a suitable binding site in the receptor protein, folded in the manner suggested for rhodopsin and other G-protein receptors. The idea that adrenergic and muscarinic receptors have

Table 8.2. Mammalian Tachykinins

Name	Sequence
Substance P	arg-pro-lys-pro-gln-gln-phe-phe-gly-leu-met-NH_2
Neurokinin A (Substance K)	his-lys-thr-asp-ser-phe-val-gly-leu-met-NH_2
Neurokinin B (Neuromedin K)	his-lys-thr-asp-ser-phe-val-gly-leu-met-NH_2
"Septide"[a] (Substance P analogue)	pglu-phe-phe-pro-leu-met-NH_2
"Senktide"[b] (Neurokinin B analogue)	succinyl-asp-phe-N-methylphe-pro-leu-met-NH_2

[a] Synthetic analog active as ligand for substance P receptor. pglu, pyroglutamic. Replacement of gly by an N-methylgly residue or pro enhances specificity toward a particular tachykinin receptor (see reference 1273).

[b] Synthetic analog active as ligand for neurokinin B receptor. Replacement of phe by an N-methylphe residue enhances specificity toward a particular tachykinin receptor. The succinyl-asp combination in the synthetic peptide increases water solubility, but is equivalent to the pglu derivative in activity and specificity (see reference 1273).

intrabilayer ligand binding sites between M2 and M3 and/or M5 provides a useful analogy, especially in view of the homologous presence of 79 asp. Comparison of the length of the "ligand-active" portion of neurokinin A, shown below, with the site-bound catecholamine in Figure 8.8 shows great similarity. The shape and charge interactions might also be similar. I suggest that the ligand binding is intrabilayer, and that the mechanism of conformational change is like those of the other G-protein receptors. The synthetic analogues of tachykinins contain a minimal number of groups in the "ligand-active" portion of the receptor. The N-methyl groups probably influence conformation at the binding sites, which are similar in the tachykinin receptors. Further analysis requires a comparison of the sequences of the three receptors.

The implications of this conclusion are profound and worth further exploration with respect to the general idea that receptor binding sites are generally accessible to the milieu that contains the prospective ligand. Another set of ligands that might penetrate protein receptors could be odorants (odorant receptor, Chap. 3).

Activation of the SKR presumably releases a G_α.GTP complex, which is thought to activate the inositol lipid second messenger system (1272, 1274).

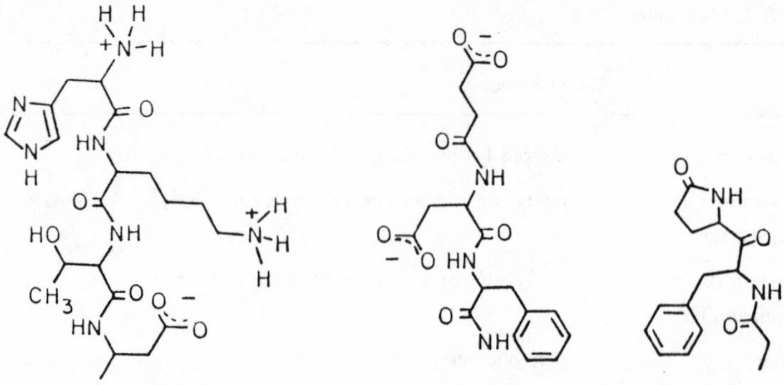

Neurokinin A Neurokinin B analogs

Ligand-active regions

Ligand-activated Phospholipase C-related Receptors

Several receptors are linked to the hydrolysis of inositol phospholipids, either indirectly via G-proteins or directly through phospholipase C, which catalyzes the hydrolysis of phosphatidylinositol 4,5-bisphosphate ($PtdIP_2$) to inositol 1,4,5-trisphosphate (IP_3) and 1,2-diacylglycerol (DAG) (538–540, 1275, 1276). Inositol 1,3,4,5-tetrakisphosphate (IP_4) is formed from IP_3 by phosphorylation with IP_3-3-kinase (1277). Both IP_3 and IP_4 are needed for activation of Ca^{++}-dependent K^+ channels in lacrimal acinar cells (1278). A functional role in the nervous system for IP_5 and IP_6 has been suggested (1279). However, some inositol phosphates are metabolic intermediates rather than second messengers (1280). A specific receptor for IP_3 has been found by radioligand binding to anterior pituitary membranes (1281). The formulas of the inositol phosphates are shown in Figure 8.9. Prolonged reduction in voltage-dependent K^+ currents (both ''A-current'' and Ca^{++}-dependent K^+ current) occur after introducing IP_3 into isolated *Hermissenda* photoreceptors, an effect also caused by injection of Ca^{++} (1282). Introduction of IP_3 into either *Musca domestica* or white-eyed *Drosophila* photoreceptor cells mimicked the effects of light on electrophysiological responses (1283).

Muscarinic agonists activate a G-protein which promotes phospholipase C-catalyzed hydrolysis of $PtdIP_2$ to IP_3 and DAG. The DAG stimulates protein kinase C which inactivates a G-protein (G_i?) by phosphorylation. The neurotransmitter adenosine produces an outward current in rat hippocampal cells by activating the G_i, and it is this response that is blocked by muscarinic agonists. Phorbol-12,13-diacetate, which activates protein kinase C, also blocks the action of adenosine (1284). Bovine and human protein kinase Cs have been sequenced, and represent a well-conserved family of proteins (1285, 1286) involved in intracellular signal transfers. Serotonin (5-HT) causes appearance

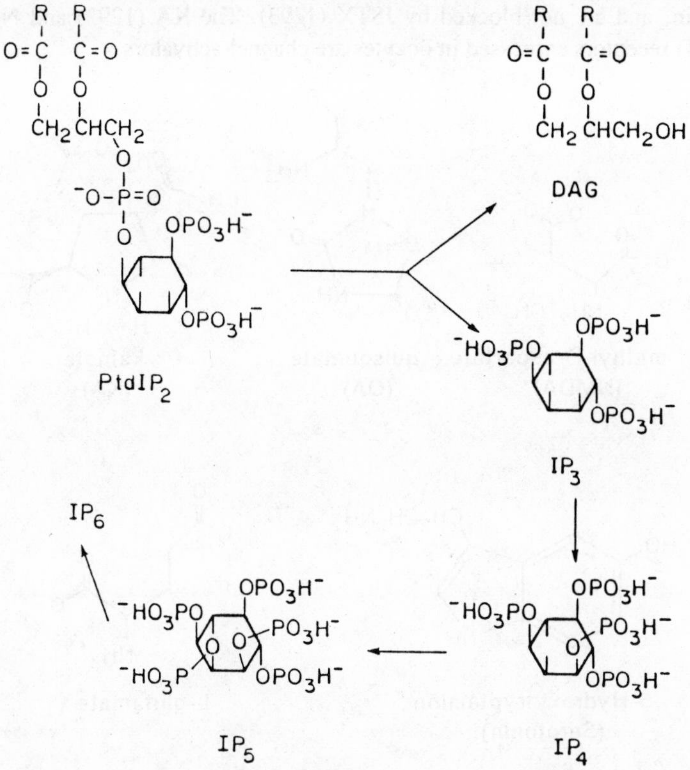

Fig. 8.9. Relationships between phosphatidylinositol diphosphate ($PtdIP_2$), diacylglycerol (DAG), and inositol phosphates (IP_3, IP_4, IP_5, and IP_6).

of Cl^- currents after injection of rat brain mRNA into *Xenopus* öocytes via G-proteins and IP_3 formation (1287). Serotonin receptor-rich mRNA is fractionated by following expression in *Xenopus* öocytes, eventually yielding the 460-amino-acid sequence of the 5-HTR1c (1241, 1288–1290). The roles of G-protein subunits has been reviewed (1291). A "Ca^{++}-mobilizing" receptor from human polymorphonuclear leucocyte (PMN) plasma membranes responds to the chemoattractant tripeptide, fmet-leu-phe, by formation of activated G-protein and subsequent phospholipase C hydrolysis of $PtdIP_2$ (1292).

Excitatory amino acid receptors have thus far escaped sequencing and modeling. Three major subtypes are recognized in the central nervous system by pharmacological means, according to their response to the agonists, N-methyl-D-aspartate (NMDA), quisqualate (QA), and kainate (KA) [cf. (1238)]. Excitatory amino receptors have been classified as channel activators, analogous to the nAChR. Fast QA/glu receptor channels are blocked by Joro spider toxin (JSTX). However, the QA receptors expressed from rat brain mRNA injected into *Xenopus* öocytes activate the formation of IP_3 via a G-protein, and are not blocked by JSTX (1293). The KA (1293) and NMDA (1294) receptors expressed in öocytes are channel activators.

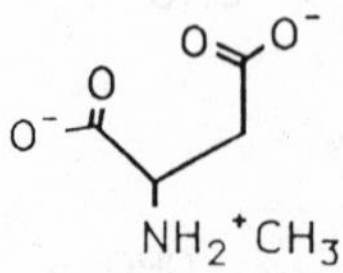

N-methyl-D-aspartate (NMDA) quisqualate (QA) kainate (KA)

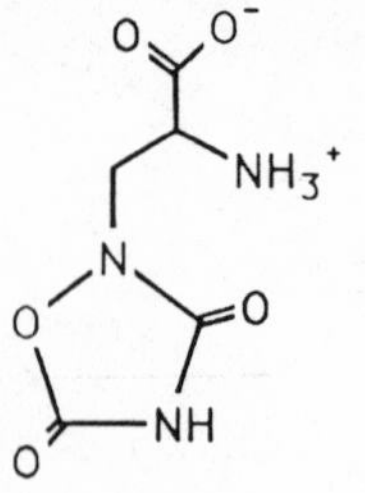

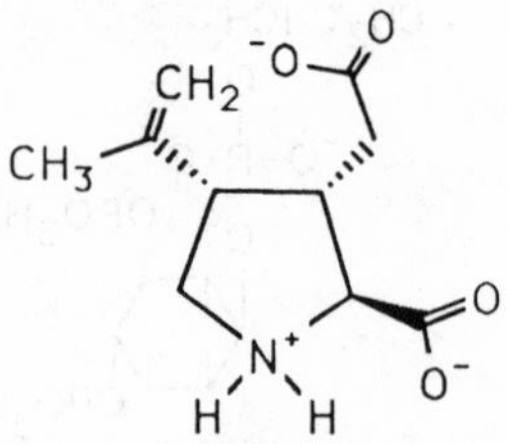

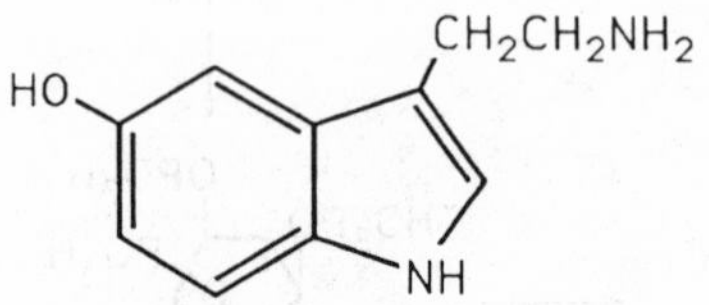

5-Hydroxytryptamine (Serotonin) L-glutamate

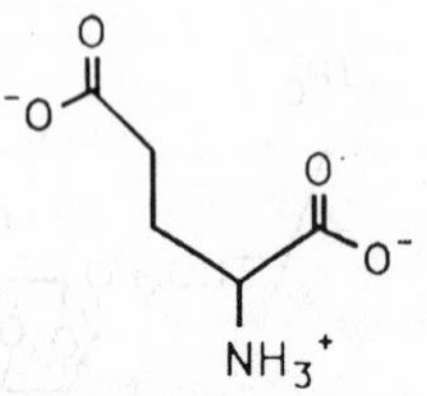

Acetylcholine, glutamate, and γ-aminobutyrate are neurotransmitters, which can activate either a ligand-activated ion channel receptor or a ligand-activated G-protein receptor. The biological and evolutionary significance of these multiple activities might arise from the need to activate responses over different time scales.

Ligand-activated Phospholipase C-related Receptors Involving Arachidonate Metabolites

Serotonin (5-HT) receptors decrease the activity of "S-type" K^+-channels in sensory neurons of the marine mollusc *Aplysia californica* through phosphorylation by cAMP generated by G-protein activation. A tetrapeptide, phe-met-arg-phe-NH_2 (FMRFamide), increases the activity of the channels, acting via a complex pathway involving G-protein stimulation of phospholipase C, which releases arachidonic acid from phospholipid. The arachidonic acid is oxygenated by 12-lipoxygenase, 5-lipoxygenase, or cyclooxygenase, the last leading to prostaglandins. Although both 5- and 12-hydroxyeicosatetraenoic acid (5- and 12-HETE) could be detected after application of FMRFamide to *Aplysia* tissue, the 12-hydroperoxy derivative (12-HPETE) had an effect on the action potential even greater than the FMRFamide. Since 12-HPETE can diffuse away from the site of generation, it might affect other ion channels in the local region (1295).

Two important points are clear. First, a molecular pathway is defined for the action of a neuropeptide. Second, a novel second messenger appears to be produced. How can 12-HPETE affect the operation of a channel, and how can its action counter the effect of phosphorylation? The model for the Na^+ channel provides a way of understanding the effect. First, the hydrophobic 12-HPETE is probably active within the bilayer. Second, the OOH group is more strongly acidic than an OH group. Thus, if the OOH group encountered an NH_3^+, it is possible that an ion-pair, NH_3^+-OO^-, would be formed. The net effect would be to prevent the positively charged group from interfering with the passage of K^+ ions through the channel. (The superfamily relationship of ion channels suggests some similarity in mechanism.) Enzymatic transformation of the 12-HPETE to 12-HETE would greatly decrease the effect, since an OH group is much less acidic than OOH. The formulas for arachidonic acid and the 12-hydroperoxy derivative indicate the hydrophobic character of this class of compounds:

R-CH=CH-CH_2-CH=CH-CH_2-CH=CH-CH_2-CH=CH-X

arachidonic acid (all-*cis*-5,8,11,14-eicosatetraenoic acid)

R-CH=CH-CH_2-CH(OOH)CH=CH-CH=CH-CH_2-CH=CH-X

12-hydroperoxy-all-*cis*-5,8,10,14-eicosatetraenoic acid (12-HPETE)

R=$CH_3(CH_2)_5$, X=$(CH_2)_3$COOH

Neuropeptides

A large number of peptides (1296,1297), of which some are illustrated, have neurophysiological or related activities. The substance K receptor has been described as being related to rhodopsin. The FMRFamide receptor seems to be a G-protein receptor (1295). Another neuropeptide, neurotensin, has been shown to share common response on *Xenopus* öocytes in activating a G_i protein and in producing IP_3 (1298). Neuropeptide receptors appear to belong to the ligand-gated G-protein receptor superfamily.

Channel Superfamilies

The dihydropyridine binding protein (DHP) is thought to be a calcium channel. The sequence of the 170K DHP protein from rabbit skeletal muscle is

Neuroactive Peptides*

Name of neuropeptide (No. of amino acid residues) Amino acid sequence

Carnosine (2) ala-his

Thyrotropin Releasing Hormone (TRH) (3) pglu-his-proNH2

met-Enkephalin (5) tyr-gly-gly-phe-met

leu-Enkephalin (5) tyr-gly-gly-phe-leu

angiotensin II (8) asp-arg-val-tyr-ile-his-pro-pheNH2

cholecystokinin-like peptide (9) asp-tyr-met-gly-trp-met-asp-pheNH2
|
SO3H

oxytocin (9) ile-tyr-cys
| |
gln-asn-cys-pro-leu-glyNH2

vasopressin (9) phe-tyr-cys
| |
gln-asn-cys-pro-arg-glyNH_2

luteinizing-hormone
releasing hormone (LHRH) (10) pglu-his-trp-ser-tyr-gly-leu-arg-pro-glyNH2

neurotensin (13) pglu-leu-tyr-glu-asn-lys-pro-arg-arg-pro-tyr-ile-leu

bombesin (14) pglu-gln-arg-leu-gly-asn-gln-trp-ala-val-gly-his-leu-metNH2

somatostatin (14) ala-gly-cys-lys-asn-phe-phe-trp
| |
cys-ser-thr-phe-thr-lys

vasoactive intestinal
polypeptide (VIP) (28) his-ser-asp-ala-val-phe-thr-asp-asn-tyr-thr-arg-leu-arg
|
asn-leu-ile-ser-asn-leu-tyr-lys-lys-val-ala-met-gln-lys
|
NH2

β-Endorphin (31) tyr-gly-gly-phe-met-thr-ser-glu-lys-ser-gln-thr-pro-leu
|
lys-his-ala-asn-lys-val-ile-ala-asn-lys-phe-leu-thr-val
|
lys-gly-gln

ACTH (Corticotropin) (39) ser-tyr-ser-met-glu-his-phe-arg-tyr-gly-lys-pro-val-gly
|
glu-ala-gly-asp-pro-tyr-val-lys-val-pro-arg-arg-lys-lys
|
asp-glu-leu-ala-glu-ala-phe-pro-leu-glu-pheNH2

*refs.1296,1297

almost 60% homologous with the rat sodium channel protein in amino acid sequence, including both identical and functionally plausible substitutions. The domain distribution is also similar in two respects, in that there are four repeats with substantial homology, and each repeat has five hydrophobic segments and one positively charged segment, called S4. The positively charged segments were first identified as important in a sodium channel model (902). The similarity in sequence and domain distribution between the sodium and calcium channels suggest that these arose from a common ancestor and constitute a channel superfamily of voltage-gated ion channels (1299).

Sodium Channels

The amino acid sequence of a sodium channel was first determined for the channel from an electric eel (1100), and later for rat brain sodium channels I, II, and III (1101, 1102). Using an eel cDNA probe to screen a *Drosophila melanogaster* genomic library, a gene was isolated and subjected to DNA sequencing. The putative sodium channel of the fruit fly is highly homologous in amino acid sequence (49% [53%] identity and FPS to eel [rat]) and domain distribution to the sodium channels of eel and rat, suggesting that the channel protein evolved before the separation of the vertebrates from the invertebrates in evolution more than 600 million years ago (1300).

Potassium Channels

On physiological grounds, the voltage-dependent K^+ channel which conducts the transient A current is formed from a protein encoded at the Shaker locus of *Drosophila melanogaster*. The sequence of cloned cDNA from this region suggests that a channel component might have been identified. The presence of hydrophobic segments and a highly conserved (eel/fly) positively charged segment indicates strongly that a channel protein might have been identified. The protein is only 70K, and probably occurs as a polymer (dimer or tetramer being the most likely) (1301). The K^+ channel system from the Shaker locus of *Drosophila* (1302–1304) seems to be remarkably complex, with at least five different subunits that may form channels in different combinations. If the same situation exists in the brains of higher vertebrates, stimulating protein synthesis by cell depolarization or other means might lead to changes in the concentrations of hundreds of different K^+ channels. There are at least 11 recognizable types of K^+ channels, grouped into four classes: voltage-sensitive, Ca^{++}-activated, receptor-coupled, and "other" (1305). A structurally new class of K^+ channel protein has been detected electrophysiologically after expression of sucrose gradient fractionated rat kidney mRNA in *Xenopus* öocytes. The 130-amino-acid sequence has no similarity to other proteins, has a highly hydrophobic (transmembrane) central portion flanked by a few charged groups, and probably functions as a channel in polymeric form (1306).

Calcium Channels

The sequence and domain resemblances between Na^+ and Ca^{++} channels gave rise to the concept of channel superfamilies. A variety of Ca^{++} channel proteins have been detected by antibodies (1307, 1308). Further information on the structures would aid in constructing models, since a preliminary examination of the DHP-binding protein sequence suggests some similarities and some differences from the Na^+ channel. A number of Ca^{++} channels are active in neurons (1309).

Conclusions

The fact that receptors and channels can be grouped into families promises to be a powerful aid to analysis of the mechanisms of action. This is especially true for the construction of preliminary models for new receptors or channels (1205), as I have shown for the case of the $GABA_A$ receptor (856).

9 Learning and Memory

In the first chapter, learning by higher organisms was defined as a multicellular representation of the input for the system, consisting of the changes occurring in the individual cells comprising the multicellular set. The initial representation will normally consist of transient changes. The local changes can disappear without further effect, or the representation can be retained by being recycled or reinforced. At some point, some molecules or molecular systems are permanently transformed into a memory. Memory is the record of the representation. There may be several stages in the molecular storage or memory formation, one short-term and one long-term. A complete theory for learning and memory is expected to describe both the molecular transformations and the intercellular connectivity.

Model Systems for Learning

On the wet sand were Descartes' words, "*Cogito ergo sum*" (I think, therefore I am). An unusually large wave crashed near at hand, and the swirling water erased the letters. The salty smell of the ocean filled the air. I closed my eyes, saw the words, heard the wave, smelled the ocean, and said out loud, "What is 'I'?"

Suppose I had imbibed the necessary ^{18}F-containing cocktail, and placed my head into a PET (positron emission tomography) scanner. One would have seen beautiful false-color images of activity in the visual cortex (looking at the words and waves), the auditory cortex (hearing the wave), the speech center (speaking), and the forebrain (searching for meaning) (1310, 1311).

Many hours, or days, or years later, there would have been no problem in my remembering the scene and the words. How can such a transient conjunction of sight, sound, and smell be learned so rapidly, retained, and recalled?

In view of the obvious complexity and difficulty in working with the human brain, a variety of model systems have been developed to study learning. Some are the marine snails *Aplysia californica* and *Hermissenda crassicornis*, the fruit fly *Drosophila* (1312–1314), the macaque monkey, *Macaca* (1315), the cat (1316), the goldfish (1317, 1318), chick embryonic myotubes (1319), the day-old chick (1320), and others. Marine snails have been especially well studied as models for higher organisms.

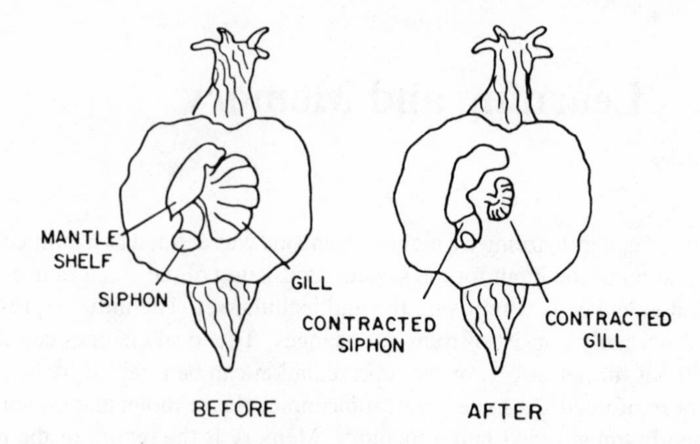

Fig. 9.1. The response of the sea hare *Aplysia californica* to a tactile stimulus. A tactile stimulus to the siphon, such as a sudden jet of water or a light touch, produces a contraction of both the siphon and the gill. (Adapted with permission from *Cellular Basis of Behavior* by Eric R. Kandel, copyright © 1976, W. H. Freeman and Company [reference 1321].)

Aplysia

The sea slugs (marine hind-gilled snails) are effective model systems for probing the connection between behavior and nerve cell properties (1321). Among the relatively small number of 10^5 nerve cells in the nervous system of the sea hare *Aplysia californica*, there are many large and readily identifiable neurons, highly suitable for neurophysiological measurements. Moreover, certain aspects of the well-defined behavioral responses can be learned, and the structural, electrophysiological, and biochemical changes followed in identified neurons (1322).

A defensive withdrawal reflex involves the exposed external organs, the siphon, gill, and mantle shelf, as illustrated in Figure 9.1. One touch causes withdrawal of the siphon and gill (1323); as few as four repeated touches cause less response to the stimulus (habituation). Stimulating a different area of the body fully restores the withdrawal response (dishabituation).

Reflex Circuit

A partial neural circuit for the transformation of the stimulus to the *Aplysia* siphon into a motor response is shown in Figure 9.2. A tactile stimulus changes the physical state of a mechanoreceptor. The change in the receptor opens ion channels, thus depolarizing the dendrites and then the cell body of a sensory neuron. With a sufficiently large depolarization, axon depolarization

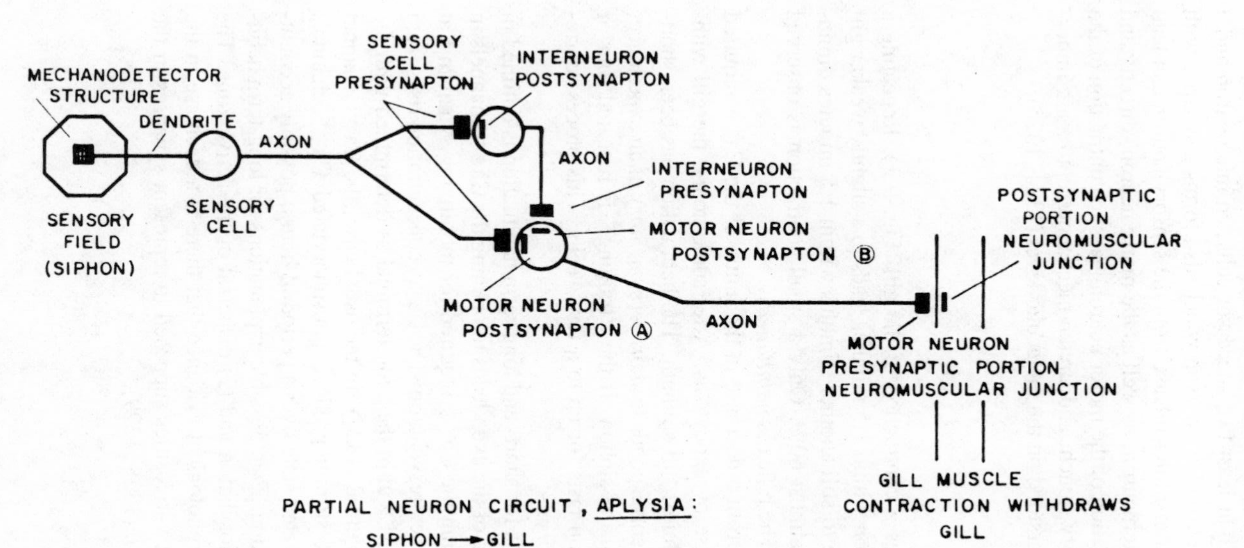

Fig. 9.2. Partial neural circuit for the transformation of a tactile stimulus into gill withdrawal in *Aplysia californica*. The stimulus changes the physical state of a mechanoreceptor. The change in the receptor opens ion channels, depolarizing the dendrites and then the cell body of a sensory neuron. With a sufficiently large depolarization, axon depolarization is triggered, transferring a signal to the axon terminals. The sensory axon terminals form synapses with at least two successor cells, an interneuron and a motor neuron. The interneuron has an axon that also forms a synapse with the motor neuron. Neurotransmitter release from the sensory neuron can lead to the excitation of the interneuron, as well as the motor neuron, directly and via the interneuron. Excitation of the motor neuron leads to contraction of the muscle in the gill. The synaptic connections between cells are composed of a presynapton and a postsynapton (Chap. 1).

is triggered, transferring a signal to the axon terminals. The sensory axon terminals form synapses with at least two successor cells, an interneuron and a motor neuron. The interneuron has an axon which also forms a synapse with the motor neuron. Neurotransmitter release from the sensory neuron can lead to the excitation of the interneuron, as well as the motor neuron, directly and via the interneuron. Excitation of the motor neuron leads to contraction of the muscle in the gill. Auditory, touch, and stretch mechanoreceptors are important, but not yet well understood on the molecular level (1324, 1325).

Habituation

A weak stimulus causes a contraction of the gill (Fig. 9.1). Repeating a weak stimulus to the siphon within 1.5 minutes leads to a slightly weaker gill withdrawal. The response to still another stimulus within 1.5 minutes diminishes the contraction by almost 60%. Only a small contraction is observed after still further stimuli. The reflex is *habituated*.

The habituation arises from a decrease in the influx of the Ca^{++} required for neurotransmitter release by exocytosis. After some hours, the gill withdrawal reflex returns to almost full strength. The habituation has been short-term. The overall process corresponds to *short-term memory* of an experience, that of several touches on the siphon. If the "training" is repeated four or more times, the habituation lasts for up to a few weeks. This process corresponds to *long-term memory*.

A partial molecular basis for short- and long-term habituation is outlined in Figure 9.3. Depolarization of the axon leads to opening the Ca^{++} channels in the axon terminal, admitting the Ca^{++} required for mobilizing synapsin I at the active zone (1326) of the presynapton. The Ca^{++} concentration rises from 90–100 nM to 110–140 nM within the axon terminal as determined with the fluorescent Ca^{++} agent, Fura-2 (1327). In the usual case, the Ca^{++} channel is inactivated. Groups of stimuli produce phosphorylated Ca^{++} channels which return to the inactivated state over a period of hours, possibly accounting for short-term habituation. Two possible explanations for long-term habituation are (1) polyphosphorylation and (2) dispersal of the active zone. The latter is supported by morphological evidence that there is a decrease in the number of active zones at varicosities supposed to represent synapses in the related case of *sensitization* (1328, 1329).

Sensitization

A strong stimulus to the tail of the *Aplysia* can *sensitize* the creature to a weak stimulus at the siphon. Furthermore, sensitization can restore the gill response to animals that have been habituated to the long-term siphon stimulus. The neural circuit that influences the reflex pathway is shown in Figure

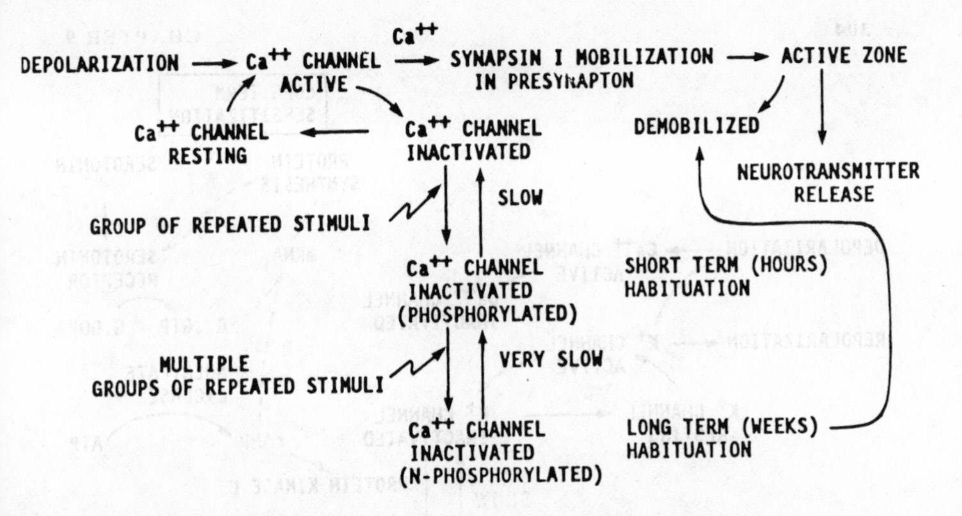

Fig. 9.3. Molecular processes in habituation. Depolarization of the axon opens Ca^{++} channels in the axon terminal, admitting the Ca^{++} required for mobilizing synapsin I at the active zone of the presynapton. In the usual case, the Ca^{++} channel is inactivated. Groups of stimuli produce a phosphorylated Ca^{++} channel which returns to the inactivated state over a period of hours, possibly accounting for short-term habituation. Two possible explanations for long-term habituation are (1) polyphosphorylation and (2) dispersal of the active zone.

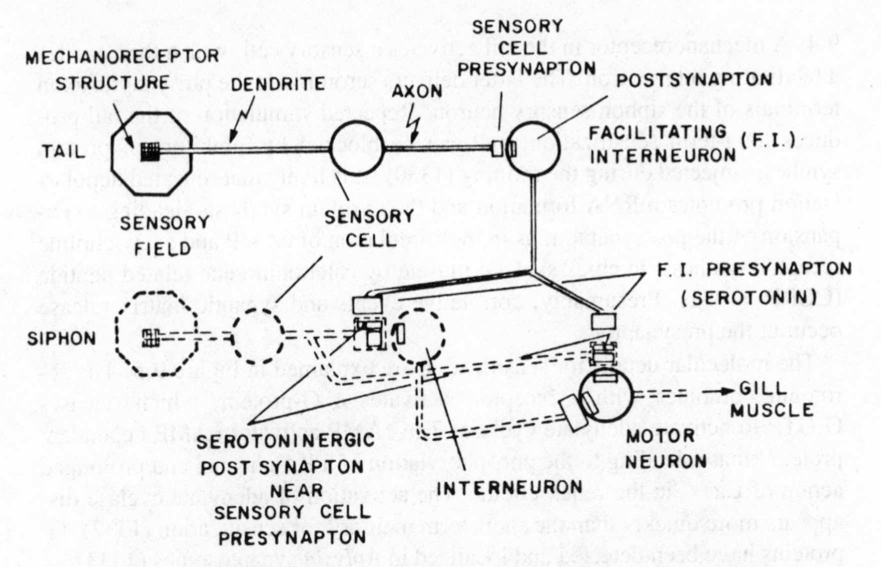

Fig. 9.4. The neural circuit that sensitizes the siphon reflex pathway in *Aplysia californica*. A mechanoreceptor in the tail activates a sensory cell, which then excites a facilitating interneuron. The latter delivers serotonin to the presynaptic axon terminals of the siphon sensory neuron. The serotonin activates a secondary messenger system, which prolongs excitation at the terminals and facilitates (or sensitizes) the activity at the synapses.

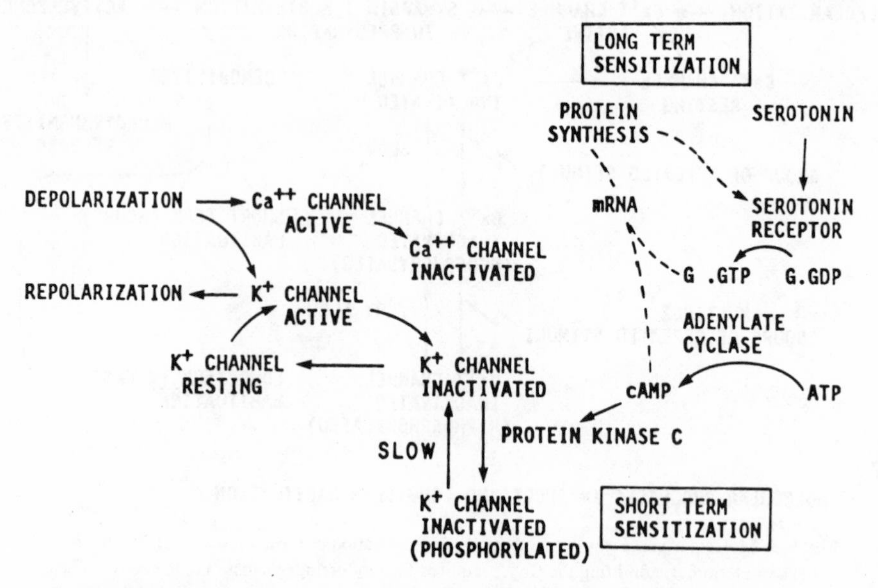

Fig. 9.5. Molecular processes in sensitization. The serotonin combines with a receptor and activates a G-protein, which releases G_{α}.GTP that activates adenylate cyclase. The cAMP activates cAMP-stimulated protein kinase, leading to the phosphorylation of a K^+ channel and prolonged action of Ca^{++} in the reflex circuit.

9.4. A mechanoreceptor in the tail activates a sensory cell, which then excites a facilitating interneuron. The latter delivers serotonin to the presynaptic axon terminals of the siphon sensory neuron. Repeated stimulation of the tail produces long-term sensitization, and can be blocked by inhibitors of protein synthesis injected during the training (1330). It is likely that repeated depolarization promotes mRNA formation and then protein synthesis, leading to expansion of the postsynapton, as in the stimulation of cAMP and acetylcholine receptor synthesis in chick skeletal muscle by calcitonin-gene related peptide (CGRP) (1331). Presumably, correlative events and synaptic matrix release occur at the presynapton.

The molecular details for sensitization are explained in Figure 9.5. The serotonin combines with a receptor, activates a G-protein, which releases G.GTP to activate adenylate cyclase. The cAMP activates cAMP-dependent protein kinase, leading to the phosphorylation of a K^+ channel and prolonged action of Ca^{++} in the reflex circuit. The activation of adenylate cyclase disappears more quickly than the short-term memory for sensitization (1332). G-proteins have been detected and localized in *Aplysia* synaptosomes (1333).

The simplified circuits shown in the Figures 9.2 and 9.4 are actually more complex, with 24 sensory cells in the siphon and 6 motor cells for the gill (1334). Together with 24 interneurons, one would expect 54 cells to be active on stimulation of the siphon. Optical measurements of neuron activity show that at least 90 neurons are active during habituation. Sensitization involves at least 150 neurons rather than the 70–80 cells that might have been predicted (1335). Other neurons involved in the response are inhibiting interneurons which release the peptide FMRFamide in response to tail shock (1336). Optical measurements depend on the change in fluorescence or absorption by potential-sensitive dyes added to the neural preparation (1337–1340).

Hermissenda

The marine snail *Hermissenda crassicornis*, can be trained to associate a conditioned stimulus (CS) with an unconditioned stimulus (US). The snail moves toward light naturally (phototaxis) near the surface of the ocean to consume hydroids, the branching colonial coelenterates related to hydra. Turbulence causes the snail to move more slowly. Alkon (1341) was inspired by these natural attributes to use light as the CS and rotation on a turntable as the US in training the snails. Animals that had been exposed to CS 1 second before US moved less than 1/3 as fast toward the light as compared with control animals. The timing of the process may not be critical (1342). Genetically uniform snails responded in the same way but with less variability. A genetic defect in the statocyst organ, which detects motion with hairs in contact with small stones, leaves the snails unable to associate light and rotation.

A thorough mapping of the neural system allowed the convergent pathways to be be identified by neurophysiological measurements, with the most essential elements shown in Figure 9.6. The two A-type cells of the eye respond to light by signalling the motor neurons via interneurons to move the animal toward the light. The three B-type cells inhibit the signal from the A-cell to the motor neurons.

Conditioning the snail with the head facing the rotation origin excites the caudal hair cells of the statocyst, increases the sensitivity of B-type cells in the snail's eye, and results in a reduced speed in phototaxis. Conditioning the snail with the head facing outward excites the cephalad hair cells, decreases the excitability of the B-type cells, and leads to an increased speed in phototaxis. The learned response implicit in the conditioning is based on a change in the B-cell.

The complex connections in the neural system responsible for the convergence of the phototaxis and rotation responses makes it difficult to trace the changes in B-cells. However, certain features of the components can be elucidated. Exciting a B-cell directly during exposure to light increases B-cell excitability, thereby simulating the effects of rotation. Animals subjected to this procedure in place of rotation move more slowly toward light, confirming that the neurophysiological effects had a behavioral counterpart.

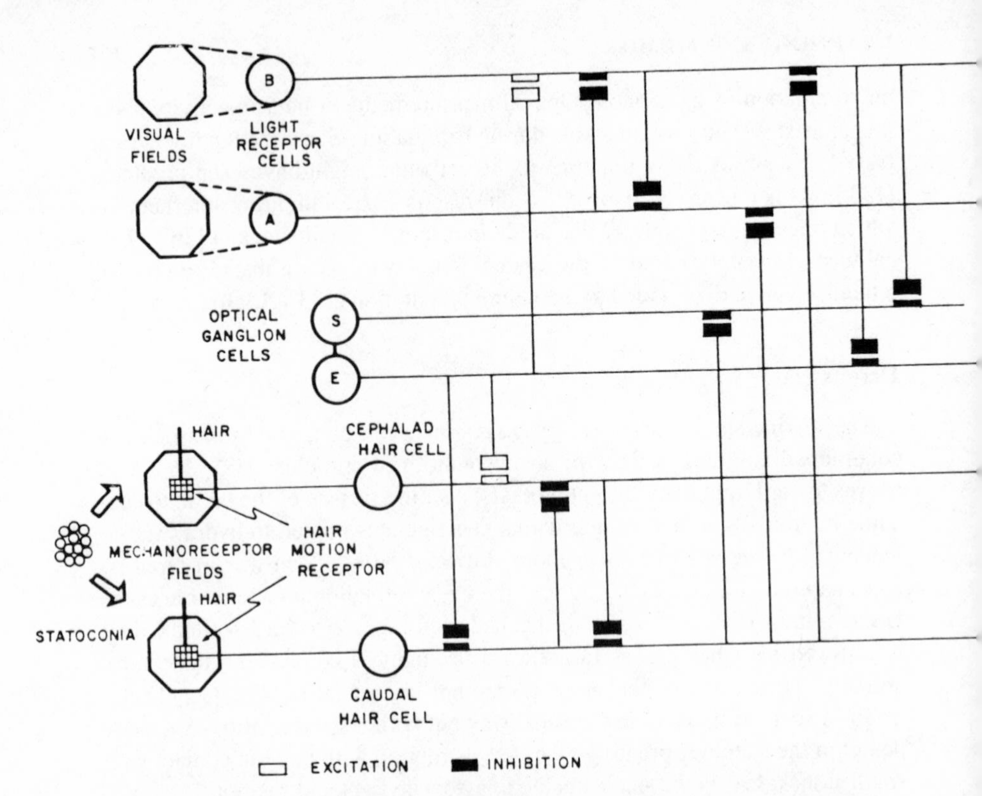

Fig. 9.6. The neural system of the marine snail *Hermissenda*, which couples light stimulation to the detection of motion. The system has three elements; the eye, the optic ganglion, and the statocyst. There are two types of cells in the eye, A-type and B-type. The A-type cells respond to light by signalling the motor neurons to move the animal toward the light. The B-type cells inhibit this motion. There is mutual inhibition between A- and B-type cells. Exciting the B-type cells inhibits S- and E-cells in the optic ganglion. These cells in turn act to excite B-type visual cells and cephalad hair cells and to inhibit caudal hair cells. In the statocyst, there are two types of hair cells with neural output, between which there is mutual inhibition. Conditioning with the head facing the rotation origin excites the caudal hair cells of the statocyst, increases the sensitivity of the B-type cell in the snail's eye, and produces a reduced speed in phototaxis. Conditioning with the head facing outward excites the cephalad hair cells, decreases the sensitivity of the B-type cells, and produces an increased speed in phototaxis. The caudal hair cells inhibit the optical ganglion and both A- and B-type visual cells. (Adapted from reference 1341.)

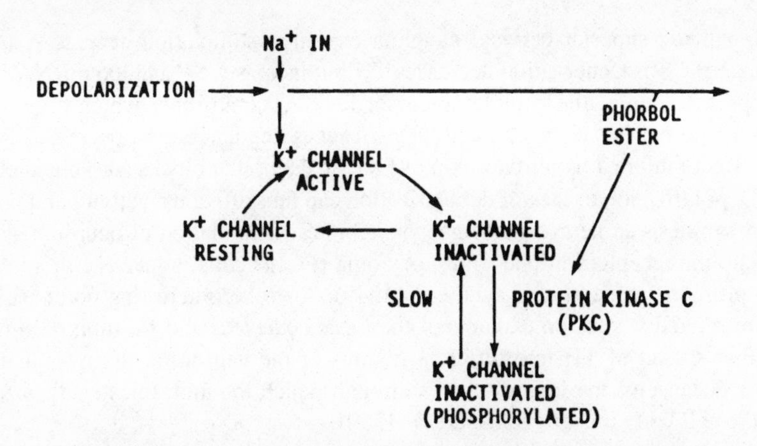

Fig. 9.7. Molecular processes in *Hermissenda* B-cell excitability. Depolarization involves an influx of Na^+, which is subsequently counterbalanced by the exit of K^+. Protein kinase C (PKC) catalyzes the phosphorylation of the K^+ channel, and thus inactivation. The B-cell under these conditions stays active (depolarized) for a longer time.

Molecular Basis of Hermissenda *Learning*

Injection of exogenous protein kinase C (PKC) or bath application of the tumor promoter, phorbol ester, into a B-cell causes an increase in excitability which mimics the effects produced by associative training. The responses of two types of K^+ channel (the ''A-channel'' and a calcium-dependent K channel) are diminished and that of a calcium channel is enhanced after PKC addition or activation (1343, 1344). The molecular processes are diagrammed in Figure 9.7. Imipramine, a tricyclic antidepressant, is a relatively specific inhibitor of PKC-dependent effects and prevents the in-vitro conditioning-induced changes in B-cells (1345). Pairing light with the application of serotonin produces a long-term reduction in positive phototactic behavior (1346). Long-term changes in B-cell responses, but not short-term responses, are prevented by a protein synthesis inhibitor, desacetylanisomycin (1347). Introduction of cAMP-dependent protein kinase into the B-cell diminishes a K^+ current (I_K) different from those changed by training (1348).

Molecular Basis of Learning in Rat Peripheral Nervous System

Environmental stress stimulates a discharge in the peripheral sympathetic nervous system, releasing noradrenaline (norepinephrine) and adrenaline (epinephrine). Reserpine or direct nerve stimulation for 30 to 90 minutes leads to an increase in tyrosine hydroxylase (TH) due to protein synthesis from increased mRNA for TH. Biosynthesis of catecholamines proceeds through the sequence shown in Figure 9.8. The increase in TH lasts at least three days. In sympathetic superior cervical ganglion explants, stimulation decreases substance P (SP). Denervation decreases TH but increases SP and the mRNA for preprotachykinin, the peptide precursor for SP. Veratridine, which prolongs the action of Na^+ channels, and thus enhances depolarization, decreases SP, an effect inhibited by tetrodotoxin (TTX), an agent that blocks Na^+ channels.

Depolarization or lack of depolarization can thus affect the activity and protein synthesis in neurons of higher vertebrates. Stimulation of neuronal acetylcholine receptors in pheochromocytoma (PC12) cells induces actin and *c-fos* protooncogene protein synthesis (1349). A further interesting point arises from related studies on explants of the *locus coeruleus* and *substantia nigra*, in that spread of TH into different regions of the neuron occurs over many days because of the long distances through which the molecules synthesized in the cell body must be transported (1350).

Some of the complications encountered in defining the neurophysiological behavior of neural systems are illustrated by the finding that a quisqualate-sensitive dipeptidase, NAALADase, in rat synaptosomes converts the neurotransmitter peptide N-acetyl-aspartyl-glutamate into N-acetylaspartate and glutamate (1351). The glutamate is subject to reuptake (1352), but might also act as an excitatory neurotransmitter. Quisqualate (QUIS) is an agonist of one type of glutamate receptor (the QUIS receptor) (1353), the other two being the kainate (KA) and N-methyl-D-aspartate (NMDA) receptors. A dipeptide derived from KA, N-kainyl-L-glutamate, inhibits NMDA receptors (1354). Long-term exposure to excitatory amino acids (KA or NMDA) induces pathology in rat hippocampal cells, apparently by promoting the entry of Ca^{++}, which activates the thiol protease, calpain I (1355). Cells normally contain an inhibitor of calpain, calpastatin, as well as insufficient Ca^{++} for extensive activation of the protease (1356). The removal of inactivation for Na^+ channels by pronase (Chap. 7; 1357) raises the possibility that proteases may influence the time course of depolarization. Non-inactivating Na^+ channels have been found in Purkinje (1358) and other cells; their influence on excitability has been simulated (1359).

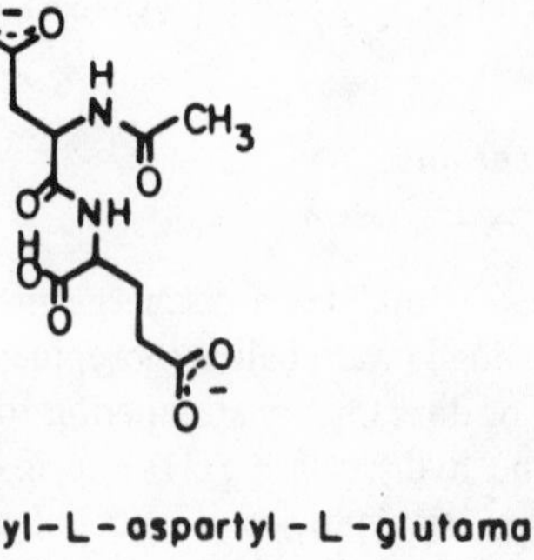

N-acetyl-L-aspartyl-L-glutamate

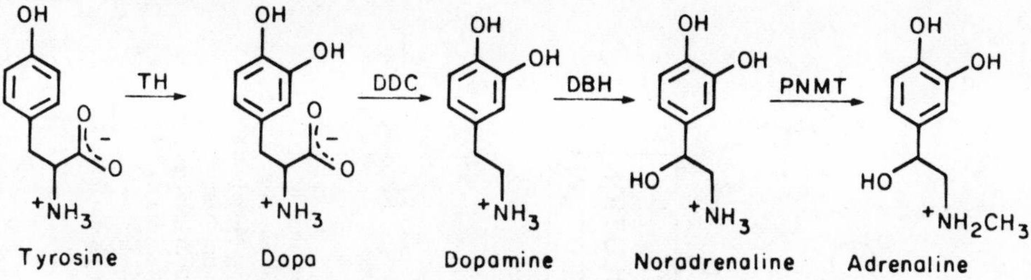

Fig. 9.8. Catecholamine biosynthesis. The enzymes catalyzing the individual steps are TH, tyrosine hydroxylase; DDC, aromatic amino acid decarboxylase; DBH, dopamine β-hydroxylase; PNMT, phenylethanolamine-N-methyl transferase.

Substance P (SP) may be related to scotophobin, the controversial "fear peptide" isolated from rat brains (1360).

The turnover of synaptic proteins like acetylcholine receptor (AChR) in denervated mouse muscle (1–2 days) occurs much more rapidly than that of AChR in innervated cells (≈8 days), the intra- or extrajunctional status having less influence (1361). Cyclic AMP may regulate the number of functional receptors on chicken ciliary ganglion neurons (1362).

Biological Basis of Learning

The effects of highly specific lesions on systems that can be defined in part by electrophysiological means suggests that critical brain structures for learning include the cerebellum, the hippocampus, the amygdala, and the cerebral cortex. The central point is that neuronal circuits involved in any particular response are limited in number and extent (1363).

Circuits for visual stimuli are initiated from the cells in the retina of the eye. The color-sensitive cone cells activate via bipolar cells two distinct types, small and large, of ganglion cells. The large ganglion cells signal magnocellular cells in the lateral geniculate body, and these in turn activate M-cells in visual area 1. The M-cells stimulate visual-area-2 cells, which then excite cells in the middle temporal area, the excitation being traceable to movement or stereo perception. The small ganglion cells signal parvocellular cells in the lateral geniculate body, and these in turn activate P-cells in visual area 1. There are small groups of cells, called "blobs," present in visual area 1. Between the blobs are the interblob cells. The P-cells excite blob and interblob cells, the combination being responsible for high-resolution perception. Further excitation by interblob and blob cells in visual areas 2 and 4 occurs in response to color. A much simplified diagram illustrating the partial circuits for perception of color, depth, and movement is shown in Figure 9.9 (1364).

More macroscopic features in visual circuits are the ocular dominance col-

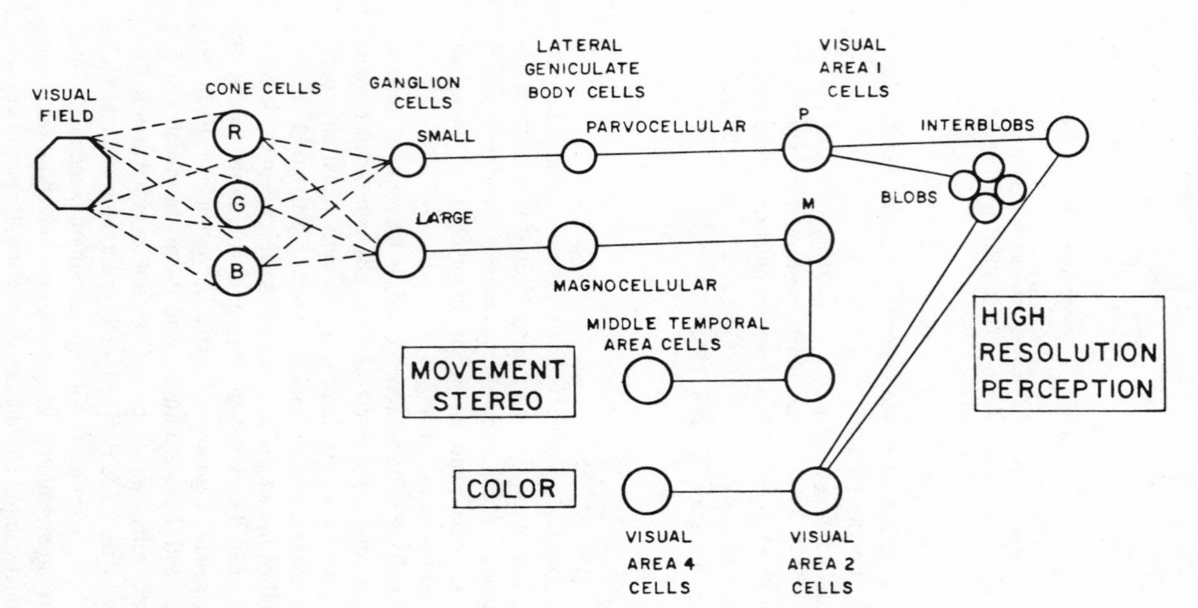

Fig. 9.9. A simplified diagram of the neural circuits in a part of the visual system. Circuits for visual stimuli are initiated from the cells in the retina of the eye. The red (R), green (G), and blue (B) sensitive cone cells activate via bipolar cells both small and large ganglion cells. The large ganglion cells signal magnocellular cells in the lateral geniculate body, and these in turn activate M-cells in visual area 1. The M-cells stimulate cells in visual area 2, which then excite cells in the middle temporal area, the excitation being traceable to movement or stereo perception. The small ganglion cells signal parvocellular cells in the lateral geniculate body, and these in turn activate P-cells in visual area 1. There are small groups of cells, called ''blobs,'' present in visual area 1. Between the blobs are the interblob cells. The P-cells excite blob and interblob cells, the combination being responsible for high-resolution perception. Further excitation by interblob and blob cells in visual areas 2 and 4 occurs in response to color. This is a much simplified diagram illustrating the partial circuits for perception of color, based on a more complete diagram in reference 1364.

umns of the visual cortex, in which vertically organized aggregates of cells respond to similar stimuli (1365). Orientation preference columns are also found, and have been confirmed by optical imaging of the signals (1366, 1367). Columns that extend through all or most of the six cortical layers are present in many parts of the cortex, and include between 1000 and 10,000 cells (1368).

Another major target of the optic nerve is the *superior collicus* (SC), part of the midbrain (Table 1.4). A cell in the SC starts firing 50 milliseconds after a localized visual stimulus appears to a monkey. Eye motion (a jump called a saccade) starts 150 milliseconds afterward (1369). The location of the visual stimulus can be remembered, with cells in the *substantia nigra para reticulata* (SNr) and SC changing their firing pattern within 100 milliseconds after the stimulus (1370). Other brain areas activated by visual stimuli are the anterior ventral temporal cortex (1371), the posterior parietal cortex, and the frontal eye field, the latter two clearly visible in PET scans of humans (1372).

The neuronal circuits that subserve fast responses must involve a reasonably small number of neurons, between 5 and 50. The minimum is based on Figure 9.10, with SC being considered in place of the lateral geniculate body. The maximum is based on 0.5 millisecond as the time for both synaptic delays and passage through a single neuron (21). Given that the time for eliciting a complex visual memory can be a small number of seconds, one can conclude that a basic circuit for producing elements of such memories probably contains between 100 and 2000 neurons.

At the lowest level, the minimal group of neurons involved in an elementary learning act constitutes a local circuit. More complex learning acts would involve multiple local circuits, the number varying with the complexity of the stimulus (1368, 1371). There is no contradiction between this concept and the participation in local circuits of individual cells that respond to complex stimuli such as vertical bars moving in a horizontal direction.

A Generalized View of Learning

Based on the experiments with the marine snails and information from many other systems, one can develop a generalized view of learning. Exciting a neural system with a stimulus leads to three types of changes: short-term (milliseconds), medium-term (minutes to hours), and long-term (days to years). Short-term changes involve depolarizations initiated by ligands opening ion channels or propagated by channels opening in response to the depolarization. Medium-term changes arise from phosphorylation of K^+ or Ca^{++} channels, an alteration that prolongs the inactivated state of these channels. Long-term changes depend upon an increase in specific mRNAs and protein synthesis, possibly of synaptic components. The turnover of synaptic components occurs

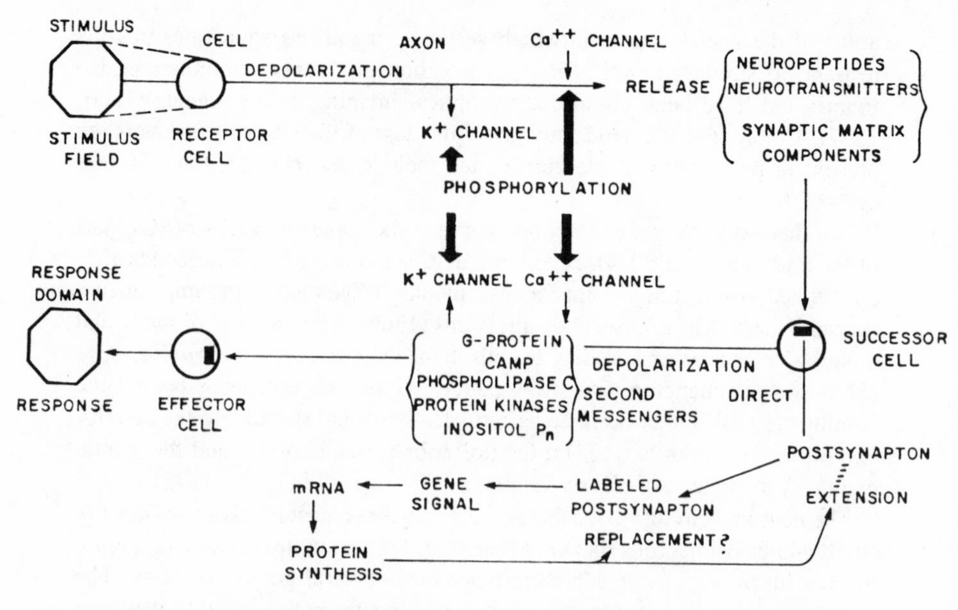

Fig. 9.10. Molecular stimulus-response scheme for a multicellular system. A stimulus (light, stress, neurotransmitter, etc.) combines with or changes a receptor in a receptor cell, depolarizing the cell and then, if the depolarization is sufficiently large, the axon. The depolarization is effected by the entry of Na^+ ion; repolarization occurs with the exit of K^+ ion. Phosphorylation decreases the rate at which the inactivated K^+ channel may be reactivated, and thus prolongs the depolarization. Another channel activated by depolarization is the Ca^{++} channel, especially in the axon terminals, in which Ca^{++} promotes neurotransmitter release. The reactivation of Ca^{++} channels can be delayed by phosphorylation. The exit of neuropeptides or neurotransmitters from the cell may be accompanied by synaptic matrix components. The successor cell can be depolarized by the neurotransmitter, directly or via one of a number of second messenger systems. The activation of the successor cell leads to a signal to a gene, which produces mRNA and then protein (Fig. 9.12). The successor cell can (eventually) activate an effector cell, which produces a response.

on a shorter time scale than the lifetime of the behavior that is supported by the neurons. Systems for marking, recognizing, removing, and replacing synaptic components presumably exist. A simplified—but more specific—view of some of the molecular processes connected to learning is shown in Figure 9.10. In some cases, proteases within neurons may be activated by neurotransmitters.

A model for the spatial and temporal aspects of cell activation involving Ca^{++} emphasizes the balance between phosphorylation and dephosphorylation in second messenger systems (1373). The time course for the spread of new molecules synthesized in the cell body may have bearing on the rhythm of learning. The formation of new synapses at neuromuscular junctions is diminished in vector-free gravity fields, implying that artificial gravity fields may be required in long-term space travel (1373). Repetition may have greater effectiveness if the stimuli are distributed over appropriate time periods, an idea that could be extremely important in the planning of educational programs.

A Generalized View of Memory

The time scales in learning are well suited to serve as the precursors for the memory scales observed in behavior. In the case of marine snails, a single stimulus is not sufficient to create a prolonged change within the excited cell. Several stimuli within a short time are needed for the learning experience to be expressed in a change that survives on the medium time scale. The analogue of the requirement for multiple stimuli in a higher nervous system would be a repetition of the stimulus, a process called reverberation. It seems reasonable to suppose that the hippocampus and other brain tissues implicated in memory formation are involved in reverberation events.

Although no direct evidence has been obtained for reverberation (1374), there are at least three indications that the idea should continue to be considered. First, the time within which the reiterative signal is introduced into the original trace would be very short, 5–15 milliseconds. Only a few neurons, including one or more in the hippocampus and related structures, might be involved. Only a thorough, multicell electrophysiological study might distinguish a signal produced by the original stimulus and one induced or promoted by a reverberative stimulus. Second, multiple stimuli are required to produce short-term habituation in marine snails. Although there are no precise analogues for complex memories in humans, I suspect that reiteration might be important. Third, there seems little doubt that brain structures like the hippocampus and amygdala contribute to memory formation without being a repository for the particular memory. In addition, return connections between the visual cortex and the lateral geniculate nucleus appear to be important in controlling visual responses (48). Reinforcing the trace at the initial stages might be one possible contribution to learning.

Repeated multiple stimuli are needed for long-term habituation in marine snails, a process that is accompanied by protein synthesis. The proteins synthesized include nAChR and possibly other synaptic and channel components. The K^+ channel system related to the Shaker locus of *Drosophila* (1302–1304) seems to be remarkably complex, with at least five different subunits that may form channels in different combinations. If the same situation exists in the brains of higher vertebrates, stimulating protein synthesis by cell depolarization or other means might lead to changes in the concentrations of hundreds of different K^+ channels. Novel proposals for other learning induced changes are (1) short-term alterations in cytoskeletal components of dendrites (which could cause twitching) (1375) and (2) a bistable autophosphorylating kinase (1376).

Stimuli ⟶ Short term changes ⟶ Long term changes

Ionic and second messenger concentrations

mRNA and protein synthesis synapse and channel alterations

Scheme 9.1. Short- and long-term changes arising from stimuli to the nervous system.

The rich variety of species and events that may occur within a nervous system is expressed in its barest form in Scheme 9.1. The local circuits and neuronal assemblies (a collection of circuits) are altered by these changes, which generate very small to very large effects on responses. The catalogue of possible changes is quite large when both spatial and temporal variations in concentrations are taken into account. Nevertheless, the commonality of much human experience suggests that the net results are quite similar, even though perceptions may differ from one individual to another. Probing the circuit or assembly can be done from any connection. A partial stimulus can elicit the whole response much as the taste of the madeleine soaked in lime-flower tea brought forth Proust's "Remembrance of Things Past" (1377).

Higher-order Representations

Learning and memory has its molecular basis in short- and long-term chemical changes which in turn are generated by the operations of receptors and channels within individual cells. Neuronal circuits and neuronal assemblies are the biological basis for the expression of the chemical changes. The ways in which the cooperative behavior among the cells is developed, organized, and utilized must be analyzed if one is to understand, in human terms, learning and memory. One must explain how the representations of the traces produced by internal or external stimuli can produce or be elicited by higher-order representations.

Learning by higher organisms is defined as a multicellular representation of the input for the system, consisting of the changes occurring in the individual cells comprising the multicellular set. An association with another representation, such as the verbal name for a visually observed object, would be via a higher-order representation. A set of such higher-order representations might constitute a still higher-order representation.

A model for association is derived from the behavior of the Belousov-Zhabotinskii (B-Z) reaction. The latter is a metal-catalyzed reaction of bromide, bromate, and malonic acid, for which red-to-blue-to-red color changes take place periodically and are distributed in space and in time. Periodic spiral oscillations can be initiated in a covered planar Petri dish. Exposure to the atmosphere for the ferroin-catalyzed reaction produces "hot spots" which prove to be convection cells induced by cooling of the surface as a result of

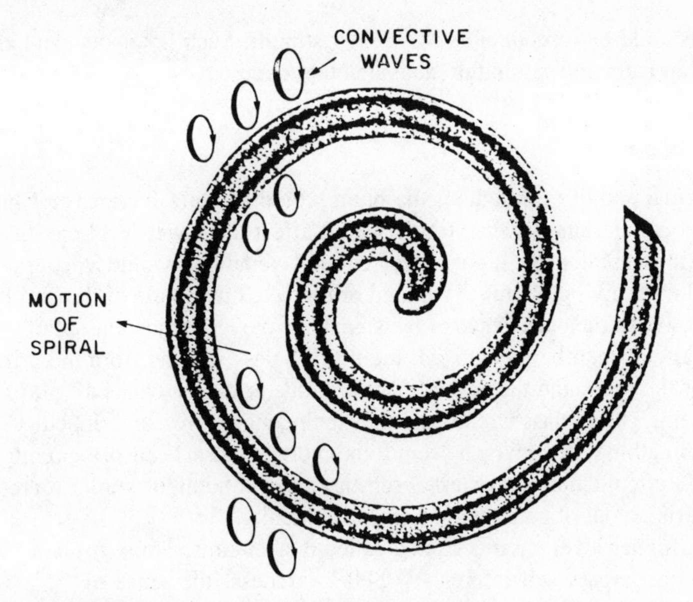

Fig. 9.11. Spiral Belousov-Zhabotisky (B-Z) wave. The convective cells on the sides result from a perturbation, usually by exposing to the atmosphere the thin layer in which the chemical reactions occur. The change in temperature brought about by evaporation of the solvent induces transport of material in the direction normal to the surface and the plane in which the spiral waves are moving.

the evaporation of water. The convected material flows in directions different from those of the initial original spiral waves; the color changes occur within the convection cells (1378–1380) (Fig. 9.11). Similar material distributions and convective inhomogeneities are found in thin layers of dihydronicotinamide adenine dinucleotide (NADH) solutions (1381) and yeast extracts, the latter involving oscillations in glycolysis (1382).

The B-Z model can be translated into neural terms, as the discoverer of chemical oscillations (1383) pointed out in 1906! Waves move through each layer of arbitrary thickness in a B-Z spiral arm in concert with neighboring layers. Each layer is equivalent to a local circuit, with waves of successive activity passing through sets of local volumes or "cells." The B-Z "cells" are analogous to neurons, with the degree of excitation rising or falling in accordance with the balance among neurotransmitters, receptor and channel modifiers, second messengers, ion fluxes, and synthetic and degradative processes. The hot spots could represent higher-level local circuits, arising near the most active portion of the original local circuits in a natural way. The structure of the higher-level local circuits would depend upon the connectivity of available neurons; spatial relationships among the neurons of the local circuit may not be as contiguous as in a convection cell. The formation of hot spots should be reproducible for similar stimuli. Such hot spots can expand, i.e., can learn and can in turn activate other circuits.

Metacircuits

Normal activity throughout the brain would create a hierarchy of higher-level local circuits. Inverse traffic would affect the lower-level circuits. The conditions created by higher-level circuit excitations would encourage the stimulus-receiving circuit to respond or to make other parts of the brain more responsive to the occurrence of the signal. Two overlapping classes of normal local circuits can be recognized, the first, P-class, arising from the primitive part of the brain and the second, A-class, due to the ''advanced'' portions of the brain. The P-class circuits monitor temperature, pressure, appetite, heartbeat, breathing, etc., which seem to be mostly internal sensory circuits. The A-class circuits monitor external sensory events and higher-order representations arising out of combinations of local circuits.

The higher-level circuits can be called metacircuits. Some A-class circuits might encompass self-references. Self-awareness, the sense of ''I,'' would include local external and internal sensory (EIS) circuits, and a running repertoire of reactivated records of previous EIS. Paying attention might involve a generalized internal representation or one focused on certain sensory circuits. Clearly, both hardware (genetically based and developmentally expressed organization of neurons) and software (connections augmented by facilitation, sensitization, etc.) are involved.

Major areas in cognitive neuroscience are perception, attention, memory, and emotion. The circuits that carry the object perceived are specific and hard-wired by genetics or development, i.e., ''labeled.'' Labeled circuits are delimited by the necessity for reiteration, either via the hippocampus or by some other pathway. Synaptic reorganization in the hippocampus (1384) can influence the response of labeled circuits. With sufficient signal strength in the labeled circuit and sufficient excitation in target neurons, higher-level local circuits can be induced. The position in the hierarchy of higher-level local circuits will influence the extent to which attention is excited by the perception. The attention circuits should be unlabeled, and require continual excitation from other representations. This description might account for the fact that attention can easily wander, but does not contradict the possibility of both short-term and long-term storage processes for attention. Records of the attention rather than the perceptions can affect interaction between perception and attention. A stimulus to almost any point in a labeled circuit can elicit a trace very similar to that originally induced by the stimulus, i.e., a memory. Reinstituting the memory for an unlabeled circuit should involve more specific entry points. Emotion can be understood as a hierarchy of unlabeled P-class circuits arising from the more primitive parts of the brain, modulated by a variety of neuropeptides for strength, length, and nature (Scheme 9.2).

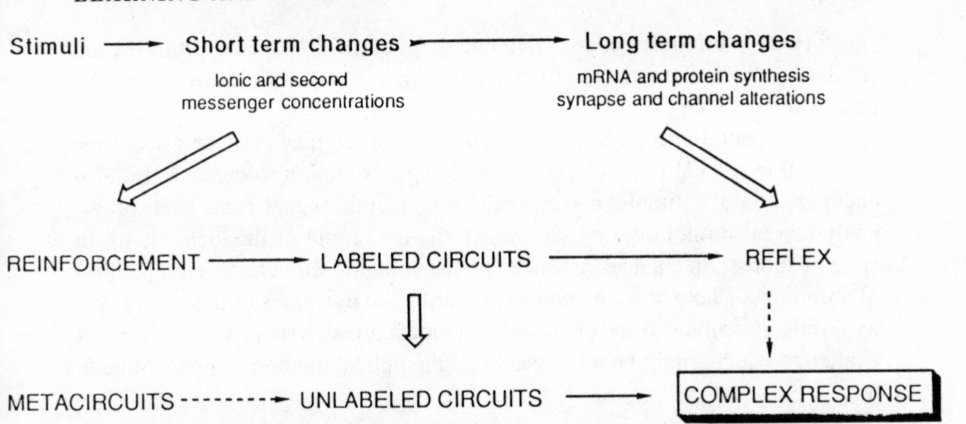

Scheme 9.2. Short- and long-term changes leading to complex responses from the nervous system. Labeled circuits of excited neurons are composed of groups fixed by genetics or development, such as those that give rise to a reflex. Metacircuits are higher-order local circuits of excited neurons (setting a level of attention), which can activate unlabeled arrangements. Together, the labeled and unlabeled circuits yield a complex response.

Neural Network Models

P. S. Churchland (32) discusses the need for constructing theories of brain function, and for theories and models to guide experiments within the complex field of neurobiology. A number of neural network models have been developed and several translated into electrical circuitry. The reasons for such models are twofold: first, one can advance our theoretical understanding of how neural networks learn, memorize, and compute; and second, there are highly practical reasons, such as translating text into speech (1385) and solving certain ordinarily difficult computational problems.

Caution is, however, advised in using neural networks as models for how the brain works rather than more specific molecular and biochemical facts (1385).

The connection between stimuli and learning and memory (Scheme 9.1) is expanded in Scheme 9.2 to express the role of the metacircuits and the computational outcome. Many questions can be asked even with this simplified picture, which I do not discuss further. How is a complex response computed from an input? How is an old memory recognized? An exposition (32) of the tensor network theory of Pellionisz and Llinás (1386) models how a visual input may be transformed into a motor action by way of the cerebellum. The use of energy minimization to map memories onto a neural network is due to Hopfield (36), in an approach that has been applied to the computation of solutions to the traveling salesman problem and the construction of hardware (1387). The model has been developed in various ways by Hopfield (1388, 1389) and others (1390, 1391). Real neural networks can be simulated (1392). An electronic sensor module (‘‘HYMOSS’’) has been produced and has the capacity for photon detection, edge enhancement (for shape detection), time resolution (for motion detection), and scanning (for visual field coverage), the latter similar to the eye's saccadic motion (1393).

With respect to recognition of stimuli, new stimuli may produce disequilibrium within a local circuit, leading to enough excitation to create long-term changes. "Old" stimuli, corresponding to memories that have been previously stored, do not alter by very much the net output of the local circuit in spite of short-term changes produced by the stimulus. Nevertheless, repetition of stimuli does help to fix memories. Relating the dynamics of the neural system to the dynamics of the chemical and biochemical systems is an important challenge which might be addressable by the optical methods already alluded to.

A Molecular Scheme for Memory

The information now available can be assembled into a coherent scheme for memory. I differentiate between internal stimuli and external stimuli through control of the amount of Ca^{++} entry. Synapses are activated by neurotransmitters. If synapse precursors are available, an activated synapse can grow by utilizing a combination of externally supplied components and internally supplied components, carried by the organelle transport on microtubules from the cell body. The conveyor belt is filled in the cell body if it has been emptied at the synapse. In this manner, one can understand the role of mRNA and protein synthesis in relation to synapse activation. The scheme is diagrammed in Figure 9.12.

Representations in Culture

Strong parallels exist between learning and memory by single individuals and by a community of individuals. A weak stimulus to a person can create a short-term memory which decays within a short time; a stronger stimulus leads to a long-term memory. If the memory stimulates the individual to communicate the stimulus to others, a short- or long-term social representation is created.

Table 1.3 lists language as the program for interaction among individuals in family groups. The family group is based on proximity, pheromones, and progeny. The interaction of family groups in clans involves the same factors ("residual language"), with the long-term memory more important than the actual proximity. The connections among such individuals can be regarded as labeled social circuits.

To create and prolong the association of clans, a belief system is needed. The contents of such belief systems have developed rapidly in content and sophistication over the past 10,000 years, especially after the development of

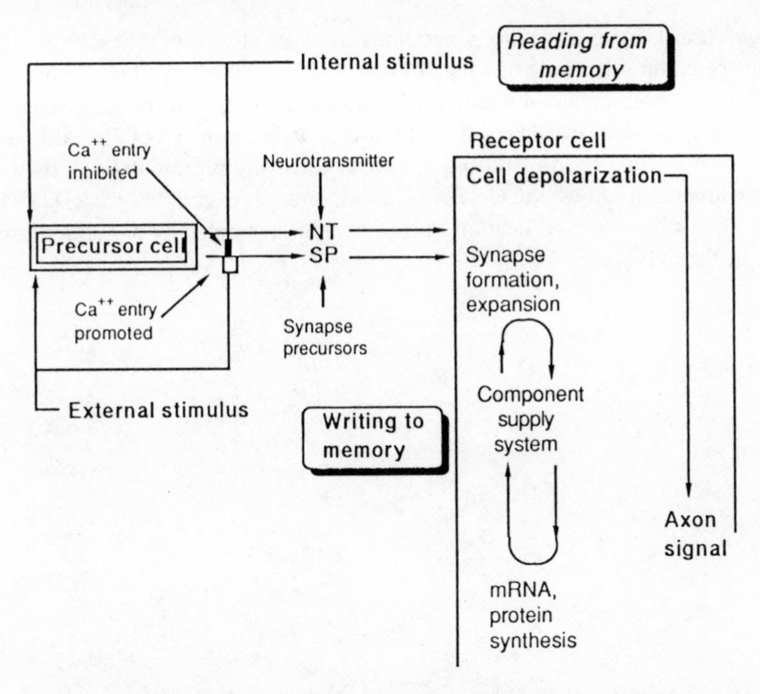

Fig. 9.12. A molecular scheme for learning and memory. An external stimulus depolarizes the receptor cell, leading to entry of calcium and release of both neurotransmitter and synapse precursors (e.g., agrin). The successor cell is depolarized, producing an axon signal. With sufficient stimulus (and once may be enough!), internal synapse precursors are acquired from the component supply system (the microtubules that carry organelles from the cell body to the periphery and back) (30, 31). Removing material from the component supply system (a "conveyor belt") stimulates (a) protein synthesis and (b) mRNA synthesis. The external stimulus has written to memory. An internal stimulus activates the same part of the nervous system but inhibits the release of calcium. Learning and writing to memory is not easily achieved with an internal stimulus. The control is not absolute, and continued activation of the same set of cells can lead to increased synapse formation and writing to memory.

writing. Cultural evolution involves the introduction of economic systems, which with time become historical systems, reinforced by cultural networks. Belief, economic, historical, and cultural representations influence the responses of individuals who may be involved simultaneously at all levels. Perturbing systems by exposing them to new influences, as in the case of the B-Z model, can lead to "hot spots," local circuits that can have short- and long-term effects.

The parallels to individual learning and memory can be expressed even more strongly. To analyze the cultural representations, one must understand the dynamics of individual behavior thoroughly in the same way that the molecular dynamics of receptors and channels have been probed in order to understand learning and memory within the individual.

One must leave to future generations the task of outlining the individual basis of cultural learning and memory. A vast repertoire of cultural representations has been created by humans in art, literature, music, history, science, and lifestyles as well as in political, religious, and economic beliefs. Although persistence (long-term memory) is evident for many representations such as those among the Jews, the Greeks, and the Chinese (over 3000 years) (1394), these are subject to modification by new insights generated by our understanding of the universe.

References

1. Changeux, J.-P. *Neuronal Man: The Biology of Mind*. New York: Pantheon, 1985. 348 pp.
2. Bear, M. F.; Cooper, L. N.; and Ebner, F. F. *Science* 237: 42–48 (1987). "A physiological basis for a theory of synapse modification."
3. Olsen, S. J., and Olsen, J. W. *Science* 197: 533–535 (1977). "The Chinese wolf, ancestor of new world dogs."
4. Bond, J. R., and van den Bergh, S. *Nature* 320: 489–490 (1986). "Galaxy distances and deviations from universal expansion." Peebles, P.J.E., and Silk, J. *Nature* 335: 601–606 (1988). "A cosmic book."
5. Groth, E. J.; Peebles, P.J.E.; Seldner, M.; and Soneira, R. M. *Sci. Amer.* 237(5): 76–98 (1977). "The clustering of galaxies." Peebles, P.J.E. *Science* 224: 1385–1391 (1984). "The origin of galaxies and clusters of galaxies." Silk, J.; Szalay, A. S.; and Zel'dovich, Y. B. *Sci. Amer.* 249(4): 56–64 (1983), "The large-scale structure of the universe." Tully, R. B. Cited in *Sci. Amer.* 258(1): 13–14 (1988). "Cosmic complex."
6. Schulman, L. S., and Seiden, P. E., *Science* 233: 425–430 (1986). "Percolation and galaxies."
7. Iberall, A., and Soodak, H. *Collective Phenomena* 3: 9–24 (1978). "Physical basis for complex systems: Some propositions relating levels of organization."
8. Miller, J. G. *Living Systems*. New York: McGraw-Hill, 1978. 1102 pp.
9. Haken, H. *Synergetics*. Berlin: Springer, 1977. 325 pp.
10. Thom, R. *Structural Stability and Morphogenesis*. Reading, Mass.: Benjamin, 1975. 348 pp.
11. Woodcock, A., and Davis, M. *Catastrophe Theory*. New York: Dutton, 1978. 152 pp.
12. Stewart, I. *New Scientist*, 20 Nov. 1975, pp. 447–454. "The seven elementary catastrophes."
13. Koestler, A. *Janus*. New York: Random House, 1978. 354 pp.
14. Wilson, D. *General Systems* 14: 3–15 (1969). Yrbk. Soc. of Gen. Systems Research. "Forms of hierarchy: A selected bibliography."
15. Lee, R. B. *The !Kung San*. Cambridge: Cambridge University Press, 1979. 526 pp.
16. Clarke, E., and O'Malley, C. D. *The Human Brain and Spinal Cord*. Berkeley: University of California Press, 1968. 926 pp.
17. Kandel, E. R., and Schwartz, J. H., eds. *Principles of Neural Science*. New York: Elsevier, 1985. 979 pp.
18. Williams, P. L., and Warwick, R. *Functional Neuroanatomy of Man*. Neurology sec., in *Gray's Anatomy*, pp. 746–1194 (35th ed.). London: Churchill Livingstone, 1975.
19. Netter, F. H. *The Nervous System*. Vol. 1, Ciba Coll. Med. Illustrations. Summit, New Jersey: CIBA, 1972. 168 pp.

20. Lewis, A. J. *Mechanisms of Neurological Disease*. Boston: Little, Brown, 1976. 540 pp.
21. Kuffler, S. W.; Nicholls, J. G.; and Martin, A. R. *From Neuron to Brain*. 2nd ed. Sunderland: Sinauer, 1984. 651 pp. (a) pp. 157–160. (b) pp. 246–248.
22. Bullock, T. H.; Orkand, R.; and Grinnell, A. *Introduction to Nervous Systems*. San Francisco: W. H. Freeman, 1977. 559 pp.
23. McGeer, P. L.; Eccles, J. C.; and McGeer, E. G. *Molecular Neurobiology of the Mammalian Brain*. New York: Plenum, 1978. 644 pp.
24. Shepherd, G. M. *The Synaptic Organization of the Brain*. 2nd ed. New York: Oxford University Press, 1979. 436 pp.
25. (a) Quarton, G. C.; Melnechuk, T.; and Schmitt, F. O. *The Neurosciences*. New York: Rockefeller University, 1962. 962 pp. (b) Schmitt, F. O.; Quarton, G. C.; Melnechuk, T.; and Adelman, G. *The Neurosciences—Second Study Program*. New York: Rockefeller University, 1970. 1068 pp. (c) Schmitt, F. O.; Worden, F. G.; and Adelman, G. *The Neurosciences—Third Study Program*. Cambridge: MIT, 1974. 1107 pp. (d) Schmitt, F. O.; Worden, F. G.; Adelman, G.; and Smith, B. H. *The Neurosciences—Fourth Study Program*. Cambridge: MIT, 1979. 1185 pp.
26. Bloom, F. E.; Lazerson, A.; and Hofstadter, L. *Brain, Mind and Behavior*. New York: W. H. Freeman, 1985. 323 pp.
27. Grisolia, S.; Guerri, C.; Samson, F.; Norton, S.; and Reinoso-Suarez, F., eds. *Ramon y Cajal's Contribution to the Neurosciences*. Amsterdam: Elsevier, 1983. 267 pp.
28. Allen, R. D. *Sci. Amer.* 256(2): 26–33 (1987). "The microtubule as an intracellular engine."
29. Schnapp, B. J.; Vale, R. D.; Sheetz, M. P.; and Reese, T. S. *Ann. N.Y. Acad. Sci.* 466: 909–918 (1986). "Microtubules and the mechanism of directed organelle movement." Schroer, T. A.; Schnapp, B. J.; Reese, T. S.; and Sheetz, M. P. *J. Cell. Biol.* 107: 1785–1792 (1988). "The role of kinesin and other soluble factors in organelle movement along microtubules."
30. Lasek, R. J., "Studying the intrinsic determinants of neuronal form and function." In *Intrinsic Determinants of Neuronal Form and Function*, ed. R. J. Lasek and M. M. Black, pp. 3–58, New York: A. R. Liss, 1988.
31. Shotton, D. M. *J. Cell Sci.* 89: 129–150 (1988). "Review: Video-enhanced light microscopy and its applications in cell biology."
32. Churchland, P. S. *Neurophilosophy: Towards a Unified Science of the Mind/Brain*. Cambridge: MIT Press, 1986. 546 pp. Churchland, P. S., and Sejnowski, T. J. *Science* 242: 741–745 (1988). "Perspectives on cognitive neuroscience."
33. Squire, L. R. *Science* 232: 1612–1619 (1986). "Mechanisms of memory."
34. Zola-Morgan, S.; Squire, L. R.; and Mishkin, M. *Science* 218: 1337–1339 (1982). "The neuroanatomy of amnesia: Amygdala-hippocampus versus temporal stem."
35. Rawlins, J.N.P. *The Behavioral and Brain Sciences* 8: 479–496 (1985). "Associations across time: The hippocampus as a temporary memory store."
36. Hopfield, J. J. *Proc. Nat'l. Acad. Sci. USA* 79: 2554–2558 (1982). "Neural networks and physical systems with emergent collective computational abilities."
37. Hopfield, J. J., and Tank, D. W., "Collective computation with continuous variables, " In *Disordered systems and biological organization*, ed. E. Bienenstock, F. Fogelman Soulie, and G. Weisbuch. NATO ASI-Series, Vol. F20. Berlin: Springer 1986. 405 pp.

38. Hopfield, J. J., and Tank, D. W. *Science* 233: 625–633 (1986). ''Computing with neural circuits: A model.''

39. Stanton, P. K.; Sarvey, J. M.; and Moskal, J. R. *Proc. Nat'l. Acad. Sci. USA* 84: 1684–1688 (1987). ''Inhibition of the production and maintenance of long-term potentiation in rat hippocampal slices by a monoclonal antibody.''

40. Sastry, B. R.; Goh, J. W.; and Auyeung, A. *Science* 232: 988–990 (1986). ''Associative induction of posttetanic and long-term potentiation in CA1 neurons of rat hippocampus.''

41. Larson, J., and Lynch, G. *Science* 232: 985–988 (1986). ''Induction of synaptic potentiation in hippocampus by patterned simulation involves two events.''

42. White, G.; Levy, W. B.; and Steward, O. *Proc. Nat'l. Acad. Sci. USA* 85: 2368–2372 (1988). ''Evidence that associative interactions between synapses during the induction of long-term potentiation occur within local dendritic domains.''

43. Rosenzweig, M. R., and Bennett, E. L., eds. *Neural Mechanisms of Learning and Memory*. Cambridge: MIT Press, 1976. 637 pp.

44. Changeux, J.-P., and Konishi, M., eds. *The Neural and Molecular Bases of Learning*. S. Bernhard, Dahlem Konferenzen. Chichester: Wiley, 1987.

45. Cotman, C. W.; Gibbs, R. G.; and Nieto-Sampiedro, M. ''Synapse turnover in the central nervous system, '' Pages 375–398 in ref. 44.

46. Ullman, S. *Ann. Rev. Neurosci.* 9: 1–26 (1986). ''Artificial intelligence and the brain: Computational studies of the visual system.''

47. Dudai, Y. ''On neuronal assemblies and memories.'' Pages 399–410 in ref. 44.

48. Altman, J. *Nature* 328: 572–573 (1987). ''A quiet revolution in thinking.'' Report on Neurobiology of Neocortex, 50th Dahlem Conference, Berlin, 17–22 May 1987.

49. Maddox, J. *Nature* 328: 571 (1987). ''Modelling for its own sake.''

50. Baudry, M. ''Activity-dependent regulation of synaptic transmission and its relationship to learning.'' Pages 153–176 in ref. 44.

51. Singer, W. ''Activity-dependent self-organization of synaptic connections as a substrate of learning.'' Pages 301–336 in ref. 44.

52. Klarsfeld, A., and Changeux, J.-P. *Proc. Nat'l. Acad. Sci. USA* 82: 4558–4562 (1985). ''Activity regulates the level of acetylcholine receptor α-subunit mRNA in cultured chick myotubes.''

53. Fontaine, B.; Klarsfeld, A.; Hökfelt, T.; and Changeux, J.-P. *Neurosci. Lett.* 71: 59–65 (1986). ''Calcitonin gene-related peptide, a peptide present in spinal cord motoneurons, increases the number of acetylcholine receptors in primary cultures of chick myotubes.''

54. Fontaine, B.; Klarsfeld, A.; and Changeux, J.-P. *J. Cell. Biol.* 105: 1337–1342 (1987). ''Calcitonin gene-related peptide and muscle activity regulate acetylcholine receptor α-subunit mRNA levels by distinct intracellular pathways.''

55. Abstracts, Int'l. Workshop on *Structural and Functional Aspects of the Cholinergic Synapse*, Neve Ilan, Israel, 30 Aug.–4 Sept. 1987.

56. Changeux, J.-P.; Klarsfeld, A.; Fontaine, B.; and Laufer, R. ''Long-term evolution of the acetylcholine receptor during synapse formation.'' Pages 17–22 in ref. 55.

57. Thompson, W. J. ''Activity-dependent regulation of gene expression.'' Pages 13–30 in ref. 44.

58. McMahan, U. J. ''Agrin: An extracellular synaptic organizing molecule.'' Page 27 in ref. 55.

59. Wallace, B. G. "Agrin: Mechanism of action." Page 28 in ref. 55.
60. The Compact Edition of the *Oxford English Dictionary*, p. 1768. Oxford: Oxford University Press, 1971. Supplement, p. 4108.
61. Schopf, J. W., and Packer, B. M. *Science* 237: 70–73 (1987). "Early Archean (3.3-Billion- to 3.5-Billion-Year-Old) microfossils from Warrawoona group, Australia."
62. Foster, K. W.; Saranak, J.; Patel, N.; Zarilli, G.; Okabe, M.; Kline, T.; and Nakanishi, K. *Nature* 311: 756 (1984). "A rhodopsin is the functional photo-receptor for phototaxis in the unicellular eukaryote, *Chlamydomonas*."
63. Horridge, G. A. *Interneurons*. San Francisco: W. H. Freeman, 1968. 436 pp.
64. Stewart, R. C., and Dahlquist, F. W. *Chem. Rev.* 87: 997–1025 (1987). "Molecular components of bacterial chemotaxis."
65. Fox, G. E.; Stackebrandt, E.; Hespell, R. B.; Gibson, J.; Maniloff, J.; Dyer, T. A.; Wolfe, R. S.; Balch, W. E.; Tanner, R. S.; Magrum, L. J.; Zablen, L. B.; Blakemore, R.; Gupta, R.; Bonen, R.; Bonen, L.; Lewis, B. J.; Stahl, T.; Luehrsen, K. R.; Chen, K. N.; Woese, C. R. *Science* 209: 457–463 (1980). "The phylogeny of prokaryotes."
66. Achenbach-Richter, L.; Stetter, K. O.; and Woese, C. R. *Nature* 327: 348–349 (1987). "A possible biochemical missing link among archaebacteria."
67. Field, K. G.; Olsen, G. J.; Lane, D. J.; Giovannoni, S. J.; Ghiselin, M. T.; Raff, E. C.; Pace, N. R.; and Raff, R. A. *Science* 239: 748–753 (1988). "Molecular phylogeny of the animal kingdom."
68. Adler, J. *Cold Spring Harbor Symp. Quant. Biol.* 30: 289–292 (1965). "Chemotaxis in *Escherichia coli*."
69. Koshland, D. E., Jr. *Bacterial Chemotaxis as a Model Behavioral System*. New York: Raven Press, 1980.
70. Macnab, R. M. "Motility and chemotaxis in *Escherichia coli* and *Salmonella typhimurium*." In *Cellular and Molecular Biology*, ed. J. Ingraham, K. B. Low, B. Magasanik, M. Schaechter, H. E. Umbarger, and F. C. Neidhardt, pp. 732–759. Washington, D.C.: ASM Publications, 1987.
71. Ordal, G. W. *Crit. Revs. Microbiol.* 12: 95–130 (1985). "Bacterial chemotaxis: Biochemistry of behavior in a single cell."
72. Berg, H. C. *Ann. Rev. Biophys. Bioeng.* 4: 119–136 (1975). "Chemotaxis in bacteria."
73. Adler, J. *Ann. Rev. Biochem.* 44: 341–356 (1975). "Chemotaxis in bacteria."
74. (a) Koshland, D. E., Jr. *Adv. in Neurochem.* 2: 277–341 (1977). "Sensory response in bacteria." (b) Hazelbauer, G. L., and Harayama, S. *Int'l. Rev. Cytol.* 81: 33–70 (1983). "Sensory transduction in bacterial chemotaxis." (c) Taylor, B. L., and Panasenko, S. M. "Biochemistry of chemosensory behavior in prokaryotes and unicellular eukaryotes." In *Membranes and Sensory Transduction*, ed. G. Colombetti and F. Lenci, pp. 71–112. New York: Plenum Press, 1984.
75. Koshland, D. E., Jr. *Ann. Rev. Neurosci.* 3: 43–75 (1980). "Bacterial chemotaxis in relation to neurobiology."
76. Macnab, R. M. *Crit. Revs. Biochem.* 5: 291–341 (1978). "Bacterial mobility and chemotaxis."
77. Bonner, J. T. *Sci. Amer.* 248(4): 106–114 (1983). "Chemical signals of social amboebae."

78. Dworkin, M., and Kaiser, D. *Science* 230: 18–24 (1985). ''Cell interactions in myxobacterial growth and development.''
79. Rosenberg, E., ed. *Myxobacteria: Development and cell interactions*. New York: Springer, 1984.
80. Paerl, H. W., and Gallucci, K. W. *Science* 227: 647–649 (1985). ''Role of chemotaxis in establishing a specific nitrogen-fixing cyanobacterial-bacterial association.''
81. Adler, J. *Science* 166: 1588–1597 (1969). ''Chemoreceptors in bacteria.''
82. Adler, J. *Science* 153: 708–716 (1966). ''Chemotaxis in bacteria.''
83. Hazelbauer, G. L.; Mesibov, R. E.; and Adler, J. *Proc. Nat'l. Acad. Sci. USA* 64: 1300–1307 (1969). ''*Escherichia coli* mutants defective in chemotaxis toward specific chemicals.''
84. Adler, J.; Hazelbauer, G. L.; and Dahl, M. M. *J. Bacteriol.* 115: 824–847 (1973). ''Chemotaxis towards sugars in *Escherichia coli*.''
85. Adler, J. *Harvey Lectures* 72: 195–230 (1978). ''Chemotaxis in bacteria.''
86. Imae, Y.; Oosawa, K.; Mizuno, T.; Kihara, M.; and Macnab, R. M. *J. Bacteriol.* 169: 371–379 (1987). ''Phenol: A complex chemoeffector in bacterial chemotaxis.''
87. Oosawa, K., and Imae, Y. *J. Bacteriol.* 154: 104–112 (1983). ''Glycerol and ethylene glycol: members of a new class of repellants of *Escherichia coli* chemotaxis.''
88. Hazelbauer, G. L., and Adler, J. *Nature New Biology* 230: 101–104 (1971). ''Role of the galactose binding protein in chemotaxis of *Escherichia coli* towards galactose.''
89. Clarke, S., and Koshland, D. E., Jr. *J. Biol. Chem.* 254: 9695–9702 (1979). ''Membrane receptors for aspartate and serine in bacterial chemotaxis.''
90. Aksamit, R., and Koshland, D. E., Jr. *Biochemistry* 13: 4473–4478 (1974). ''Identification of the ribose binding protein as the receptor for ribose chemotaxis in *Salmonella typhimurium*.''
91. Parnes, J. R., and Boos, W. *J. Biol. Chem.* 248: 4436–4445 (1973). ''Unidirectional transport activity mediated by the galactose-binding protein of *Escherichia coli*.''
92. Miller, D. M., III; Olson, J. S.; and Quiocho, F. A. *J. Biol. Chem.* 255: 2465–2471 (1980). ''The mechanism of sugar binding to the periplasmic receptor for galactose chemotaxis and transport in *Escherichia coli*.''
93. Brass, J. M., and Manson, M. D. *J. Bacteriol.* 157: 881–890 (1984). ''Reconstitution of maltose chemotaxis in *Escherichia coli* by addition of maltose-binding protein to calcium treated cells of maltose regulon mutants.''
94. Berg, H. C., and Brown, D. A. *Nature* 239: 500–504 (1972). ''Chemotaxis in *Escherichia coli* analysed by three-dimensional tracking.''
95. Berg, H. C. *Nature* 254: 389–392 (1975). ''Bacterial behavior.''
96. Macnab, R. M., and Koshland, D. E., Jr. *Proc. Nat'l. Acad. Sci. USA* 69: 2509–2512 (1972). ''The gradient-sensing mechanism in bacterial chemotaxis.''
97. Macnab, R. M. *Proc. Nat'l. Acad. Sci. USA* 74: 221–225 (1977). ''Bacteria flagella rotating in bundles: A study in helical geometry.''
98. Asakura, S.; Eguchi, G.; and Iino, T. *J. Mol. Biol.* 16: 302 (1966). ''*Salmonella* flagella: in vitro reconstruction and overall shapes of flagellar filaments.''
99. Hotani, H. *J. Mol. Biol.* 106: 151–166 (1976). ''Light microscope study of mixed helices in reconstructed *Salmonella* flagella in vitro.''

100. Khan, S.; Macnab, R. M.; DeFranco, A. L.; and Koshland, D. E., Jr. *Proc. Nat'l. Acad. Sci. USA* 75: 4150–4154 (1978). "Inversion of a behavioral response in bacterial chemotaxis: Explanation at the molecular level."
101. Silverman, M., and Simon, M. *Nature* 249: 73–74 (1974). "Flagellar rotation and the mechanism of bacterial motility."
102. Springer, M. S.; Goy, M. F.; and Adler, J. *Proc. Nat'l. Acad. Sci. USA* 74: 3312–3316 (1977). "Sensory transduction in *Escherichia coli*: Two complimentary pathways of information processing that involve methylated proteins."
103. Kondoh, H.; Ball, C. B.; and Adler, J. *Proc. Nat'l. Acad. Sci. USA* 76: 260–264 (1979). "Identification of a methyl-accepting chemotaxis protein for the ribose and galactose chemoreceptors of *Escherichia coli*."
104. Manson, M. D.; Blank, V.; Brade, G.; and Higgins, C. F. *Nature* 321: 253–256 (1986). "Peptide chemotaxis in *E.coli* involves the Tap signal transducer and the dipeptide permease."
105. Berg, H. C. *Nature* 321: 200–201 (1986). "Chemotaxis gene unveiled."
106. Kleene, S. J.; Hobson, A. C.; and Adler, J. *Proc. Nat'l. Acad. Sci. USA* 76: 6309–6313 (1979). "Attractants and repellants influence methylation and demethylation of methyl-accepting chemotaxis proteins in an extract of *Escherichia coli*."
107. Springer, M. S.; Goy, M. F.; and Adler, J. *Nature* 280: 279–284 (1979). "Protein methylation in behavioural control mechanisms and in signal transduction."
108. Mowbray, S. L., and Koshland, D. E., Jr. *Cell* 50: 171–180 (1987). "Additive and independent responses in a single receptor: Aspartate and maltose stimuli on the Tar protein."
109. Goldman, D. J.; Worobec, S. W.; Siegel, S. W.; Hecker, R. V.; and Ordal, G. W. *Biochemistry* 21: 915–920 (1982). "Chemotaxis in *Bacillus subtilis* . Effects of attractants on the level of methylation of methyl-accepting chemotaxis proteins and the role of demethylation in the adaptation process."
110. Goldman, D. J.; Nettleton, D. O.; and Ordal, G. W. *Biochemistry* 23: 675–680 (1984). "Purification and characterization of chemotactic methylesterase from *Bacillus subtilis*."
111. Goldman, D. J., and Ordal, G. W. *Biochemistry* 23: 2600–2606 (1984). "In vitro methylation and demethylation of methyl-accepting chemotaxis proteins in *Bacillus subtilis*."
112. Van der Werf, P., and Koshland, D. E., Jr. *J. Biol. Chem.* 252: 2793–2795 (1977). "Identification of a γ-glutamyl methyl ester in bacterial membrane protein involved in chemotaxis."
113. Kleene, S. J.; Toews, M. L.; and Adler, J. *J. Biol. Chem.* 252: 3214–3218 (1977). "Isolation of a glutamic acid methyl ester from an *Escherichia coli* membrane protein involved in chemotaxis."
114. Paoni, N. F., and Koshland, D. E., Jr. *Proc. Nat'l. Acad. Sci. USA* 76: 3693–3697 (1979). "Permeabilization of cells for studies on the *Biochemistry* of bacterial chemotaxis."
115. DeFranco, A. L., and Koshland, D. E., Jr. *Proc. Nat'l. Acad. Sci. USA* 77: 2429–2433 (1980). "Multiple methylation in processing of sensory signals during bacterial chemotaxis."
116. Chelsky, D., and Dahlquist, F. W. *Proc. Nat'l. Acad. Sci. USA* 77: 2434–2438 (1980). "Structural studies of methyl-accepting chemotaxis proteins of *Escherichia coli*: Evidence for multiple methylation sites."

117. Armitage, J. *Nature* 289: 121–122 (1981). ''Multiple methylation and bacterial adaptation.''
118. Kehry, M., and Dahlquist, F. W. *J. Biol. Chem.* 257: 10378–10386 (1982). ''The methyl-accepting chemotaxis proteins of *Escherichia coli.*''
119. Boyd, A., and Simon, M. I. *J. Bacteriol.* 143: 809–815 (1980). ''Multiple electrophoretic forms of methyl-accepting chemotaxis proteins generated by stimulus elicited methylation in *Escherichia coli.*''
120. Engström, P., and Hazelbauer, G. L. *Cell* 20: 165–171 (1980). ''Multiple methylation of methyl-accepting chemotaxis proteins during adaptation of *E. coli* to chemical stimuli.''
121. Kehry, M. R., and Dahlquist, F. W. *Cell* 29: 761–772 (1982). ''Adaptation in bacterial chemotaxis: CheB-dependent modification permits additional methylations of sensory transducer proteins.''
122. Terwilliger, T. C., and Koshland, D. E., Jr. *J. Biol. Chem.* 259: 7719–7725 (1984). ''Sites of methyl esterification and deamination on the aspartate receptor involved in chemotaxis.''
123. Kehry, M. R.; Doak, T. G.; and Dahlquist, F. W. *J. Bacteriol.* 161: 105–112 (1985). ''Aberrant regulation of methylesterase activity in cheD chemotaxis mutants of *Escherichia coli.*''
124. Kehry, M. R.; Doak, T. G.; and Dahlquist, F. W. *J. Bacteriol.* 163: 983–990 (1985). ''Sensory adaptation in bacterial chemotaxis: Regulation of demethylation.''
125. Clarke, S. *Ann. Rev. Biochem.* 54: 479–506 (1985). ''Protein carboxyl methyltransferases: Two distinct classes of enzymes.''
126. Stock, J.; Borczuk, A.; Chiou, F.; and Burchenal, J. *Proc. Nat'l. Acad. Sci. USA* 82: 8364–8368 (1986). ''Compensatory mutations in receptor function: A reevaluation of the role of methylation in bacterial chemotaxis.''
127. Hirota, N. *J. Biochem.* 99: 349–356 (1986). ''Methylation and demethylation of solubilized chemoreceptors from thermophilic bacterium PS–3.''
128. Terwilliger, T. C.; Wang, J. Y.; and Koshland, D. E., Jr. *J. Biol. Chem.* 261: 10814–10820 (1986). ''The multiply methylated aspartate receptors involved in bacterial chemotaxis.''
129. Parkinson, J. S. *Cell* 4: 183–188 (1975). ''Genetics of chemotactic behavior in bacteria.''
130. (a) Parkinson, J. S., and Revello, P. T. *Cell* 15: 1221–1230 (1978). ''Sensory adaptation mutants of *E. coli.*'' (b) Slocum, M. K., and Parkinson, J. S. *J. Bacteriol.* 155: 565–577 (1983). ''Genetics of methyl-accepting chemotaxis proteins in *Escherichia coli*: Organization of the tar region.''
131. (a) Springer, W. R., and Koshland, D. E., Jr. *Proc. Nat'l. Acad. Sci. USA* 74: 533–537 (1977). ''Identification of a protein methyltransferase as the cheR gene product in a bacterial sensing system.'' (b) Clarke, S.; Sparrow, K.; Panasenko, S.; and Koshland, D. E., Jr. *J. Supramol. Struct.* 13: 315–318 (1980). ''In vitro methylation of bacterial chemotaxis proteins: Characterization of protein methyltransferase activity in crude extracts of *Salmonella typhimurium.*''
132. Stock, J. B., and Koshland, D. E., Jr. *Proc. Nat'l. Acad. Sci. USA* 75: 3659–3663 (1978). ''A protein methylesterase involved in bacterial sensing.''
133. Chelsky, D., and Dahlquist, F. W. *Biochemistry* 19: 4633–4639 (1980). ''Chemotaxis in *Escherichia coli*: Associations of protein components.''
134. Milligan, D. L., and Koshland, D. E., Jr. *J. Biol. Chem.* 263: 6268–6275 (1988).

"Site-directed cross-linking. Establishing the dimeric structure of the aspartate receptor of bacterial chemotaxis."

135. Falke, J. J., and Koshland, D. E., Jr. *Science* 237: 1596–1600 (1987). "Global flexibility in a sensory receptor: A site-directed cross-linking approach."
136. Kehry, M. R.; Doak, T. G.; and Dahlquist, F. W. *J. Biol. Chem.* 259: 11828–11835 (1984). "Stimulus-induced changes in methylesterase activity during chemotaxis in *Escherichia coli*."
137. Goy, M. F.; Springer, M. S.; and Adler, J. *Cell* 15: 1231–1240 (1978). "Failure of sensory adaptation in bacterial mutants that are defective in a protein methylation reaction."
138. Rollins, C., and Dahlquist, F. W. *Cell* 25: 333–340 (1981). "The methyl-accepting chemotaxis proteins of *E. coli*: A repellent-stimulated, covalent modification, distinct from methylation."
139. Kehry, M. R.; Bond, M. W.; Hunkapiller, M. W.; and Dahlquist, F. W. *Proc. Nat'l. Acad. Sci. USA* 80: 3599–3603 (1983). "Enzymatic deamidation of methyl-accepting chemotaxis proteins in *Escherichia coli* catalyzed by the cheB gene product."
140. Hazelbauer, G. L., and Engström, P. *J. Bacteriol.* 145: 35–42 (1981). "Multiple forms of methyl-accepting chemotaxis proteins distinguished by a factor in addition to multiple methylation."
141. Moulton, R. C., and Montie, T. C. *J. Bacteriol.* 137: 274–280 (1979). "Chemotaxis by *Pseudomonas aeruginosa*."
142. Stinson, M. W.; Cohen, M. A.; and Merrick, J. M. *J. Bacteriol.* 131: 672–681 (1977). "Purification and properties of the periplasmic glucose binding protein of *Pseudomonas aeruginosa*."
143. Craven, R. C., and Montie, T. C. *J. Bacteriol.* 154: 780–786 (1983). "Chemotaxis of *Pseudomonas aeruginosa*: Involvement of methylation."
144. Ahlgren, J. A., and Ordal, G. W. *Biochem. J.* 213: 759–763 (1983). "Methyl esterification of glutamic acid residues of methyl-accepting chemotaxis proteins in *Bacillus subtilis*."
145. Bedale, W. A.; Nettleton, D. O.; Sopata, C. S.; Thoelke, M. S.; and Ordal, G. W. *J. Bacteriol.* 170: 223–227 (1988). "Evidence for methyl group transfer between the methyl-accepting chemotaxis proteins in *Bacillus subtilis*."
146. Schimz, A. *FEBS Lett.* 125: 205–207 (1981). "Methylation of membrane proteins is involved in chemosensory and photosensory behavior of *Halobacterium halobium*."
147. Bibikov, S. I.; Baryshev, V. A.; and Glagolev, A. N. *FEBS Lett.* 146: 255–258 (1982). "The role of methylation in the taxis of *Halobacterium halobium* to light and chemoeffectors."
148. Shaw, P.; Gomes, S. L.; Sweeney, K.; Ely, B.; and Shapiro, L. *Proc. Nat'l. Acad. Sci. USA* 80: 5261–5265 (1983). "Methylation involved in chemotaxis is regulated during *Caulobacter* differentiation."
149. Kathariou, S., and Greenberg, E. P. *J. Bacteriol.* 156: 95–100 (1983). "Chemoattractants elicit methylation of specific polypeptides in *Spirochaeta aurantia*."
150. Nowlin, D. M.; Nettleton, D. O.; Ordal, G. W.; and Hazelbauer, G. L. *J. Bacteriol.* 163: 262–266 (1983). "Chemotactic transducer proteins of *Escherichia coli* exhibit homology with methyl-accepting proteins from distantly related bacteria."

151. Greenberg, E. P., and Canale-Parola, E. *J. Bacteriol.* 130: 485–494 (1977). ''Chemotaxis in *Spirochaeta aurantia*.''
152. Janson, C. A., and Clarke, S. A. *J. Biol. Chem.* 255: 11640–11643 (1980). ''Identification of aspartic acid as a site of methylation in human erythrocyte membrane proteins.''
153. O'Connor, C. M., and Clarke, S. *J. Biol. Chem.* 258: 8485–8492 (1983). ''Methylation of erythrocyte membrane proteins at extracellular and intracellular D-aspartyl sites in vitro.''
154. Galletti, P.; Paik, W. K.; and Kim, S. *Eur. J. Biochem.* 97: 221–227 (1979). ''Methyl acceptors for protein methylase II from human erythrocyte membrane.''
155. Johnson, B. A.; Murray, E. D., Jr.; Clarke, S.; Glass, D. B.; and Aswad, D. W. *J. Biol. Chem.* 262: 5622–5629 (1987). ''Protein carboxyl methyltransferase facilitates conversion of atypical L-isoaspartyl peptides to normal L-aspartyl peptides.''
156. Paik, W. K., and Kim, S. *Protein Methylation*. New York: Wiley, 1980. 282 pp.
157. Chelsky, D.; Olson, J. F.; and Koshland, D. E., Jr. *J. Biol. Chem.* 262: 4303–4309 (1987). ''Cell-cycle dependent methyl esterification of Lamin B.''
158. Galletti, P.; DeRosa, M.; Gambacorta, A.; Manna, C.; Festinese, R.; and Zappia, V. *FEBS Lett.* 124: 62–66 (1981). ''Protein methylation in *Calderiella acidophila*, an extreme thermo-acidophilic archaebacterium.''
159. Saier, M. H., Jr.; Grenier, F. C.; Lee, C. A.; and Waygood, E. B. *J. Cellular Biochem.* 27: 43–56 (1985). ''Evidence for the evolutionary relatedness of the proteins of the bacterial phosphoenolpyruvate:sugar phosphotransferase system.''
160. Postma, P. W., and Lengeler, J. W. *Microbiol. Revs.* 49: 232–269 (1985). ''Phosphoenolpyruvate: Carbohydrate phosphotransferase system of bacteria.''
161. Adler, J., and Epstein, W. *Proc. Nat'l. Acad. Sci. USA* 71: 2895–2899 (1974). ''The phosphotransferase-system enzymes as chemoreceptors for certain sugars in *Escherichia coli* chemotaxis.''
162. Saier, M. H., Jr.; Newman, M. J.; and Rephaeli, A. W. *J. Biol. Chem.* 252: 8890–8898 (1977). ''Properties of a phosphoenolpyruvate:mannitol phosphotransferase system in *Spirochaeta aurantia*.''
163. Saier, M. H., Jr.; Feucht, B. U.; and Mora, W. K. *J. Biol. Chem.* 252: 8899–8907 (1977). ''Sugar phosphate:sugar transphosphorylation and exchange group translocation catalyzed by the enzyme II complexes of the bacterial phosphoenolpyruvate:sugar phosphotransferase system.''
164. Saier, M. H., Jr.; Cox, D. F.; and Moczydlowski, E. G. *J. Biol. Chem.* 252: 8908–8916 (1977). ''Sugar phosphate:sugar transphosphorylation coupled to exchange group translocation catalyzed by the enzyme II complexes of the phosphoenolpyruvate:sugar phosphotransferase system in membrane vesicles of *Escherichia coli*.''
165. Saier, M. H., Jr. *J. Supramol. Struct.* 14: 281–294 (1980). ''Catalytic activities associated with the enzymes II of the bacterial phosphotransferase system.''
166. Pecher, A.; Renner, I.; and Lengeler, J. W. ''The phosphoenolpyruvate-dependent carbohydrate:phosphotransferase system enzymes II, a new class of chemosensors in bacterial chemotaxis,'' In *Mobility and Recognition in Cell Biology*, ed. H. Sund, and K. Veeger, pp. 517–531. Berlin: Walter de Gruyter, 1983.
167. (a) Niwano, M., and Taylor, B. L. *Proc. Nat'l. Acad. Sci. USA* 79: 11–15 (1982). ''Novel sensory adaptation mechanism in bacterial chemotaxis to oxygen and phos-

photransferase substrates.'' (b) Taylor, B. L. *Ann. Rev. Microbiol.* 37: 551–573 (1983). ''Role of protonmotive force in sensory transduction in bacteria.''

168. Bouma, C. L.; Meadow, N. D.; Stover, E. W.; and Roseman, S. *Proc. Nat'l. Acad. Sci. USA* 84: 930–934 (1987). ''II-Bglc, a glucose receptor of the bacterial phosphotransferase system: Molecular cloning of ptsG and purification of the receptor from an overproducing strain of *Escherichia coli.*''
169. Grenier, F. C.; Waygood, E. B.; and Saier, M. H., Jr. *Biochemistry* 24: 4872–4876 (1985). ''Bacterial phosphotransferase system: Regulation of the glucose and mannose enzymes II by sulfhydryl oxidation.''
170. Macnab, R. M., and Koshland, D. E., Jr. *J. Mol. Biol.* 84: 399–406 (1974). ''Bacterial motility and chemotaxis: light-induced tumbling response and visualization of individual flagella.''
171. Vogler, A. P., and Lengeler, J. W. *J. Bacteriol.* 169: 593–599 (1987). ''Indirect role of adenylate cyclase and cyclic AMP in chemotaxis to phosphotransferase system carbohydrates in *Escherichia coli* K-12.''
172. Fillingame, R. H. *Ann. Rev. Biochem.* 49: 1079–1113 (1980). ''The proton-translocating pumps of oxidative phosphorylation.''
173. Wikström, M. *Nature* 308: 558–560 (1984). ''Pumping of protons from the mitochondrial matrix by cytochrome oxidase.''
174. Glagolev, A. N., and Sherman, M. Y. *FEMS Microbiol. Lett.* 17: 147–150 (1983). ''The glucose phosphotransferase system is involved in *Escherichia coli* oxygen taxis.''
175. Repaske, D. R., and Adler, J. *J. Bacteriol.* 145: 1196–1208 (1981). ''Change in intracellular pH of *Escherichia coli* mediates the chemotactic response to certain attractants and repellents.''
176. Kihara, M., and Macnab, R. M. *J. Bacteriol.* 145: 1209–1221 (1981). ''Cytoplasmic pH mediates pH taxis and weak-acid repellent taxis of bacteria.''
177. Slonczewski, J. L.; Alger, J. R.; and Macnab, R. M. ^{31}P-NMR results cited in ref. 176.
178. Taylor, B. L.; Miller, J. B.; Warrick, H. M.; and Koshland, D. E., Jr. *J. Bacteriol.* 140: 567–573 (1979). ''Electron acceptor taxis and blue light effect on bacterial chemotaxis.''
179. Clancy, M.; Madill, K. A.; and Wood, J. M. *J. Bacteriol.* 146: 902–906 (1981). ''Genetic and biochemical requirements for chemotaxis to L-proline in *Escherichia coli.*''
180. Omirbekova, N. G.; Gabai, V. L.; Sherman, M. Yu.; Vorobyeva, N. V.; and Glagolev, A. N. *FEMS Microbiol. Lett.* 28: 259–263 (1985). ''Involvement of Ca^{++} and cGMP in bacterial taxis.''
181. Eisenbach, M.; Margolin, Y.; and Ravid, S. ''Excitatory signaling in microorganisms,'' In *Sensing and Response in Microorganisms*, ed. M. Eisenbach, and M. Balaban, pp. 43–61. Amsterdam: Elsevier, 1985.
182. Miller, J. B., and Koshland, D. E., Jr. *J. Mol. Biol.* 111: 183–201 (1977). ''Membrane fluidity and chemotaxis: effects of temperature and membrane lipid composition on the swimming behavior of *Salmonella typhimurium* and *Escherichia coli.*''
183. Parkinson, J. S., and Houts, S. E. *J. Bacteriol.* 151: 106–113 (1982). ''Isolation and behavior of *Escherichia coli* deletion mutants lacking chemotaxis functions.''
184. Imae, Y.; Mizuno, T.; and Maeda, K. *J. Bacteriol.* 159: 368–374 (1984). ''Che-

mosensory excitation and thermosensory excitation in adaptation-deficient mutants of *Escherichia coli*.''

185. Imae, Y.; Mizuno, T.; Yamamoto, K.; and Lee, L. Abstracts, p. 73, 9th Int'l. Biophys. Congr., Jerusalem, Israel, 1987. ''Mechanism of thermosensory transduction in bacteria.''
186. Baryshev, V. A.; Glagolev, A. N.; and Skulachev, V. P. *J. Gen. Microbiol.* 129: 367–373 (1983). ''The interrelation of phototaxis, membrane potential and K^+/Na^+ gradient in *Halobacterium halobium*.'' Cf. Glagolev, A. N. *Mobility and Taxis in Prokaryotes*. London: Harwood, 1984.
187. Brown, I. G.; Galperin, M. Y.; Glagolev, A. N.; and Skulachev, V. P. *Eur. J. Biochem.* 134: 345–349 (1983). ''Utilization of energy stored in the form of K^+ and Na^+ ion gradients by bacterial cells.''
188. Spudich, J. L., and Bogomolni, R. A. *Nature* 312: 509–513 (1984). ''Mechanism of colour discrimination by a bacterial sensory rhodopsin.'' Spudich, J. L. ''Bacterial sensory rhodopsin (SR), a dual attractant and repellent phototaxis receptor.'' In *Sensing and Response in Microorganisms*, ed. M. Eisenbach, and M. Balaban, pp. 119–127. Amsterdam: Elsevier, 1985.
189. Berg, H. C. *Nature* 315: 354–355 (1985). ''Bacterial diplomacy.''
190. Anderson, R. A. ''Formation of the bacterial flagellar bundle.'' In *Swimming and Flying in Nature*, ed. T.Y.T. Wu, C. Brokaw, and C. Brennen, pp. 45–56. New York: Plenum, 1975.
191. Segall, J. E.; Ishihara, A.; and Berg, H. C. *J. Bacteriol.* 161: 51–59 (1985). ''Chemotactic signaling in filamentous cells of *Escherichia coli*.'' Hess, J. F.; Bourret, R. M.; and Simon, M. I. *Nature* 336: 139–143 (1988). ''Hidstidine phosphorylation and phosphoryl group transfer in bacterial chemotaxis.'' Hazelbauer, G. L. *Can. J. Microbiol.* 34: 466–474 (1988). ''The bacterial chemosensory system.'' Wolfe, A. J.; Conley, M. P.; and Berg, H. C. *Proc. Nat'l. Acad. Sci. USA* 85: 6711–6715 (1988). ''Acetyladenylate plays a role in controlling the direction of flagellar rotation.'' Borkovich, K. A.; Kaplan, N.; Hess, J. F.; and Simon, M. I. *Proc. Nat'l. Acad. Sci. USA* 86: 1208–1212 (1989). ''Transmembrane signal transduction in bacterial chemotaxis involves ligand-dependent activation of phosphate group transfer.''
192. Ravid, S., and Eisenbach, M. *J. Bacteriol.* 158: 222–230 (1984). ''Direction of flagellar rotation in bacterial cell envelopes.''
193. Szupica, C. J., and Adler, J. *J. Bacteriol.* 162: 451–453 (1985). ''Cell envelopes of chemotaxis mutants of *Escherichia coli* rotate their flagella counterclockwise.''
194. Stock, J. B.; Maderis, A. M.; and Koshland, D. E., Jr. *Cell* 27: 37–44 (1981). ''Bacterial chemotaxis in the absence of receptor carboxymethylation.''
195. Stock, J. B.; Kersulis, G.; and Koshland, D. E., Jr. *Cell* 42: 683–690 (1985). ''Neither methylating nor demethylating enzymes are required for bacterial chemotaxis.''
196. Callahan, A. M., and Parkinson, J. S. *J. Bacteriol.* 161: 96–104 (1985). ''Genetics of methyl-accepting chemotaxis proteins in *Escherichia coli*: cheD mutations affect the structure and function of the tsr transducer.''
197. Ravid, S.; Matsumura, P.; and Eisenbach, M. *Proc. Nat'l. Acad. Sci. USA* 83: 7157–7161 (1986). ''Restoration of flagellar clockwise rotation in bacterial envelopes by insertion of the chemotaxis protein CheY.''

198. Eisenbach, M., and Matsumura, P. *Botan. Acta* 101: 105–110 (1988). "*In vitro* approach to bacterial chemotaxis."

199. Volz, K.; Beman, J.; and Matsumura, P. *J. Biol. Chem.* 261: 4723–4725 (1986). "Crystallization and preliminary characterization of CheY, a chemotaxis control protein from *Escherichia coli.*"

200. Matsumura, P.; Rydel, J. J.; Linzmeier, R.; and Vacante, D. *J. Bacteriol.* 160: 36–41 (1984). "Overexpression and sequence of the *Escherichia coli* cheY gene and biochemical activities of the cheY protein."

201. Stock, A.; Koshland, D. E., Jr.; and Stock, J. *Proc. Nat'l. Acad. Sci. USA* 82: 7989–7993 (1985). "Homologies between the *Salmonella typhimurium* cheY protein and proteins involved in the regulation of chemotaxis, membrane protein synthesis and sporulation."

202. Simms, S. A.; Keane, M. G.; and Stock, J. *J. Biol. Chem.* 260: 10161–10168 (1985). "Multiple forms of the cheB methylesterase in bacterial chemosensing."

203. Borczuk, A.; Staub, A.; and Stock, J. *Biochem. Biophys. Res. Commun.* 141: 918–923 (1986). "Demethylation of bacterial chemoreceptors is inhibited by attractant stimuli in the complete absence of the regulatory domain of the demethylating enzyme."

204. Sherman, M. Y. *FEBS Lett.* 148: 192–197 (1982). "Hypothesis: Interrelation between taxis and sporulation systems."

205. Tribhuwan, R. C.; Johnson, M. S.; and Taylor, B.L. *J. Bacteriol.* 168: 624–630 (1986). "Evidence against direct involvement of cyclic GMP or cyclic AMP in bacterial chemotactic signalling."

206. Engström, P. Ph.D. Thesis, University of Uppsala, Sweden (1982). Cited in ref. 205.

207. Stock, A.; Mottonen, J.; Chen, T.; and Stock, J. *J. Biol. Chem.* 262: 535–537 (1987). "Identification of a possible nucleotide binding site in cheW, a protein required for sensory transduction in bacterial chemotaxis."

208. Kuo, S. C., and Koshland, D. E., Jr. *J. Bacteriol.* 169: 1307–1314 (1987). "Roles of cheY and cheZ gene products in controlling flagellar rotation in bacterial chemotaxis of *Escherichia coli.*"

209. Park, C., and Hazelbauer, G. L. *J. Bacteriol.* 168: 1378–1383 (1986). "Mutation plus amplification of a transducer gene disrupts general chemotactic behavior in *Escherichia coli.*"

210. Nowlin, D. M.; Bollinger, J.; and Hazelbauer, G. L. *J. Biol. Chem.* 262: 6039–6045 (1987). "Site of covalent modification in Trg, a sensory transducer of *Escherichia coli.*"

211. Thoelke, M. W.; Bedale, W. A.; Nettleton, D. O.; and Ordal, G. W. *J. Biol. Chem.* 262: 2811–2816 (1987). "Evidence for an intermediate methyl-acceptor for chemotaxis in *Bacillus subtilis.*"

212. Segall, J. E.; Block, S. M.; and Berg, H. C. *Proc. Nat'l. Acad. Sci. USA* 83: 8987–8991 (1986). "Temporal comparisons in bacterial chemotaxis."

213. DePamphilis, M. L., and Adler, J. *J. Bacteriol.* 105: 384–395 (1971). "Fine structure and isolation of the hook-basal body complex of flagella from *Escherichia coli* and *Bacillus subtilis.*"

214. DePamphilis, M. L., and Adler, J. *J. Bacteriol.* 105: 396–407 (1971). "Attachment of flagellar basal bodies to the cell envelope: specific attachment to the outer membrane and the cytoplasmic membrane."

215. Block, S. M., and Berg, H. C. *Nature* 309: 470–472 (1984). ''Successive incorporation of force-generating units in the bacterial rotary motor.''

216. Coulton, J. W., and Murray, R.G.E. *J. Bacteriol.* 136: 1037–1049 (1978). ''Cell envelope associations of *Aquaspirillum serpens* flagella.''

217. Dean, G. E.; Macnab, R. M.; Stader, J.; Matsumura, P.; and Burks, C. *J. Bacteriol.* 159: 991–999 (1984). ''Gene sequence and predicted amino acid sequence of the motA protein, a membrane associated protein required for flagellar rotation in *Escherichia coli*.''

218. Stader, J.; Matsumura, P.; Vacante, D.; Dean, G. E.; and Macnab, R. M. *J. Bacteriol.* 166: 244–252 (1986). ''Nucleotide sequence of the *Escherichia coli* motB gene and site-limited incorporation of its product into the cytoplasmic membrane.''

219. Kato, S.; Okamoto, M.; and Asakura, S. *J. Mol. Biol.* 173: 463–476 (1984). ''Polymorphic transition of the flagellar polyhook from *Escherichia coli* and *Salmonella typhimurium*.''

220. Spencer, M. *Nature* 309: 404–405 (1984). ''Bacterial motion: progress in flagellation.''

221. Berg, H. C., and Anderson, R. A. *Nature* 245: 380–382 (1973). ''Bacteria swim by rotating their flagellar filaments.''

222. Berg, H. C.; Manson, M. D.; and Conley, M. P. *Symp. Soc. Exp. Biol.* 35: 1–35 (1982). ''Dynamics and energetics of flagellar rotation in bacteria.''

223. Macnab, R. M., and Aizawa, S.-I. *Ann. Rev. Biophys. Bioeng.* 13: 51–83 (1984). ''Bacterial motility and the bacterial flagellar motor.''

224. Lowe, G.; Meister, M.; and Berg, H. C. *Nature* 325: 637–640 (1987). ''Rapid rotation of flagellar bundles in swimming bacteria.''

225. Shimada, K., and Berg, H. C. *J. Mol. Biol.* 193: 585–589 (1987). ''Response of the flagellar rotary motor to abrupt changes in extracellular pH.''

226. Manson, M. D.; Tedesco, P.; Berg, H. C.; Harold, F. M.; and van der Drift, C. *Proc. Nat'l. Acad. Sci. USA* 74: 3060–3064 (1977). ''A protonmotive force drives bacteria flagella.''

227. Calladine, C. R. *Sci. Prog. (Oxf.)* 68: 365–385 (1983). ''Construction and operation of bacterial flagella.''

228. Ritchie, R. J. *Progr. Biophys. Molec. Biol.* 43: 1–32 (1984). ''A critical assessment of the use of lipophilic cations as membrane potential probes.''

229. Mitchell, P. *FEBS Lett.* 176: 287–293 (1984). ''Bacterial flagellar motors and osmoelectric molecular rotation by an axially transmembrane well and turnstile mechanism.''

230. Manson, M. D.; Tedesco, P.; and Berg, H. C. *J. Mol. Biol.* 138: 541–561 (1980). ''Energetics of flagellar rotation in bacteria.''

231. Khan, S., and Macnab, R. M. *J. Mol. Biol.* 138: 563–597 (1980). ''The steady state counterclockwise/clockwise ratio of bacterial flagellar motors is regulated by protonmotive force.''

232. Khan, S., and Macnab, R. M. *J. Mol. Biol.* 138: 599–614 (1980). ''Proton chemical potential, proton electrical potential and bacterial motility.''

233. Shioi, J.-I.; Matsuura, S.; and Imae, Y. *J. Bacteriol.* 144: 891–7 (1980). ''Quantitative measurements of proton motive force and motility in *B.subtilis*.''

234. Goulbourne, E. A., Jr., and Greenberg, E. P. *J. Bacteriol.* 143: 1450–1457 (1980). ''Relationship between protonmotive force and motility in *Spirochaeta aurantia*.''

235. Glagolev, A. N., and Skulachev, V. P. *Nature* 272: 280–282 (1978). ''The proton pump is a molecular engine of motile bacteria.''

236. Ravid, S., and Eisenbach, M. *J. Bacteriol.* 158: 1208–1210 (1984). ''Minimal requirements for rotation of bacterial flagella.''

237. Larsen, S. H.; Adler, J.; Gargus, J. J.; and Hogg, R. W. *Proc. Nat'l. Acad. Sci. USA* 71: 1239–1243 (1974). ''Chemomechanical coupling without ATP: The source of energy for motility and chemotaxis in bacteria.''

238. Thipayathaana, P., and Valentine, R. C. *Biochim. Biophys. Acta* 347: 464–468 (1974). ''Requirement for energy-transducing ATPase for anaerobic motility in *Escherichia coli*.''

239. Hirota, N.; Matsuura, S.; Mochizuki, N.; Mutoh, N.; and Imae, Y. *J. Bacteriol.* 148: 399–405 (1981). ''Use of lipophilic cation-permeable mutants for measurement of transmembrane electrical potential in metabolizing cells of *Escherichia coli*.''

240. Meister, M.; Lowe, G.; and Berg, H. C. *Cell* 49: 643–650 (1987). ''The proton flux through the bacterial flagellar motor.''

241. Läuger, P. *Nature* 268: 360–362 (1977). ''Ion transport and rotation of bacterial flagella.''

242. Läuger, P. *Biophys. J.* 53: 53–65 (1988). ''Torque and rotation rate of the bacterial flagellar motor.''

243. Meister, M., and Berg, H. C. *Biophys. J.* 52: 413–419 (1987). ''The stall torque of the bacterial flagellar motor.''

244. Macnab, R. M., and Han, D. P. *Cell* 32: 109–117 (1983). ''Asynchronous switching of flagellar motors on a single bacterial cell.''

245. Ishihara, A.; Segall, J. E.; Block, S. M.; and Berg, H. C. *J. Bacteriol.* 155: 228–237 (1983). ''Coordination of flagella on filamentous cells of *Escherichia coli*.''

246. Segall, J. E.; Manson, M. D.; and Berg, H. C. *Nature* 296: 855–857 (1982). ''Signal processing times in bacterial chemotaxis.''

247. Lapidus, I. R.; Welch, M.; and Eisenbach, M. *J. Bacteriol.* 170: 3627–3632 (1988). ''Pausing of flagellar rotation is a component of bacterial motility and chemotaxis.''

248. Ordal, G. W. *Nature* 270: 66–67 (1977). ''Calcium ion regulates chemotactic behavior in bacteria.''

249. Khan, S., and Berg, H. C. *Cell* 32: 913–919 (1983). ''Isotope and thermal effects in chemiosmotic coupling to the flagellar motor of *Streptococcus*.''

250. Oosawa, F., and Masai, J. *J. Phys. Soc. Japan* 51: 631–641 (1982). ''Mechanism of flagellar motor rotation in bacteria.''

251. Szmelcman, S., and Adler, J. *Proc. Nat'l. Acad. Sci. USA* 73: 4387–4391 (1976). ''Changes in membrane potential during chemotaxis.''

252. Miller, J. B., and Koshland, D. E., Jr. *Proc. Nat'l. Acad. Sci. USA* 74: 4752–4756 (1977). ''Sensory electrophysiology of bacteria: Relationship of the membrane potential to motility and chemotaxis in *Bacillus subtilis*.''

253. Snyder, M. A.; Stock, J. B.; and Koshland, D. E., Jr. *J. Mol. Biol.* 149: 241–257 (1981). ''Role of membrane potential and calcium in chemotactic sensing by bacteria.''

254. Eisenbach, M. *Biochemistry* 21: 6818–6825 (1982). ''Changes in membrane potential of *Escherichia coli* in response to temporal gradients of chemicals.''

255. Eisenbach, M.; Raz, T.; and Ciobotariu, A. *Biochemistry* 22: 3293–3298 (1983).

"A process related to membrane potential involved in bacterial chemotaxis to galactose."

256. Eisenbach, M.; Margolin, Y.; Ciobotariu, A.; and Rottenberg, H. *Biophys. J.* 45: 463–467 (1984). "Distinction between changes in membrane potential and surface charge upon chemotactic stimulation of *Escherichia coli*."
257. Margolin, Y., and Eisenbach, M. *J. Bacteriol.* 159: 605–610 (1984). "Voltage clamp effects on bacterial chemotaxis."
258. Goulbourne, E. A., Jr., and Greenberg, E. P. *J. Bacteriol.* 148: 837–844 (1981). "Chemotaxis of *Spirochaeta aurantia*: Involvement of membrane potential in chemosensory signal transduction."
259. Goulbourne, E. A., Jr., and Greenberg, E. P. *J. Bacteriol.* 153: 916–920 (1983). "A voltage clamp inhibits chemotaxis of *Spirochaeta aurantia*."
260. Goulbourne, E. A., Jr., and Greenberg, E. P. *J. Bacteriol.* 155: 1443–1445 (1983). "Inhibition of *Spirochaeta aurantia* chemotaxis by neurotoxins."
261. Plonsey, R. *Bioelectric Phenomena*. New York: McGraw-Hill, 1969. p. 124.
262. Dahl, M. K., and Manson, M. D. *J. Bacteriol.* 164: 1057–1063 (1985). "Interspecific reconstitution of maltose transport and chemotaxis in *Escherichia coli* with maltose-binding protein from various enteric bacteria."
263. Manson, M. D., and Kossmann, M. *J. Bacteriol.* 165: 34–40 (1986). "Mutations in tar suppress defects in maltose chemotaxis caused by specific malE mutations."
264. Axelrod, D. *J. Membrane Biol.* 75: 1–10 (1983). "Lateral motion of membrane proteins and biological function."
265. Mowbray, S. L., and Petsko, G. A. *J. Biol. Chem.* 258: 7991–7997 (1983). "The x-ray structure of the periplasmic galactose binding protein from *Salmonella typhimurium* at 3.0Å resolution."
266. Mahoney, W. C.; Hogg, R. W.; and Hermodson, M. A. *J. Biol. Chem.* 256: 4350–4356 (1981). "The amino acid sequence of the D-galactose binding protein from *Escherichia coli* B/r."
267. Argos, P.; Mahoney, W. C.; Hermodson, M. A.; and Hanei, M. *J. Biol. Chem.* 256: 4357–4361 (1981). "Structural prediction of sugar-binding proteins functional in chemotaxis and transport."
268. Vyas, N. K.; Vyas, M. N.; and Quiocho, F. A. *Nature* 327: 635–8 (1987). "A novel calcium binding site in the galactose-binding protein of bacterial transport and chemotaxis." Vyas, N. K.; Vyas, M. N.; and Quiocho, F. A. *Science* 242: 1290–1295 (1988). "Sugar and signal-transducing sites of the *Escherichia coli* galactose chemoreceptor protein."
269. Wang, E. A., and Koshland, D. E., Jr. *Proc. Nat'l. Acad. Sci. USA* 77: 7157–7160 (1980). "Receptor structure in the bacterial sensing system."
270. Russo, A. F., and Koshland, D. E., Jr. *Science* 220: 1016–1020 (1983). "Separation of signal transduction and adaptation functions of the aspartate receptor in bacterial sensing."
271. Boyd, A.; Kendall, K.; and Simon, W. *Nature* 301: 623–626 (1983). "Structure of the serine chemoreceptor in *Escherichia coli*."
272. Bollinger, J.; Park, C.; Harayama, S.; and Hazelbauer, G. L. *Proc. Nat'l. Acad. Sci. USA* 81: 3287–3291 (1984). "Structure of the trg protein: Homologies with and differences from other sensory transducers of *Escherichia coli*."
273. Park, C., and Hazelbauer, G. L. *J. Bacteriol.* 167: 101–109 (1986). "Mutations specifically affecting ligand interaction of the Trg chemosensory transducer."

274. Simms, S. A.; Stock, A. M.; and Stock, J. B. *J. Biol. Chem.* 262: 8537–8543 (1987). ''Purification and characterization of the S-adenosylmethionine: glutamyl transferase that modifies chemoreceptor proteins in bacteria.''

275. Kosower, E. M. *Biochem. Biophys. Res. Commun.* 115: 648–652 (1983). ''Selection of ion channel elements in the serine and aspartate methyl-accepting chemotaxis proteins of bacteria.''

276. Schimz, A. *FEBS Lett.* 139: 283–286 (1982). ''Localization of the methylation system involved in sensory behavior of *Halobacterium Halobium* and its dependence on calcium.''

277. Harwood, C. S.; Rivelli, M.; and Ornston, L. N. *J. Bacteriol.* 160: 622–628 (1984). ''Aromatic acids are chemoattractants for *Pseudomonas putida*.''

278. Parke, D.; Rivelli, M.; and Ornston, L. N. *J. Bacteriol.* 163: 417–422 (1985). ''Chemotaxis to aromatic and hydroaromatic acid: Comparison of *Bradyrhizobium japonicum* and *Rhizobium trifolii*.''

279. Barreau, C., and van der Wel, H. *Chem. Senses* 8: 71–79 (1983). ''The effect of thaumatins on the chemotactic behaviour of *Escherichia coli*.''

280. Law, J. H., and Regnier, F. E. *Ann. Rev. Biochem.* 40: 533–548 (1971). ''Pheromones.''

281. Lewin, R. *Science* 225: 153–156 (1984). ''The continuing tale of a small worm.''

282. Golden, J. W., and Riddle, D. L. *Proc. Nat'l. Acad. Sci. USA* 81: 819–824 (1984). ''A pheromone-induced developmental switch in *Caenorhabditis elegans*: Temperature-sensitive mutants reveal a wild-type temperature-dependent process.''

283. Prestwich, G. D. *Quart. Rev. Biol.* 60: 437–456 (1985). ''Communication in insects. II. Molecular communication of insects.''

284. Hecker, E., and Butenandt, A. ''Bombykol revisited—Reflections on a pioneering period and on some of its consequences.'' In *Techniques in Pheromone Research*, ed. H.E. Hummel, T. A. Miller, chap. 1, pp. 1–44. New York: Springer, 1984.

285. Karlson, P., and Lüscher, M. *Nature* 183: 55–56 (1959). ''Pheromones, a new term for a class of biologically active substances.''

286. Attygalle, A. B., and Morgan, E. D. *Angew. Chem. Int'l. Edn.* 27: 460–478 (1988). ''Pheromones in nanogram quantities: Structural determination by microchemical and gas chromatographic methods.''

287. Roelofs, W. L., and Brown, R. L. *Ann. Rev. Ecol. Syst.* 13: 395–422 (1982). ''Pheromones and evolutionary relationships of *Tortricidae*.''

288. Haynes, K. F.; Gaston, I. K.; Mistrot-Pope, M.; and Baker, T. C. *J. Chem. Ecol.* 10: 1551–1565 (1984). ''Potential for evolution of resistance to pheromones.''

289. O'Connell, R. J., and Grant, A. J. In Olfaction and Taste IX, ed. S. D. Roper, and J. Atema, *Ann. N.Y. Acad. Sci.* 510: 79–85 (1987). ''Electrophysiological responses of olfactory receptor neurons to stimulation with mixtures of individual pheromone components.''

290. O'Connell, R. J. *Experientia* 42: 232–241 (1986). ''Chemical communication in invertebrates.''

291. Slessor, K. N.; Kaminski, L. -A.; King, G.G.S.; Borden, J. H.; and Winston, M. L. *Nature* 332: 354–355 (1988). ''Semiochemical basis of the retinue response to queen honey bees.''

292. Altner, H., and Prillinger, L. *Int'l. Rev. Cytol.* 67: 69–139 (1980). ''Ultrastruc-

ture of Invertebrate Chemo-, Thermo- and Hygroreceptors and its Functional Significance.''

293. Vogt, R. G.; Riddiford, L. M.; and Prestwich, G. D. *Proc. Nat'l. Acad. Sci. USA* 82: 8827–8831 (1985). ''Kinetic properties of a sex pheromone-degrading enzyme: The sensillar esterase of *Antheraea polyphemus*.''
294. Vogt, R. G.; Prestwich, G. D.; and Riddiford, L. M. *J. Biol. Chem.* 263: 3952–3959 (1988). ''Sex pheromone receptor proteins. Visualization using a radiolabeled photoaffinity analog.''
295. Kaissling, K.-E. *Ann. Rev. Neurosci.* 9: 121–145 (1986). ''Chemo-electrical transduction in insect olfactory receptors.''
296. Vogt, R. G., and Riddiford, L. M. *Nature* 293: 161–163 (1981). ''Pheromone binding and inactivation by moth antennae.''
297. Kaissling, K.-E. ''Temporal characteristics of pheromone receptor cell responses in relation to orientation behaviour of moths.'' In *Mechanisms in Insect Olfaction*, ed., T. L. Payne, M. C. Birch, and C.E.J. Kennedy, pp. 193–199. Oxford: Clarendon Press, 1985.
298. Vogt, R. G. ''The molecular basis of pheromone reception: Its influence on behavior.'' In *Pheromone Biochemistry*, ed. G. D. Prestwich, G. J. Blomquist, chap. 12, pp. 385–431. New York: Academic Press, 1987.
299. Bignetti, E.; Cavaggioni, A.; Pelosi, P.; Persaud, K. C.; Sorbi, R. T.; and Tirindelli, R. *Eur. J. Biochem.* 149: 227–231 (1985). ''Purification and characterization of an odorant-binding protein from cow nasal tissue.''
300. Pevsner, J.; Sklar, P. B.; and Snyder, S. H. *Proc. Nat'l. Acad. Sci. USA* 83: 4942–4946 (1986). ''Odorant-binding protein: Localization to nasal glands and secretions.''
301. Pevsner, J.; Hwang, P. M.; Sklar, P. B.; Venable, J. C.; and Snyder, S. H. *Proc. Nat'l. Acad. Sci. USA* 85: 2383–2387 (1988). ''Odorant-binding protein and its mRNA are localized to lateral nasal gland implying a carrier function.''
302. Lee, K.-H.; Wells, R. G.; and Reed, R. R. *Science* 235: 1053–1056 (1987). ''Isolation of an olfactory cDNA: Similarity to retinol-binding protein suggests a role in olfaction.''
303. Aoqvist, J.; Sandblom, P.; Jones, T. A.; Newcomer, M. E.; van Gunsteren, W. F.; and Tapia, O., *J. Mol. Biol.* 192: 593–604 (1986). ''Molecular dynamics simulations of the holo and apo forms of retinol binding protein. Structural and dynamical changes induced by retinol removal.''
304. Prestwich, G. D. *Science* 237: 999–1006 (1987). ''Chemistry of pheromone and hormone metabolism in insects.''
305. Vogt, R. G., and Riddiford, L. M. ''Pheromone reception.'' In *Mechanisms in Insect Olfaction*, ed. T. L. Payne, M. C. Birch, and C.E.J. Kennedy, chap. 23, pp. 201–208. Oxford: Clarendon Press, 1986.
306. Prestwich, G. D. ''Chemical studies of pheromone reception and catabolism.'' In *Pheromone Biochemistry*, ed. G. D. Prestwich, and G. J. Blomquist, chap. 14, pp. 473–527. New York: Academic Press, 1987.
307. Prestwich, G. D. *Pure & Applied Chem.* 61: 551–554 (1989). ''Bio-organic studies of insect olfaction.''
308. Vogt, R. G., and Prestwich, G. D. In Olfaction and Taste IX, ed. S. D. Roper, and J. Atema, *Ann. N.Y. Acad. Sci.* 510: 689–691 (1987). ''Variation in olfactory proteins. Evolvable elements encoding insect behavior.''

309. Smith, J. M., *Nature* 332: 311–312 (1988). "Punctuation in perspective."

310. Dobzhansky, T. In *Genetics of the Evolutionary Process*, Chaps. 10 & 11, pp. 311–390. New York: Columbia University Press, 1970.

311. Smith, J. M., ed. In *Evolution Now: A century after Darwin*, chaps. 4 and 5, pp. 107–181. San Francisco: W. H. Freeman, 1982. May, R. M. *Science* 241: 1441–1449 (1988). "How many species are there on Earth?" Partridge, L., and Harvey, P. H. *Science* 241: 1449–1455 (1988). "The ecological context of life history evolution." Lande, R. *Science* 241: 1455–1459 (1988). "Genetics and demography in biological conservation." Eigen, M.; McCaskill, J.; and Schuster, P. *J. Phys. Chem.* 92: 6881–6891 (1988). "Molecular quasi-species."

312. Kimura, M., In *Evolution of Genes and Proteins*, ed. M. Nei, and R. K. Koehn, chap. 11, pp. 208–233. Sunderland, Mass.: Sinauer, 1983. See also Kimura, M. *The Neutral Theory of Molecular Evolution*. Cambridge: Cambridge University Press, 1983. 366pp, and remarks by Smith, J. M. *Nature* 306: 713–714 (1983).

313. Selander, R. K., and Whittam, T. S. In *Evolution of Genes and Proteins*, ed. M. Nei, and R. K. Koehn, chap. 5, pp. 89–114. Sunderland, Mass.: Sinauer, 1983.

314. Boppré, M. *Naturwissen.* 73: 17–26 (1986). "Insects pharmacophagously utilizing defensive plant chemicals (pyrrolizine alkaloids)."

315. Edgar, J. A. *Trans. R. Soc. (Lond.)* B272: 467–476 (1975). "*Danainae* (Lep.) and 1, 2-dehydropyrrolizidine alkaloid-containing plants—with reference to observations made in the New Hebrides."

316. Boppré, M.; Petty, R. L.; Schneider, D.; and Meinwald, J. *J. Comp. Physiol.* 126: 97–103 (1978). "Behaviorally mediated contacts between scent organs: Another prerequisite for pheromone production in *Danaus chrysippus* males (Lepidoptera)."

317. Culvenor, C.C.J., and Edgar, J. A. *Experientia* 28: 627–628 (1972). "Dihydropyrrolizine secretions associated with coremata of *Utetheisa* moths (Family Arctiidae)."

318. Conner, W. E.; Eisner, T.; Vander Meer, R. K.; Guerrero, A.; and Meinwald, J. *Behav. Ecol. Sociobiol.* 9: 227–235 (1981). "Precopulatory sexual interaction in an arctiid moth (*Utethesia ornatrix*"): role of a pheromone derived from dietary alkaloids."

319. Schneider, D.; Boppré, M.; Zweig, J.; Horsley, S. B.; Bell, T. W.; Meinwald, J.; Hansen, K.; and Diehl, E. W. *Science* 215: 1264–1265 (1982). "Scent organ development in *Creatonotos* moths: Regulation by pyrrolizidine alkaloids."

320. Wunderer, H. J.; Hansen, K.; Bell, T. W.; Schneider, D.; and Meinwald, J. *Exp. Biol.* 46: 11–27 (1986). "Sex pheromones of two Asian moths (*Creatonotos transiens*, *C.gangis*; Lepidoptera-Arctiidae): behavior, morphology, chemistry and electrophysiology."

321. Bell, T. W., and Meinwald, J. *J. Chem. Ecol.* 12: 385–409 (1986). "Pheromones of two Arctiid moths (*Creatonos transiens* and *C.gangis*): Chiral components from both sexes and achiral female components."

322. Williams, N. H., and Whitten, W. M. *Biol. Bull.* 164: 355–395 (1983). "Orchid floral fragrances and male euglossine bees. Methods and advances in the last sesquidecade." Dressler, R. L. *Ann. Rev. Ecol. Syst.* 13: 373–394 (1982). "Biology of the orchid bees (*Euglossini*)."

323. Metcalf, R. L.; Metcalf, E. R.; and Mitchell, W. C. *Proc. Nat'l. Acad. Sci. USA*

83: 1549–1553 (1986). ''Benzyl acetates as attractants for the male oriental fruit fly, *Dacus dorsalis*, and the male melon fly, *Dacus cucurbitae*.''

324. Metcalf, R. L.; Metcalf, E. R.; Mitchell, W. C.; and Lee, L.W.Y. *Proc. Nat'l. Acad. Sci. USA* 76: 1561–1575 (1979). ''Evolution of olfactory receptor in oriental fruit fly *Dacus dorsalis*.''
325. Metcalf, R. L.; Mitchell, W. C.; and Metcalf, E. R. *Proc. Nat'l. Acad. Sci. USA* 80: 3143–3147 (1983). ''Olfactory receptors in the melon fly *Dacus cucurbitae* and the oriental fruit fly, *Dacus dorsalis*.''
326. Festenstein, H., and Demant, P. *HLA and H-2: Basic Immunogenetics, Biology and Clinical Relevance*. London: E. Arnold, 1978, 212 pp.
327. Thomas, L. *The Youngest Science: Notes of a Medicine-Watcher*, pp. 208–219. New York: Viking Press, 1983.
328. Yamazaki, K.; Yamaguchi, M.; Beauchamp, G. K.; Bard, J.; Boyse, E. A.; and Thomas, L. In *Biochemistry of Taste and Olfaction*, ed. R. H. Cagan, and M. R. Kare, chap. 5, pp. 85–92. New York: Academic Press, 1981.
329. Yamazaki, K.; Beauchamp, G. K.; Egorov, I. K.; Bard, J.; Thomas, L.; and Boyse, E. A. *Proc. Nat'l. Acad. Sci. USA* 80: 5685–5688 (1983). ''Sensory distinction between H-2b and H-2bml mutant mice.''
330. Yamazaki, K; Beauchamp, G. K.; Thomas, L.; and Boyse, E. A. *J. Mol. Cell. Immunol.* 1: 79–82 (1984). ''Chemosensory identity of H-2 heterozygotes.''
331. Schwende, F. J.; Jorgenson, J. W.; and Novotny, M. *J. Chem. Ecol.* 10: 1603–1615 (1984). ''Possible chemical basis for histocompatibility-related mating preference in mice.''
332. Matzinger, P., and Zamoyska, R. *Nature* 297: 628 (1982). ''A beginner's guide to major histocompatibility complex function.''
333. Yamazaki, K; Beauchamp, G. K.; Matsuzaki, O.; Bard, J.; Thomas, L.; and Boyse, E. A. *Proc. Nat'l. Acad. Sci. USA* 83, 4438–4440 (1986). ''Participation of the murine X and Y chromosomes in genetically determined chemosensory identity.''
334. Robertson, M. *Nature* 297: 629–632 (1982). ''The evolutionary past of the major histocompatibility complex and the future of cellular immunology.''
335. Steinmetz, M., and Hood, L. *Science* 222: 727–733 (1983). ''Genes of the major histocompatibility complex in mouse and man.''
336. Singh, P. B.; Brown, R. E.; and Roser, B. *Nature* 327: 161–164 (1987). ''MHC antigens as olfactory recognition cues.''
337. Ivanyi, P. *Proc. Roy. Soc. (Lond.)* B202: 117–158 (1978). ''Some aspects of the H-2 system, the major histocompatibility system in the mouse.''
338. Hall, P. F. *Int. Rev. Cytol.* 86: 53–96 (1984). ''Cellular organization for steroidogenesis.''
339. Stowe, M. K.; Tumlinson, J. H.; and Heath, R. R. *Science* 236: 964–967 (1987). ''Chemical mimicry: Bolas spiders emit component d of moth prey species sex pheromones.''
340. Cairns, J.; Overbaugh, J.; and Miller, S. *Nature* 335: 142–145 (1988). ''The origin of mutants.'' Cairns, J. *Nature* 336: 527–528 (1988).
341. Yamazaki, K.; Beauchamp, G. K.; Wysocki, C. J.; Bard, J.; Thomas, L.; and Boyse, E. A. *Science* 221: 186–188 (1983). ''Recognition of H-2 types in relation to the blocking of pregnancy in mice.''
342. Hall, B. G. ''Evolution of new metabolic functions in laboratory organisms.'' In

Evolution of Genes and Proteins, ed. M. Nei, and R. K. Koehn, chap. 12, pp. 234–257. Sunderland, Mass.: Sinauer, 1983.

343. Lewin, R. *Science* 216: 1212–1213 (1982). "Adaptation can be a problem for evolutionists."

344. Remsen, J. V., Jr. *Science* 224: 171–173 (1984). "High incidence of 'leapfrog' pattern of geographic variation in Andean birds: Implications for the speciation process."

345. Anon., Ed. comment. *Lancet*: Feb. 6, pp. 279–280 (1971). "A human pheromone?"

346. Singer, A. G.; Macrides, F.; Clancy, A. N.; and Agosta, W. C. *J. Biol. Chem.* 261: 13323–13326 (1986). "Purification and analysis of a proteinaceous aphrodisiac pheromone from hamster vaginal discharge."

347. Bartoshuk, L. M. "History of taste research." In *Handbook of Perception, Vol. VIA, Tasting and Smelling*, ed. E. C. Carterette, and M. P. Friedman, chap. 1, pp. 1–18. New York: Academic Press, 1978. 321 pp.

348. Pfaffman, C. "The vertebrate phyogeny, neural code and integrative processes of taste." In *Handbook of Perception, Vol. VIA, Tasting and Smelling*, ed. E. C. Carterette, and M. P. Friedman, chap. 3, pp. 51–123. New York: Academic Press, 1978. 321 pp.

349. Faurion, A. *Progr. in Sensory Physiol.* 8: 129–201 (1987). "Physiology of the sweet taste."

350. Scott, T. R., and Mark, G. P. *Progr. in Neurobiol.* 27: 293–317 (1986). "Feeding and taste."

351. Sato, T. *Prog. in Sensory Physiol.* 6: 1–38 (1986). "Receptor potential in rat taste cells."

352. Arvidson, K., and Friberg, U. *Science* 209: 807–808 (1980). "Human taste: Response and taste bud number in fungiform papillae."

353. Miller, I. *J. Gen. Physiol.* 57: 1–25 (1971). "Peripheral interactions among single papilla inputs to gustatory nerve fibers."

354. Miller, I. *J. Comp. Neurol.* 158: 155–166 (1974). "Branched *chorda tympani* neurons and interactions among taste receptors."

355. Sato, T., and Beidler, L. M. *Compar. Biochem. Physiol.* 73A: 1–10 (1982). "The response characteristics of rat taste cells to four basic taste stimuli."

356. Pfaff, D. W., ed. *Taste, Olfaction and the Central Nervous System*. New York: Rockefeller University Press, 1985. 346 pp.

357. Halpern, B. P. "Time as a factor in gustation: Temporal patterns of taste stimulation and response." Pages 181–209 in ref. 356. Kelling, S. T., and Halpern, B. P. *Science* 219: 412–414 (1983). "Taste flashes: Reaction times, intensity and quality."

358. Creighton, T. E. *Proteins: Structure and Molecular Principles*, p. 184, Table 5–3, New York: W. H. Freeman, 1983. 515 pp.

359. Vernin, G. *The Chemistry of heterocyclic flavoring and aroma compounds*. Chichester, England: Ellis Horwood, 1982. 375 pp.

360. Schiffman, S. S. *New Engl. J. Med.* 308: 1273–1279, 1337–1343 (1983). "Taste and smell in disease."

361. Pfaffman, C. "De Gustibus: Praeteriti, Praesentis, Futuri." Pages 19–44 in ref. 356.

362. Beidler, L. M., and Tomosaki, K. ''Multiple Sweet Receptor Sites and Taste Theory.'' Pages 47–64 in ref. 356.

363. Erickson, R. P. ''Definitions: A matter of taste.'' Pages 129–150 in ref. 356.

364. Bartoshuk, L. M., and Gent, J. F. ''Taste mixtures: An analysis of synthesis.'' Pages 210–232 in ref. 356.

365. Boudreau, J. C. *Naturwissen*. 67: 14–20 (1980). ''Taste and the taste of foods. A review and a report on a symposium.''

366. *Chem. & Engr. News* 65(19): 26 (1987). ''Biotechnology project aids fish farming.''

367. Jakinovich, W., Jr. ''Comparative study of sweet taste specificity.'' In *Biochemistry of Taste and Olfaction*, ed. R. H. Cagan, and M. R. Kare, chap. 7, pp. 117–138. New York: Academic Press, 1981. 539 pp.

368. Hough, L. *Chem. Soc. Revs*. 14: 357–374 (1985). ''The sweeter side of chemistry.''

369. Crammer, B., and Ikan, R. *Chem. Soc. Revs*. 6: 431–466 (1977); 7: 164–165 (corrections) (1978). ''Properties and syntheses of sweetening agents''; (b) Crammer, B.; Ikan, R., *Chemistry in Britain* 22: 915–918 (1986). ''Sweet glycosides from the stevia plant.''

370. Compadre, C. M.; Pezzuto, J. M.; Kinghorn, A. D.; and Kamath, S. K. *Science* 227: 417–419 (1985). ''Hernandulcin: An intensely sweet compound discovered by review of ancient literature.''

371. DuBois, G. E.; Dietrich, P. S.; Lee, J. F.; McGarraugh, G. V.; and Stephenson, R. A. *J. Med. Chem*. 24: 1269–1271 (1981). ''Diterpenoid sweeteners. Synthesis and sensory evaluation of stevioside analogues nondegradable to steviol.''

372. DuBois, G. E., and Stephenson, R. A. *J. Med. Chem*. 28: 93–98 (1985). ''Diterpenoid sweeteners. Synthesis and sensory evaluation of stevioside analogues with improved organoleptic properties.''

373. Pezzuto, J. M.; Compadre, C. M.; Swanson, S. M.; Nanayakkara, N.P.D.; and Kinghorn, A. D. *Proc. Nat'l. Acad. Sci. USA* 82: 2478–2482 (1985). ''Metabolically activated steviol, the aglycone of stevioside, is mutagenic.''

374. Moncrieff, R. W. *The Chemical Senses*, 3d ed. Cleveland, Ohio: CRC Press, 1967. 760 pp.

375. Clauss, K., and Jensen, H. *Angew. Chem. Int'l. Edn*. 12: 869–876 (1973). ''Oxathiazinone dioxides—A new group of sweetening agents.''

376. Boehm, M. F., and Bada, J. L. *Proc. Nat'l. Acad. Sci. USA* 81: 5263–5266 (1984). ''Racemization of aspartic acid and phenylalanine in the sweetener aspartame at 100°C.''

377. Fuller, W. D.; Goodman, M.; and Verlander, M. S. *J. Am. Chem. Soc*. 107: 582–583 (1985). ''A new class of amino acid based sweeteners.''

378. Goodman, M.; Coddington, J.; Mierke, D. F.; and Fuller, W. D. *J. Am. Chem. Soc*. 109: 4712–4714 (1987). ''A model for the sweet taste of stereoisomeric retro-inverso and dipeptide amides.''

379. Goodman, M.; Mierke, D. F.; and Fuller, W. D. In *Peptide Chemistry* 1987, ed. T. Shiba, and S. Sakakibara, pp. 699–704. Osaka: Protein Research Foundation, 1988. ''A stereoisomeric approach to the molecular basis of taste of retro-inverso and dipeptide amides.''

380. Tsang, J. W.; Schmied, B.; Nyfeler, R.; and Goodman, M. *J. Med. Chem*. 27:

1663–1668 (1984). "Peptide sweeteners. 6. Structural studies on the C-terminal amino acid of L-aspartyl dipeptide sweeteners."

381. Goodman, M. *Biopolymers* 24: 137–155 (1985). "Peptide homologs, isosteres and isomers: A general approach to structure-activity relationships."

382. van der Wel, H. *Trends in Biochem. Sci.* 5: 122–124 (1980). "Sweet-tasting proteins."

383. Cagan, R. H., and Morris, R. W. *Proc. Nat'l. Acad. Sci. USA* 76: 1692–1696 (1979). "Biochemical studies of taste sensation: Binding to taste tissue of ^{3}H-labeled monellin, a sweet-tasting protein."

384. du Villard, X. D.; van der Wel, H.; and Brouwer, J. N. *Chem. Senses* 5: 93–98 (1980). "Enhancement of the perceived sucrose sweetness in the rat by thaumatin."

385. van der Wel, H., and Arvidson, K. *Chem. Senses and Flavour* 3: 291–297 (1978). "Qualitative psychophysical studies on the gustatory effects of the sweet tasting proteins thaumatin and monellin."

386. van der Wel, H., and Bel, W. *Chem. Senses* 3: 99–104 (1978).

387. Hough, C., and Edwardson, J. *Nature* 271: 381–383 (1978). "Antibodies to thaumatin as a model of the sweet taste receptor."

388. de Vos, A. M.; Hatada, M.; van der Wel, H.; Krabbendam, H.; Peeredeman, A. F.; and Kim, S.-H. *Proc. Nat'l. Acad. Sci. USA* 82: 1406–1409 (1985). "Three-dimensional structure of thaumatin I, an intensely sweet protein."

389. Richardson, M.; Valdes-Rodriguez, S.; and Blanco-Labra, A. *Nature* 327: 432–434 (1987). "A possible function for thaumatin and a TMV-induced protein suggested by homology to a maize inhibitor."

390. Cornelissen, B.J.C.; Hooft van Huijsduijnen, R.A.M.; and Bol, J. F. *Nature* 321: 531–532 (1986). "A tobacco mosaic virus-induced tobacco protein is homologous to the sweet-tasting protein thaumatin."

391. Ogata, C. M., and Kim, S.-H. *Trends in Biochem. Sci.* 13: 13–15 (1988). "Crystal structures of two intensely sweet tasting proteins."

392. Ogata, C. M.; Hatada, M.; Tomlinson, G.; Shin, W.-C.; and Kim, S.-H. *Nature* 328: 739–742 (1987). "Crystal structure of the intensely sweet protein monellin."

393. Hatada, M.; Jancarik, J.; Graves, B.; and Kim, S.-H. *J. Am. Chem. Soc.* 107: 4279–4282 (1985). "Crystal structure of Aspartame, a peptide sweetener."

394. Burley, S. K., and Petsko, G. A. *Science* 229: 23–28 (1985). "Aromatic-aromatic stabilization: A mechanism of protein structure stabilization."

395. Görbitz, C. H. *Acta Chem. Scand.* B41: 87–92 (1987). "Crystal and molecular structure of Aspartame.HCl.$2H_2O$."

396. Temussi, P. A.; Lelj, F.; Tancredi, T.; Castiglione-Morelli, M. A.; and Pastore, A. *Int. J. Quantum Chem.* 26: 889–906(1984). "Soft agonist-receptor interactions: Theoretical and experimental simulation of the active site of the receptor of sweet molecules."

397. Wong, R. Y., and Horowitz, R. M. *J. Chem. Soc. Perkin Trans*; I: 843–848 (1986). "The x-ray crystal and molecular structure of neohesperidin dihydrochalcone sweetener."

398. Bartoshuk, L. M.; Lee, C.-H.; and Scarpellino, R. *Science* 178: 988–990 (1972). "Sweet taste of water induced by artichoke (*Cynara scolymus*)."

399. Edgeworth, P., Letter (Banda [India], 30 August 1847) to Linnean Society, 7 Dec. 1847, *Proc. Linn. Soc. Lond.* [1838–1848] 1: 353 (1847).

400. Kurihara, K. ''Taste modifiers.'' In *Handbook of Sensory Physiology, Chemical Senses 2, Taste*, ed. L. M. Beidler, pp. 363–378. Berlin: Springer, 1971.
401. Chakravanti, D., and Debnath, N. B. *J. Inst. Chemists (India)* 53: 155–158 (1981). ''Isolation of gymnemagenin, the sapogenin from *Gymnema sylvestre* R. Br. (*Asclepiadaceae*).''
402. Kurihara, Y.; Ookubo, K.; Tasaki, H.; Kodama, H.; Akiyama, Y.; Yagi, A.; and Halpern, B. *Tetrahedron* 44: 61–68 (1988). ''Studies on the taste modifiers. I. Purification and structure determination of sweetness inhibiting substance in leaves of *Ziziphus jujuba*.''
403. Sefecka, R., and Kennedy, L. M. In Olfaction and Taste IX, ed. S. D. Roper, and J. Atema, *Ann. N.Y. Acad. Sci.* 510: 602–605 (1987). ''Chemical analyses of hodulcin, the sweetness-suppressing principle from *Hovenia dulcis* leaves.''
404. Jakinovich, W., Jr. ''Sugar taste reception in the gerbil.'' Chapter 5, pages 65–91 in ref. 356.
405. Jakinovich, W., Jr. *Science* 219: 408–410 (1983). ''Methyl 4,6-dichloro-4,6-dideoxy-α-D-galactopyranoside: An inhibitor of sweet taste responses in gerbils.''
406. Glaser, D.; Hellekant, G.; Brouwer, J. N.; and van der Wel, H. *Chem. Senses* 8: 367–374 (1984). ''Effects of gymnemic acid on sweet taste perception in primates.''
407. Kurihara, K., and Beidler, L. M. *Science* 161: 1241–1243 (1968). ''Taste modifying protein from miracle fruit.''
408. Theerasilp, S., and Kurihara, Y. *J. Biol. Chem.* 263: 11536–11539 (1988). ''Complete purification and characterization of the taste-modifying protein, miraculin, from miracle fruit.'' Theerasilp, S.; Hitotsuya; Nakajo, S.; Nakaya, K.; Nakamura, Y.; and Kurihara, Y. *J. Biol. Chem.* 264: 6655–6659 (1989). ''Complete amino acid sequence and structure characterization of the taste-modifying protein, miraculin.''
409. Otagiri, K.; Shigenaga, T.; Kanehisa, H.; and Okai, H. *Bull. Chem. Soc. Japan* 57: 90–96 (1984). ''Studies of bitter peptides from casein hydrolysate IV. Relationship between bitterness and hydrophobic amino acids moiety in the C-terminal of BPIa (arg-gly-pro-pro-phe-ile-val).''
410. Kanehisa, H., and Okai, H. *Bull. Chem. Soc. Japan* 57: 301-302 (1984). ''Studies of bitter peptides from casein hydrolysate V. Bitterness of the synthetic C-terminal analogs of des-gly-BPIa (arg-pro-pro-phe-ile-val).''
411. Kanehisa, H. *Bull. Chem. Soc. Japan* 57: 97–102 (1984). ''Studies of bitter peptides from casein hydrolysate VI. Syntheses and bitter taste of BPIc (val-try-pro-phe-pro-pro-gly-ile-asn-his) and its analogs and fragments.''
412. Shigenaga, T.; Otagiri, K.; Kanehisa, H.; and Okai, H. *Bull. Chem. Soc. Japan* 57: 103–107 (1984). ''Studies of bitter peptides from casein hydrolysate VII. Bitterness of the retro-BPIa (val-ile-phe-pro-pro-gly-arg) and its fragments.''
413. Merck Index, 10th edn., entries 2863, 3046, 5596 and 6163. Rahway, New Jersey: Merck, 1983.
414. Akabas, M. H.; Dodd, J.; and Al-Awqati, Q. *Neurosci. Abstr.* 17: 361 (1987). ''Mechanism of transduction of bitter taste in rat taste bud cells.'' Akabas, M. H.; Dodd, J.; Al-Awqati, Q. *Science* 242: 1047–1050 (1988). ''A bitter substance induces a rise in intracellular calcium in a subpopulation of rate taste cells.''
415. Teeter, J., and Gold, G. H. *Nature* 331: 298–299 (1988). ''A taste of things to come.''

416. Tonosaki, K., and Funakoshi, M. *Nature* 331: 354–356 (1988). ''Cyclic nucleotides may mediate taste transduction.''
417. Avenet, P.; Hofmann, F.; and Lindemann, B. *Nature* 331: 351–354 (1988). ''Transduction in taste receptor cells requires cAMP-dependent protein kinase.''
418. Lancet, D.; Striem, B. J.; Pace, U.; Zehavi, U.; and Naim, N. *Neurosci. Abstr.* 17: 361 (1987). ''Adenylate cyclase and GTP-binding protein in rat sweet taste transduction.''
419. Striem, B. J.; Pace, U.; Zehavi, U.; Naim, N.; and Lancet, D. *Chem. Senses* 11: 669 (1986). ''Is adenylate cyclase involved in sweet taste transduction?''
420. Metcalf, R. L.; Metcalf, R. A.; and Rhodes, A. M. *Proc. Nat'l. Acad. Sci. USA* 77: 3769–3772 (1980). ''Cucurbitacins as kairomones for diabroticite beetles.''
421. Jabloner, H.; Dunbar, B. I.; and Hopfinger, A. J. *J. Poly. Sci., Polymer Chem.* 18: 2933–2940 (1980). ''A molecular approach to flavor synthesis. I. Menthol esters of varying size and polarity.''
422. Kawamura, Y., and Klare, M., eds. *Umami: A basic taste*. New York: Dekker, 1985.
423. Huang, T.-C.; Ho, C.-T.; and Chang, S. S. Abstr. 33, Agric. & Food Chem. Div. (AGFD), Amer. Chem. Soc. 191st mtg., New York, April 13–18, 1986. ''Characterization of two peptides with delicious taste.''
424. Goodman, M., and Temussi, P. A. *Biopolymers* 24: 1629–1633 (1985). ''Structure-activity relationship of a bitter diketopiperazine revisited.''
425. Kinnamon, S. C., and Roper, S. D. *Biophys. J.* 49: 21a (1986). ''Evidence for the role of voltage-sensitive channels in taste transduction.'' Kinnamon, S. C. *Trends in Neurosci.* 11: 491–496 (1988). ''Taste transduction: A diversity of mechanisms.'' Kinnamon, S. C.; Dionne, V. E.; and Beam, K. G. *Proc. Nat'l. Acad. Sci. USA* 85: 7023–7027 (1988). ''Apical localization of K^+ channels in taste cells provides the basis for sour taste transduction.''
426. Frank, M. E. ''On the neural code for sweet and salty tastes.'' Pages 107–128 in ref. 356.
427. Suami, A.; Ogawa, S.; Takata, M.; Yasuda, K.; Suga, A.; Takei, K.; and Uematsu, Y. *Chem. Lett.*: 719–722 (1985). ''Synthesis of sweet-tasting pseudo-β-fructopyranose.''
428. Shallenberger, R. S., and Acree, T. E. *Nature* 216: 480–482 (1967). ''Molecular theory of sweet taste.''
429. Shallenberger, R. S. *Advanced Sugar Chemistry. Principles of Sugar Stereochemistry*. Chichester, England: E. Horwood, 1982.
430. Bartoshuk, L. M. *Science* 205: 934–935 (1979). ''Bitter taste of saccharin related to the genetic ability to taste the bitter substance, 6-n-propylthiouracil.''
431. Henkin, R. I., and Shallenberger, R. S. *Nature* 227: 965–966 (1970). ''Aglycogeusia-inability to recognize sweetness and its possible molecular basis.''
432. Ninomiya, Y.; Higashi, T.; Mizukoshi, T.; and Funakoshi, M. In Olfaction and Taste IX, ed., S. D. Roper, and J. Atema, *Ann. N.Y. Acad. Sci.* 510: 527–529 (1987). ''Genetics of the ability to perceive sweetness of D-phenylalanine in mice.''
433. Arora, K.; Rodrigues, V.; Joshi, S.; Shanbhag, S.; and Siddiqui, O. *Nature* 330: 62–63 (1987). ''A gene affecting the specificity of the chemosensory neurons of *Drosophila*.''
434. Henkin, R. I., and Bradley, D. F. *Proc. Nat'l. Acad. Sci. USA* 62: 30–37 (1969). ''Regulation of taste acuity by thiols and metal ions.''

435. Beidler, L. M., ed. *Handbook of Sensory Physiology*, vol. 4, pt. 1, *Olfaction*. Heidelberg: Springer, 1971. 518 pp.
436. (a) Cagan, R. H., and Kare, M. R., eds. *Biochemistry of Taste and Olfaction*, New York: Academic Press, 1981. 539 pp. (b) Bestmann, H. J., and Vostrowsky, O. *Naturwissen*. 69: 457–71 (1982). ''Insektenpheromone.''
437. Daeniker, H. U.; Gubler, B. A.; Schlegel, W.; and Schudel, P. *Chimia* 31: 146–162 (1977). ''Aromen.''
438. Theimer, E. T., ed., *Fragrance Chemistry*, New York: Academic Press, 1982. 635 pp.
439. Moskowitz, H. R., and Warren, C. B., eds. *Odor Quality and Chemical Structure*. Amer. Chem. Soc., Symp. 148: 1981. 243 pp.
440. Amoore, J. E. *Molecular Basis of Odor*. Springfield, Ill.: Thomas, 1970.
441. Amoore, J. E. ''Odor theory and odor classification.'' *Fragrance Chemistry*, ed. E. T. Theimer, pp. 28–76. New York: Academic Press, 1982.
442. Amoore, J. E. *J. Am. Water Assn*. 78: 70–76 (1986). ''The chemistry and physiology of odor sensitivity.''
443. Lancet, D. *Ann. Rev. Neurosci*. 9: 329–355 (1986). ''Vertebrate olfactory reception.''
444. Holley, A., and MacLeod, P. *J. Physiol. Paris* 73: 725–828 (1977). ''Transduction et codage des informations olfactives chez les vertébrés.''
445. Gibbons, B. *Nat'l. Geographic* 170: 324–360 (1986). ''The intimate sense of smell.''
446. Wright, R. H.; Hughes, J. R.; and Hendrix, D. E. *Nature* 216: 404–406 (1967). Wright, R. H. *Chem. Senses* 8: 103–106 (1983). ''Molecular vibration and odour blending.''
447. Friedman, L., and Miller, J. G. *Science* 172: 1044–1046 (1971). ''Odor incongruity and chirality.''
448. Russell, G. F., and Hills, J. I. *Science* 172: 1043–1044 (1971). ''Odor differences between enantiomeric isomers.''
449. Davies, J. T. ''Olfactory Theories.'' Pages 322–350 in ref. 435.
450. Getchell, T. L.; Margolis, F. L.; and Getchell, M. L. *Progr. in Neurobiology* 23: 317–345 (1984). ''Perireceptor and receptor events in vertebrate olfaction.''
451. Graziadei, P.P.C., and Monti Graziadei, G. A. *J. Neurocytol*. 8: 1–18 (1979). ''Neurogenesis and neuron regeneration in the olfactory system of mammals. I. Morphological aspects of differentiation and structural organization of the olfactory sensory neurons.''
452. Hinds, J. W.; Hinds, P. L.; and McNelly, N. A. *The Anatomical Record* 210: 375–383 (1984). ''An audioradiographic study of the mouse olfactory epithelium: Evidence for long-lived receptors.''
453. Simmons, P. A., and Getchell, T. L. *J. Neurophysiol*. 45: 516–528 (1981). ''Neurogenesis in olfactory epithelium: Loss and recovery of transepithelial voltage transients following olfactory nerve section.''
454. Shepherd, G. M. *The Synaptic Organization of the Brain*, 2d ed., pp. 153–155, 319. New York: Oxford University Press, 1979.
455. Ottoson, D. *Physiology of the Nervous System*. New York: Oxford University Press, 1983. 528 pp.
456. van Meer, G.; Gambiner, B.; and Simons, K. *Nature* 322: 639–641 (1986). ''The

tight junction does not allow lipid molecules to diffuse from one epithelial cell to the next.''

457. Menco, B.P.M.; Dodd, G. H.; Davey, M.; and Bannister, L. H. *Nature* 263: 597–599 (1976). ''Presence of membrane particles in freeze-etched bovine olfactory cilia.''
458. Menco, B.P.M. *Cell Tissue Res.* 211: 5–29 (1980). ''Qualitative and quantitative freeze-fracture studies on olfactory and nasal respiratory epithelial surfaces of frog, ox, rat and dog.''
459. Pevsner, J.; Trifiletti, R. R.; Strittmatter, S. M.; and Snyder, S. H. *Proc. Nat'l. Acad. Sci. USA* 82: 3050–3054 (1985). ''Isolation and characterization of an olfactory receptor protein for odorant pyrazines.''
460. Pevsner, J.; Reed, R. R.; Feinstein, P. G.; and Snyder, S. H. *Science* 241: 336–339 (1988). ''Molecular cloning of an odorant-binding protein: Member of a ligand carrier family.''
461. Rogers, K. E.; Dasgupta, P.; Gubler, U.; Grillo, M.; Khew-Goodall, Y. S.; and Margolis, F. L. *Proc. Nat'l. Acad. Sci. USA* 84: 1704–1708 (1987). ''Molecular cloning and sequencing of a cDNA for olfactory marker protein.''
462. Sydor, W.; Teitelbaum, Z.; Blacher, R.; Sun, S.; Benz, W.; and Margolis, F. L. *Arch. Biochem. Biophys.* 249: 351–362 (1986). ''Amino acid sequence of a unique neuronal protein: Rat olfactory marker protein.''
463. Margolis, F. L. *Scand. J. Immunol.* 15 (Suppl.9), 181–199 (1982). ''Olfactory marker protein (OMP).''
464. Baker, H., and Margolis, F. L. *Neurosci. Abstr.* 17: 363 (1987). ''Olfactory marker protein (OMP)-like immunoreactivity in discrete nuclei of rodent brain.''
465. Chen, Z., and Lancet, D. *Proc. Nat'l. Acad. Sci. USA* 81: 1859–1863 (1984). ''Membrane proteins unique to vertebrate olfactory cilia: Candidates for sensory receptor molecules.''
466. Chen, Z.; Pace, U.; Ronen, D.; and Lancet, D. *J. Biol. Chem.* 261: 1299–1305 (1986). ''Polypeptide gp95. A unique glycoprotein of olfactory cilia with transmembrane receptor properties.''
467. Kropf, R.; Lancet, D.; and Lazard, D. *Neurosci. Abstr.* 17: 1410 (1987). ''A bovine olfactory cilia preparation: Specific transmembrane glycoproteins and phosphoproteins.''
468. Fesenko, E. E.; Novoselov, V. I.; and Bystrova, M. F. *FEBS Lett.* 219: 224–226 (1987). ''The subunits of specific odor-binding glycoproteins from rat olfactory epithelium.''
469. Chen, Z.; Ophir, D.; and Lancet, D. *Brain Res.* 368: 329–338 (1986). ''Monoclonal antibodies to ciliary glycoproteins of frog olfactory neurons.''
470. Harper, R.; Bate-Smith, E. C.; and Land, E. G. *Odour description and odour classification*. New York: American Elsevier, 1968.
471. Harper, R.; Bate-Smith, E. C.; Land, E. G.; and Griffiths, N. M. *Perfum. Essent. Oil Rec.* 59: 22-37 (1968).
472. Harper, R.; Land, E. G.; Griffiths, N. M.; and Bate-Smith, E. C. *Brit. J. Psychol.* 59: 231–252 (1968). ''Odour qualities—a glossary of usage.''
473. Ohloff, G.; Vial, C.; Wolf, H. R.; Job, K.; Jegou, E.; Polonsky, J.; and Lederer, E. *Helv. Chim. Acta* 63: 1932–1946 (1980). ''Stereochemistry-odor relationships in enantiomeric ambergris fragrances.''
474. Ohloff, G. In Olfaction and Taste IV, *Proc. Fourth Int'l. Symp. on Olfaction and*

Taste, Starnberg, Fed. Rep. Ger., Aug. 2–4, 1971, ed. D. Schneider, Stuttgart: Wissenschaftl. Verlagsges., 1972. Pp. 156–160.

475. Meiselman, H. L.; Rivlin, R. S., eds. *Clinical Measurement of Taste and Smell*. New York: Macmillan, 1986. 602 pp.

476. Graedel, T. E. *J. Chem. Ed.* 61: 681–686 (1984). "The perceived intensity of natural odors."

477. Ohloff, G. *Progr. Chem. Organic Natural Products* 35: 431–527 (1978). "Recent Developments in the Field of Naturally-Occurring Aroma Components."

478. Amoore, J. E. *Nature* 214: 1095–1096 (1967). "Specific anosmia: a clue to the olfactory code."

479. Labows, J. N., and Shushan, B. *American Laboratory*, March, 1983. "Direct analysis of food aromas."

480. Dravnieks, A. *Science* 218: 799–801 (1982). "Odor quality: Semantically generated multidimensional profiles are stable."

481. Berglund, B.; Berglund, U.; Lindvall, T. *Experientia* 42: 280–287 (1986). "Theory and methods for odor evaluation."

482. Brower, K. R.; Schaefer, R. *J. Chem. Ed.* 52: 538–540 (1975). "The recognition of chemical types by odor. The effect of steric hindrance at the functional group."

483. Schafer, R.; Brower, K. R. "Psychophysical recognition of functional groups in odorants." In *Olfaction and Taste V*. London: Academic Press, 1974, pp. 313–316.

484. Doty, R. L.; Shaman, P.; Applebaum, S. L.; Giberson, R.; Siksorski, L.; and Rosenberg, L. *Science* 226: 1441–1443 (1984). "Smell identification ability: Changes with age."

485. McKusick, V. A. *Mendelian Inheritance in Man: Catalogs of autosomal dominant, autosomal recessive and X-linked phenotypes*. Baltimore: Johns Hopkins University Press, 1986. 1741 pp. Anosmias listed are 10557, androst-16-en-3-one smelling ability (ref. 486), 10720, congenital anosmia (includes butanethiol anosmia [ref.487] and complete anosmia [ref. 488]), 25415, musk-smelling ability (pentadecanolactone [ref. 489], musk ambrette [ref. 490]), 30430, inability to smell cyanide (hydrogen cyanide) (ref. 491) and 22925, inability to smell freesia flowers (ref. 492). The genetics of the inability to smell isovaleric acid 24345 have not been studied (ref. 493).

486. Wysocki, C. J., and Beauchamp, G. K. *Proc. Nat'l. Acad. Sci. USA* 81: 4899–4902 (1984). "Ability to smell androstenone is genetically determined."

487. Patterson, P. M., and Lauder, B. A. *J. Hered.* 39: 295–297 (1948). "The incidence and probable inheritance of 'smell blindness' to normal butyl mercaptan."

488. Lygonis, C. S. *Hereditas* 61: 413–415 (1969). "Familial absence of olfaction."

489. Whissell-Buechy, D., and Amoore, J. E. *Nature* 242: 271–273 (1973). "Odour-blindness to musk: simple recessive inheritance."

490. Kalmus, H., and Seedburgh, D. *Ann. Human Genetics* 38: 495–499 (1975). "Correlated odour threshold bimodality in two out of three synthetic musks."

491. Brown, K. S.; MacLean, C. M.; and Robinette, R. R. *Human Biol.* 40: 456–472 (1968). "The distribution of the sensitivity to chemical odors in man."

492. McWhirter, K. G. *Canad. J. Genet. Cytol.* 11: 479 (1969). "Ethnography of specific anosmia."

493. Hirth, L.; Abadamian, D.; and Goedde, H. W. *Human Heredity* 36: 1–5 (1986). "Incidence of specific anosmia in northern Germany."

494. Hubert, H. B.; Fabsitz, R. R.; Feinlib, M.; and Brown, K. S. *Science* 208: 607–

609 (1980). "Olfactory sensitivity in humans: genetic versus environmental control."

495. Kjaer, A. *Pure and Applied Chem.* 49: 137–152 (1977). "Low molecular weight sulphur-containing compounds in nature: A survey."
496. Andersen, K. K., and Bernstein, D. T. *J. Chem. Ecol.* 1: 493–499 (1975). "Some chemical constituents of the scent of the striped skunk, *Mephitis mephitis*."
497. Getchell, M. L., and Gesteland, R. C. *Proc. Nat'l. Acad. Sci. USA* 69: 1494–1498 (1972). "The chemistry of olfactory reception: Stimulus-specific protection from sulfhydryl reagent inhibition."
498. Delaleu, J. C., and Holley, A. *Chem. Senses and Flavour* 5: 205–218 (1980). "Modification of transduction mechanisms in the frog's olfactory mucosa using a thiol reagent as olfactory stimulus."
499. Criswell, D. W.; McClure, F. L.; Schafer, R.; and Brower, K. R. *Science* 210: 425–426 (1980). "War gases as olfactory probes."
500. Schafer, R.; Fracek, S. P., Jr.; Criswell, D. W.; and Brower, K. R. *Chem. Senses* 9: 55–72 (1984). "Protection of olfactory responses from inhibition by ethyl bromoacetate, diethylamine, and other chemically active odorants by certain esters and other compounds."
501. Fracek, S. P., Jr., and Schafer, R. *Neurosci. Abstr.* 13: 1019 (1983). "Chemically active odorants and vaporous protectants as chemical probes of olfaction."
502. Shirley, S.; Polak, E.; and Dodd, G. H. *Eur. J. Biochem.* 132: 485–494 (1983). "Chemical modification studies on rat olfactory mucosa using a thiol-specific reagent and enzymatic iodination."
503. Mason, J. R.; Clark, L.; and Morton, T. H. *Science* 226: 1092–1094 (1984). "Selective deficits in the sense of smell caused by chemical modification of the olfactory epithelium."
504. Delaleu, J. C., and Holley, A. *Neurosci. Lett.* 37: 251–266 (1983). "Investigations of the discriminative properties of the frog's olfactory mucosa using a photoactivatable odorant."
505. Dodd, G., and Persaud, K. "Biochemical mechanisms in vertebrate primary olfactory neurons." In *Biochemistry of Taste and Olfaction*, ed. R. H. Cagan, and M. R. Kare, chap. 16, pp. 333–357. New York: Academic Press, 1981. 539 pp.
506. Doty, R. L. *Experientia* 42: 257–271 (1986). "Odor-guided behavior in mammals."
507. Albone, E. S. *Mammalian semiochemistry: an investigation of chemical signals between mammals*. Chichester, England: Wiley-Interscience, 1984. 360 pp.
508. Ferris, A. M.; Schlitzer, J. L.; and Schierberl, M. J. "Nutrition and taste in smell deficits: A risk factor or an adjustment?" In *Clinical Measurement of Taste and Smell*, ed. H. L. Meiselman, and R. S. Rivlin, chap. 16, pp. 264–278. New York: Macmillan, 1986. 602 pp.
509. Poziomek, E. J.; Boshart, G. L.; Crabtree, E. V.; Dehn, R.; Green, J. M.; Hoy, D. J.; Mackay, R. A.; Pryor, G. T.; Stone, G.; and Tanabe, M. *Microchem. J.* 16: 136–144 (1971). "Olfactory detection of chemical warfare agents."
510. Schleppnik, A. A. In Abstr., Eur. Chemoreception Res. Org., 3rd Congr., p. 76 (1975). Cf. *Science* 190: 870 (1975); *Chem. & Engr. News* 53(41): 24–25 (1975).
511. Marcus, B. In *Chem. & Engr. News* 66(22): 30–33 (1988). "New molecular sieves eliminate odors."
512. Takagi, S. F. "Biophysics of Smell." In *Handbook of Perception, Vol. VIA*,

Tasting and Smelling, ed. E. C. Carterette, and M. P. Friedman, Chap. 7, pp. 233–243. New York: Academic Press, 1978.

513. Masukawa, L. M.; Hedlund, B.; and Shepherd, G. M. *J. Neurosci.* 5: 128–135 (1985). ''Electrophysiological properties of identified cells in the olfactory epithelium of the tiger salamander.''

514. Anderson, P.A.V., and Hamilton, K. A. *Biophys. J.* 49: 556a (1986). ''Intracellular recordings from olfactory receptor neurons isolated from the salamander.''

515. Gesteland, R. C. ''The neural code: integrative neural mechanisms.'' In *Handbook of Perception, Vol. VIA, Tasting and Smelling*, ed. E. C. Carterette, M. P. Friedman, chap. 9, pp. 259–276. New York: Academic, 1978.

516. Persaud, K. C.; DeSimone, J. A.; Getchell, M. L.; Heck, G. L.; and Getchell, T. V. *Biochim. Biophys. Acta* 902: 65–79 (1987). ''Ion transport across the frog olfactory mucosa: the basal and odorant-stimulated states.''

517. Adrian, E. D. *Brit. Med. J.* 1: 287–290 (1954). ''The basis of sensation—some recent studies of sensation.''

518. Mozell, M. M., and Hornung, D. E. ''Peripheral mechanisms in the olfactory process.'' Pages 253–279 in ref. 356.

519. Sokoloff, L.; Revich, M.; Kennedy, C.; Des Rosiers, M. H.; Patlak, C. S.; Pettigrew, K. D.; Sakurada, O.; and Shinohara, M. *J. Neurochem.* 28: 897–916 (1977). ''The [^{14}C]deoxyglucose method for the measurement of local cerebral glucose utilization: Theory, procedure and normal values in the conscious and anesthetized albino rat.''

520. Lancet, D.; Greer, C. A.; Kauer, J. S.; and Shepherd, G. M. *Proc. Nat'l. Acad. Sci. USA* 79: 670–674 (1982). ''Mapping of odor-related neuronal activity in the olfactory bulb by high-resolution 2-deoxyglucose autoradiography.''

521. Shepherd, G. M. ''Are there labeled lines in the olfactory pathway?'' Chapter 13, pages 307–321 in ref. 356.

522. Maue, R. A., and Dionne, V. E. *Biophys. J.* 49: 556a (1986). ''Membrane conductance mechanisms in neonatal and embryonic mouse olfactory neurons.''

523. Jastreboff, P. J.; Pedersen, P. E.; Greer, C. A.; Stewart, W. B.; Kauer, J. S.; Benson, T. E.; and Shepherd, G. M. *Proc. Nat'l. Acad. Sci. USA* 81: 5250–5254 (1984). ''Specific olfactory receptor populations projecting to identified glomeruli in the rat olfactory bulb.''

524. Matsumoto, S. G., and Hildebrand, J. G. *Proc. Roy. Soc. London, Ser. B* 213: 249–277 (1981). ''Olfactory mechanisms in the moth *Manduca sexta*: Response characteristics and morphology of central neurons in the antennal lobes.''

525. Tolbert, L. P., and Hildebrand, J. G. *Proc. Roy. Soc. London, Ser. B* 213: 279–301 (1981). ''Organization and synaptic ultrastructure of glomeruli in the antennal lobes of the moth *Manduca sexta*: A study using thin sections and freeze fracture.''

526. Schwob, J. E., and Gottlieb, D. I. *J. Neurosci.* 6: 3393–3404 (1986). ''The primary olfactory projection has two chemically distinct zones.''

527. Lancet, D., and Pace, U. *Trends in Biochem. Sci.* 12: 63–66 (1987). ''The molecular basis of odor recognition.''

528. Lancet, D.; Chen, Z.; Ciobotariu, A.; Eckstein, F.; Khen, M.; Heldman, J.; Ophir, D.; Shafir, I.; and Pace, U. In Olfaction and Taste IX, ed. S. D. Roper, and J. Atema, *Ann. N.Y. Acad. Sci.* 510: 27–32 (1987). ''Toward a comprehensive molecular analysis of olfactory transduction.''

529. Kurihara, K., and Koyama, N. *Biochem. Biophys. Res. Commun.* 48: 30–34 (1972). ''High activity of adenyl cyclase in olfactory and gustatory organs.''
530. Pace, U.; Hanski, E.; Salomon, Y.; and Lancet, D. *Nature* 316: 255–258 (1985). ''Odorant-sensitive adenylate cyclase may mediate olfactory reception.''
531. Shirley, S. G.; Robinson, C. J.; Dickinson, K.; Aujla, R.; and Dodd, G. H. *Biochem. J.* 240: 605–607 (1986). ''Olfactory adenylate cyclase of the rat.''
532. Shirley, S. G.; Robinson, C. J.; and Dodd, G. H. *Biochem. Soc. Trans.* 15: 503 (1987). ''Olfactory adenylate cyclase of the rat.''
533. Shepherd, G. M. *Nature* 316: 214–215 (1985). ''Welcome whiff of biochemistry.''
534. Menevse, A.; Dodd, G.; and Poynder, T. M. *Biochem. Biophys. Res. Commun.* 77: 671–677 (1977). ''Evidence for the specific involvement of cyclic AMP in the olfactory transduction mechanism.''
535. Minor, A. V., and Sakina, N. L. *Neurofysiologiya* 5: 415–422 (1973). (*Biol. Abs.* 57: 5567 [Abstr. 52193] [1974]). ''Role of cyclic adenosine 3′,5′-monophosphate in olfactory reception.''
536. Nakamura, T., and Gold, G. H. *Nature* 325: 442–444 (1987). ''A cyclic nucleotide-gated conductance in olfactory receptor cilia.''
537. Sklar, P. B.; Anholt, R.R.H.; and Snyder, S. H. *J. Biol. Chem.* 261: 15538–15543 (1986). ''The odorant-sensitive adenylate cyclase of olfactory receptor cells.''
538. Berridge, M. J., and Irvine, R. F. *Nature* 312: 315–321 (1984). ''Inositol trisphosphate, a novel second messenger in cellular signal transduction.''
539. Michell, B. *Nature* 324: 613 (1986). ''A second messenger function for inositol tetrakisphosphate.''
540. Michell, R. H. *Trends in Biochem. Sci.* 4: 128–131 (1979). ''Inositol phospholipids in membrane function.''
541. Henkin, R. I. *J. Clin. Endocrin. Metab.* 28: 624–628 (1968). ''Impairment of olfaction and of the tastes of sour and bitter in pseudohypoparathyroidism.''
542. Weinstock, R. S.; Wright, H. N.; Spiegel, A. M.; Levine, M. A.; Moses, A. M. *Nature* 322: 635–666 (1986). ''Olfactory dysfunction in humans with guanine-nucleotide binding protein.'' Cf. entry 30080, ref. 485 concerning pseudohypoparathyroidism (PHP).
543. Wright, H. N.; Weinstock, R. S.; Spiegel, A. M.; Levine, M. A.; Moses, A. M. In Olfaction and Taste IX, ed. S. D. Roper, and J. Atema, *Ann. N.Y. Acad. Sci.* 510: 719–722 (1987). ''Guanine nucleotide-binding stimulatory protein. A requisite for human odorant perception.''
544. Pace, U., and Lancet, D. *Proc. Nat'l. Acad. Sci. USA* 83: 4947–4951 (1986). ''Olfactory GTP-binding signal transducing polypeptide of vertebrate chemosensory neurons.''
545. Ueda, K., and Hayaishi, O. *Ann. Rev. Biochem.* 54: 73–100 (1985). ''ADP-Ribosylation.''
546. Anholt, R.R.H.; Mumby, S. M.; Stoffers, D. A.; Girard, P. R.; Kuo, J. F.; and Snyder, S. H. *Biochemistry* 26: 788–795 (1987). ''Transduction proteins of olfactory receptor cells: Identification of guanine nucleotide binding proteins and protein kinase C.''
547. Nairn, A. C.; Hemmings, H. C., Jr.; and Greengard, P. *Ann. Rev. Biochem.* 54: 931–976 (1985). ''Protein kinases in the brain.''

548. Gergel, M. G. *Excuse me sir, would you like to buy a kilo of isopropyl bromide*, p. 26, Pierce Chemical Co., 1979.

549. Getchell, T. V.; Heck, G. L.; DeSimone, J. A.; and Price, S. *Biophys. J.* 29: 397–411 (1980). ''The location of olfactory receptor sites: Inferences from latency measurements.''

550. Masukawa, L. M.; Hedlund, B.; and Shepherd, G. M. *J. Neurosci.* 5: 136–141 (1985). ''Changes in the electrical properties of olfactory epithelial cells in the tiger salamander after olfactory nerve transection.''

551. Fudenberg, H. H.; Pink, J.R.L.; Wang, A.-C.; and Ferrara, G. B. *Basic Immunogenetics*, pp. 190–201. 3d ed. New York: Oxford University Press, 1984. 302 pp.

552. Goodman, C. S.; Bestiani, M. J.; Doe, C. Q.; du Lac, S.; Helfand, S. L.; Kuwada, J. Y.; and Thomas, J. B. *Science* 225: 1271–1279 (1984). ''Cell recognition during neuronal development.''

553. Dixon, R.A.F.; Kobilka, B. K.; Strader, D. J.; Benovic, J. L.; Dohlman, H. G.; Frielle, T.; Bolanowski, M. A.; Bennett, C. D.; Rands, E.; Diehl, R. E.; Mumford, R. A.; Slater, E. E.; Sigal, I. S.; Caron, M. G.; Lefkowitz, R. J.; and Strader, C. D. *Nature* 321: 75–79 (1986). ''Cloning of the gene and cDNA for mammalian β-adrenergic receptor and homology with rhodopsin.''

554. Dohlman, H. G.; Caron, M. G.; and Lefkowitz, R. J. *Biochemistry* 26: 2657–2664 (1987). ''A family of receptors coupled to guanine nucleotide regulatory proteins.''

555. Kubo, T.; Fukuda, K.; Mikami, A.; Maeda, A.; Takahashi, H.; Mishina, M.; Haga, T.; Haga, K.; Ichiyama, A.; Kangawa, K.; Kojima, M.; Matsuo; H.; Hirose, T.; and Numa, S. *Nature* 323: 411–416 (1986). ''Cloning, sequencing and expression of complementary DNA encoding the muscarinic acetylcholine receptor.''

556. Kubo, T.; Maeda, A.; Sugimoto, K.; Akiba, I.; Mikami, A.; Takahashi, H.; Haga, T.; Haga, K.; Ichiyama, A.; Kangawa, K.; Matsuo, H.; Hirose, T.; and Numa, S. *FEBS Lett.* 209: 367–372 (1986). ''Primary structure of porcine cardiac muscarinic acetylcholine receptor deduced from the cDNA sequence.''

557. Bonner, T. I.; Buckley, N.J.; Young, A. C.; and Brann, M. R. *Science* 237: 527–532 (1987). ''Identification of a family of muscarinic acetylcholine receptor genes.''

558. Bourne, H. R. *Nature* 321: 814–816 (1986). ''One molecular machine can transduce diverse signals.'' *Nature* 337: 504–505 (1989). ''G-protein subunits: Who carries what message?''

559. Hargrave, P., and Applebury, M. L. *Vision Res.* 26: 1881–1895 (1986). ''Molecular biology of the visual pigments.''

560. Kebabian, J. W., and Caine, D. B. *Nature* 277: 93–96 (1979). ''Multiple receptors for dopamine.''

561. Stoof, J. C., and Kebabian, J. W. *Life Sci.* 35: 2281 (1984). ''Dopamine receptors—Biochemistry, physiology and pharmacology.''

562. Hargrave, P. A.; McDowell, J. H.; Curtis, D. R.; Wang, J. K.; Juszczak, E.; Fong, S.-L.; Mohana Rao, J. K.; and Argos, P. *Biophys. Struct. Mech.* 9: 235–244 (1983). ''The structure of bovine rhodopsin.''

563. Hargrave, P. A.; McDowell, J. H.; Feldmann, R. J.; Atkinson, P. H.; Mohana Rao, J. K.; and Argos, P. *Vision Res.* 24: 1487–1499 (1984) . ''Rhodopsin's protein and carbohydrate structure: Selected aspects.''

564. Kosower, E. M. *Proc. Nat'l. Acad. Sci. USA* 85: 1076–1080 (1988). ''Assign-

ment of the Groups Responsible for the "Opsin shift" and the Light Absorptions of Rhodopsin and Red, Green and Blue Iodopsins (Cone Pigments)."

565. Ohloff, G.; Maurer, B.; Winter, B.; and Giersch, W. *Helv. Chim. Acta* 66: 192–217 (1983). "Structural and configurational dependence of the sensory process in steroids."

566. Ohloff, G.; Giersch, W.; Thommen, W.; and Willhalm, B. *Helv. Chim. Acta* 66: 1343–1354 (1983). "Conformationally controlled odor perception 'steroid-type' molecules."

567. Ohloff, G.; Giersch, W.; Pickenhagen, W.; Furrer, A.; and Frei, B. *Helv. Chim. Acta* 68: 2022–2029 (1985). "Significance of the geminal dimethyl group in the odor principle of Ambrox."

568. Schulte-Elte, K. H.; Giersch, W.; Winter, B.; Pamingle, H.; and Ohloff, G. *Helv. Chim. Acta* 68: 1961–1985 (1985). "Diastereoselektivität der Geruchswahrnehmung von Alkoholen der Ionen reihe."

569. Ohloff, G., and Giersch, W. *Helv. Chim. Acta* 63: 76–94 (1980). "Stereochemistry-activity relationships in olfaction. Odorants containing a proton donor/proton acceptor unit."

570. Ohloff, G. *Experientia* 42: 271–279 (1986). "Chemistry of odor stimuli."

571. Sicard, G., and Holley, A. *Brain Res.* 292: 283–296 (1984). "Receptor cell responses to odorants. Similarities and differences among odorants."

572. Firestein, S., and Werblin, F., cited in Shepherd, G. M.; Getchell, T. V.; and Mistretta, C. M. *Nature* 324: 17–18 (1986). "Questions of taste and smell."

573. Hughes, J. R.; Hendrix, D. E.; Wetzel, N.; Johnston, J. W., Jr. In *Olfaction and Taste III*, ed. C. Pfaffman, pp. 172–191. New York: Rockefeller University Press, 1969. "Correlations between electrophysiological activity from the human olfactory bulb and the subjective response to odoriferous stimuli."

574. Hara, T. J.; Freese, M.; and Scott, K. R. *Jap. J. Physiol.* 23: 325–333 (1973). "Spectral analysis of olfactory bulbar responses in rainbow trout."

575. Satou, M. *Jap. J. Physiol.* 24: 389–402 (1974). "Electrical responses at various levels of the olfactory pathway in Himé salmon, *Oncorhynchus nerka*."

576. Hara, T. J., *Comp. Biochem. Physiol.* 54A: 31–36 (1976). "Structure-activity relationships of amino acids in fish olfaction."

577. Hara, T. J. *Comp. Biochem. Physiol.* 54A: 37–39 (1976). "Effects of pH on the olfactory responses to amino acids in rainbow trout, *Salmo gairdneri*."

578. Rhein, L. D., and Cagan, R. H. *Proc. Nat'l. Acad. Sci. USA* 77: 4412–4416 (1980). "Biochemical studies of olfaction: Isolation, characterization and odorant binding activity of cilia from rainbow trout olfactory rosettes."

579. Fein, A., and Szuts, E. Z. *Photoreceptors: their role in vision*. Cambridge: Cambridge University Press, 1982. 212 pp.

580. Barlow, H. B., and Mollon, J. D., eds. *The Senses*. Cambridge: Cambridge University Press, 1982. 490 pp.

581. Schmidt, R. F., ed. *Fundamentals of Sensory Physiology*. 2d ed. New York: Springer Verlag, 1981. 286 pp.

582. Kaufman, L. *Perception*. New York: Oxford University Press, 1979. 416 pp.

583. Chabré, M. *Ann. Rev. Biophys. Biophys. Chem.* 14: 331–360 (1985). "Trigger and amplification mechanisms in visual phototransduction."

584. Masland, R. H. *Sci. Amer.* 255(6): 90–99 (1986). "The functional architecture of the retina."

585. Attwell, D.; Borges, S.; Wu, S. M.; and Wilson, M. *Nature* 328: 522 (1987). "Signal clipping by the rod output synapse."

586. Besharse, J. C., and Dunis, D. A. *Science* 219: 1341–1343 (1983). "Methoxyindoles and photoreceptor metabolism: activation of rod shedding."

587. Curcio, C. A.; Sloan, K. J., Jr.; Packer, O.; Hendrickson, A. E.; Kalina, R. E. *Science* 236: 579–582 (1987). "Distribution of cones in human and monkey retina: Individual variability and radial asymmetry."

588. Brann, M. R., and Cohen, L. V. *Science* 235: 585–587 (1987). "Diurnal expression of transducin mRNA and translocation of transducin in rods of rat retina."

589. Olive, J. *Int'l. Rev. Cytol.* 64: 107–166 (1980). "The structural organization of the mammalian retinal disk membrane."

590. Saibil, H.; Chabré, M.; and Worcester, D. *Nature* 262: 266-270 (1976). "Neutron diffraction studies of retinal rod outer segment membranes."

591. Anderson, R. A.; Benolken, R. M.; Dudley, P. A.; Landis, D. J.; and Wheeler, T. G. *Exp. Eye Res.* 18: 205–213 (1974). "Polyunsaturated fatty acids of photoreceptor membranes."

592. Aveldaño, M. I. *J. Biol. Chem.* 262: 1172–1179 (1987). "A novel group of very long chain polyenoic fatty acids in dipolyunsaturated phosphatidylcholines from vertebrate retina."

593. Aveldaño, M. I., and Sprecher, H. *J. Biol. Chem.* 262: 1180–1186 (1987). "Very long chain (C_{24} to C_{36}) polyenoic fatty acids of the n-3 and n-6 series in dipolyunsaturated phosphatidylcholines from bovine retina."

594. Aveldaño, M. I. *Biochemistry* 27: 1229–1239 (1988). "Phospholipid species containing long and very long chain fatty acids remain with rhodopsin after hexane extraction."

595. Rotstein, N. P., and Aveldaño, M. I. *Biochem. J.* 249: 191–200 (1988). Synthesis of very long chain fatty acids (up to 36 carbon) tetra-, penta- and hexaenoic fatty acids in retina."

596. Kemp, C. M. "Rhodopsin and its role in visual transduction." In *Biological Membranes* 5: 145–196, ed. D. Chapman, New York: Academic Press, 1985.

597. Pfister, C.; Chabré, M.; Plouet, J.; Tuyen, V. V.; DeKozak, Y.; Faure, J. P.; and Kühn, H. *Science* 228: 891–893 (1985). "Retinal S antigen identified as the 48K protein regulating light-dependent phosphodiesterase in rods."

598. Fung, B.K.-K., and Hubbell, W. L. *Biochemistry* 17: 4403–4410 (1978). "Organization of rhodopsin in photoreceptor membranes. 2. Transmembrane organization of bovine rhodopsin: Evidence from proteolysis and lactoperoxidase catalyzed iodination of native and reconstituted membranes."

599. Liebman, P. A.; Parker, K. R.; and Dratz, E. A. *Ann. Rev. Physiol.* 49:a 765–791 (1987). "The molecular mechanism of visual excitation and its relation to the structure and composition of the rod outer segment."

600. Pugh, E. N., Jr., and Cobbs, W. H. *Vision Res.* 26: 1613–1643 (1986). "Visual transduction in vertebrate rods and cones: A tale of two transmitters, calcium and cyclic GMP."

601. Hagins, W. A., and Yoshikami, S. *Exp. Eye Res.* 18: 299–305 (1974). "A role for Ca^{++} in excitation of retinal rods and cones."

602. Brown, J. E., and Blinks, J. R. *J. Gen. Physiol.* 64: 643–665 (1974). "Changes in intracellular free calcium concentration during illumination of invertebrate photoreceptors."

603. Yau, K.-W., and Nakatani, K. *Nature* 311: 661–663 (1984). "Electrogenic Na-Ca exchange in the retinal rod outer segment."

604. Yau, K.-W., and Nakatani, K. *Nature* 313: 579–582 (1985). "Light-induced reduction of cytoplasmic free calcium in retinal rod outer segment."

605. Pugh, E. N., and Altman, J. *Nature* 334: 16–17 (1988). "Phototransduction: A role for calcium in adaptation."

606. Matthews, H. R.; Murphy, R.L.W.; Fain, G. L.; and Lamb, T. D. *Nature* 334: 68–69 (1988). "Photoreceptor light adaptation is mediated by cytoplasmic calcium concentration."

607. Nakatani, K., and Yau, K.-W. *Nature* 334: 69-71 (1988). "Calcium and light adaptation in retinal rods and cones."

608. Hodgkin, A. L., and Nunn, B. J. *J. Physiol. (Lond.)* 403: 439–471 (1988). "Control of light-sensitive current in salamander rods."

609. Yau, K.-W., cited by Altman, J. *Nature* 313: 264–265 (1985) . "New visions in photoreception." Lewin, R. *Science* 227: 500–503 (1985). "Unexpected progress in photoreception."

610. Matthews, G. *Proc. Nat'l. Acad. Sci. USA* 84: 299–302 (1987). "Single channel recordings demonstrate that cGMP opens the light-sensitive ion channel of the rod photoreceptor."

611. Zimmerman, A. L., and Baylor, D. A. *Nature* 321: 70–72 (1986). "Cyclic GMP-sensitive conductance of retinal rods consists of aqueous pores."

612. Tanaka, J. C.; Furman, R. E.; Cobbs, W. H.; and Mueller, P. *Proc. Nat'l. Acad. Sci. USA* 84: 724–728 (1987). "Incorporation of a retinal rod cGMP-dependent conductance into planar bilayers."

613. Stern, J. H.; Knutsson, H.; and MacLeish, P. R. *Science* 236: 1674–1678 (1987). "Divalent cations directly affect the conductance of excised patches of rod photoreceptor membrane."

614. Cook, N. J.; Hanke, W.; and Kaupp, U. B. *Proc. Nat'l. Acad. Sci. USA* 84: 585–589 (1987). "Identification, purification and functional reconstitution of the cyclic GMP-dependent channel from rod photoreceptors." Cf. *Nature* 342: 762–766 (1989).

615. Fung, B.K.-K., and Stryer, L. *Proc. Nat'l. Acad. Sci. USA* 77: 2500–2504 (1980). "Photolyzed rhodopsin catalyzes the exchange of GTP for bound GDP in retinal rod outer segments."

616. Stryer, L. *Biopolymers* 24: 29–47 (1985). "Molecular design of an amplification cascade in vision." Stryer, L. *Sci. Amer.* 257(1): 32–40 (1987). "The molecules of visual excitation."

617. Kühn, H. *Nature* 283: 587–589 (1980). "Light- and GTP-regulated interaction of GTPase and other proteins with bovine photoreceptor membranes."

618. Fung, B.K-K.; Hurley, J. B.; and Stryer, L. *Proc. Nat'l. Acad. Sci. USA* 78: 152–156 (1981). "Flow of information in the light-triggered cyclic nucleotide cascade of vision."

619. Vuong, T. M.; Chabré, M.; and Stryer, L. *Nature* 311: 659–661 (1984). "Millisecond activation of transducin in the cyclic nucleotide cascade of vision."

620. Bitensky, M. W.; Wheeler, G. L.; Yamazaki, G.; Rasenick, M. M.; and Stein, P. J. *Curr. Topics Membrane Transp.* 15: 237–271 (1981). "Cyclic nucleotide metabolism in vertebrate photoreceptors: A remarkable analogy and an unraveling enigma."

621. Hurley, J. B., and Stryer, L. *J. Biol. Chem.* 257: 11094–11099 (1982). "Purification and characterization of the γ-regulatory subunit of the cyclic GMP phosphodiesterase from retinal rod outer segments." Eckstein, F.; Karpen, J. W.; Critchfield, J. M.; and Stryer, L. *J. Biol. Chem.* 263: 14080–14085 (1988). "Stereochemical course of the reaction catalyzed by the cyclic GMP phosphodiesterase from retinal rod outer segments."

622. Pober, J. S., and Bitensky, M. W. *Adv. in Cyclic Nucleotide Res.* 11: 265–301 (1979). "Light-regulated enzymes of vertebrate retinal rods."

623. Sitaramayya, A.; Harkness, J.; Parkes, J. H.; Gonzalez-Oliva, C.; and Liebman, P. A. *Biochemistry* 25: 651–656 (1986). "Kinetic studies suggest that light-activated cyclic GMP phosphodiesterase is a complex with G-protein subunits."

624. Weishaar, R. E.; Cain, M. H.; and Briston, J. A. *J. Med. Chem.* 28: 537–545 (1985). "A new generation of phosphodiesterase inhibitors: Multiple molecular forms of phosphodiesterase and the potential for drug selectivity."

625. Oprian, D. D.; Molday, R. S.; Kaufman, R. J.; and Khorana, H. G. *Proc. Nat'l. Acad. Sci. USA* 84: 8874–8878 (1987). "Expression of a synthetic bovine rhodopsin gene in monkey kidney cells."

626. Franke, R. R.; Sakmar, T. P.; Oprian, D. D.; and Khorana, H. G. *J. Biol. Chem.* 263: 2119–2123 (1988). "A single amino acid substitution in rhodopsin (lysine 248 → leucine) prevents activation of transducin."

627. Palczewski, K.; McDowell, J. H.; and Hargrave, P. A. *Biochemistry* 27: 2306–2313 (1988). "Rhodopsin kinase: Substrate specificity and factors that influence activity." Palczewski, K.; McDowell, J. H.; Hargrave, P. A.; Ingebritsen, T. S. *Biochemistry* 28: 415–419 (1989). "The catalytic subunit of phosphatase 2A dephosphorylates phosphoopsin."

628. Bitensky, M. W.; Wheeler, M. A.; Rasenick, M. M.; Yamazaki, A.; Stein, P. J.; Halliday, K. R.; and Wheeler, G. L. *Proc. Nat'l. Acad. Sci. USA* 79: 3408–3412 (1982). "Functional exchange of components between light-activated photoreceptor phosphodiesterase and hormone-activated adenylate cyclase systems."

629. Weiss, E. R.; Hadcock, J. R.; Johnson, G. L.; and Malbon, C. C. *J. Biol. Chem.* 262: 4319–4323 (1987). "Antipeptide antibodies directed against cytoplasmic rhodopsin sequences recognize the β-adrenergic receptor."

630. Lefkowitz, R. J., and Caron, M. G. *J. Biol. Chem.* 263: 4993–4996 (1988). "Adrenergic receptors. Models for the study of receptors coupled to guanine nucleotide regulatory proteins."

631. McCormick, F.; Clark, B.F.C.; LaCour, T.F.M.; Kjeldgaard, M.; Norskov-Lauritsen, L.; and Nyborg, J. *Science* 230: 78–82 (1985). "A model for the tertiary structure of p21, the product of the ras oncogene."

632. LaCour, T.F.M.; Nyborg, J.; Thirup, S.; and Clark, B.F.C. *EMBO J.* 4: 2385–2388 (1985). "Structural details of the binding of guanosine diphosphate to elongation factor Tu from E.coli as studied by X-ray crystallography." Jurnak, F. *Science* 230: 32–36 (1985). "Structure of the GDP domain of EF-Tu and location of the amino acids homologous to ras oncogene proteins."

633. Marx, J. L. *Science* 239: 863 (1988). "First portrait of an oncogene product."

634. De Vos, A. M.; Tong, L.; Milburn, M. V.; Matias, P. M.; Jancarik, J.; Noguchi, S.; Nishimura, S.; Miura, K.; Ohtsuka, E.; and Kim, S.-H. *Science* 239: 863 (1988). "Three-dimensional structure of an oncogene protein: Catalytic domain of human c-H-*ras* p21."

635. Beckner, S. K.; Hattori, S.; and Shih, T. Y. *Nature* 317: 71–72 (1983). "The ras oncogene product is not a regulatory component of adenylate cyclase."
636. Medynski, D. C.; Sullivan, K.; Smith, D.; Van Dop, C.; Chang, F.-H.; Fung, B.K.K.; Seeburg, P. H.; and Bourne, H. R. *Proc. Nat'l. Acad. Sci. USA* 82: 4311–4315 (1985). "Amino acid sequence of the α-subunit of transducin deduced from the cDNA sequence."
637. Hurley, J. B.; Simon, M. I.; Teplow, D. B.; Robishaw, J. D.; and Gilman, A. G. *Science* 226: 860–862 (1984). "Homologies between signal transducing G proteins and ras gene products."
638. Tanabe, T.; Nukada, T.; Nishikawa, Y.; Sugimoto, K.; Suzuki, H.; Takahashi, H.; Noda, M.; Haga, T.; Ichiyama, A.; Kangawa, K.; Minamino, N.; Matsuo, H.; Numa, S. *Nature* 315: 242–245 (1985). "Primary structure of the α-subunit of transducin and its relationship to ras proteins."
639. Yatsunami, K., and Khorana, H. G. *Proc. Nat'l. Acad. Sci. USA* 82: 4316–4320 (1985). "GTPase of bovine rod outer segments. The amino acid sequence of the a-subunit as derived from the cDNA sequence."
640. Lochrie, M. A.; Hurley, J. B.; and Simon, M. I. *Science* 228: 96–99 (1985). "Sequence of the alpha subunit of photoreceptor G protein: Homologies between transducin, ras and elongation factors."
641. Benovic, J. L.; Kühn, H.; Weyand, I.; Codina, J.; Caron, M. G.; and Lefkowitz, R. *Proc. Nat'l. Acad. Sci. USA* 84: 8879–8882 (1987). "Functional desensitization of the isolated β-adrenergic receptor by the β-adrenergic receptor kinase: Potential role of an analog of the retinal protein arrestin (48-kDa protein)."
642. Stryer, L., Thomas, D. D.; and Meares, C. F. *Ann. Rev. Biophys. Bioeng.* 11: 203–222 (1982). "Diffusion-enhanced fluorescence energy transfer."
643. Michel-Villaz, M.; Roche, C.; and Chabré, M. *Biophys. J.* 37: 603–616 (1982). "Orientational changes of the absorbing dipole of retinal upon the conversion of rhodopsin to bathorhodopsin, lumirhodopsin and isorhodopsin."
644. Mathies, R., and Stryer, L. *Proc. Nat'l. Acad. Sci. USA* 73: 2169–2173 (1976). "Retinal has a highly dipolar vertically excited singlet state: Implications for vision."
645. Cone, R. A. *Nature New Biol.* 236: 39–43 (1972). "Rotational diffusion of rhodopsin in the visual receptor membrane."
646. Brown, P. K. *Nature New Biol.* 236: 35–38 (1972). "Rhodopsin rotates in the visual receptor membrane."
647. Träuble, H., and Sackmann, E. *Nature* 245: 210–211 (1973). "Lipid motion and rhodopsin motion."
648. Liebman, P. A., and Entine, G. *Science* 185: 457–459 (1974). "Lateral diffusion of visual pigment in photoreceptor disk membranes."
649. Poo, M.-M., and Cone, R. A. *Nature* 247: 438–441(1974). "Lateral diffusion of rhodopsin in the photoreceptor membrane."
650. Takezoe, H., and Yu, H. *Biochemistry* 20: 5275–5281 (1981). "Lateral diffusion of photopigments in photoreceptor disk membrane by the dynamic Kerr effect."
651. Dale, R. E. *FEBS Lett.* 192: 255–259 (1985). "Interpretation of fluorescence photobleaching recovery experiments on oriented cell membranes."
652. Wald, G. *Science* 162: 230–239 (1968). "Molecular basis of visual excitation."
653. Ottolenghi, M. *Adv. Photochem.* 12: 97–200 (1980). "The photochemistry of rhodopsins."

654. Birge, R. R. *Ann. Rev. Biophys. Bioeng.* 10: 315–354 (1981). ''Photophysics of light transduction in rhodopsin and bacteriorhodopsin.''

655. Busch, G. E.; Applebury, M. L.; Lamola, A.; and Rentzepis, P. M. *Proc. Nat'l. Acad. Sci. USA* 69: 2802–2806 (1972). ''Formation and decay of prelumirhodopsin at room temperature.''

656. Peters, K.; Applebury, M. L.; and Rentzepis, P. M. *Proc. Nat'l. Acad. Sci. USA* 74: 3119–3123 (1977). ''Primary photochemical event in vision: Proton translocation.''

657. Shichida, Y.; Kobayashi, T.; Ohtani, H.; Yoshizawa, T.; and Nagakura, S. *Photochem. Photobiol.* 27: 335–341 (1978). ''Picosecond laser photolysis of squid rhodopsin at room and low temperatures.''

658. Renk, G. E.; Crouch, R. K.; and Feix, J. B. *Biophys. J.* 53: 361–365 (1988). ''Lack of interaction of rhodopsin chromophore with membrane lipids. An electron-electron double resonance study using ^{14}N:^{15}N pairs.''

659. Wiedmann, T. S.; Pates, R. D.; Beach, J. M.; Salmon, A.; and Brown, M. F. *Biophys. J.* 51: 15a (1987). ''Influences of phospholipid polar headgroups and acyl chain polyunsaturation on photochemical function of rhodopsin in recombinant membranes.''

660. Yoshizawa, T., and Wald, G. *Nature* 197: 1279–1286 (1965). ''Prelumirhodopsin and the bleaching of visual pigments.''

661. Aton, B.; Callender, R. H.; and Honig, B. *Nature* 273: 784–786 (1978). ''Photochemical *cis-trans* isomerization of bovine rhodopsin at liquid helium temperatures.''

662. Eyring, G.; Curry, B.; Mathies, R.; Fransen, R.; Palings, I.; and Lugtenburg, J. *Biochemistry* 19: 2410–2418 (1980). ''Resonance Raman studies of bathorhodopsin based on visual pigment analogues.''

663. Eyring, G.; Curry, B.; Broek, A.; Lugtenburg, J.; and Mathies, R. *Biochemistry* 21: 384–393 (1982). ''Resonance Raman studies of bathorhodopsin.'' Loppnow, G. R.; Barry, B. A.; and Mathies, R. A. *Proc. Nat'l. Acad. Sci. USA* 86: 1515–1518 (1989). ''Why are blue visual pigments blue? A resonance Raman microprobe study.''

664. Bagley, K. A.; Balogh-Nair, V.; Croteau, A. A.; Dollinger, G.; Ebrey, T. G.; Eisenstein, L.; Hong, M. K.; Nakanishi, K.; and Vittitow, J. *Biochemistry* 24: 6055–6071 (1985). ''Fourier-transform infrared difference spectroscopy of rhodopsin and its photoproducts at low temperature.''

665. Kawamura, S.; Tokunaga, F.; Yoshizawa, T.; Sarai, A.; and Kakitani, T. *Vision Res.* 19: 879–884 (1979). ''Orientational changes of the transition dipole moment of retinal chromophore due to the conversion of rhodopsin to bathorhodopsin and to isorhodopsin.''

666. Kawamura, S.; Wakabayashi, S.; Maeda, A.; and Yoshizawa, T. *Vision Res.* 18: 457–462 (1978). ''Isorhodopsin: Conformation and orientation of its chromophore in frog disk membrane.''

667. Kobayashi, T. *Photochem. Photobiol.* 32: 207–215 (1980). ''Existence of hypsorhodopsin as the first intermediate in the primary photochemical process of cattle rhodopsin.''

668. Schick, G. A.; Cooper, T. M.; Holloway, R. A.; Murray, L. P.; and Birge, R. R. *Biochemistry* 26: 2556–2562 (1987). ''Energy storage in the primary photochemical events of rhodopsin and isorhodopsin.''

669. Cooper, A. *Nature* 282: 531–533 (1979). ''Energy uptake in the first step of visual excitation.''
670. Abdulaev, N. G.; Artamonov, I. D.; Bogachuk, A. S.; Feigina, M. Yu; Kostina, B.; Kudelin, A. B.; Martynov, V. I.; Miroshnikov, A. I.; Zolotarev, A. S.; and Ovchinnikov, Yu. A. *Biochemistry Int'l.* 5: 693–703 (1982). ''Structure of light-activated proteins: Visual rhodopsin.''
671. Wang, J. K.; McDowell, J. H.; and Hargrave, P. A. *Biochemistry* 19: 5111–5117 (1980). ''Site of attachment of 11-*cis*-retinal in bovine rhodopsin.''
672. Nathans, J., and Hogness, D. S. *Cell* 34: 807–814 (1983). ''Isolation, sequence analysis and intron-exon arrangement of the gene encoding bovine rhodopsin.''
673. Nathans, J., and Hogness, D. S. *Proc. Nat'l. Acad. Sci. USA* 81: 4851–4855 (1984). ''Isolation and nucleotide sequence of the gene encoding human rhodopsin.''
674. Nathans, J.; Thomas, D.; and Hogness, D. S. *Science* 232: 193–202 (1986). ''Molecular genetics of human color vision: the genes encoding blue, green, and red pigments.''
675. Nathans, J.; Piantanida, T. P.; Eddy, R. L.; Shows, T. B.; and Hogness, D. S. *Science* 232: 203–210 (1986). ''Molecular genetics of inherited variation in human color vision.'' Vollrath, D.; Nathans, J.; and Davis, R. W. *Science* 240: 1669–1672 (1988). ''Tandem array of human visual pigment genes at Xq28.''
676. Hargrave, P. A. ''Molecular dynamics of the rod cell.'' In *The Retina*, Part 1, pp. 207–237. New York: Academic Press, 1986.
677. Shichida, Y.; Matuoka, S.; and Yoshizawa, T. *Photobiochem. Photobiophys.* 7: 221–228 (1984). ''Formation of photorhodopsin, a precursor of bathorhodopsin, detected by picosecond laser photolysis at room temperature.''
678. Mathies, R. A.; Cruz, C.H.B.; Pollard, W. T.; and Shank, C. V. *Science* 240: 777–779 (1988). ''Direct observation of the femtosecond excited state *cis-trans* isomerization in bacteriorhodopsin.''
679. Mathies, R. A., Univ. Calif., Berkeley. Personal communication.
680. Bennett, N.; Michel-Villaz, M.; and Kühn, H. *Eur. J. Biochem.* 127: 97–103 (1982). ''Light-induced interaction between rhodopsin and the GTP-binding protein. Metarhodopsin II is the major photoproduct involved.''
681. Stryer, L. *Ann. Rev. Neurosci.* 9: 87–119 (1986). ''Cyclic GMP cascade of vision.''
682. Dratz, E. A., and Hargrave, P. A. *Trends in Biochem.* 8: 128–131 (1983). ''The structure of rhodopsin and the rod outer segment disk membrane.''
683. Honig, B.; Dinur, B.; Nakanishi, K.; Balogh-Nair, V.; Gawinowicz, M. A.; Arnaboldi, M.; and Motto, M. G. *J. Am. Chem. Soc.* 101: 7084–7086 (1979). ''An external point charge model for wavelength regulation in visual pigments.''
684. Sheves, M., and Nakanishi, K. *J. Am. Chem. Soc.* 105: 4033–4039 (1983). ''Factors affecting the absorption maxima of polyene iminium systems: A model study for rhodopsin and bacteriorhodopsin.''
685. Kakitani, H.; Kakitani, T.; Rodman, H.; and Honig, B. *Photochem. Photobiol.* 41: 471–479 (1985). ''On the mechanism of wavelength regulation in visual pigments.''
686. Derguini, F.; Dunn, D.; Eisenstein, L.; Nakanishi, K.; Odashima, K.; Rao, V. J.; Sastry, L.; and Termini, J. *Pure & Applied Chem.* 58: 719–724 (1986).

"Studies with retinal pigments: modified point charge model for bacteriorhodopsin and difference FTIR (Fourier transform infrared) studies."

687. Fukuda, M. N.; Papermaster, D. F.; Hargrave, P. A. *J. Biol. Chem.* 254: 8201–8207 (1979). "Rhodopsin carbohydrate. Structure of small oligosaccharides attached at two site near the NH_2 terminus."
688. Liang, C.-J.; Yamashita, K.; Muellenberg, C. G.; Shichi, H.; and Kobata, A. *J. Biol. Chem.* 254: 6414–6418 (1979). "Structure of the carbohydrate moieties of bovine rhodopsin."
689. Adams, A. J.; Tanaka, M.; and Shichi, H. *Exp. Eye Res.* 27: 595–605 (1978). "Concanavalin A binding to rod outer segment membranes-usefulness for preparation of intact disks."
690. Clark, S. P., and Molday, R. S. *Biochemistry* 18: 5868–5873 (1979). "Orientation of membrane glycoproteins in sealed rod outer segment disks."
691. Hargrave, P. A.; Fong, S.-L.; McDowell, J. H.; Mas, M. T.; Curtis, D. R.; Wang, J. K.; Juszczak, E.; and Smith, D. P. *Neurochem. Int'l.* 1: 231–244 (1980). "The partial primary structure of bovine rhodopsin and its topography in the retinal rod disk membrane."
692. Findlay, J.B.C.; Barclay, P. L.; Brett, M.; Davison, M.; Pappin, D.J.C.; and Thomson, P. *Vision Res.* 24: 1501–1508 (1984). "The structure of mammalian rod opsins."
693. O'Tousa, J. E.; Baehr, W.; Martin, R. L.; Hirsh, J.; Pak, W. L.; and Applebury, M. L. *Cell* 40: 839–850 (1985). "The *Drosophila* ninaE gene encodes an opsin."
694. Zuker, C. S.; Montell, C.; Jones, K.; Laverty, T.; and Rubin, G. M. *J. Neurosci.* 7: 1550–1557 (1987). "A rhodopsin gene expressed in photoreceptor cell R7 of the *Drosophila* eye: Homologies with other signal-transducing molecules."
695. Montell, C.; Jones, K.; Zuker, C. S.; and Rubin, G. M. *J. Neurosci.* 7: 1558–1566 (1987). "A second opsin gene expressed in the ultraviolet-sensitive R7 photoreceptor cells of *Drosophila melanogaster*."
696. Zuker, C. S.; Cowman, A. F.; and Rubin, G. M. *Cell* 40: 851–858 (1985). "Isolation and structure of a rhodopsin gene from *Drosophila melanogaster*."
697. Cowman, A. F.; Zuker, C. S.; and Rubin, G. M. *Cell* 44: 705–710 (1986). "An opsin gene expressed in only one photoreceptor cell type of the *Drosophila* eye."
698. Findlay, J.B.C.; Brett, M.; and Pappin, D.J.C. *Nature* 293: 314–316 (1981). "Primary structure of C-terminal functional sites in ovine rhodopsin."
699. Henderson, R., and Unwin, P.N.T. *Nature* 257: 28–32 (1975). "Three-dimensional model of purple membrane obtained by electron microscopy."
700. Unwin, P.N.T., and Henderson, R. *J. Mol. Biol.* 94: 425–440 (1975). "Molecular structure determination by electron microscopy of unstained crystalline specimens.
701. Seiff, F.; Wallat, I.; Westerhausen, J.; and Heyn, M. P. *Biophys. J.* 50: 629–635 (1986). "Location of chemically modified lysine 41 in the structure of bacteriorhodopsin by neutron diffraction."
702. Trewhalla, J.; Popot, J.-L.; Zaccaï, G.; and Engelman, D. M. *EMBO J.* 5: 3045–3050 (1986). "Localization of two chymotryptic fragments in the structure of renatured bacteriorhodopsin by neutron diffraction."
703. Heyn, M. P.; Westerhausen, J.; Wallat, I.; and Seiff, F. *Proc. Nat'l. Acad. Sci. USA* 85: 12146–2150 (1988). "High-sensitivity neutron diffraction of membranes: Location of the Schiff base end of the chromophore of bacteriorhodopsin."

704. Lanyi, J. K.; Zimanyi, L.; Nakanishi, K.; Derguini, K.; Okabe, K.; and Honig, B. *Biophys. J.* 53: 185–191 (1988). ''Chromophore/protein and chromophore/anion interactions in halorhodopsin.''

705. Kosower, E. M. *Biochem. Biophys. Res. Commun.* 111: 1022–1026 (1983). ''Partial tertiary structure assignment for the acetylcholine receptor on the basis of the hydrophobicity of amino acid sequences and channel location using single group rotation theory.''

706. Gilson, M. K.; Rashin, M.; Fine, R.; and Honig, B. *J. Mol. Biol.* 183: 503–516 (1985). ''On the calculation of electrostatic interactions in proteins.''

707. Matthew, J. B. *Ann. Rev. Biophys. Biophys. Chem.* 14: 387–417 (1985). ''Electrostatic effects in proteins.''

708. Warshel, A.; Russell, S. T.; and Churg, A. K. *Proc. Nat'l. Acad. Sci. USA* 81: 4785–4789 (1984). ''Macroscopic models for studies of electrostatic interactions in proteins: Limitations and applicability.''

709. Warshel, A.; Russell, S. T. *Quart. Rev. Biophysics* 17: 283–422 (1984). ''Calculations of electrostatic interactions in biological systems and in solutions.''

710. Honig, B. H.; Hubbell, W. L.; Flewelling, R. F. *Ann. Rev. Biophys. Biophys. Chem.* 15: 163–193 (1986). ''Electrostatic interactions in membranes and proteins.''

711. Baasov, T., and Sheves, M. *J. Am. Chem. Soc.* 107: 7524–7533 (1985). ''Model compounds for the study of spectroscopic properties of visual pigments and bacteriorhodopsin.''

712. Harbison, G. S.; Smith, S. O.; Pardoen, J. A.; Courtin, J.M.L.; Lugtenburg, J.; Herzfeld, J.; Mathies, R. A.; Griffin, R. G. *Biochemistry* 24: 6955–6962 (1985). ''Solid state ^{13}C-NMR detection of a perturbed 6-s-*trans* chromophore in bacteriorhodopsin.''

713. Lugtenburg, J.; Muradin-Szweykowska, M.; Harbison, G. S.; Smith, S. O.; Heeremans, C.; Pardoen, J. A.; Herzfeld, J.; Griffin, R. G.; Mathies, R. A. *Biochemistry* 25: 3104 (1986). ''Mechanism for the opsin shift of retinal's absorption in bacteriorhodopsin.''

714. Okabe, M.; Balogh-Nair, V.; and Nakanishi, K. *Biophys. J.* 45: 272a (1984). ''Proton translocation by synthetic analogs of bacteriorhodopsin. The role of polyene side chain length and point charge at the ring binding site.''

715. Rodman, H.; Honig, B.; Nakanishi, K.; Okabe, M.; Shimizu, N.; Spudich, J. L.; and McCain, D. A. *Biophys. J.* 49: 210a (1986). ''Retinal protein interactions in bacteriorhodopsin and bacterial sensory rhodopsin.''

716. Mollevanger, C.P.J.; Kentgens, A. P.; Pardoen, J. A.; Courtin, J.M.L.; Veeman, W. S.; Lugtenburg, J.; and de Grip, W. J. *Eur. J. Biochem.* 163: 9–14 (1987). ''High resolution solid-state ^{13}C-NMR study of carbons C-5 and C-12 of the chromophore of bovine rhodopsin. Evidence for a 6-s-*cis* conformation with negative charge perturbation near C-12.''

717. Smith, S. O.; Palings, I.; Copie, V.; Raleigh, D. P.; Courtin, J.; Pardoen, J. A.; Lugtenburg, J.; Mathies, R. A.; and Griffin, R. G. *Biochemistry* 26: 1606–1611 (1987). ''Low-temperature solid-state ^{13}C-NMR studies of the retinal chromophore in rhodopsin.''

718. Conrad, M. P., and Strauss, H. L. *Biophys. J.* 48: 117–124 (1985). ''The vibrational spectrum of water in liquid alkanes.''

719. Gray, T. M., and Matthews, B. W. *J. Mol. Biol.* 175: 75–81 (1984). ''Intrahelical

hydrogen bonding of serine, threonine and cysteine residues within α-helices and its relevance to membrane-bound proteins.''

720. Neitz, J., and Jacobs, G. H. *Nature* 323: 623–625 (1986). ''Polymorphism of the long-wavelength cone in normal human colour vision.''

721. Mollon, J. D. and *Nature* 323: 578–579 (1986). ''Perception: Questions of sex and colour.''

722. Harris, W. A.; Stark, W. S.; and Walker, J. A. *J. Physiol. (Lond.)* 256: 415–439 (1976). ''Genetic dissociation of the photoreceptor system in the compound eye of *Drosophila melanogaster*.''

723. Chabré, M., and Breton, J. *Photochem. Photobiol.* 30: 295 (1979). ''Orientation of aromatic residues in rhodopsin. Rotation of one tryptophan upon the meta I–meta II transition after illumination.''

724. Ovchinnikov, Yu. A.; Abdulaev, N. G.; Zolotarev, A. S.; Artamonov, I. D., Bespalov, I. A.; Dergachev, A. E.; and Tsuda, M. *FEBS Lett.* 232: 69–72 (1988). ''Octopus rhodopsin. Amino acid sequence deduced from cDNA.''

725. Ohtani, H.; Kobayashi, T.; Tsuda, M.; and Ebrey, T. G. *Biophys. J.* 53: 17–34 (1988). ''Primary processes in photolysis of octopus rhodopsin.''

726. Yoshizawa, T. *Adv. Biophys.* 17: 5–67 (1984). ''Photophysiological functions of visual pigments.''

727. Sheves, M.; Albeck, A.; Ottolenghi, M.; Bovee-Geurts, P.H.M.; De Grip, W. J.; Einterz, C. M.; Lewis, J. W.; Schaechter, L. E.; and Kliger, D. S. *J. Am. Chem. Soc.* 108: 6440–6441 (1986). ''An artificial visual pigment with restricted C_{9-11} motion forms normal photolysis intermediates.''

728. Liu, R.S.H., and Asato, A. E. *Proc. Nat'l. Acad. Sci. USA* 82: 259–263 (1985). ''The primary process of vision and the structure of bathorhodopsin: A mechanism for photoisomerization of polyenes.''

729. Liu, R.S.H.; Matsumoto, H.; Asato, A. E.; and Mead, D. *J. Am. Chem. Soc.* 108: 3796–3799 (1986). ''A mechanistic model study of processes in the vertebrate and invertebrate visual cycles. Bioorganic studies of visual pigments. 4.''

730. Warshel, A. *Nature* 260: 679–683 (1976). ''Bicycle-pedal model for the first step in the vision process.''

731. Warshel, A., and Barboy, N. *J. Am. Chem. Soc.* 104: 1469–1476 (1982). ''Energy storage and reaction pathways in the first step of the vision process.''

732. Palings, I.; Pardoen, J. A.; van den Berg, E.; Winkel, C.; Lugtenburg, J.; and Mathies, R.A. *Biochemistry* 26: 2544–2556 (1987). ''Assignment of finger print vibrations in the resonance Raman spectra of rhodopsin, isorhodopsin and bathorhodopsin: Implications for chromophore structure and environment.''

733. Asato, A. E.; Denny, M.; and Liu, R.S.H. *J. Am. Chem. Soc.* 108: 5032–5033 (1986). ''Retinal and rhodopsin analogues directed toward a better understanding of the H.T.-n model of the primary process of vision.''

734. Kobilka, B. K.; Dixon, R.A.F.; Frielle, T.; Dohlman, H. G.; Bolanowski, M. A.; Sigal, I. S.; Yang-Feng, T. L.; Francke, U.; Caron, M. G.; and Lefkowitz, R. J. *Proc. Nat'l. Acad. Sci. USA* 84; 46–50 (1987). ''cDNA for the human ϑ_2-adrenergic receptor: A protein with multiple membrane-spanning domains and encoded by a gene whose chromosomal location is shared with that of the receptor for platelet-derived growth factor.''

735. Maelicke, A., ed. *Nicotinic Acetylcholine Receptor: Structure and Function*, NATO ASI Series, Vol. H3. Berlin: Springer, 1986.

736. Kosower, E. M. "A structural and dynamic model for the nicotinic acetylcholine receptor." Pages 465–483 in ref. 735.

737. Kosower, E. M. *Eur. J. Biochem.* 168: 431–449 (1987). "A structural and dynamic model for the nicotinic acetylcholine receptor."

738. Longstaff, C.; Calhoon, R. D.; and Rando, R. R. *Biochemistry* 25: 6311–6319 (1986). "Chemical modification of rhodopsin and its effect on regeneration and G-protein activation."

739. Cobbs, W. H.; Barkdoll, A. E., III; and Pugh, E. N., Jr. *Nature* 317: 64–67 (1985). "Cyclic GMP increases photocurrent and light sensitivity of retinal cones."

740. Haynes, L., and Yau, K.-W. *Nature* 317, 61–64 (1985). "Cyclic GMP-sensitive conductance in outer segment membrane of catfish cones."

741. Grunwald, G. B.; Gierschik, P.; Nirenberg, M.; and Spiegel, A. *Science* 231: 856–859 (1986). "Detection of α-transducin in retinal rods but not cones."

742. Lerea, C. L.; Somers, D. E.; Hurley, J. B.; Klock, I. B.; and Bunt-Milam, A. H. *Science* 234: 77–80 (1986). "Identification of specific transducin α-subunits in retinal rod and cone photoreceptors."

743. Ho, Y.-K., and Fung, B.K.-K. *J. Biol. Chem.* 259: 6694–6699 (1984). "Characterization of transducin from bovine retinal outer segments: Role of sulfhydryl groups."

744. Hildebrandt, J. D.; Codina, J.; Rosenthal, W.; Brinbaumer, L.; Neer, E. J.; Yamazaki, A.; and Bitensky, M. W. *J. Biol. Chem.* 260: 14867–14872 (1985). "Characterization by 2-dimensional peptide mapping of the γ-subunits of N_s and N_i, the regulatory proteins of adenylate cyclase and of transducin, the guanine nucleotide binding protein of rod outer segments of the eye."

745. Yamanaka, G.; Eckstein, F.; and Stryer, L. *Biochemistry* 25: 6149–6153 (1986). "Interaction of retinal transducin with guanosine triphosphate analogues: Specificity of the γ-phosphate binding region."

746. Kühn, H.; Hall, S. W.; and Wilden, U. *FEBS Lett.* 176: 473–478 (1984). "Light-induced binding of 48-kDa protein to photoreceptor membranes is highly enhanced by phosphorylation of rhodopsin."

747. Sitaramayya, A. *Biochemistry* 25: 5460–5468 (1986). "Rhodopsin kinase prepared from bovine rod disk membranes quenches light activation of cGMP phosphodiesterase in a reconstituted system."

748. Wilden, U.; Hall, S. W.; and Kühn, H. *Proc. Nat'l. Acad. Sci. USA* 83: 1174–1178 (1986). "Phosphodiesterase activation by photoexcited rhodopsin is quenched when rhodopsin is phosphorylated and binds the intrinsic 48-kDa protein of rod outer segments."

749. Hamm, H. E.; Deretic, D.; Arendt, A.; Hargrave, P. A.; Koenig, B.; and Hofmann, K. P. *Science* 241: 832–835 (1988). "Site of G-protein binding to rhodopsin mapped with synthetic peptides from the α subunit."

750. Martin, R. L.; Wood, C.; Baehr, W.; Applebury, M. L. *Science* 232: 1266–1269 (1986). "Visual pigment homologies revealed by DNA hybridization."

751. In ref. 485, Secs. 30380, 30390, pp. 1001–1003.

752. Rushton, W.A.H. *Sci. Amer.* 232(3): 64–74 (1975). "Visual pigments and color blindness."

753. Jacobs, G. H., and Neitz, J. *Proc. Nat'l. Acad. Sci. USA* 84: 2545–2549 (1987). "Inheritance of color vision in a New World monkey (*Saimiri sciureus*)."

754. Bolton, J. R., and Hall, D. O. *Ann. Rev. Energy* 4: 353–401, Fig. 1, p. 355 (1979). "Photochemical conversion and storage of solar energy."

755. Foster, K. W.; Saranak, J.; Derguini, F.; Rao, V. J.; Zarilli, G. R.; Okabe, M.; Fang, J.-M.; Shimizu, N.; and Nakanishi, K. *J. Am. Chem. Soc.* 110: 6588–6589 (1988). "Rhodopsin activation: A novel view suggested by in vivo *Chlamydomonas* experiments."

756. Jaffé, H. H., and Orchin, M. *Theory and Applications of Ultraviolet Spectroscopy*. New York: J. Wiley, 1962. 624 pp.

757. Das, P. K., and Becker, R. S. *J. Phys. Chem.* 82: 2081–2093 (1978). "Spectroscopy of polyenes. 1. Comprehensive investigation of absorption spectra of polyenals and polyenones related to visual chromophores."

758. Das, P. K., and Becker, R. S. *J. Phys. Chem.* 82: 2093–2105 (1978). "Spectroscopy of polyenes. 2. Comprehensive investigation of emission spectral properties of polyenals and polyenones related to visual chromophores."

759. Nakanishi, K.; Balogh-Nair, V.; Arnaboldi, M.; Tsujimoto, K.; and Honig, B. *J. Am. Chem. Soc.* 102: 7945–7947 (1980). "An external point-charge model for bacteriorhodopsin to account for its purple color."

760. Rowan, R., III; Warshel, A.; Sykes, B. D.; and Karplus, M. *Biochemistry* 13: 970–981 (1974). "Conformation of retinal isomers."

761. Birge, R. R.; Berge, C. T.; Noble, L. L.; and Neuman, R. C., Jr. *J. Am. Chem. Soc.* 101: 5162–5170 (1979). "Effect of external pressure on the spectroscopic and conformational properties of the visual chromophores."

762. Callender, R. H.; Doukas, A.; Crouch, R.; and Nakanishi, K. *Biochemistry* 15: 1621–1629 (1976). "Molecular flow resonance Raman effect from retinal and rhodopsin."

763. Cookingham, R., and Lewis, A. *J. Mol. Biol.* 119: 569–577 (1978). "S-*trans* in 11-*cis*-rhodopsin by resonance Raman spectroscopy."

764. Cookingham, R. E.; Lewis, A.; and Lemley, A. T. *Biochemistry* 17: 4699–4711 (1978). "A vibrational analysis of rhodopsin and bacteriorhodopsin chromophore analogues: Resonance Raman and infrared spectroscopy of chemically modified retinals and Schiff bases."

765. Gilardi, R. D.; Karle, I. L.; and Karle, J. *Acta Cryst.* B28: 2605–11 (1972). "The crystal and molecular structure of 11-*cis*-retinal."

766. Irving, C. S.; Byers, G. W.; and Leermakers, P. A. *Biochemistry* 9: 858–864 (1970). "Spectroscopic model for the visual pigments: Influence of microenvironmental polarizability."

767. Irving, C. S.; Byers, G. W.; and Leermakers, P. A. *J. Am. Chem. Soc.* 91: 2141–2143 (1969). "Effect of solvent polarizability on the absorption spectrum of all-*trans*-retinylpyrrolidiminium perchlorate."

768. Myers, A. B., and Birge, R. R. *J. Am. Chem. Soc.* 103: 1881–1885 (1981). "The ground-state dipole moments of all-*trans*- and 9-*cis*-retinal."

769. Ponder, M., and Mathies, R. *J. Phys. Chem.* 87: 5090–5098 (1983). "Excited state polarizabilities and dipole moments of diphenylpolyenes and retinal."

770. Grunwald, E., and Winstein, S. *J. Am. Chem. Soc.* 70: 846–854 (1948). "The correlation of solvolysis rates."

771. Kosower, E. M. *J. Am. Chem. Soc.* 80: 3253–3261 (1958). "The effect of solvent on spectra I. A new empirical measure of solvent polarity: Z-values."

772. Kosower, E. M. *An Introduction to Physical Organic Chemistry*. New York: John Wiley, 1968. 503 pp.

773. Reichardt, C.; and Dimroth, K. *Fortschr. Organ. Forsch.* 11: 1–73 (1968). ''Solvents and empirical parameters for characterization of their polarity.''

774. Reichardt, C. *Solvents and Solvent Effects in Organic Chemistry*. 2d ed. Weinheim: VCH, 1988. 534 pp.

775. Kosower, E. M. *Accts. Chem. Res.* 15: 259–266 (1982). ''Photophysics of phenylaminonaphthalenesulfonates: A paradigm for excited state intramolecular electron transfer.''

776. Kosower, E. M., and Huppert, D. *Ann. Rev. Phys. Chem.* 37: 127–156 (1986). ''Excited state electron and proton transfers.''

777. Kropf, A., and Hubbard, R. *Ann. N.Y. Acad. Sci.* 74: 266–280 (1958). ''The mechanism of bleaching rhodopsin.''

778. Kosower, E. M., and Remy, D. C. *Tetrahedron* 5: 281–289 (1959). ''The effect of a positive charge on an $\pi \rightarrow \pi^*$ transition.''

779. Tabushi, I., and Shimokawa, K. *J. Am. Chem. Soc.* 102: 5400–5402 (1980). ''Model approach to retinal pigments. Remarkable red shift due to proximal ammonium ion.''

780. Kropf, A.; Whittenberger, A.; Gott, S.; and Waggoner, Y. *Exp. Eye Res.* 17: 591 (1973). ''Circular dichroism of 5, 6-dihydrorhodopsin.''

781. Azuma, M.; Azuma, K.; Kito, Y., *Biochim. Biophys. Acta* 295, 520–527 (1973). ''Circular dichroism of visual pigment analogues containing 3-dehydroretinal and 5, 6-epoxy-3-dehydroretinal as the chromophore.''

782. Blatchly, R. A.; Carriker, J. D.; Balogh-Nair, V.; and Nakanishi, K. *J. Am. Chem. Soc.* 102: 2495–2497 (1980). ''Adamantyl allenic rhodopsin. Leniency of the ring binding site in bovine opsin.''

783. Crouch, R. K. *J. Am. Chem. Soc.* 104: 4946–4948 (1982). ''A synthetic isorhodopsin formed with a retinal derivative lacking an inert ring.''

784. Wald, G. *Exp. Eye Res.* 18: 333–343 (1974). ''Visual pigments and photoreceptors—Review and outlook.''

785. Chen, J. G.; Nakamura, T.; Ebrey, T. G.; Odashima, K.; Koono, K.; Derguini, F.; Nakanishi, K.; and Honig, B. *Biophys. J.* 51: 267a (1987). ''Wavelength regulation in a cone pigment, chicken iodopsin.''

786. Barry, B.; Mathies, R. A.; Pardoen, J. A.; and Lugtenburg, J. *Biophys. J.* 52: 603–610 (1987). ''Raman microscope and quantum yield studies on the primary photochemistry of A_2-visual pigments.''

787. Hardie, R. C. *Trends in Neurosci.* 9: 419–423 (1986). ''The photoreceptor array of the dipteran retina.''

788. Kirschfeld, K.; Franeschini, N.; and Minke, B. *Nature* 269: 386–390 (1977). ''Evidence for a sensitising pigment in fly receptors.''

789. Rath, P.; Pande, C.; Deng, F. H.; Callender, R.; and Schwemer, J. *Biophys. J.* 53: 385a (1988). ''The nature of chromophore protein linkage in the u.v. absorbing visual pigment of owl fly.''

790. Becker, R. S., and Freedman, K. *J. Am. Chem. Soc.* 107: 1477–1485 (1985). ''A comprehensive investigation of the mechanism and photophysics of isomerization of a protonated and unprotonated Schiff base of 11-*cis*-retinal.''

791. Becker, R. S.; Freedman, K.; Hutchinson, J. A.; and Noe, L. J. *J. Am. Chem.*

Soc. 107: 3942–3944 (1985). "Kinetic study of the photoisomerization of a protonated Schiff base of 11-*cis*-retinal over the picosecond-to-second time regimes."

792. Rotmans, J. P.; Bonting, S. L.; and Daemen, F.J.M. *Vision Res.* 12: 337–341 (1972). "On the chromophoric group of rhodopsin."

793. Cooper, A.; Dixon, S. F.; Nutley, M. A.; and Robb, J. L. *J. Am. Chem. Soc.* 109: 7254–7263 (1987). "Mechanism of retinal Schiff base formation and hydrolysis in relation to visual pigment hydrolysis and regeneration: Resonance Raman spectroscopy of a tetrahedral carbinolamine intermediate and oxygen-18 labeling of retinal at the metarhodopsin stage in photoreceptor membranes."

794. Saari, J. C., and Bredberg, L. *Biochim. Biophys. Acta* 716: 266-272 (1982). "Enzymatic reduction of 11-*cis*-retinal bound to cellular retinal-binding protein."

795. Bridges, C. D., and Alvarez, R. A. *Science* 236: 1678–1680 (1987). "The visual cycle operates via an isomerase acting on all-*trans*-retinol in the pigment epithelium."

796. Changeux, J-P. *Harvey Lectures* 75: 85–254 (1981). "The Acetylcholine Receptor: An 'Allosteric' Membrane Protein."

797. Langley, J. N. *J. Physiol.(Lond.)* 36: 347–384 (1907). "On the contraction of muscle, chiefly in relation to the presence of "receptive" substances."

798. Witkop, B. "Natural Products as Medicinal Agents." In *Natural Products, Receptors and Ligands*, ed. J. L. Beal, and E. Reinhard, pp. 151–183. Stuttgart: Hippokrates Verlag, 1980.

799. Loewi, O. *Pflugers Arch.* 189: 239–242 (1921). "Über humorale Übertragbarkeit der Herzenwirkung I. Mitteilung" ("On the Humoral Propagation of Cardiac Nerve Action").

800. Loewi, O., and Navratil, E. *Pflugers Arch.* 214: 678–688 (1926). "Über humorale Übertragbarkeit der Herznervenwirkung. X. Mitteilung. Über das Schicksal des Vagusstoffs." ("On the Humoral Propagation of Cardiac Nerve Action. Communication X. The Fate of the Vagus Substance.")

801. Dale, H. H.; Feldberg, W.; and Vogt, M. *J. Physiol. (Lond.)* 86: 353–380 (1936). "Release of Acetylcholine at Voluntary Motor Nerve Endings."

802. Baeyer, A. *Justus Liebig's Ann. Chem.* 142: 322–326 (1867). "I. Ueber das Neurin" (On p. 325 is the first synthesis of acetylcholine ["acetylneurine"]).

803. MacIntosh, F. C. "Cholinergic transmission: Variations on a theme." In *Progress in Cholinergic Biology: Model Cholinergic Synapses*, ed. I. Hanin, and A. M. Goldberg, pp. 1–22. New York: Raven Press, 1982.

804. Filbin, M. T.; Lunt, G. G.; and Donellan, J. F. *Eur. J. Biochem.* 132: 151–156 (1983). "Partial purification and characterization of an acetylcholine receptor with nicotinic properties from the supraoesophageal ganglion of the locust (*Schistocerca gregaria*).

805. Triggle, D. J., and Triggle, C. R. *Chemical Pharmacology of the Synapse.* (a) Table 1.1, p. 4 (b) p. 309 (c) pp. 362–363. New York: Academic Press, 1976. 654 pp.

806. Dale, H. H. *J. Pharmac. exper. Therap.* 6: 147–190 (1914). "The action of certain esters and ethers of choline, and their relation to muscarine."

807. Birdsall, N.J.M.; Hulme, E. C.; and Burgen, A.S.V. *Proc. Royal Soc. Lond. (Biol.)* 207: 1–12 (1980). "The character of muscarinic receptors in the rat brain."

808. Popot, J.-L., and Changeux, J.-P. *Physiol. Revs.* 64: 1162–1239 (1984). "The

nicotinic receptor of acetylcholine: Structure of an oligomeric integral membrane protein.''

809. Cartaud, J.; Benedetti, E. L.; Sobel, A.; and Changeux, J.-P. *J. Cell. Sci.* 29: 313–337 (1978). ''A morphological study of the cholinergic receptor protein from *Torpedo marmorata* in its membrane environment and in its detergent-extracted form.''
810. Neubig, R. R.; Krodel, E. K.; Boyd, N. D.; and Cohen, J. B. *Proc. Nat'l. Acad. Sci. USA* 76: 690–694 (1979). ''Acetylcholine and local anesthetic binding to *Torpedo* nicotinic postsynaptic membranes after removal of non-receptor peptides.''
811. Dolly, J. O.; and Barnard, E. A. *Biochem. Pharmacol.* 33: 841–858 (1984). ''Nicotinic Acetylcholine Receptors: An Overview.''
812. Gonzalez-Ros, J. M.; Paraschos, A.; Favach, M. C.; and Martinez-Carrion, M. *Biochim. Biophys. Acta* 643: 407–420 (1981). ''Characterization of acetylcholine receptor isolated from *Torpedo californica* electroplax through the use of an easily removable detergent, octyl β-D-glucopyranoside.''
813. Heidmann, T., and Changeux, J.-P. *Ann. Rev. Biochem.* 47: 317–357 (1978). ''Structure and functional properties of the acetylcholine receptor protein in its purified and membrane states.''
814. Gershoni, J. M.; Hawrot, E.; and Lentz, T. L. *Proc. Nat'l. Acad. Sci. USA* 80: 4973–4977 (1983). ''Binding of α-bungarotoxin to isolated subunit of the acetylcholine receptor of *Torpedo californica*: Quantitative analysis with protein blots.''
815. Dolly, J. O. *Int'l. Rev. Biochem.* 26: 257–309 (1979). ''Biochemistry of acetylcholine receptors from skeletal muscle.''
816. Momoi, M. Y., and Lennon, V. A. *J. Biol. Chem.* 257: 12757 (1982). ''Purification and biochemical characterization of nicotinic acetylcholine receptors of human muscle.''
817. Ross, M. J.; Klymkowsky, M. W.; Agard, D. A.; Stroud, R. M. *J. Mol. Biol.* 116: 635–659 (1977). ''Structural studies of a membrane-bound acetylcholine receptor from *Torpedo californica*.''
818. Klymkowsky, M. W., and Stroud, R. M. *J. Mol. Biol.* 128: 319–334 (1979). ''Immunospecific identification and three-dimensional structure of a membrane-bound acetylcholine receptor from *Torpedo californica*.''
819. Cartaud, J.; Benedetti, E. L.; Cohen, J. B.; Meunier, J-C.; Changeux, J.-P. *FEBS Lett.* 33: 109–113 (1973). ''Presence of a lattice structure in membrane fragments rich in nicotinic receptor protein from the electric organ of *Torpedo marmorata*.''
820. Zingsheim, H. P.; Neugebauer, D.-Ch.; Barrantes, F. J.; Frank, J. *Proc. Nat'l. Acad. Sci. USA* 77: 952–956 (1980). ''Structural details of membrane-bound acetylcholine receptor from *Torpedo marmorata*.''
821. Kistler, J.; Stroud, R. M.; Klymkowsky, M. W.; LaLancette, R.; and Fairclough, R. H. *Biophys. J.* 37: 371–383 (1982). ''Structure and function of an acetylcholine receptor.''
822. Stroud, R. M. Proc. Second SUNYA Conversation in the *Discipline of Biomolecular Stereodynamics*, vol. 2, pp. 55–73. New York: Academic Press, 1981. ''Structure of an acetylcholine receptor, a hypothesis for a dynamic mechanism of its action.''
823. Zingsheim, H. P.; Neugebauer, D.-C.; Frank, J.; Hänicke, W.; and Barrantes, F. J. *EMBO J.* 1: 541–547 (1982). ''Dimeric arrangement and structure of the membrane bound acetylcholine receptor studied by electron microscopy.''

824. Zingsheim, H. P.; Barrantes, F. J.; Frank, J.; Hänicke, W.; and Neugebauer, D.-C. *Nature* 299: 81–84 (1982). "Direct structural localization of two toxin recognition sites on an ACh receptor protein."
825. Barrantes, F. J. *Int'l. Rev. Neurobiol.* 24: 259–341 (1983). "Recent developments in the structure and function of the acetylcholine receptor."
826. Brisson, A., and Unwin, P.N.T. *Nature* 315: 474–477 (1985). "Quaternary structure of the acetylcholine receptor." Toyoshima, C., and Unwin, N. *Nature* 336: 247–250 (1988). "Ion channel of acetylcholine receptor reconstructed from images of postsynaptic membranes." Unwin, N.; Toyoshima, C.; and Kubalek, E. *J. Cell. Biol.* 107: 1123–1138 (1988). "Arrangement of the acetylcholine receptor subunits in the resting and desensitized states, determined by cryoelectron microscopy of crystallized *Torpedo* postsynaptic membranes."
827. Hucho, F. *Eur. J. Biochem.* 158: 211–226 (1986). "The nicotinic acetylcholine receptor and its ion channel."
828. Giersig, M.; Kunath, W.; Sack-Kongehl, H.; and Hucho, F. "Ultrastructural analysis of the native acetylcholine receptor." Pages 7–17 in ref. 735. Hertling-Jaweed, S.; Bandini, G.; Müller-Fahrnow, A.; Dommes, V.; and Hucho, F. *FEBS Lett.* 241: 29–32 (1988). "Rapid preparation of the nicotinic acetylcholine receptor for crystallization in detergent solution."
829. Kunath, W.; Sack-Kongehl, H.; Giersig, M.; and Hucho, F. In *Receptors and Ion Channels*, pp. 13–21. "Ultrastructural analysis of the native acetylcholine receptor."
830. Neubig, R., and Cohen, J. B. *Biochemistry* 18: 5464–5475 (1979). "Equilibrium binding of [^{3}H]tubocurarine and [^{3}H]acetylcholine by *Torpedo* postsynaptic membranes: Stoichiometry and ligand interactions."
831. Changeux, J.-P.; Devillers-Thiery, A.; and Chemouilli, P. *Science* 225: 1335–1345 (1984). "Acetylcholine receptor, an allosteric protein."
832. Burden, S. J.; DePalma, R. L.; and Gottesman, G. S. *Cell* 35: 687–692 (1983). "Crosslinking of proteins in acetylcholine receptor-rich membranes: association between the β-subunit and the 43kD subsynaptic protein."
833. Gordon, A. S.; Milfay, D.; and Diamond, I. *Neurosci. Abstr.* 13: 159 (1983). "Identification of an M_r 43,000 protein kinase in acetylcholine receptor-enriched membranes."
834. Davis, C. G.; Gordon, A. S.; and Diamond, I. *Proc. Nat'l. Acad. Sci. USA* 79: 3666–3670 (1982). "Specificity and localization of the acetylcholine receptor kinase."
835. Gysin, R.; Yost, B.; Flanagan, S. D. *Biochemistry* 22: 5781–5789 (1983). "Immunochemical and molecular differentiation of 43,000 molecular weight proteins associated with *Torpedo* neuroelectrocyte synapses."
836. Barrantes, F. J.; Mieskes, G.; and Wallimann, T. *Proc. Nat'l. Acad. Sci. USA* 80: 5440–5444 (1983). "Creatine kinase activity in the Torpedo electrocyte and in the nonreceptor peripheral ν-proteins from acetylcholine receptor-rich membranes."
837. Giraudat, J.; Devillers-Thierry, A.; Perriard, J.-C.; and Changeux, J.-P. *Proc. Nat'l. Acad. Sci. USA* 81: 7313–7317 (1984). "Complete nucleotide sequence of *Torpedo marmorata* mRNA coding for the 43,000-dalton ν2 protein: Muscle-specific creatine kinase."
838. Hamilton, S. I.; McLaughlin, M.; and Karlin, A. *Biochemistry* 18: 155–163

(1979) . ''Formation of disulfide-linked oligomers of acetylcholine receptor in membrane from *Torpedo* electric tissue.''
839. Wise, D. S.; Wall, J.; and Karlin, A. *J. Biol. Chem.* 256: 12624–12627 (1981). ''Relative locations of the β- and δ-chains of the acetylcholine receptor determined by electron microscopy of isolated receptor trimer.''
840. Wise, D. S.; Schoenborn, B. P.; and Karlin, A. *J. Biol. Chem.* 256: 4124–4126 (1981). ''Structure of acetylcholine receptor dimer determined by neutron scattering and electron microscopy.''
841. Bon, F.; LeBrun, E.; Gomel, J.; Van Rapenbusch, R.; Cartaud, J.; Popot, J.-L.; and Changeux, J.-P. *J. Mol. Biol.* 176: 205–237 (1984). ''Image analysis of the heavy form of the acetylcholine receptor from *Torpedo marmorata*.''
842. Cartaud, J.; Popot, J.-L.; and Changeux, J.-P. *FEBS Lett.* 121; 327-332 (1980). ''Light and heavy forms of the acetylcholine receptor from *Torpedo marmorata* electric organ: morphological identification using reconstituted vesicles.''
843. Young, S. H., and Poo, M.-M. *J. Neurosci.* 3: 225–231 (1983). ''Rapid lateral diffusion of extrajunctional acetylcholine receptors in the developing muscle membrane of *Xenopus* tadpole.'' Stolberg, J., and Fraser, S. E. *J. Cell Biol.* 107: 1397–1408 (1988). ''Acetylcholine receptors and concanavalin A-binding sites on cultured Xenopus muscle cells: Electrophoresis, diffusion and aggregation.''
844. Merlie, J. P.; Isenberg, K.; Carlin, B.; and Olson, E. N. *Trends in Pharmacol. Sci.* 5: 377–379 (1984). ''Regulation of the synthesis of acetylcholine receptors.''
845. Merlie, J.; Sebbane, R.; Tzartos, S.; and Lindstrom, J. *J. Biol. Chem.* 257: 2694–2701 (1982). ''Inhibition of glycosylation with tunicamycin blocks assembly of newly synthesized acetylcholine receptor subunits in muscle cells.''
846. Nestler, E. J., and Greengard, P. *Nature* 305: 583–588 (1983). ''Protein phosphorylation in the brain.''
847. Huganir, R.; Miles, K.; and Greengard, P. *Proc. Nat'l. Acad. Sci. USA* 81: 6968–6972 (1984). ''Phosphorylation of the nicotinic acetylcholine receptor by an endogenous tyrosine-specific protein kinase.''
848. Lindstrom, J.; Merlie, J.; and Yogeeswaran, G. *Biochemistry* 18: 4465–4470 (1979). ''Biochemical properties of acetylcholine receptor subunits from *Torpedo californica*.''
849. Fong, T. M., and McNamee, M. G. *Biochemistry* 26: 3871–3880 (1987). ''Stabilization of acetylcholine receptor secondary structure by cholesterol and negatively charged phospholipids in membranes.''
850. Middlemas, D. S., and Raftery, M. A. *Biochem. Biophys. Res. Commun.* 115: 1075–1082 (1983). ''Exposure of acetylcholine receptor to the lipid bilayer.''
851. Olson, E. N.; Glaser, L.; and Merlie, J. P. *J. Biol. Chem.* 259: 5364–5367 (1984). ''α and β-subunits of the nicotinic acetylcholine receptor contain covalently bound lipid.''
852. Strader, C. D., and Raftery, M. A. *Proc. Nat'l. Acad. Sci. USA* 77: 5807–5811 (1980). ''Topographic studies of *Torpedo* acetylcholine receptor subunits as a transmembrane complex.''
853. Conti-Tronconi, B. M.; Dunn, S.M.J.; and Raftery, M. A. *Biochemistry* 21: 893–899 (1982). ''Functional stability of *Torpedo* acetylcholine receptor. Effects of protease treatment.''
854. Raftery, M. A.; Hunkapiller, M. W.; Strader, C. D.; and Hood, L. E. *Science*

208: 1454–1457 (1980). "Acetylcholine receptor: Complex of homologous subunits."

855. Schofield, P. R.; Darlison, M. G.; Fujita, N.; Burt, D. R.; Stephenson, F. A.; Rodriguez, H.; Rhee, L. M.; Ramachandran, J.; Reale, V.; Glemcorse, T. A.; Seeburg, P. H.; and Barnard, E. A. *Nature* 328: 221–227 (1987). "Sequence and functional expression of the $GABA_A$ receptor shows a ligand-gated receptor superfamily."

856. Kosower, E. M. *FEBS Lett.* 231: 5–10 (1988). "A partial structure for the γ-aminobutyric acid ($GABA_A$) receptor is derived from the model for the nicotinic acetylcholine receptor. The anion-exchange protein of cell membranes is related to the $GABA_A$ receptor."

857. Noda, M.; Takahashi, H.; Tanabe, T.; Toyosato, M.; Furutani, Y.; Hirose, T.; Asai, M.; Inayama, S.; Miyata, T.; and Numa, S. *Nature* 299: 793–797 (1982). "Primary structure of α-subunit precursor of *Torpedo californica* acetylcholine receptor deduced from cDNA sequence."

858. Devillers-Thiery, A.; Giraudat, J.; Bentaboulet, M.; and Changeux, J.-P. *Proc. Nat'l. Acad. Sci. USA* 80: 2067–2071 (1983). "Complete mRNA coding sequence of the acetylcholine binding α-subunit from *Torpedo marmorata* acetylcholine receptor. A model for the transmembrane organization of the polypeptide chain."

859. Noda, M.; Takahashi, H.; Tanabe, T.; Toyosato, M.; Kikyotani, S.; Hirose, T.; Asai, M.; Takashima, H.; Inayama, S.; Miyata, T.; and Numa, S. *Nature* 301: 251–255 (1983). "Primary structures of β- and δ-subunit precursors of *Torpedo californica* acetylcholine receptor deduced from cDNA sequences."

860. Noda, M.; Takahashi, H.; Tanabe, T.; Toyosato, M.; Kikyotani, S.; Furutani, Y.; Hirose, T.; Takashima, H.; Inayama, S.; Miyata, T.; and Numa, S. *Nature* 302: 528–532 (1983). "Structural homology of *Torpedo californica* acetylcholine receptor subunits."

861. Claudio, T.; Ballivet, M.; Patrick, J.; and Heinemann, S. *Proc. Nat'l. Acad. Sci. USA* 80: 1111–1115 (1983). "Nucleotide and deduced amino acid sequences of *Torpedo californica* acetylcholine receptor γ-subunit."

862. Okayama, H., and Berg, P. *Mol. Cell. Biol.* 2: 161–170 (1982). "High-efficiency cloning of full-length cDNA."

863. Numa, S.; Noda, M.; Takahashi, H.; Tanabe, T.; Toyosato, M.; Furutani, Y.; and Kikyotani, S. *Cold Spring Harbor Symp. Quant. Biol.* 48: 57–69 (1983). "Molecular structure of the nicotinic acetylcholine receptor."

864. Noda, M.; Furutani, Y.; Takahashi, H.; Toyosato, M.; Tanabe, T.; Shimuzu, S.; Kikyotani, S.; Kayano, T.; Hirose, T.; Inayama, S.; and Numa, S. *Nature* 305: 818–823 (1983). "Cloning and sequence analysis of calf cDNA and human genomic DNA encoding α-subunit precursor of muscle acetylcholine receptor."

865. Maxam, A. M., and Gilbert, W. *Methods in Enzymol.* 65: 499–560 (1980). "Sequencing end-labeled DNA with base-specific chemical cleavages."

866. de Duve, C. *Nature* 333: 117–118 (1988). "The second genetic code." Schulman, L. H., and Abelson, J. *Science* 240: 1591–1592 (1988). "Recent excitement in understanding transfer RNA identity."

867. Schulz, G. E., and Schirmer, R. H. *Principles of Protein Structure*, pp. 73–78. Heidelberg: Springer Verlag, 1979. 314 pp.

868. Cohen, F. E.; Abarbanel, R. M.; Kuntz, I. D.; and Fletterick, R. J. *Biochemistry*

22: 4894–4904 (1983). ''Secondary structure assignment for α/ϑ proteins by a combinatorial approach.''

869. Kyte, J., and Doolittle, R. F. *J. Mol. Biol.* 157: 105–132 (1982). ''A simple method for displaying the hydropathic character of a protein.''

870. Eisenberg, D.; Weiss, R. M.; and Terwilliger, T. C. *Nature* 299: 371–374 (1982). ''The helical hydrophobic moment: A measure of the amphiphilicity of a helix.''

871. Eisenberg, D.; Weiss, R. M.; Terwilliger, T. C.; and Wilcox, W. *Faraday Symp. Chem. Soc.* 17: 109–120 (1982). ''Hydrophobic moments and protein structure.''

872. Eisenberg, D. *Ann. Rev. Biochem.* 53: 595–624 (1984). ''Three-dimensional structure of membrane and surface proteins.''

873. Guy, H. R. *Biophys. J.* 45: 249–261 (1984). ''A Structural model of the acetylcholine receptor channel based on partition energy and helix packing calculations.''

874. Schiffer, M., and Edmundson, A. B. *Biophys. J.* 7: 121–135 (1967). ''Use of helical wheels to represent the structures of proteins and to identify segments with helical potential.''

875. Kao, P. N.; Dwork, A. J.; Kaldany, R.-R.; Silver, M. L.; Wideman, J.; Stein, S.; and Karlin, A. *J. Biol. Chem.* 259: 11662–11665 (1984). ''Identification of the α-subunit half-cystine specifically labeled by an affinity reagent for the acetylcholine binding site.''

876. Ringe, D.; Petsko, G. A.; Kerr, D. E.; and Ortiz de Montellano, P. R. *Biochemistry* 23: 2–4 (1984). ''Reaction of myoglobin with phenylhydrazine: A molecular doorstop.''

877. Jaenicke, R. *Prog. Biophys. Molec. Biol.* 49: 117–237 (1987). ''Folding and association of proteins.''

878. Richardson, J. S., and Richardson, D. C. *Science* 240: 1648–1652 (1988). ''Amino acid preferences for specific locations at the ends of α-helices.'' Presta, L. G., and Rose, G. D. *Science* 240: 1632–1641 (1988). ''Helix signals in proteins.''

879. Creighton, T. E. *Science* 240: 267 (1988). ''The protein folding problem.''

880. Williams, R.J.P. *Angew. Chem. Int'l. Edn.* 16: 766–777 (1977). ''Flexible drug molecules and dynamic receptors.''

881. Williams, R.J.P. *Chem. Soc. Revs.* 9: 281–324 (1980). ''On first looking into nature's chemistry. Part I. The role of small molecules and ions: The transport of the elements.''

882. Williams, R.J.P. *Chem.Soc. Revs.* 9: 325–364 (1980). ''On first looking into nature's chemistry. Part II. The role of large molecules, especially proteins.''

883. Gund, P.; Andose, J. D.; Rhodes, J. B.; and Smith, G. M. *Science* 208: 1425–1431 (1980). ''Three-dimensional molecular modeling and drug design.''

884. Dietrich, S. W.; Boger, M. B.; Kollmann, P. A.; and Jorgensen, E. C. *J. Med. Chem.* 20: 863–880 (1977). ''Thyroxine analogues 23. Quantitative structure-activity correlation studies of in vivo and in vitro thyromimetic activities.'' See also the similar approaches used for drug-receptor interactions in refs. 885–889.

885. Humber, L. G.; Bruderlein, F. T.; Philipp, A. H.; Götz, M.; and Voith, K. *J. Med. Chem.* 22: 761–767 (1979). ''Mapping the dopamine receptor 1. Features derived from modifications in ring E of the neuroleptic butaclamol.''

886. Humber, L. G.; Philipp, A. H.; Voith, K.; Pugsley, T.; Lippmann, W.; Ahmed, F. R.; and Przybylska, M. *J. Med. Chem.* 22: 899–901 (1979). ''(+)-Isobutaclamol: A crystallographic, pharmacological and biochemical study.''

887. Philipp, A. H.; Humber, L. G.; and Voith, K. *J. Med. Chem.* 22: 768–773 (1979). "Mapping the dopamine receptor 2. Features derived from modifications in the rings A/B Region of the neuroleptic butaclamol.

888. Randall, W. C.; Anderson, P. S.; Cresson, E. L.; Hunt, C. A.; Lyon, T. F.; Rittle, K. E.; Remy, D. C.; Springer, J. P.; Hirshfield, J.; Hoogsteen, K.; Williams, M.; Risley, E. A.; and Totaro, J. A. *J. Med. Chem.* 22: 1222–1230 (1979). "Synthesis, assignment of absolute configuration, and receptor binding studies relevant to the neuroleptic activities of a series of chiral 3-substituted cyproheptadine atropisomers."

889. Andrea, T. A.; Dietrich, S. W.; Murray, W. J.; Kollman, P.; Jorgensen, E. C.; and Rothenberg, S. *J. Med. Chem.* 22: 221–232 (1979). "A model for thyroid hormone-receptor interactions."

890. Ondetti, M. A.; Rubin, B.; and Cushman, D. W. *Science* 196: 441–444 (1977). "Design of specific inhibitors of angiotensin-converting enzyme: New class of orally active antihypertensive agents."

891. Ondetti, M. A., and Cushman, D. W. *J. Med. Chem.* 24: 355–361 (1981). "Inhibition of the renin-angiotensin system. A new approach to the therapy of hypertension."

892. Gund, P. *Ann. Repts. Med. Chem.* 14: 299–308 (1979). "Pharmacophoric pattern searching and receptor mapping."

893. Waters, J. A.; Spivak, C. E.; Hermsmeier, M.; Yadav, J. S.; Liang, R. F.; and Gund, T. M. *J. Med. Chem.* 31: 545–554 (1988). "Synthesis, pharmacology, and molecular modeling of semirigid, nicotinic agonists."

894. Behling, R. W.; Yamane, T.; Navon, G.; and Jelinski, L. W. *Proc. Nat'l. Acad. Sci. USA* 85: 6721–6725 (1988). "Conformation of acetylcholine bound to the nicotinic acetylcholine receptor."

895. Urry, D. W. *Proc. Nat'l. Acad. Sci. USA* 69: 1610–1614 (1972). "Molecular theory of ion-conducting channels. Field-dependent transition between conducting and non-conducting conformations."

896. Mackay, D.H.J.; Berens, P. H.; Wilson, K. R.; and Hagler, A. T. *Biophys. J.* 45: 229–248 (1984). "Structure and dynamics of ion transport through gramicidin A."

897. Hille, B. *Ionic channels of excitable membranes*. Sunderland, Mass.: Sinauer, 1984. 426 pp.

898. Ho, S.-P.; and DeGrado, W. F. *J. Am. Chem. Soc.* 109: 6751–6758 (1987). "Design of a 4-helix bundle protein: Synthesis of peptides which self-associate into a helical protein." Lear, J. D.; Wasserman, Z. R.; and DeGrado, W. F. *Science* 240: 1177–1181 (1988). "Synthetic amphiphilic peptide models for protein channels."

899. Salemme, F. R. *Science* 241: 145 (1988). "Structural polymorphism in transmembrane channels."

900. Langs, D. A. *Science* 241: 188–191 (1988). "Three-dimensional structure at 0.86Å of the uncomplexed form of the transmembrane ion channel peptide gramicidin A."

901. Wallace, B. A., and Ravikumar, K. *Science* 241: 182–187 (1988). "The gramicidin pore: Crystal structure of a cesium complex."

902. Kosower, E. M. *FEBS Lett.* 182: 234–242 (1985). "A structural and dynamic model for the sodium channel of the electric eel, *Electrophorus electricus*."

903. Kosower, E. M. "Single Group rotation." Abstracts, Int'l. Symposium, *Struc-*

ture and Dynamics of Nucleic Acids and Proteins, San Diego, Calif. Sept. 5–9, pp. 52–3, 1982.

904. Finer-Moore, J., and Stroud, R. M. *Proc. Nat'l. Acad. Sci. USA* 81: 155–159 (1984). "Amphipathic analysis and possible formation of an ion channel in an acetylcholine receptor."

905. Kosower, E. M., and Petsko, G. A. "Probing a model acetylcholine receptor site with the agonists, acetylcholine and anatoxin-a, and the antagonist, *d*-tubocurarine." unpublished results.

906. Takai, T.; Noda, M.; Furutani, Y.; Takahashi, H.; Notake, M.; Shimizu, S.; Kayano, T.; Tanabe, T.; Tanaka, K.; Hirose, T.; Inayama, S.; and Numa, S. *Eur. J. Biochem.* 143: 109–115 (1984). "Primary structure of ÿ subunit precursor of calf-muscle acetylcholine receptor deduced from the cDNA sequence."

907. Tanabe, T.; Noda, M.; Furutani, Y.; Takai, T.; Takahashi, H.; Tanaka, K.-i.; Hirose, T.; Inayama, S.; and Numa, S. *Eur. J. Biochem.* 144: 11–17 (1984). "Primary structure of beta-subunit precursor of calf muscle acetylcholine receptor deduced from cDNA sequence."

908. Boulter, J.; Luyten, W.; Evans, K.; Mason, P.; Ballivet, M.; Goldman, D.; Stengelin, S.; Martin, G.; Heinemann, S.; and Patrick, J. *J. Neurosci.* 5: 2545–2552 (1985). "Isolation of a clone coding for the α-subunit of a mouse acetylcholine receptor."

909. Nef, P.; Mauron, A.; Stalder, R.; Alliod, C.; and Ballivet, M. *Proc. Nat'l. Acad. Sci. USA* 81: 7975–7979 (1984). "Structure, linkage and sequence of the two genes encoding the δ and ÿ subunits of the nicotinic acetylcholine receptor."

910. Takai, T.; Noda, M.; Mishina, M.; Shimizu, S.; Furutani, Y.; Kayano, T.; Ikeda, T.; Kubo, T.; Takahashi, H.; Takahashi, T.; Kuno, M.; and Numa, S. *Nature* 315: 761–764 (1985). "Cloning, sequencing and expression of cDNA for a novel subunit of acetylcholine receptor from calf muscle."

911. Boulter, J.; Evans, K.; Martin, G.; Mason, P.; Stengelin, S.; Goldman, D.; Heinemann, S.; and Patrick, J. *J. Neurosci. Res.* 16: 37–49 (1986). "Isolation and sequence of cDNA clones coding for the precursor to the γ-subunit of mouse muscle nicotinic acetylcholine receptor."

912. LaPolla, R. J.; Mixter-Mayne, K.; Davidson, N. *Proc. Nat'l. Acad. Sci. USA* 81: 7970–7974 (1984). "Isolation and characterization of a cDNA clone for the complete protein coding region of the delta-subunit of the mouse acetylcholine receptor."

913. Heinemann, S.; Boulter, J.; Connolly, J.; Goldman, D.; Evans, K.; Treco, D.; Ballivet, M.; Patrick, J. "Molecular biology of muscle and neural acetylcholine receptors." Pages 359–387 in ref. 735.

914. Numa, S. *Biochem. Soc. Symp.* 52: 119–143 (1986). "Molecular basis for the function of ionic channels."

915. Boulter, J.; Evans, K.; Goldman, D.; Martin, G.; Treco, D.; Heinemann, S.; Patrick, J. *Nature* 319: 368–372 (1986). "Isolation of a cDNA clone coding for a possible neural nicotinic acetylcholine receptor α-subunit."

916. Maelicke, A. *Angew. Chem. Int'l. Edn.* 23: 195–221 (1984). "Biochemical aspects of cholinergic excitation."

917. Maelicke, A. "Structure and function of the nicotinic acetylcholine receptor." In *The Cholinergic Synapse*, ed. V. P. Whittaker, *Handbook Exp. Pharmacol.*, vol. 86, pp. 267–313. Berlin: Springer, 1988.

918. Conti-Tronconi, B. M., and Raftery, M. A. *Ann. Rev. Biochem.* 51: 491–530 (1982). "The nicotinic cholinergic receptor: Correlation of molecular structure with functional properties."

919. Karlin, A. "Molecular properties of nicotinic acetylcholine receptors." In *The Cell Surface and Neuronal Function*, ed. C. W. Cotman, G. Poste, and G. L. Nicolson, pp. 191–260. Amsterdam: Elsevier-North-Holland, 1980.

920. Hamilton, S. L.; Pratt, D. R.; and Eaton, D. C. *Biochemistry* 24: 2210–2219 (1985). "Arrangement of the subunits of the nicotinic acetylcholine receptor of *Torpedo californica* as determined by α-neurotoxin cross-linking." Kubalek, E.; Ralston, S.; Lindstrom, J.; and Unwin, N. *J. Cell. Biol.* 105: 9–18 (1987). "Location of subunits within the acetylcholine receptor by electron image analysis of tubular crystals from *Torpedo marmorata*."

921. Holtzman, E.; Wise, D.; Wall, J.; and Karlin, A. *Proc. Nat'l. Acad. Sci. USA* 79: 310–314 (1982) . "Electron microscopy of complexes of isolated acetylcholine receptor, biotinyl-toxin and avidin."

922. Karlin, A.; Cox, R.; Kaldany, R.-R.; Lobel, P.; Holtzman, E. *Cold Spring Harbor Symp. Quant. Biol.* 48: 1–8 (1983). "The arrangement and functions of the chains of the acetylcholine receptor of *Torpedo* electric tissue."

923. Karlin, A.; Holtzman, E.; Yodh, N.; Lobel, P.; Wall, J.; and Hainfield, J. *J. Biol. Chem.* 258: 6678–6681 (1983). "The arrangement of the subunits chains of the acetylcholine receptor of *Torpedo californica*."

924. Brisson, A. "Three-dimensional structure of the acetylcholine receptor." Pages 1–6 in ref. 735.

925. Chou, P. Y., and Fasman, G. D. *Ann. Rev. Biochem.* 47: 251 (1978). "Empirical predictions of protein conformation." Cf. Fasman, G.D. *Nature* 316: 22 (1985).

926. Eisenberg, D.; Weiss, R. M.; and Terwilliger, T. C. *Proc. Nat'l. Acad. Sci. USA* 8: 140–144 (1984). "The hydrophobic moment detects periodicity in protein hydrophobicity."

927. Guy, H. R. *Biophys. J.* 47: 61–70 (1985). "Amino acid side-chain partition energies and distribution of residues in soluble proteins."

928. Mohana Rao, J. K., and Argos, P. *Biochim. Biophys. Acta* 869: 197–214 (1986). "A conformational preference parameter to predict helices in integral membrane proteins."

929. Cohen, F. E.; Abarbarnel, R. M.; Kuntz, I. D.; and Fletterick, R. J. *Biochemistry* 25: 266–275 (1986). Turn prediction in proteins using a pattern-matching approach."

930. Chothia, C. *Ann. Rev. Biochem.* 53: 537–572 (1984) . "Principles that determine the structure of proteins."

931. Creighton, T. E. *J. Phys. Chem.* 89: 2452–2459 (1985). "The problem of how and why proteins adopt folded conformations."

932. Wallace, B., Cascio, M.; and Mielke, D. L. *Proc. Nat'l. Acad. Sci. USA* 83: 9423–9427 (1986). "Evaluation of methods for the prediction of membrane protein secondary structures."

933. Argos, P., and Mohana Rao, J. K. *Methods in Enzymol.* 130: 185–207 (1986). "Prediction of protein structure."

934. Thornton, J. M. *Nature* 335: 10–11 (1988). "The shape of things to come?"

935. Rooman, M. J., and Wodak, S. J. *Nature* 335: 4549 (1988). "Identification of predictive sequence motifs limited by protein structure data base size."

936. Kellis, J. T., Jr.; Nyberg, K.; Sali, D.; and Fersht, A. *Nature* 333: 784–786 (1988). "Contribution of hydrophobic interactions to protein stability."

937. Feldmann, R. J.; Brooks, B. R.; Lee, B. Div. Computer Research and Technology, Nat'l. Inst. Health, Bethesda, Md., USA , Oct. 1986. "Understanding protein architecture through simulated unfolding."

938. McCarthy, M. P.; Earnest, J. P.; Young, E. F.; Choe, S.; and Stroud, R. M. *Ann. Rev. Neurosci.* 9: 383–413 (1986). "The molecular biology of the acetylcholine receptor."

939. Bash, P. A.; Langridge, R.; and Stroud, R. M. *Biophys. J.* 47: 43a (1985). "Model for the ion channel in the acetylcholine receptor." Stroud, R. M., and Finer-Moore, J. *Ann. Rev. Cell Biol.* 1: 317–351 (1985). "Acetylcholine receptor structure, function and evolution."

940. Kosower, E. M. *FEBS Lett.* 155: 245–247 (1983). "Partial tertiary structure assignments for the beta-, gamma- and delta-subunits of the acetylcholine receptor on the basis of the hydrophobicity of amino acid sequences and channel location using single group rotation theory."

941. Kosower, E. M. *FEBS Lett.* 172: 1–5 (1984). "Revised assignments for the β-, γ- and δ-subunits of the acetylcholine structural model."

942. Mishina, M.; Tobimatsu, T.; Imoto, K.; Tanaka, K.-i.; Fujita, Y.; Fukuda, K.; Kurasaki, M.; Takahashi, H.; Morimoto, Y.; Hirose, T.; Inayama, S.; Takahashi, T.; Kuno, M.; and Numa, S. *Nature* 313: 364–369 (1984). "Location of functional regions of acetylcholine receptor α-subunit by site directed mutagenesis."

943. Wolosin, J. M.; Lyddiatt, A.; Dolly, J. O.; and Barnard, E. A. *Eur. J. Biochem.* 109: 495–505 (1980). "Stoichiometry of the ligand-binding sites in the acetylcholine receptor oligomer from muscle and from electric organ. Measurement by affinity alkylation with bromoacetylcholine."

944. Blanchard, S. G.; Dunn, S.M.J.; and Raftery, M.A. *Biochemistry* 21: 6258–6264 (1982). "Effects of reduction and alkylation on ligand binding and cation transport by *Torpedo californica* acetylcholine receptor."

945. Dunn, S.M.J., and Raftery, M. A. *Proc. Nat'l. Acad. Sci. USA* 79: 6757–6761 (1982). "Activation and desensitization of *Torpedo* acetylcholine receptor: Evidence for separate binding sites."

946. Kao, P. N., and Karlin, A. *Biophys. J.* 49: 5a (1986). Kao, P. N., and Karlin, A. *J. Biol. Chem.* 261: 8085–8088 (1986). "The acetylcholine receptor binding site contains a disulfide crosslink between adjacent half-cystinyl residues."

947. Thornton, J. M. *J. Mol. Biol.* 151: 261–287 (1981). "Disulphide bridges in globular proteins."

948. Ovchinnikov, Y. A. *Pure and Applied Chem.* 58: 725–736 (1986). "Bioorganic chemistry of rhodopsins."

949. Ovchinnikov, Y. A.; Lipkin, V. M.; Shuvaeva, T. M.; Bogachuk, A. P.; and Shemyakin, V. V. *FEBS Lett.* 179: 107–110 (1985). "Complete amino acid sequence of the γ-subunit of the GTP-binding protein from cattle retina."

950. Capasso, S.; Mazzarella, L.; and Tancredi, T. *Biopolymers* 18: 1555–1558 (1979). "Conformational analysis of the cyclic disulfide L-cysteinyl-L-cysteine."

951. Capasso, S.; Mattia, C.; Mazzarella, L.; and Puliti, R. *Acta Cryst.* B33: 2080–2083 (1977). "Structure of a cis-peptide unit: Molecular conformation of the cyclic disulfide, L-cysteinyl-L-cysteine."

952. Bodzansky, M., and Stahl, G. L. *Proc. Nat'l. Acad. Sci. USA* 71: 2791–2794 (1974). "The structure and synthesis of Malformin A."

953. Chang, E. L.; Yager, P.; Williams, R. W.; and Dalziel, A. W. *Biophys. J.* 41: 65a (1983). "The secondary structure of reconstituted acetylcholine receptor as determined by Raman spectroscopy."

954. Earnest, T., and Rothschild, K. *Biophys. J.* 49: 294a (1986). "Secondary structure orientation of bacteriorhodopsin and the nicotinic acetylcholine receptor."

955. Fong, T. M., and McNamee, M. G. *Biochemistry* 25: 830–840 (1986). "Correlation between acetylcholine receptor function and structural properties of membranes."

956. Chothia, C. *J. Mol. Biol.* 75: 295–302 (1973). "Conformation of twisted β-pleated sheets in proteins."

957. Mishina, M.; Kurosaki, T.; Tobimatsu, T.; Morimoto, Y.; Noda, M.; Yamamoto, T.; Terao, M.; Lindstrom, J.; Takahashi, T.; Kuno, M.; and Numa, S. *Nature* 307: 604–608 (1984). "Expression of functional acetylcholine receptor from cloned cDNAs."

958. Fairclough, R. H.; Finer-Moore, J.; Love, R. A.; Kristofferson, D.; Desmeules, P. J.; and Stroud, R. M. *Cold Spring Harbor Symp. Quant. Biol.* 48: 9–20 (1983). "Subunit organization and structure of an acetylcholine receptor."

959. Stroud, R. M. *Neurosci. Comment.* 1: 124–138 (1983). "Acetylcholine receptor structure."

960. Huganir, R. L.; Delcour, A. H.; Greengard, P.; and Hess, G. P. *Nature* 321: 774–777 (1986). "Phosphorylation of the nicotinic acetylcholine receptor regulates its rate of desensitization." Hopfield, J. F.; Tank, D. W.; Greengard, P. W.; and Huganir, R. L. *Nature* 336: 677–680 (1988). "Functional modification of the nicotinic acetylcholine receptor by tyrosine phosphorylation."

961. Huganir, R. L. "Regulation of the nicotinic acetylcholine receptor by protein phosphorylation." Pages 88–89 in ref. 55.

962. Hess, G. P. "New approaches to studies of neuronal receptors." Page 86 in ref. 55.

963. Udgaonkar, J. B., and Hess, G. P. *J. Membrane Biol.* 93: 93–109 (1986). "Acetylcholine receptor kinetics: Chemical kinetics."

964. Pasquale, E. B.; Udgaonkar, J. B.; and Hess, G. P. *J. Membrane Biol.* 93: 195–204 (1986). "Single-channel current recordings of acetylcholine receptors in electroplax isolated from the *Electrophorus electricus* Main and Sachs' electric organs."

965. Deshpande, S. S.; Aracava, Y.; Alkondon, M.; Daly, J. W.; and Albuquerque, E. X. *Neurosci. Abstr.* 16: 738 (1986). "Possible involvement of cAMP in desensitization of the nicotinic acetylcholine receptor (AChR)—a mechanism for autoregulatory function revealed by studies with forskolin."

966. Albuquerque, E. X.; Deshpande, S. S.; Aracava, Y.; Alkondon, M.; and Daly, J. W. *FEBS Lett.* 199: 113–120 (1986). "Possible involvement of cAMP in desensitization of the nicotinic acetylcholine receptor (AChR)—a mechanism for autoregulatory function revealed by studies with forskolin."

967. Seamon, K. B.; Padgett, W.; and Daly, J. W. *Proc. Nat'l. Acad. Sci. USA* 78: 3363–3367 (1981). "Forskolin: Unique diterpene activator of adenylate cyclase in membranes and in intact cells."

968. Middleton, P.; Jaramillo, F.; and Schuetze, S. M. *Proc. Nat'l. Acad. Sci. USA*

83: 4967–4971 (1986). "Forskolin increases the rate of acetylcholine receptor desensitization at rat *soleus* endplates."

969. Eusebi, F.; Molinaro, M.; and Zani, B. M. *J. Cell. Biol.* 100: 1339–1342 (1985). "Agents that activate protein kinase C reduce acetylcholine sensitivity in cultured myotubes."

970. Saitoh, T., and Changeux, J.-P. *Proc. Nat'l. Acad. Sci. USA* 78: 4430–4434 (1981). "Changes in state of phosphorylation of acetylcholine receptor during maturation of the electromotor synapse in *Torpedo marmorata* electric organ."

971. Ross, A. F.; Rapuano, M.; Schmidt, J. H.; and Prives, J. M. *J. Biol. Chem.* 262: 14640–14647 (1987). "Phosphorylation and assembly of nicotinic acetyl-choline receptor subunits in cultured chick muscle cells."

972. Kelly, P. T.; McGuiness, T. L.; and Greengard, P. *Proc. Nat'l. Acad. Sci. USA* 81: 945–949 (1984). "Evidence that the major postsynaptic density protein is a component of a Ca^{++}/calmodulin dependent protein kinase."

973. Bursztain, S., and Fischbach, G. D. *J. Cell. Biol.* 98: 498–506 (1984). "Evidence that coated vesicles transport acetylcholine receptors to the surface membrane of chick myotubes."

974. Kalcheim, C.; Vogel, Z.; and Duskin, D. *Proc. Nat'l. Acad. Sci. USA* 79: 3077–3081 (1982). "Embryonic brain extract induces collagen biosynthesis in cultured muscle cells: Involvement in acetylcholine receptor aggregation."

975. Nemethy, G., and Scheraga, H. A. *Biopolymers* 23: 2781–2799 (1984). "Role of proline-proline interactions in the packing of collagenlike poly(tripeptide) triple helices."

976. Sealock, R.; Wray, B. E.; and Froehner, S. C. *J. Cell. Biol.* 98: 2239–2244 (1984). "Ultrastructural localization of the Mr 43,000 protein and the acetylcholine receptor in *Torpedo* postsynaptic membranes using monoclonal antibodies."

977. Froehner, S. C. *J. Cell. Biol.* 99: 88–96 (1984). "Peripheral proteins of postsynaptic membranes from Torpedo electric organ identified with monoclonal antibodies."

978. Walker, J. H.; Boustead, C. M.; and Witzemann, V. *EMBO J.* 3: 2287–2290 (1984). "The 43-K protein, $\nu 1$, associated with acetylcholine receptor containing membrane fragments is an actin-binding protein."

979. LaRochelle, W. J., and Froehner, S. C. *J. Biol. Chem.* 261: 5270–5274 (1986). "Determination of tissue distributions and relative concentrations of the postsynaptic 43-kDa protein and the acetylcholine receptor in *Torpedo*."

980. Bloch, R. J., and Resneck, W. G. *J. Cell. Biol.* 99: 984–983 (1984). "Isolation of acetylcholine receptor clusters in substrate-associated material from cultured rat myotubes using saponin."

981. Froehner, S. C.; LaRochelle, W. J.; and Murnane, A. A. *J. Cell. Biol.* 103: 376a (1986). "Quantitative comparison of the postsynaptic 43K protein and acetylcholine receptor in muscle cells that differ in receptor organization."

982. Froehner, S. C.; LaRochelle, W. J.; and Murnane, A. A. "Association of the postsynaptic 43K protein with sites of clustered and diffusely-distributed acetylcholine receptors." Pages 281–289 in ref. 735.

983. Porter, S., and Froehner, S. C. *Biochemistry* 24: 425–432 (1985). "Interaction of the 43K protein with components of *Torpedo* postsynaptic membranes."

984. Froehner, S. C. *Trends in Neurosci.* 9: 37–41 (1986). "The role of the postsynaptic cytoskeleton in AChR organization."

985. Fallon, J. R.; Nitkin, R. M.; Reist, N. E.; Wallace, B.G.; and McMahan, U. J. *Nature* 315: 571–574 (1985). ''Acetylcholine receptor-aggregating factor is similar to molecules concentrated at neuromuscular junctions.''

986. Wallace, B. G.; Nitkin, R. M.; Reist, N. E.; Fallon, J. R.; Nicole-Moayeri, N.; and McMahan, U. J. *Nature* 315: 574–577 (1985). ''Aggregates of acetylcholinesterase induced by acetylcholine receptor-aggregating factor.''

987. Salpeter, M. M., and Loring, R. H. *Progr. in Neurobiol.* 25: 297–325 (1985). ''Nicotinic acetylcholine receptors in vertebrate muscle: Properties, distribution and neural control.''

988. Giraudat, J.; Dennis, M.; Heidmann, T.; Chang, J.-Y.; and Changeux, J.-P. *Proc. Nat'l. Acad. Sci. USA* 83: 2719–2723 (1986). ''Structure of the high affinity binding site for noncompetitive blockers of the acetylcholine receptor: Serine-262 of the δ subunit is labeled by [^{3}H]chlorpromazine.''

989. Giraudat, J.; Dennis, M.; Heidmann, T.; Haumont, P.-Y.; Lederer, F.; and Changeux, J.-P. *Biochemistry* 26: 2410–2418 (1987). ''Structure of the high affinity binding site for noncompetitive blockers of the acetylcholine receptor: [^{3}H]chlorpromazine labels homologous residues in the ϑ and δ chains.''

990. Oberthür, W.; Muhn, P.; Baumann, H.; Lottspeich, F.; Wittmann-Liebold, B.; and Hucho, F. *EMBO J.* 5: 1815–1819 (1986). ''The reaction site of a non-competitive antagonist in the δ-subunit of the nicotinic acetylcholine receptor.''

991. Hucho, F.; Oberthür, W.; and Lottspeich, F. *FEBS Lett.* 205: 137–142 (1986). ''The ion channel of the nicotinic acetylcholine receptor is formed by the homologous helices M II of the receptor subunits.''

992. Hucho, F. ''The helix-II model of the nicotinic acetylcholine receptor.'' Page 73 in ref. 55.

993. Hucho, F.; Oberthür, W.; Lottspeich, F.; Wittmann-Liebold, B. ''A structural model of the nicotinic acetylcholine receptor.'' Pages 115–127 in ref. 735.

994. DiPaola, M.; Kao, P. N.; and Karlin, A. *Neurosci. Abstr.* 16: 961 (1986). ''Residues at the acetylcholine binding site may be contained in a single functional domain of the receptor α-subunit.''

995. Herz, J. M.; Johnson, D. A.; Brown, R. D.; and Taylor, P. *Biophys. J.* 51: 59a (1987). ''Topography of the acetylcholine receptor binding sites revealed by fluorescence energy transfer.''

996. Bolger, M. B.; Dionne, V.; Chrivia, J.; Johnson, D. A.; and Taylor, P. *Mol. Pharmacol.* 26: 57–69 (1984). ''Interaction of a fluorescent dicholine with the nicotinic receptor and acetylcholinesterase.''

997. Johnson, D. A.; Brown, R. D.; Herz, J. M.; Berman, H. A.; Andreasen, G. L.; and Taylor, P. *J. Biol. Chem.* 262: 14022–14029 (1987). ''Decidium. A novel fluorescent probe of the agonist/antagonist and noncompetitive inhibitor sites on the nicotinic acetylcholine receptor.''

998. Herz, J. M.; Johnson, D.; and Taylor, P. *Biophys. J.* 49: 5a (1986) ''Distance between the local anesthetic and agonist binding sites on the acetylcholine receptor (AChR) determined by fluorescence energy transfer.''

999. Taylor, P. ''Topography of the acetylcholine receptor revealed by fluorescence energy transfer.'' Pages 61–74 in ref. 735.

1000. Oswald, R. E.; Bamberger, M. J.; and Mclaughlin, J. T. *Mol. Pharmacol.* 25: 360–368 (1984). ''Mechanism of phencyclidine binding to the acetylcholine receptor from *Torpedo* electroplaque.''

1001. Honig, B. H.; Hubbell, W. L.; Flewelling, R. F. *Ann. Rev. Biophys. Biophys. Chem.* 15: 163–193 (1986). "Electrostatic interactions in membranes and proteins."

1002. Yee, A. S.; Corley, D. E.; and McNamee, M. G. *Biochemistry* 25: 2110–2119 (1986). "Thiol-group modification of *Torpedo californica* acetylcholine receptor: Subunit localization and effects on function."

1003. Clarke, J. H., and Martinez-Carrion, M. *J. Biol. Chem.* 261: 10063–10072 (1986). "Labeling of functionally sensitive sulfhydryl containing domains of acetylcholine receptor from *Torpedo californica* membranes."

1004. Imoto, K.; Methfessel, C.; Sakmann, B.; Mishina, M.; Mori, Y.; Konno, T.; Fukuda, K.; Kurasaki, M.; Bujo, H.; Fujita, Y.; and Numa, S. *Nature* 324: 670–674 (1986). "Location of a δ-subunit region determining ion transport through the acetylcholine receptor channel."

1005. Imoto, K.; Busch, C.; Sakmann, B.; Mishina, M.; Konno, T.; Nakai, J.; Bujo, H.; Mori, Y.; Fukuda, K.; and Numa, S. *Nature* 335: 645–648 (1988). "Ring of negatively charged amino acids determnine the acetylcholine receptor channel conductance."

1006. Linse, S.; Brodin, P.; Johansson, C.; Thulin, E.; Grundström, T.; and Forsen, S. *Nature* 335: 651–652 (1988). "The role of protein surface charges in ion binding."

1007. Zabrecky, J. R., and Raftery, M. A. *J. Receptor Res.* 5: 397–417 (1985). "The role of lipids in the function of the acetylcholine receptor."

1008. Fong, T. M. *Biophys. J.* 49: 153a (1986). "Correlation between *Torpedo californica* acetylcholine receptor function and structural properties of membranes."

1009. Jones, O. T., and McNamee, M. G. *Biochemistry* 27: 2364–2374 (1988). "Annular and nonannular binding sites for cholesterol associated with the nicotinic acetylcholine receptor."

1010. McNamee, M. G., "Biophysical studies of acetylcholine receptor in reconstituted membranes: Role of lipids in regulating function." Pages 84–85 in ref. 55.

1011. Leibel, W. S.; Firestone, L. L.; Legler, D. C.; Braswell, L. M.; Miller, K. W. *Biochim. Biophys. Acta* 897: 249–260 (1987). "Two pools of cholesterol in acetylcholine receptor rich membranes from *Torpedo*."

1012. Barrantes, F. J. "The membrane environment of the acetylcholine receptor." Page 83 in ref. 55.

1013. Earnest, J. P.; Shuster, M. J.; McCarthy, M. P.; Stroud, R. M. "Structural studies aimed at understanding the function of acetylcholine receptor." Page 71 in ref. 55.

1014. McNamee, M. G.; Fong, T. M.; Jones, O. T.; Earnest, J. P. "Lipid-protein interactions and acetylcholine receptor function in reconstituted membranes." Pages 147–157 in ref. 735.

1015. Chang, H. W., and Neumann, E. *Proc. Nat'l. Acad. Sci. USA* 73: 3364–3368 (1976). "Dynamic properties of isolated acetylcholine receptor proteins: Release of calcium ions caused by acetylcholine binding."

1016. Fairclough, R. H.; Miake-Lye, R.C.; Stroud, R. M.; Hodgson, K. O.; and Doniach, S. *J. Mol. Biol.* 189: 673–680 (1986). "Location of terbium-binding sites on acetylcholine receptor-enriched membranes."

1017. Blatt, Y.; Montal, M. S.; Lindstrom, J.; and Montal, M. *J. Neurosci.* 6: 481–486 (1986). "Monoclonal antibodies specific to the β- and γ-subunits of the *Torpedo* acetylcholine receptor inhibit single channel activity."

1018. Gundelfinger, E. D.; Hermans-Borgmeyer, I.; Zopf, D.; Sawruk, E.; Betz, H. ''Characterization of the mRNA and the gene of a putative neuronal nicotinic acetylcholine receptor protein from *Drosophila*.'' Pages 437–446 in ref. 735.

1019. White, M. M.; Mayne, K. M.; Lester, H. A.; and Davidson, N. *Proc. Nat'l. Acad. Sci. USA* 82: 4852–4856 (1986). ''Mouse-*Torpedo* hybrid acetylcholine receptors: Functional homology does not equal sequence homology.''

1020. Sakmann, B.; Methfessel, C.; Mishina, M.; Takahashi, T.; Takai, T.; Kurasaki, M.; Fukuda, K.; and Numa, S. *Nature* 318: 538–543 (1985). ''Role of acetylcholine receptor subunits in gating of the channel.''

1021. Mishina, M.; Takai, T.; Imoto, K.; Noda, M.; Takahashi, T.; Numa, S.; Methfessel, C.; and Sakmann, B. *Nature* 321: 406–411 (1986). ''Molecular distinction between fetal and adult forms of muscle acetylcholine receptor.''

1022. Hochschwender, S.; Lindstrom, J.; Froehner, S.; and Hall, Z. *Neurosci. Abstr.* 16: 146 (1986). ''Subunit structure of the acetylcholine receptor in developing, and in normal and denervated adult, mammalian muscle.''

1023. Heidmann, O.; Buonanno, A.; Geoffroy, B.; Robert, B.; Guenet, J.-L.; Merlie, J. P.; and Changeux, J.-P. *Science* 234: 866–868 (1986). ''Chromosomal localization of muscle nicotinic acetylcholine receptor genes in the mouse.''

1024. Young, E. F.; Ralston, E.; Blake, J.; Ramachandran, J.; Hall, Z. W.; and Stroud, R. M. *Proc. Nat'l. Acad. Sci. USA* 82: 626–630 (1985). ''Topological mapping of the acetylcholine receptor model: Evidence for a model with five transmembrane segments and a cytoplasmic C-terminal peptide.''

1025. Lindstrom, J.; Criado, M.; Hochschwender, S.; Fox, J. L.; and Sarin, V. *Nature* 311: 573–575 (1984). ''Immunochemical tests of acetylcholine receptor subunit models.''

1026. Froehner, S. C.; Douville, K.; Klink, S.; and Culp, W. *J. Biol. Chem.* 258: 7112–7120 (1983). ''Monoclonal antibodies to cytoplasmic domains of the acetylcholine receptor.''

1027. LaRochelle, W. J., and Froehner, S. C. *Neurosci. Abstr.* 14: 734 (1984). ''Synthetic nonadecapeptide of Torpedo AChR gamma subunit recognized by monoclonal antibodies specific for cytoplasmic domains.''

1028. LaRochelle, W. J.; Wray, B. E.; Sealock, R.; and Froehner, S. C. *J. Cell. Biol.* 100: 684–691 (1985). ''Immunochemical demonstration that amino acids 360–377 of the acetylcholine receptor gamma-subunit are cytoplasmic.''

1029. Ratnam, M., and Lindstrom, J. *Biochem. Biophys. Res. Commun.* 122: 1225–1233 (1984). ''Structural features of the nicotinic acetylcholine receptor revealed by antibodies to synthetic peptides.''

1030. Eldefrawi, A. T.; Bakry, N. M.; Eldefrawi, M. E.; Tsai, M-C.; and Albuquerque, E. X. *Mol. Pharmacol.* 17: 172–179 (1980). ''Nereistoxin interaction with the acetylcholine receptor-ionic channel complex.''

1031. Spivak, C. E.; Waters, J.; Witkop, B.; and Albuquerque, E. X. *Mol. Pharmacol.* 23, 337–343 (1983). ''Potencies and channel properties induced by semi-rigid agonists at frog nicotinic acetylcholine receptors.''

1032. Kanne, D. B.; Ashworth, D. J.; Cheng, M. T.; and Mutter, L. C. *J. Am. Chem. Soc.* 108: 7864–7865 (1986). ''Synthesis of the first highly potent bridged nicotinoid, 9-azabicyclo[4.2.1]nona[2, 3-c]pyridine (pyrido[3, 4b]norhomotropane).''

1033. Kanne, D. B., and Abood, L. G. *J. Med. Chem.* 31: 506–509 (1988). ''Synthesis and biological characterization pyridohomotropanes. Structure-activity relationships of conformationally restricted nicotinoids.''

1034. Taylor, P.; Brown, R. D.; and Johnson, D. A. *Curr. Topics in Membranes and Transport* 18: 407–444 (1983). ''The linkage between ligand occupation and response of the nicotinic acetylcholine receptor.''

1035. Sheridan, R. E., and Lester, H. A. *J. Gen. Physiol.* 80: 499–515 (1982). ''Functional stoicheiometry at nicotinic receptor.''

1036. Lukas, R. J. *Biophys. J.* 49: 3a (1986). ''Heterogeneity of functional nicotinic acetylcholine receptors on clonal cell lines is revealed by the use of ^{86}rubidium ion efflux studies.''

1037. Meyers, H-W.; Jürss, R.; Brenner, H. R.; Fels, G.; Prinz, H.; Watzke, H.; and Maelicke, A. *Eur. J. Biochem.* 137: 399–404 (1983) . ''Synthesis and properties of NBD-*n*-acylcholines, fluorescent analogues of acetylcholine.''

1038. Prinz, H., and Maelicke, A. *J. Biol. Chem.* 258: 10273–10282 (1983). ''Interaction of cholinergic ligands with the purified acetylcholine receptor protein.''

1039. Prinz, H. ''A general treatment of ligand binding to the acetylcholine receptor.'' Pages 129–146 in ref. 735.

1040. Neely, A., and Lingle, C. J. *Biophys. J.* 50: 981–986 (1986). ''Trapping of an open-channel blocker at the frog neuromuscular acetylcholine channel.''

1041. Siemen, D.; Hellmann, S.; Maelicke, A. ''Single channel studies of acetylcholine receptors covalently alkylated with acetylcholine.'' Pages 233–241 in ref. 735.

1042. Colquhoun, D., and Sakmann, B. *Nature* 294: 464–466 (1981). ''Fluctuations in the microsecond time range of the current through single acetylcholine receptor channels.''

1043. Nerbonne, J. M.; Sheridan, R. E.; Chabala, L. D.; and Lester, H. A. *Mol.* Pharmacol. 23: 344–349 (1983). ''*cis*-3,3'-Bis-[trimethylammoniomethyl]azobenzene (*cis*-Bis-Q).''

1044. Sine, S. M., and Steinbach, J. H. *Biophys. J.* 47: 42a (1985). ''Activation of acetylcholine receptor by low concentrations of agonist.''

1045. Jackson, M. B. *Biophys. J.* 49: 663–672 (1986). ''Kinetics of unliganded acetylcholine receptor channel gating.''

1046. Sigworth, F. J. *Biophys. J.* 47: 709–720 (1985). ''Open channel noise I. Noise in acetylcholine receptor currents suggests conformational fluctuations.''

1047. Neubig, R. R., and Cohen, J. B. *Biochemistry* 19: 2770–2779 (1980). ''Permeability control by cholinergic receptors in *Torpedo* postsynaptic membranes: Agonist-dose response relations measured at second and millisecond times.''

1048. Boyd, N. D., and Cohen, J. B. *Biochemistry* 19: 5344–5353 (1980). ''Kinetics of binding of [^{3}H]acetylcholine and [^{3}H]carbamoylcholine to *Torpedo* postsynaptic membranes: Slow conformational transitions of the cholinergic receptor.''

1049. Boyd, N. D., and Cohen, J. B. *Biochemistry* 19: 5353–5358 (1980). ''Kinetics of binding of [^{3}H]acetylcholine to *Torpedo* postsynaptic membranes: Association and dissociation rate constants by rapid mixing and ultrafiltration.''

1050. Neubig, R. R.; Boyd, N. D.; and Cohen, J. B. *Biochemistry* 21: 3460–3466 (1982). ''Conformations of *Torpedo* acetylcholine receptor associated with ion transport and desensitization.''

1051. Heidmann, T., and Changeux, J.-P. *Eur. J. Biochem.* 94: 255–279 (1978). ''Fast kinetic studies on the interaction of a fluorescent agonist with the membrane-bound acetylcholine receptor from *Torpedo marmorata*.''

1052. Aronstam, R. S., and Witkop, B. *Proc. Nat'l. Acad. Sci. USA* 78: 4639–4643 (1981). "Anatoxin-a interactions with cholinergic synaptic molecules."

1053. Koskinen, A.M.P., and Rapoport, H. *J. Med. Chem.* 28: 1301–1309 (1985). "Synthetic and conformational studies on anatoxin-a: A potent acetylcholine agonist."

1054. Damle, V. N., and Karlin, A. *Biochemistry* 19: 3924–3932 (1980). "Effects of Agonists and antagonists on the reactivity of the binding site disulfide in acetylcholine receptor from *Torpedo californica*."

1055. Tzartos, S. J., and Starzinski-Powitz, A. *FEBS Lett.* 196: 91–95 (1986). "Decrease in acetylcholine-receptor content of human myotube cultures mediated by monoclonal antibodies to α-, β- and γ-subunits."

1056. Drachman, D. *Ann. Rev. Neurosci.* 4: 195–225 (1981). "The biology of *myasthenia gravis*."

1057. Grob, D., ed. "*Myasthenia gravis*: Pathophysiology and management." *Ann. N.Y. Acad. Sci.* 377: 1–902 (1981).

1058. Drachman, D. B., ed. "*Myasthenia gravis*: Biology and treatment." *Ann. N.Y. Acad. Sci.* 505: 914 pp. (1987).

1059. McIntosh, K. R., and Drachman, D. B. *Science* 232: 401–403 (1986). "Induction of suppressor cells specific for AChR in experimental autoimmune *myasthenia gravis*."

1060. Barkas, T.; Juillerat, M.; and Tzartos, S. J. *Biochem. J.* 231: 245 (1985). "Antigenic sites of the nicotinic acetylcholine receptor."

1061. McCormick, D. J.; Lennon, V. A.; and Atassi, M. Z. *Biochem. J.* 231: 245–246 (1985). "Antigencity of synthetic peptides 159–169 and 151–169 of *Torpedo* acetylcholine receptor a chain."

1062. Lindstrom, J., and Criado, M. *Biochem. J.* 231: 247 (1985). "Antigenicity of the sequence 152–167 in α-subunits of *Torpedo* acetylcholine receptor."

1063. Neumann, D.; Gershoni, J. M.; Fridkin, M.; and Fuchs, S. *Proc. Nat'l. Acad. Sci. USA* 82: 3490–3493 (1985). "Antibodies to synthetic peptides as probes for the binding site on the α-subunit of the acetylcholine receptor."

1064. Schuetze, S. M.; Vicini, S.; and Hall, Z. W. *Proc. Nat'l. Acad. Sci. USA* 82: 2533–2537 (1985). "Myasthenic serum selectively blocks acetylcholine receptors with long channel open times at developing rat endplates."

1065. Ratnam, M.; Sargent, P.; Sarin, V.; Fox, J.; Le Nguyen, D.; Rivier, J.; Criado, M.; and Lindstrom, J. Unpublished results cited in ref. 1017.

1066. Montal, M.; Anholt, R.; and Labarca, P. In *Ion Channel Reconstitution*, ed. C. Miller, pp. 157–204. New York: Plenum Press, 1986. "The reconstituted acetylcholine receptor."

1067. Montal, M. "Functional reconstitution of membrane proteins in planar lipid bilayer membranes." In *Analysis of Membrane Proteins*, ed. C. I. Ragan, R. J. Cherry, pp. 97–128. London: Chapman and Hall, 1986.

1068. Wilson, P. T.; Lentz, T. L.; and Hawrot, E. *Proc. Nat'l. Acad. Sci. USA* 82: 8790–8794 (1985). "Determination of the primary amino acid sequence specifying the α-bungarotoxin binding site on the α-subunit of the acetylcholine receptor from *Torpedo californica*."

1069. Barkas, T.; Mauron, A.; Roth, B.; Alliod, C.; Tzartos, S. J.; and Ballivet, M. *Science* 235: 77–80 (1987). "Mapping the main immunogenic region and toxin-binding site of the nicotinic acetylcholine receptor."

1070. Neumann, D.; Barchan, D.; Safran, A.; Gershoni, J. M.; and Fuchs, S. *Proc. Nat'l. Acad. Sci. USA* 83: 3008–3011(1986). "Mapping of the α-bungarotoxin binding site within the α-subunit of the acetylcholine receptor."

1071. Neumann, D.; Barchan, D.; Fridkin, M.; and Fuchs, S. *Proc. Nat'l. Acad. Sci. USA* 83: 9250–9253 (1986). "Analysis of ligand binding to the synthetic dodecapeptide 185–196 of the acetylcholine receptor α subunit."

1072. Shipolini, R. A., and Bailey, G. S.; Edwardson, J. A.; and Banks, B.E.C. *Eur. J. Biochem.* 40: 337–344 (1973). "Separation and characterization of polypeptides from the venom of *Dendroaspis viridis*."

1073. Shipolini, R. A., and Banks, B.E.C. *Eur. J. Biochem.* 49: 399–405 (1974). "The amino acid sequence of a polypeptide from the venom of *Dendroaspis viridis*."

1074. Conti-Tronconi, B. M., and Raftery, M. A. *Proc. Nat'l. Acad. Sci. USA* 83: 6646–6650 (1986). "Nicotinic acetylcholine receptor contains multiple binding sites: Evidence from binding of α-dendrotoxin."

1075. Goldman, D.; Simmons, D.; Swanson, L. W.; Patrick, J.; Heinemann, S. *Proc. Nat'l. Acad. Sci. USA* 83: 4076–4080 (1986). "Mapping of brain areas expressing RNA homologous to two different acetylcholine receptor α-subunit cDNAs."

1076. Clarke, P.B.S. "Radioligand labeling of nicotinic receptors in mammalian brain." Pages 345–357 in ref. 735.

1077. Sugiyama, H., and Yamashita, Y. *Brain Res.* 373: 22–26 (1986). "Characterization of putative acetylcholine receptors solubilized from rat brains."

1078. Endo, T.; Nakanishi, M.; Furukawa, S.; Joubert, F. J.; Tamiya, N.; and Hayashi, K. *Biochemistry* 25: 395–404 (1986). "Stopped-flow fluorescence studies on binding kinetics of neurotoxins with acetylcholine receptor."

1079. Cheung, A. T.; Johnson, D. A.; and Taylor, P. *Biophys. J.* 45: 447–454 (1984). "Kinetics of interaction of N^{ϵ}-fluorescein isothiocyanato-lysine-23-cobra α-toxin with the acetylcholine receptor."

1080. Kang, S., and Maelicke, A. *J. Biol. Chem.* 255: 7326–7332 (1980). "Fluorescein isothiocyanate-labeled α-cobra toxin: Biochemical characterization and interaction with acetylcholine receptor from *Electrophorus electricus*."

1081. Endo, T., and Tamiya, N. *Pharmac. Ther.* 34: 403–451 (1987). "Current view on the structure-function relationship of postsynaptic neurotoxins from snake venoms."

1082. Maelicke, A. private communication.

1083. Maelicke, A., and Kosower, E. M. Unpublished results.

1084. Maelicke, A.; Watters, D.; and Fels, G. "Monoclonal antibodies as probes of acetylcholine receptor function." Pages 83–91 in ref. 735.

1085. Fels, G.; Plümer-Wilk, R.; Schreiber, M.; and Maelicke, A. *J. Biol. Chem.* 261: 15746–15754 (1986). "A monoclonal antibody interfering with binding and response of the acetylcholine receptor."

1086. Maelicke, A.; Fels, G.; Plümer-Wilk, R.; Wolff, E. K.; Covarrubias, M.; Methfessel, C. In *Ion Channels in Neural Membranes*, pp. 275–282. New York: Alan R. Liss, 1986. "Specific blockade of acetylcholine receptor function by monoclonal antibodies."

1087. Ratnam, M.; Le Nguyen, D.; Rivier, J.; Sargent, P. B.; and Lindstrom, J. *Biochemistry* 25: 2633–2643 (1986). "Transmembrane topography of nicotinic acetylcholine receptor: Immunochemical tests contradict theoretical predictions based on hydrophobicity profiles."

1088. Lindstrom, J.; Whiting, P.; Schoepfer, R.; Luther, M.; and Das, M. "Structure of nicotinic acetylcholine receptors from muscle and neurons." In *Computer-assisted modeling of receptor-ligand interactions*, ed. R. Rein, and A. Golombek, New York: 1989.
1089. Bialek, W., and Goldstein, R. F. *Biophys. J.* 48: 1027–1044 (1985). "Do vibrational spectroscopies uniquely describe protein dynamics? The case for myoglobin."
1090. Dunn, S.M.J.; Conti-Tronconi, B.; and Raftery, M. A. *Biochem. Biophys. Res. Commun.* 139: 830–837 (1986). "Acetylcholine receptors are stabilized by extracellular disulfide bonding."
1091. Fujita, N.; Nelson, N.; Fox, T. D.; Claudio, T.; and Hess, G. P. *Biophys. J.* 49: 364a (1986). "Biosynthesis of the *Torpedo californica* acetylcholine receptor α-subunit in yeast."
1092. Sweet, M. T.; Fujita, N.; Donaldson, P.; Fox, T. D.; Nelson, N.; Claudio, T.; Lindstrom, J. M.; and Hess, G. P. *Neurosci. Abstr.* 16: 147 (1986). "Orientation of acetylcholine receptor α subunit in yeast plasma membrane."
1093. Yamada, H.; Kuroki, R.; Hirata, M.; and Imoto, T. *Biochemistry* 22: 4551–4556 (1983). "Intramolecular cross-linkage of lysozyme. Imidazole catalysis of the formation of the cross-link between lysine-13 (ε-amino) and leucine 129 (α-carboxyl) by carbodiimide reaction."
1094. Ueda, T.; Yamada, H.; Hirata, M.; and Imoto, T. *Biochemistry* 24: 6316–6322 (1985). "An intramolecular cross-linkage of lysozyme. Formation of cross-links between lysine-1 and histidine-15 with bis(bromoacetamide) derivatives by a two-stage reaction procedure and properties of the resulting derivatives."
1095. Kim, H. Y.; Pilosof, D.; Dyckes, D. F.; and Vestal, M. L. *J. Am. Chem. Soc.* 106: 7304–7309 (1984). "On-line peptide sequencing by enzymatic hydrolysis, high performance liquid chromatography and thermospray mass spectrometry."
1096. Catterall, W. A. *Science* 242: 50–61 (1988). "Structure and function of voltage-sensitive ion channels."
1097. Blundell, T. L.; Sibanda, B. L.; Sternberg, M.J.E.; and Thornton, J. M. *Nature* 326: 347–352 (1987). "Knowledge-based prediction of protein structures and the design of novel structures."
1098. Kawakami, K.; Noguchi, S.; Noda, M.; Takahashi, H.; Ohta, T.; Kawamura, M.; Nojima, H.; Nagano, K.; Hirose, T.; Inayama, S.; Hayashida, H.; Miyata, T.; and Numa, S. *Nature* 316: 733–736 (1985). "Primary structure of the α-subunit of *Torpedo californica* (Na^+/K^+)ATPase deduced from the cDNA sequence."
1099. Noguchi, S.; Noda, M.; Takahashi, H.; Kawakami, K.; Ohta, T.; Nagano, K.; Hirose, T.; Inayama, S.; Kawamura, M.; and Numa, S. *FEBS Lett.* 196: 315–320 (1986). "Primary structure of the β-subunit of *Torpedo californica* (Na^+/K^+)ATPase deduced from the cDNA sequence."
1100. Noda, M.; Shimizu, S.; Tanabe, T.; Takai, T.; Kayano, T.; Ikeda, T.; Takahashi, H.; Nakayama, H.; Kanaoka, Y.; Minamino, N.; Kangawa, K.; Matsuo, H.; Raftery, M. A.; Hirose, T.; Notake, M.; Inayama, S.; Hayashida, H.; Miyata, T.; and Numa, S. *Nature* 312: 121–127 (1984). "Primary structure of *Electrophorus electricus* sodium channel deduced from cDNA sequence."
1101. Noda, M.; Ikeda, T.; Kayano, T.; Suzuki, H.; Takeshima, H.; Kurasaki, M.;

Takahashi, H.; and Numa, S. *Nature* 320: 188–192 (1986). "Existence of distinct sodium channel messenger RNAs in rat brain."

1102. Kayano, T.; Noda, M.; Flockerzi, V.; Takahashi, H.; and Numa, S. *FEBS Lett.* 228: 187–194 (1988). "Primary structure of rat brain sodium channel III deduced from the cDNA sequence."

1103. Hartshorne, R. P., and Catterall, W. A. *J. Biol. Chem.* 259: 1667–1675 (1984). "The sodium channel from rat brain. Purification and subunit composition."

1104. Hartshorne, R.; Tamkun, M.; and Montal, M. "The reconstituted sodium channel from brain." In *Ion Channel Reconstitution*, ed. C. Miller, chap. 5, pp. 337–362. New York: Plenum, 1986.

1105. Talvenheimo, J. A. *J. Membrane Biol.* 87: 77–91 (1985). "The purification of ion channels from excitable cells."

1106. Messner, D. J.; Feller, D. J.; Scheuer, T.; and Catterall, W. A. *J. Biol. Chem.* 261: 14882–14890 (1986). "Functional properties of rat brain sodium channels lacking the β1- or β2-subunit."

1107. Goldin, A. L.; Snutch, T.; Lübbert, H.; Dowsett, A.; Marshall, J.; Auld, V.; Downey, W.; Fritz, L. C.; Lester, H. A.; Dunn, R.; Catterall, W. A.; and Davidson, N. *Proc. Nat'l. Acad. Sci. USA* 83: 7503–7507 (1986). "Messenger RNA coding for only the α-subunit of the rat brain sodium channel is sufficient for expression of functional channels in *Xenopus* öocytes."

1108. Hirono, H.; Yamagishi, S.; Ohara, R.; Hisanaga, Y.; Nakayama, T.; and Sugiyama, H. *Brain Res.* 359: 57–64 (1985). "Characterization of the mRNA responsible for induction of functional sodium chanels in *Xenopus* öocytes."

1109. Noda, M.; Ikeda, T.; Suzuki, H.; Takeshima, H.; Takahashi, H.; Kuno, M.; and Numa, S. *Nature* 322: 826–828 (1986). "Expression of functional sodium channels from cloned cDNA."

1110. Krafte, D.; Leonard, J.; Snutch, T.; Davidson, N.; and Lester, H. A. *Biophys. J.* 51: 194a (1987). "Slowly inactivating sodium currents from injection of high molecular weight rat brain RNA into *Xenopus* öocytes."

1111. Agnew, W. S.; Levinson, S. R.; Brabson, J. S.; and Raftery, M. A. *Proc. Nat'l. Acad. Sci. USA* 75: 2606–2610 (1978). "Purification of the tetrodotoxin-binding component associated with the voltage sensitive sodium channel from *Electrophorus electricus* electroplax membranes."

1112. Nakayama, H.; Withy, R. M.; and Raftery, M. A. *Proc. Nat'l. Acad. Sci. USA* 79: 7575–7579 (1982). "Use of a monoclonal antibody to purify the tetrodotoxin binding component from the electroplax of *Electrophorus electricus*."

1113. Sefton, B. M., and Buss, J. E. *J. Cell. Biol.* 104: 1449–1453 (1987). "The covalent modification of eukaryotic proteins with lipid."

1114. Magee, T., and Hanley, M. *Nature* 335: 114–115 (1988). "Sticky finger and CAAX boxes."

1115. Meiri, H.; Spira, G.; Marei, S.; Namir, M.; Schwartz, A.; Komoriya, A.; Kosower, E. M.; and Palti, Y. *Proc. Nat'l. Acad. Sci. USA* 84: 5058–5062 (1987). "Mapping a region associated with sodium channel inactivation using antibodies to a synthetic peptide corresponding to a part of the channel."

1116. Rosenthal, V.; Meiri, H.; Spira, G.; Kraoz, Z.; Kosower, E. M.; and Palti, Y. Interaction of the anti-eel antibody 5F3 with OCEIII and the frog sciatic nerve, unpublished results.

1117. Greenblatt, R. E.; Blatt, Y.; and Montal, M. *FEBS Lett.* 193: 125–134 (1985).

"The structure of the voltage-sensitive sodium channel. Inferences derived from computer-aided analysis of the *Electrophorus electricus* channel primary structure."

1118. Guy, H. R., and Seetharamulu, P. *Proc. Nat'l. Acad. Sci. USA* 83: 508–512 (1986). "Molecular model of the action potential sodium channel." Guy, H. R. *Biophys. J.* 51: 193a (1987).

1119. Guy, H. R. *Current topics in membrane transport: Molecular biology of ion channels* 33: 289–308 (1988). "A model relating the sodium channel's structure to its function."

1120. Engelman, D. M.; Steitz, T. A.; and Goldman, A. *Ann. Rev. Biophys. Biophys. Chem.* 15: 321–333 (1986). "Identifying nonpolar transbilayer helices in amino acid sequences of membrane proteins."

1121. Barlow, D. J., and Thornton, J. M. *J. Mol. Biol.* 201: 601–619 (1988). "Helix geometry in proteins."

1122. Wilke, M. E., Mitra, A., Yamamoto, T.; Stroud, R. *Biophys. J.* 51: 454a (1987). "Structural study of a natural insecticide."

1123. Toniolo, C.; Bonora, G. M.; Bavoso, A.; Benedetti, E.; DiBlasio, B.; Pavone, V.; and Pedone, C. *Macromol.* 19: 472–479 (1986). "A long, regular polypeptide 3_{10}-helix." Tosteson, M. T.; Auld, D. S.; and Tosteson, D. C. *Proc. Nat'l. Acad. Sci. USA* 86: 707–710 (1989). "Voltage-gated channels formed in lipid bilayers by a positively charged segment of the Na-channel polypeptide."

1124. Roberts, R. H., and Barchi, R. L. *J. Biol. Chem.* 262: 2298–2303 (1987). "The voltage-sensitive sodium channel from rabbit skeletal muscle."

1125. Messner, D. J., and Catterall, W. A. *J. Biol. Chem.* 260: 10597–10604 (1985). "The sodium channel from rat brain. Separation and characterization of subunits."

1126. Gordon, R. D.; Fieles, W. E.; Schotland, D. L.; Hogue-Angeletti, R.; and Barchi, R. L. *Proc. Nat'l. Acad. Sci. USA* 84: 308–312 (1987). "Topographical localization of the C-terminal region of the voltage-dependent sodium channel from *Electrophorus electricus* using antibodies raised against a synthetic peptide."

1127. Hodgkin, A. L., and Huxley, A. F. *J. Physiol.(Lond.)* 117: 500–544 (1952). "A quantitative description of membrane current and its application to conduction and excitation in nerve."

1128. Vassilev, P. M.; Scheuer, T.; and Catterall, W. A. *Science* 241: 1658–1661 (1988). "Identification of an intracellular peptide segment involved in sodium channel inactivation."

1129. Armstrong, C. M., and Bezanilla, F. *J. Gen. Physiol.* 63: 533–552 (1974). "Charge movement associated with opening and closing of the activation gates of the Na channels."

1130. Stimers, J. R.; Bezanilla, F.; and Taylor, R. E. *J. Gen. Physiol.* 85: 65–82 (1985). "Sodium channel activation in the squid giant axon: Steady state properties."

1131. Bezanilla, F. *J. Membrane Biol.* 88: 97–111 (1985). "Gating of sodium and potassium channels."

1132. French, R. J., and Horn, R. *Ann. Rev. Biophys. Bioeng.* 12: 319–356 (1983). "Sodium channel gating: Models, mimics and modifiers."

1133. Stühmer, W.; Methfessel, C.; Sakmann, B.; Noda, M.; and Numa, S. *Eur. Biophys. J.* 14: 131–138 (1987). "Patch clamp characterization of sodium channels expressed from rat brain cDNA."

1134. Zell, A.; Ealick, S. E.; and Bugg, C. E. In *Molecular architecture of proteins and enzymes*, pp. 65–97. New York: Academic Press, 1985. "Three-dimensional structures of scorpion neurotoxins."

1135. Catterall, W. A. *Ann. Rev. Pharmacol. Toxicol.* 20: 15–43 (1980). "Neurotoxins that act on voltage-sensitive sodium channels in excitable membranes."

1136. Barchi, R. L. *Int'l. Rev. Neurobiol.* 23: 69–101 (1982). "Biochemical studies of the excitable membrane sodium channel."

1137. Daly, J. W.; Myers, C. W.; Warnick, J. E.; and Albuquerque, E. X. *Science* 208: 1383–1385 (1980). "Levels of batrachotoxin and lack of sensitivity to its action in poison-dart frogs (*Phyllobates*)."

1138. Rosenberg, R. L.; Tomiko, S. A.; and Agnew, W. S. *Proc. Nat'l. Acad. Sci. USA* 81: 1239–1243 (1984). "Reconstitution of neurotoxin-modulated ion transport by the voltage-regulated sodium channel isolated from the electroplax of *Electrophorus electricus*."

1139. Bar-Sagi, D., and Prives, J. *J. Biol. Chem.* 260: 4740–4744 (1985). "Negative modulation of sodium channels in cultured chick muscle cells by the channel activator batrachotoxin."

1140. Tejedor, F. J.; McHugh, E.; and Catterall, W. A. *Biochemistry* 27: 2389–2397 (1988). "Stabilization of a sodium channel state with a high affinity for saxitoxin by intramolecular cross-linking. Evidence for allosteric effects of saxitoxin binding."

1141. Kosower, E. M. *FEBS Lett.* 163: 161–164 (1983). "A hypothesis for the mechanism of action of sodium channel opening by batrachotoxin and related toxins."

1142. Leibowitz, M. D.; Sutro, J. B.; and Hille, B. *J. Gen. Physiol.* 87: 25–46 (1986). "Voltage-dependent gating of veratridine-modified Na channels."

1143. Sutro, J. B. *J. Gen. Physiol.* 87: 1–24 (1986). "Kinetics of veratridine action on Na channels of skeletal muscle."

1144. Tanguy, J.; Yeh, J. Z.; and Narahashi, T., *Biophys. J.* 51: 7a (1987). "Gating charge immobilization after removal of Na-channel inactivation."

1145. Corbett, A. M.; Zinkand, W. C.; and Krueger, B. K. *Biophys. J.* 51: 435a (1987). "Activation of single neuronal sodium channels by veratridine and polypeptide neurotoxin in planar lipid bilayers."

1146. Garber, S. S., and Miller, C. *J. Gen. Physiol.* 89: 459–480 (1987). "Single Na^+ channels activated by veratridine and batrachotoxin."

1147. Khodorov, B. I. *Progr. Biophys. Mol. Biol.* 45: 57–148 (1985). "Batrachotoxin as a tool to study voltage-sensitive sodium channels of excitable membranes."

1148. Brown, G. B. *Int'l. Rev. Neurobiol.* 29: 77–116 (1988). "Batrachotoxin: A window on the allosteric nature of the voltage-sensitive sodium channel."

1149. Seyama, I.; Yamada, K.; Kato, R.; Masutani, T.; and Hamada, M. *Biophys. J.* 53: 271–274 (1988). "Grayanotoxin opens Na channels from inside the squid axonal membrane."

1150. Rando, T. A., and Strichartz, G. R. *Biophys. J.* 49: 785–794 (1986). "Saxitoxin blocks batrachotoxin-modified sodium channels in the node of Ranvier in a voltage-dependent manner."

1151. Fontecilla-Camps, J. C.; Almassy, R. J.; Suddath, F. L.; Watt, D. D.; and Bugg, C. E. *Proc. Nat'l. Acad. Sci. USA* 77: 6496–6500 (1980). "Three-dimensional structure of a protein from scorpion: A new structural class of neurotoxins."

1152. Chinn, K., and Narahashi, T. *Biophys. J.* 49: 42a (1986). "Stabilization of sodium channel states by deltamethrin in mouse neuroblastoma cells."

1153. Brown, L. D., and Narahashi, T. *Biophys. J.* 49: 377a (1986). "Modification of squid axon sodium channels by deltamethrin."

1154. Lund, A. E., and Narahashi, T. *NeuroTox* 3: 11–24 (1982). "Dose-dependent interaction of the pyrethroid isomers with sodium channels of squid axon membranes."

1155. Narahashi, T. *Comp. Biochem. Physiol.* 72C: 411–414 (1982). "Modification of nerve membrane sodium channels by the insecticide pyrethroids."

1156. Frelin, C.; Vigne, P.; and Lazdunski, M. *Eur. J. Biochem.* 119: 437–442 (1981). "The specificity of the sodium channel for monovalent cations."

1157. Sigworth, F. J., and Spalding, B. C. *Nature* 283: 293–295 (1980). "Chemical modification reduces the conductance of sodium channels in nerve."

1158. Worley, J. F., III; Krueger, B. K.; and French, R. J. *Biophys. J.* 87: 327–349 (1986). "Trimethyloxonium modification of single batrachotoxin-activated sodium channels reduces block in planar bilayers."

1159. Kao, P. N. *Biophys. J.* 51: 438a (1987). "Modification of saxitoxin binding site of squid giant axon by carbodiimide but not by trimethyloxonium."

1160. Oxford, G. S.; Wu, C. H.; and Narahashi, T. *J. Gen. Physiol.* 71: 227–247 (1978). Removal of sodium channel inactivation in squid giant axons by N-bromoacetamide.

1161. Armstrong, C. M.; Bezanilla, F.; and Rojas, E. *J. Gen. Physiol.* 62: 375–391 (1973). "Destruction of sodium channel inactivation in squid axons perfused with pronase."

1162. Rojas, E., and Rudy, B. *J. Physiol.(Lond.)* 262: 501–531 (1976). "Destruction of the sodium conductance inactivation by a specific protease in perfused nerve fibers from *Loligo*."

1163. Eaton, D. C.; Brodwick, M. S.; Oxford, G. S.; and Rudy, B. *Nature* 271: 473–476 (1978). "Arginine-specific reagents remove sodium channel inactivation."

1164. Oxford, G. S., and Yeh, J. Z. *Biophys. J.* 37: 104a (1982). "Molecular mechanisms of sodium channel inactivation: Discriminations using temperature, octanol and ion competition."

1165. Lo, M-v. C., and Shrager, P. *Biophys. J.* 35: 31–43 (1981) . "Block and inactivation of sodium channels in nerve by amino acid derivatives I."

1166. Lo, M-v. C., and Shrager, P. *Biophys. J.* 35: 45–57 (1981) . "Block and inactivation of sodium channels in nerve by amino acid derivatives II."

1167. Yamamoto, D.; Yeh, J. Z.; and Narahashi.T. *Biophys. J.* 45: 337–344 (1984). Voltage-dependent calcium block of normal and tetramethrin modified single sodium channels."

1168. Meiri, H.; Zeitoun, I.; Grunhagen, H. H.; Lev-Ram, V.; Eshhar, Z.; and Schlessinger, J. *Brain Res.* 310: 168–173 (1984). "Monoclonal antibodies associated with sodium channel block block nerve impulse and stain nodes of Ranvier."

1169. Meiri, H.; Goren, E.; Bergmann, H.; Zeitoun, I.; Rosenthal, Y.; and Palti, Y. *Proc. Nat'l. Acad. Sci. USA* 83: 8385–8389 (1986). "Specific modulation of sodium channels in mammalian nerve by monoclonal antibodies."

1170. Maelicke, A. ed. *Molecular biology of neuroreceptors and ion channels*, NATO-ASI Series, vol. H-? Berlin: Springer Verlag, 1989.

1171. Maelicke, A.; Plümer-Wilk, R., Fels, G.; Spencer, S. R.; and Conti-Tronconi, B. M. "Interaction of monoclonal antibodies with *Torpedo* acetylcholine receptor," in ref. 1170. Maelicke, A.; Plümer-Wilk, R.; Fels, G.; Spencer, S. R.; Engelhard,

M.; Veltel, D.; and Conti-Tronconi, B. M. *Biochemistry* 28: 1396–1405 (1989). ''Epitope mapping employing antibodies raised against short synthetic peptides.''

1172. Woodbury, J. W.; D'Arrigo, J. S.; and Eyring, H. In ''Molecular Mechanisms of Anesthesia.'' *Progr. Anesthesia*, ed. B. R. Fink, vol. 1, pp. 53–87. New York: Raven Press, 1975. ''Physiological mechanism of general anesthesia: general blockade.''

1173. Hille, B. In ''Molecular Mechanisms of Anesthesia.'' *Progr. Anesthesia*, ed. B. R. Fink, vol. 2, pp. 1–5. New York: Raven Press, 1980. ''Theories of anesthesia: general perturbations versus specific receptors.''

1174. Strichartz, G., ed. ''Local Anesthetics.'' *Handbk. Exp. Pharmacol.*, vol. 81. Berlin: Springer, 1987. 292 pp.

1175. Richards, C. D.; Keightley, C. A.; Hesketh, T. R.; and Metcalfe, J. C. In ''Molecular Mechanisms of Anesthesia.'' *Progr. Anesthesia*, ed. B. R. Fink, vol. 2, pp. 337–351. New York: Raven Press, 1980. ''A critical evaluation of the lipid hypotheses of anesthetic action.''

1176. Roth, S. H. *Ann. Rev. Pharmacol. Toxicol.* 19: 159–178 (1979). ''Physical methods of anesthesia.''

1177. Hille, B. *J. Gen. Physiol.* 69: 497–515 (1977). ''Local anesthetics: Hydrophilic and hydrophobic pathways for the drug-receptor reaction.''

1178. Van Dyke, C., and Byck, R. *Sci. Amer.* 246(3): 108–119 (1982). ''Cocaine.''

1179. Banerjee, D. K.; Lutz, R. A.; Levine, M. A.; Rodbard, D.; and Pollard, H. B. *Proc. Nat'l. Acad. Sci. USA* 84: 1749–1753 (1987). ''Uptake of norepinephrine and related catecholamines by cultured chromaffin cells: Characterization of cocaine-sensitive and -insensitive plasma membrane transport sites.''

1180. Gabe, E. J., and Barnes, W. H. *Acta Cryst.* 16: 796–801 (1963). ''The crystal and molecular structure of *l*-cocaine hydrochloride.''

1181. Hyrnchuk, R. J.; Barton, R. J.; and Robertson, B. E. *Can. J. Chem.* 61: 481–487 (1983). ''The crystal structure of free base cocaine, $C_{17}H_{21}NO_4$.''

1182. Postma, S. W., and Catterall, W. A. *Mol. Pharmacol.* 25: 219–227 (1984). ''Inhibition of binding [^{3}H]Batrachotoxinin A 20α-benzoate to sodium channels by local anesthetics.''

1183. Creveling, C. R.; Lewandowsky, G. A.; and Daly, J. W. *Neurosci. Abstr.* 16: 1511 (1986). ''Effect of local anesthetics on the dissociation rate of the batrachotoxin-sodium channel complex.''

1184. McNeal, E. T.; Lewandowski, G. A.; Daly, J. W.; and Creveling, C. R. *J. Med. Chem.* 28: 381–388 (1985). ''[^{3}H]Batrachotoxinin A 20α-benzoate binding to voltage-sensitive sodium channels: A rapid and quantitative assay for local anesthetic activity in a variety of drugs.''

1185. Courtney, K. R. *J. Pharm. Exp. Therap.* 213: 114–119 (1980). ''Structure-activity relations for frequency-dependent sodium channel block in nerve by local anesthetics.''

1186. Courtney, K. R. *Biophys. J.* 45: 42–44 (1984). ''Size-dependent kinetics associated with drug block of sodium current.''

1187. Romey, G.; Quast, U.; Pauron, D.; Frelin, C.; Renaud, J. F.; and Lazdunski, M. *Proc. Nat'l. Acad. Sci. USA* 84: 896–900 (1987). ''Na^+ channels as sites of action of the cardioactive agent DPI 201–106 with agonist and antagonist enantiomers.''

1188. Cahalan, M. D., and Almers, W. *Biophys. J.* 27: 39–56 (1979). "Interaction between quaternary lidocaine, the sodium channel gates and tetrodotoxin."
1189. Evers, A. S.; Berkowitz, B. A.; and d'Avignon, D. A. *Nature* 328: 157–160 (1987). "Correlation between the anesthetic effect of halothane and saturable binding in brain."
1190. Franks, N. P., and Lieb, W. R. *Nature* 328: 113–114 (1987). "Anesthetics on the mind."
1191. Hendry, B. M.; Elliott, J. R.; and Haydon, D. A. *Biophys. J.* 47: 841–845 (1985). "Further evidence that membrane thickness influences voltage-gated channels."
1192. Levinson, S. R.; Duch, D. S.; Urban, B. W.; Recio-Pinto, E. *Ann. N.Y. Acad. Sci.* 479: 162–178 (1987). "The sodium channel from *Electrophorus electricus*."
1193. Giraudat, J.; Montecucco, C.; Bisson, R.; Changeux, J.-P. *Biochemistry* 24: 3121–3127 (1985). "Transmembrane topology of acetylcholine receptor subunits probed with photoreactive phospholipids."
1194. Costa, M.R.C., and Catterall, W. A. *J. Biol. Chem.* 259: 8210–8218 (1984). "Cyclic AMP-dependent phosphorylation of the α-subunit of the sodium channel in synaptic nerve ending particles."
1195. Rossie, S.; Gordon, D.; and Catterall, W. A. *J. Biol. Chem.* 262: 17530–17535 (1987). "Identification of an intracellular domain of the sodium channel having multiple cAMP-dependent phosphorylation sites."
1196. Elmer, L. W.; O'Brien, B. J.; Nutter, T. J.; and Angelides, K. J. *Biochemistry* 24: 8128–8137 (1985). "Physicochemical characterization of the α-peptide of the sodium channel from rat brain."
1197. Waechter, C. J.; Schmidt, J.; and Catterall, W. A. *J. Biol. Chem.* 258: 5117–5123 (1983). "Glycosylation is required for maintenance of functional sodium in neuroblastoma cells."
1198. Bar-Sagi, D., and Prives, J. *J. Cell. Physiol.* 114: 77–81 (1983). "Tunicamycin inhibits the expression of surface Na^+ channels in cultured muscle cells."
1199. Schmidt, J., and Catterall, W. A. *Cell* 46: 437–445 (1986). "Biosynthesis and processing of the α-subunit of the voltage-sensitive sodium channels in rat brain neurons."
1200. Catterall, W. A. *Trends in Neurosci.* 9: 7–10 (1986). "Voltage-dependent gating of sodium channels correlating structure and function."
1201. Gordon, R. D.; Fieles, W. E.; Schotland, D. L.; and Barchi, R. L. *Biophys. J.* 51: 437a (1987). "Immunochemical testing of current molecular models of the voltage-dependent sodium channel from *Electrophorus electricus*." Gordon, R. D.; Li, Y.; Fieles, W. E.; Schotland, D. L.; and Barchi, R. L. *J. Neurosci.* 8: 3742–3749 (1988). "Topological localization of a segment of the eel voltage-dependent sodium channel primary sequence (AA 927–938) that discriminates between models of tertiary structure."
1202. Finkelstein, A., and Peskin, C. S. *Biophys. J.* 46: 549–558 (1984). "Some unexpected consequences of a simple physical mechanism for voltage-dependent gating in biological membranes."
1203. Starmer, C. F.; Grant, A. O.; and Strauss, H. C. *Biophys. J.* 46: 15–27 (1984). "Mechanisms of use-dependent block of sodium channels in excitable membranes by local anesthetics."
1204. Venkataraghavan, B., and Feldmann, R. J., eds. *Ann. N.Y. Acad. Sci.* 439: 1–

209 (1985). "Macromolecular structure and specificity: Computer-assisted modeling and applications."

1205. Kosower, E. M., in ref. 1170. "A dynamic and structural model for the sodium channel."

1206. Werman, R. *Compar. Biochem. Physiol.* 18: 745–766 (1966). "Criteria for identification of a central nervous system transmitter."

1207. Bradford, H. F. *Chemical Neurobiology: An Introduction to Neurochemistry.* New York: W. H. Freeman, 1986. 507 pp.

1208. Grenningloh, G.; Rienitz, A.; Schmitt, B.; Methfessel, C.; Zensen, M.; Beyreuther, K.; Gundelfinger, E. D.; and Betz, H. *Nature* 328: 215–220 (1987). "The strychnine-binding subunit of the glycine receptor shows homology with nicotinic acetylcholine receptors."

1209. Stevens, C. F. *Nature* 328: 198 (1987). "Channel families in the brain."

1210. Luyten, W.H.M.L., and Heinemann, S. F. *Ann. Repts. Med. Chem.* 22: 281–291 (1987). "Molecular cloning of the nicotinic acetylcholine receptor: New opportunities in drug design?"

1211. Patrick, J.; Boulter, J.; Deneris, E.; Connolly, J.; Wada, K.; Goldman, D.; and Heinemann, S. Page 78 in ref. 55. Boulter, J.; Connolly, J.; Deneris, E.; Goldman, D.; Heinemann, S.; and Patrick, J. *Proc. Nat'l. Acad. Sci. USA* 84: 7763–7767 (1987). "Functional expression of two neuronal nicotinic acetylcholine receptors from cDNA clones identifies a gene family."

1212. Goldman, D.; Deneris, E.; Luyten, W.; Kochhar, A.; Patrick, J.; and Heinemann, S. *Cell* 48: 965–973 (1987). "Members of a nicotinic acetylcholine receptor gene family are expressed in different regions of the mammalian central nervous system."

1213. Loring, R. H., and Zigmond, R. E. *Neurosci. Abstr.* 11: 92 (1985). "Amino acid sequence of a neurotoxin that blocks neuronal nicotinic receptors and the localization of its binding sites in chick ciliary ganglion."

1214. Halvorsen, S. W., and Berg, D. K. *J. Neurosci.* 6: 3405–3412 (1986). "Identification of a nicotinic acetylcholine receptor on neurons using an α-neurotoxin that blocks receptor function."

1215. Aracava, Y.; Deshpande, S. S.; Swanson, K. L.; Rapoport, H.; Wonnacott, S.; Lunt, G.; and Albuquerque, E. X. *FEBS Lett.* 222: 63–70 (1987). "Nicotinic acetylcholine receptors in cultured neurons from the hippocampus and brain stem of the rat characterized by single channel recording."

1216. Langosch, D.; Thomas, L.; and Betz, H. *Proc. Nat'l. Acad. Sci. USA* 85: 7394–7399 (1988). "Conserved quaternary structure of ligand-gated ion channels: The postsynaptic glycine receptor is a pentamer."

1217. Blair, L.A.C.; Levitan, E. S.; Marshall, J.; Dionne, V. E.; and Barnard, E. A. *Science* 242: 577–579 (1988). "Single subunits of the $GABA_A$ receptor form ion channels with properties of the native receptor."

1218. Barnard, E. A.; Darlison, M. G.; and Seeburg, P. *Trends in Neurosci.* 10: 502–509 (1987). "Molecular biology of the $GABA_A$ receptor—the receptor/channel superfamily."

1219. Aoshima, H.; Anan, M.; Ishii, H.; Iio, H.; and Kobayashi, S. *Biochemistry* 26: 4811–4816 (1987). "Minimal model to account for the membrane conductance increase and desensitization of γ-aminobutyric acid receptors synthesized in the *Xenopus* öocytes injected with rat brain mRNA."

1220. Squires, R.; Casida, J.; Richardson, M.; and Saederup, E. *Mol. Pharmacol.* 23: 326–336 (1983). "[^{35}S]t-Butylbicyclophosphorothionate binds with high affinity to brain-specific sites coupled to γ-aminobutyric acid-A and ion-recognition sites."

1221. Havoundjian, H.; Paul, S. M.; and Skolnick, P. *Proc. Nat'l. Acad. Sci. USA* 83: 9241–9244 (1986). "The permeability of γ-aminobutyric acid-gated chloride channels is described by the binding of a "cage" convulsant, t-butylbicyclophosphoro-[^{35}S]thionate."

1222. Kaila, K.; and Voipio, J. *Nature* 330: 163–165 (1987). "Postsynaptic fall in intracellular pH induced by GABA-activated bicarbonate conductance."

1223. Thomas, R. *Nature* 330: 110–111 (1987). "Inhibition by acidification."

1224. Demuth, D. R.; Showe, L. C.; Ballantine, M.; Palumbo, A.; Fraser, P. J.; Cioe, L.; Rovera, G.; and Curtis, P. J. *EMBO J.* 5: 1205–1214 (1986). "Cloning and structural characterization of a human non-erythroid band-3 like protein."

1225. Kopito, R. R., and Lodish, H. F. *Nature* 316: 234–238 (1985). "Primary structure and transmembrane orientation of the murine anion exchange protein."

1226. Brock, C. J.; Tanner, M.J.A.; and Kempf, C. *Biochem. J.* 213: 577–586 (1983). "The human erythrocyte transport protein."

1227. Jenkins, R. E., and Tanner, M.J.A. *Biochem. J.* 161: 139–147 (1977). "The structure of the major protein of the human erythrocyte membrane."

1228. Lawrence, L. J., and Casida, J. E. *Science* 221: 1399–1401 (1983). "Stereospecific action of pyrethroid insecticides on the γ-aminobutyric acid receptor-ionophore complex."

1229. Falke, J. J., and Chan, S. I. *Biochemistry* 25: 7895–7898 (1986). "Molecular mechanisms of band 3 inhibitors. 2. Channel blockers."

1230. Falke, J. J., and Chan, S. I. *Biochemistry* 25: 7899–7906 (1986). "Molecular mechanisms of band 3 inhibitors. 3. Translocation inhibitors."

1231. Bormann, J.; Hamill, O. P.; and Sakmann, B. *J. Physiol. (Lond.)* 385: 243–286 (1987). "Mechanism of anion permeation through channels gated by glycine and γ-aminobutyric acid in mouse cultured spinal neurones."

1232. Dwyer, T. M.; Adams, D. J.; and Hille, B. *J. Gen. Physiol.* 75: 469–492 (1980). "The permeability of the endplate channel to organic cations in frog muscle."

1233. Betz, H. *Angew. Chem. Int'l. Edn.* 24: 365–370 (1985). "The glycine receptor of rat spinal cord: Exploring the site of action of the plant alkaloid strychnine."

1234. Aprison, M. H.; Lipkowitz, K. B.; and Simon, J. R. *J. Neurosci. Res.* 17: 209–213 (1987). "Identification of a glycine-like fragment on the strychnine molecule."

1235. Sullivan, K. A.; Miller, R. T.; Masters, S. B.; Beiderman, H.; Heidman, W.; and Bourne, H. R. *Nature* 330: 758–760 (1987). "Identification of receptor contact involved in receptor-G protein coupling."

1236. Gilman, A. G. *Ann. Rev. Biochem.* 56: 615–649 (1987). "G proteins: Transducers of receptor-generated signals."

1237. Hescheler, J.; Rosenthal, W.; Trautwein, W.; and Schultz, G. *Nature* 325: 445–447 (1987). "The GTP-binding protein, G_o, regulates neuronal calcium channels."

1238. Masu, Y.; Nakayama, K.; Tamaki, H.; Harada, Y.; Kuno, M.; and Nakanishi, S. *Nature* 329: 836–838 (1987). "cDNA cloning of bovine substance-K receptor through oocyte expression system." Cf. *Nature* 342: 643–8 (1989).

1239. Frielle, T.; Collins, S.; Daniel, K. W.; Caron, M. G.; Lefkowitz, R. J.; and

Kobilka, B. K. *Proc. Nat'l. Acad. Sci. USA* 84: 7920–4 (1987). "Cloning of the cDNA for the human β_1-adrenergic receptor."

1240. Kobilka, B. K.; Matsui, H.; Kobilka, T. S.; Yang-Feng, T. L.; Francke, U.; Caron, M. G.; Lefkowitz, R. J.; and Regan, J. W. *Science* 238: 650–656 (1987). "Cloning, sequencing, and expression of the gene coding for the human platelet α_2-adrenergic receptor."

1241. Julius, D.; MacDermott, A. B.; Axel, R.; and Jessell, T. M. *Science* 241: 558–564 (1988). "Molecular characterization of a functional cDNA encoding the serotonin 1c receptor."

1242. Levitzki, A. *FEBS Lett.* 211: 113–118 (1987). "Regulation of adenylate cyclase by hormones and G-proteins."

1243. Dohlman, H. G.; Caron, M. G.; and Lefkowitz, R. J. *Biochemistry* 26: 2657–2664 (1987). "A family of receptors coupled to guanine nucleotide regulatory proteins."

1244. Dohlman, H. G.; Bouvier, M.; Benovic, J. L.; Caron, M. G.; and Lefkowitz, R. J. *J. Biol. Chem.* 262: 14282–14286 (1987). "The multiple membrane topography of the β_2-adrenergic receptor."

1245. Dixon, R.A.F.; Sigal, I. S.; Rands, E.; Kobilka, B. K.; Register, R. B.; Candelore, M. R.; Blake, A. D.; and Strader, C. D. *Nature* 326: 73–77 (1987). "Ligand binding to the β-adrenergic receptor involves its rhodopsin-like core."

1246. Dixon, R.A.F.; Sigal, I. S.; Candelore, M. R.; Register, R. B.; Scattergood, W.; Rands, E.; and Strader, C. D. *EMBO J.* 6: 3269–3275 (1987). "Structural features required for ligand binding to the β-adrenergic receptor."

1247. Fraser, C. M.; Chung, F.-Z.; Wang, C.-D.; and Venter, J. C. *Proc. Nat'l. Acad. Sci. USA* 85: 5478–5482 (1988). "Site-directed mutagenesis of human β-adrenergic receptors: Substitution of aspartic acid-130 with asparagine produces a receptor with high-affinity agonist binding that is uncoupled from adenylate cyclase." Bunzow, J. R.; Van Tol, H.H.M.; Grandy, D. K.; Albert, P.; Salon, J.; Christie, M.; Machida, C. A.; Neve, K. A.; and Civelli, O. *Nature* 336: 783–787 (1988). "Cloning and expression of a rat D2 dopamine receptor cDNA."

1248. Sibley, D. R.; Benovic, J. L.; Caron, M. G.; and Lefkowitz, R. J. *Cell* 48: 913–922 (1987). "Regulation of transmembrane signalling by receptor phosphorylation."

1249. Benovic, J. L.; Mayor, F., Jr.; Staniszewski, C.; Lefkowitz, R. J.; and Caron, M. G. *J. Biol. Chem.* 262: 9026–9032. "Purification and characterization of the β-adrenergic receptor kinase."

1250. Palczewski, K.; McDowell, J. H.; and Hargrave, P. A. *J. Biol. Chem.* 263: 14067–14073 (1988). "Purification and characterization of rhodopsin kinase."

1251. Yee, G. H., and Huganir, R. L. *J. Biol. Chem.* 262: 16748–16753 (1987). "Determination of the sites of cAMP-dependent phosphorylation on the nicotinic acetylcholine receptor."

1252. Haga, K.; Haga, T.; and Ichiyama, A. *J. Biol. Chem.* 261: 10133–10140 (1986). "Reconstitution of the muscarinic acetylcholine receptor."

1253. Peralta, E. G.; Winslow, J. W.; Peterson, G. L.; Smith, D. H.; Ashkenazi, A.; Ramachandran, J.; Schimerlik, M. I.; and Capon, D. J. *Science* 236: 600–605 (1987). "Primary structure and biochemical properties of an M2 muscarinic receptor."

1254. Akiba, I.; Kubo, T.; Maeda, A.; Bujo, H.; Nakai, J.; Mishina, M.; and Numa,

S. *FEBS Lett.* 235: 257–261 (1988). ''Primary structure of porcine muscarinic acetylcholine receptor III and antagonist binding studies.''

1255. Giachetti, A.; Mittmann, U.; Brown, J. H.; Goldstein, D.; Masters, S. B.; Birdsall, N.J.M.; Hulme, E. C.; Kromer, W.; Stockton, J. W.; Engel, W.; Eberlein, W.; Trummlitz, G.; Mihm, G.; Ladinsky, H.; Giraldo, E.; Schiavi, G. B.; Monferini, E.; Hammer, R.; Montagna, E.; Micheletti, R.; Roeske, W. R.; Bahl, J.; Vickroy, T.; Watson, M.; Yamamura, H. I.; Schuessler, R.; Boineau, J.; and Watanabe, A. M. *Fed. Proc.* 46: 2523–2535 (1987). ''Symposium summary: Cardioselective muscarinic antagonists.''

1256. Fukuda, K.; Kubo, T.; Akiba, I.; Maeda, A.; Mishina, M.; and Numa, S. *Nature* 327: 623–625 (1987). ''Molecular distinction between muscarinic acetylcholine receptor subtypes.''

1257. Haga, T.; Haga, K.; Berstein, G.; Nishiyama, T.; Uchiyama, H.; and Ichiyama, A. *Trends in Pharmacol. Sci.* Suppl., pp. 12–18, *Subtypes of muscarinic receptors III*, ed. Levine, R. R.; Birdsall, N.J.M.; North, R. A.; Holman, M.; Watanabe, A.; Iversen, L. L., Elsevier, Cambridge, U.K., 1988. ''Molecular properties of muscarinic receptors.''

1258. Bonner, T. I.; Buckley, N. J.; Young, A. C.; and Brann, M. R. *Science* 237: 527–531 (1987). ''Identification of a family of muscarinic acetylcholine receptor genes.''

1259. Peralta, E. G.; Winslow, J. W.; Ashkenazi, A.; Smith, D. H.; Ramachandran, J.; and Capon, D. J. *Trends in Pharmacol. Sci.* Suppl., pp. 6–11, Subtypes of muscarinic receptors III, ed. R. R. Levine, N.J.M. Birdsall, R. W. North, M. Holman, A. Watanabe, L. L. Iversen. Cambridge, England: Elsevier, 1988. ''Structural basis of muscarinic acetylcholine receptor subtype diversity.''

1260. Hulme, E., and Birdsall, N. *Nature* 323: 396–397 (1987). ''Distinctions in acetylcholine receptor activity.''

1261. Wheatley, M.; Hulme, E. C.; Birdsall, N.J.M.; Curtis, C.A.M.; Eveleigh, P.; Pedder, E. K.; and Poyner, D. *Trends in Pharmacol. Sci.* Suppl., pp. 19–24, Subtypes of muscarinic receptors III, ed. R. R. Levine, N.J.M. Birdsall, R. A. North, M. Holman, A. Watanabe, L. L. Iversen, Cambridge, England: Elsevier, 1988.

1262. Ashkenazi, A.; Winslow, J. W.; Peralta, E. G.; Peterson, G. L.; Schimerlik, M. I.; Capon, D. J.; and Ramachandran, J. *Science* 238: 672–675 (1987). ''An M2 muscarinic receptor subtype coupled to both adenylyl cyclase and phosphoinositide turnover.''

1263. Pfaffinger, P. J.; Martin, J. M.; Hunter, D. D.; Nathanson, N. M.; and Hille, B. *Nature* 317: 536–538 (1985). ''GTP-binding proteins couple cardiac muscarinic receptors to a K channel.''

1264. Martin, J. M.; Hunter, D. D.; and Nathanson, N.M. *Biochemistry* 24: 7521–7525 (1985). ''Islet-activating protein inhibits physiological responses evoked by cardiac muscarinic receptors. Role of guanosine triphosphate-binding proteins in regulation of potassium permeability.''

1265. Holz, G. G., IV; Rane, S. G; and Dunlap, K. *Nature* 319: 670–672 (1986). ''GTP-binding proteins mediate transmitter inhibition of voltage-dependent calcium channels.''

1266. Codina, J.; Yatani, A.; Grenet, D.; Brown, A. M.; and Birnbaumer, L. *Science* 236: 442–445 (1987). ''The α-subunit of the GTP binding protein G_k opens atrial potassium channels.''

1267. Yatani, A.; Codina, J.; Imoto, Y.; Reeves, J. P.; Birnbaumer, L.; and Brown, A. M. *Science* 238: 1288–1292 (1987). "A G-protein directly regulates mammalian cardiac calcium channels."

1268. Bormann, J. *Trends in Neurosci.* 11: 112–116 (1988). "Electrophysiology of $GABA_A$ and $GABA_B$ receptor subtypes."

1269. Breitwieser, G. E., and Szabo, G. *Nature* 317: 538–540 (1985). "Uncoupling of cardiac muscarinic and β-adrenergic receptors from ion channels by a guanine nucleotide analogue."

1270. Erspamer, V. *Trends in Neurosci.* 4: 267–269 (1981). "The tachykinin peptide family."

1271. Laufer, R.; Wormser, U.; Friedman, Z. Y.; Gilon, C.; Chorev, M.; and Selinger, Z. *Proc. Nat'l. Acad. Sci. USA* 82: 7444–7448 (1985). "Neurokinin B is a preferred agonist for a neuronal substance P receptor and its action is antagonized by enkephalin."

1272. Hanley, M. R., and Jackson, T. *Nature* 329: 766–767 (1987). "Return of the magnificent seven."

1273. Wormser, U.; Laufer, R.; Hart, Y.; Chorev, M.; Gilon, C.; and Selinger, Z. *EMBO J.* 5: 2805–2808 (1986). "Highly selective agonists for substance P receptor subtypes."

1274. Hanley, M. R. *Nature* 315: 14–15 (1985). "Neuropeptides as mitogens."

1275. Hirasawa, K., and Nishizuka, Y. *Ann. Rev. Pharmacol. Toxicol.* 25: 147–170 (1985). "Phosphatidylinositol turnover in receptor in receptor mechanism and signal transduction."

1276. Abdel-Latif, A. A. *Pharmacol. Revs.* 38: 227–272 (1986). "Calcium-mobilizing receptors, polyphosphoinositides, and the generation of second messengers."

1277. Burgess, G. M.; Godfrey, P. P.; McKinney, J. S.; Berridge, M. J.; Irvine, R. F.; and Putney, J. W., Jr. *Nature* 309: 63–66 (1984). "The second messenger linking receptor activation to internal Ca release in liver."

1278. Morris, A. P.; Gallacher, D. V.; Irvine, R. F.; and Petersen, O. H. *Nature* 330: 653–655 (1987). "Synergism of inositol trisphosphate and inositol tetrakisphosphate in activating Ca^{++}-dependent K^+ channels."

1279. Vallejo, M.; Jackson, T.; Lightman, S.; and Hanley, M. R. *Nature* 330: 656–658 (1987). "Occurrence and extracellular actions of inositol pentakisphosphate and inositol pentakisphosphate."

1280. Altman, J. *Nature* 331: 119–120 (1988). "Ins and outs of cell signalling."

1281. Guillemette, G.; Balla, T.; Baukal, A. J.; and Catt, K. J. *Proc. Nat'l. Acad. Sci. USA* 84: 8195–8199 (1987). "Inositol 1, 4, 5-trisphosphate binds to a specific receptor and releases microsomal calcium in the anterior pituitary gland."

1282. Sakakibara, M.; Alkon, D. L.; Neary, J. T.; Heldman, E.; and Gould, R. *Biophys. J.* 50: 797–803 (1986). "Inositol trisphosphate regulation of photoreceptor membrane currents."

1283. Devary, O.; Heichal, O.; Blumenfeld, A.; Cassel, D.; Suss, E.; Barash, S.; Rubinstein, C. T.; Minke, B.; and Selinger, Z. *Proc. Nat'l. Acad. Sci. USA* 84: 6939–6943 (1987). "Coupling of photoexcited rhodopsin to inositol phospholipid hydrolysis in fly photoreceptors."

1284. Worley, P. F.; Baraban, J. M.; McCarren, M.; Snyder, S. H.; and Alger, B. E. *Proc. Nat'l. Acad. Sci. USA* 84: 3467–3471 (1987). "Cholinergic phosphatidylinositol modulation of inhibitory, G-protein linked, neurotransmitter actions: Electrophysiological studies in rat hippocampus.''

1285. Parker, P. J.; Coussens, L.; Totty, N.; Rhee, L.; Young, S.; Chen, E.; Stabel, S.; Waterfield, M. D.; and Ullrich, A. *Science* 233: 853–859 (1986). ''The complete primary structure of protein kinase C—the major phorbol ester receptor.''
1286. Coussens, L.; Parker, P. J.; Rhee, L.; Yang-Feng, T. L.; Chen, E.; Waterfield, M. D.; Francke, U.; and Ullrich, A. *Science* 233: 859–866 (1986). ''Multiple, distinct forms of bovine and human protein kinase C suggest diversity in cellular signalling pathways.''
1287. Dascal, N.; Ifune, C.; Hopkins, R.; Snutch, T. P.; Lübbert, H.; Davidson, N.; Simon, M. I.; and Lester, H. A. *Molecular Brain Res.* 1: 201–209 (1986). ''Involvement of a GTP-binding protein in mediation of serotonin and acetylcholine responses in Xenopus oocytes injected with rat brain messenger RNA.''
1288. Lübbert, H.; Hoffman, B. J.; Snutch, T. P.; Vandyke, T.; Levine, A. J.; Hartig, P. R.; Lester, H. A.; and Davidson, N. *Proc. Nat'l. Acad. Sci. USA* 84: 4332–4336 (1987). ''CDNA cloning of a serotonin 5-HT_{1C} receptor by electrophysiological assays of messenger RNA-injected *Xenopus* öocytes.''
1289. Levitan, E. S. *Trends in Neurosci.* 11: 41–43 (1988). ''Cloning of serotonin and substance K receptors by functional expression in öocytes.''
1290. Lübbert, H. ''Isolation by functional expression and analysis of 5-HT receptor cDNA and genomic clones,'' in ref. 1170.
1291. Neer, E. J., and Clapham, D. E. *Nature* 333: 129–134 (1988). ''Roles of G-protein subunits in transmembrane signalling.''
1292. Smith, C. D.; Cox, C. C.; and Snyderman, R. *Science* 232: 97–100 (1986). ''Receptor-coupled activation of phosphoinositide-specific phospholipase C by a N-protein.''
1293. Sugiyama, H.; Ito, I.; and Hirono, C. *Nature* 325: 531–533 (1987). ''A new type of glutamate receptor linked to inositol phospholipid metabolism.''
1294. Verdoorn, T. A.; Kleckner, N. W.; and Dingledine, R. *Science* 238: 1114–1116 (1987). ''Rat brain N-methyl-D-aspartate receptors expressed in *Xenopus* öocytes.''
1295. Piomelli, D.; Volterra, A.; Dale, N.; Siegelbaum, S. A.; Kandel, E. R.; Schwartz, J. H.; and Belardetti, F. *Nature* 328: 38–43 (1987). ''Lipoxygenase metabolites of arachidonic as second messengers for presynaptic inhibition of *Aplysia* sensory cells.''
1296. Iversen, L. *Sci. Amer.* 241(3): 134–149 (1979). ''Chemistry of the brain.''
1297. Shepherd, G. M. *Neurobiology*, p. 153. New York: Oxford University Press, 1983.
1298. Hirono, C.; Ito, I.; and Sugiyama, H. *J. Physiol. (Lond.)* 382: 523–535 (1987). ''Neurotensin and acetylcholine evoke common responses in frog öocytes injected with rat brain messenger ribonucleic acid.''
1299. Tanabe, T.; Takeshima, H.; Mikami, A.; Flockerzi, V.; Takahashi, H.; Kangawa, K.; Kojima, M.; Matsuo, H.; Hirose, T.; and Numa, S. *Nature* 328: 313–318 (1987). ''Primary structure of the receptor for calcium channel blockers from skeletal muscle.'' Tanabe, T.; Beam, K. G.; Powell, K. A.; and Numa, S. *Nature* 336: 134–139 (1988). ''Restoration of excitation-contraction coupling and slow calcium current in dysgenic muscle by dihydropyridine receptor complementary DNA.'' Reuter, H., and Porzig, H. *Nature* 336: 113–114 (1988). ''Calcium channels: Diversity and complexity.''

1300. Salkoff, L.; Butler, A.; Wei, A.; Scavarda, N.; Giffen, K.; Ifune, C.; Goodman, R.; and Mandel, G. *Science* 237: 744–753 (1987). "Genomic organization and deduced amino acid sequence of a putative sodium channel gene in *Drosophila*."

1301. Tempel, B. L.; Papazian, D. M.; Schwarz, T. L.; Jan, Y. N.; and Jan, L. Y. *Science* 237: 770–775 (1987). "Sequence of a probable potassium channel component encoded at Shaker locus of *Drosophila*."

1302. Agnew, W. S. *Nature* 331: 114–115 (1988). "A Rosetta stone for K^+ channels."

1303. Schwarz, T. L.; Tempel, B. L.; Papazian, D. M.; Jan, Y. N.; and Jan, L. Y. *Nature* 331: 137–142 (1988). "Multiple potassium channel components are produced by alternative splicing at the Shaker locus in *Drosophila*."

1304. Timpe, L. C.; Schwarz, T. L.; Tempel, B. L.; Papazian, D. M.; Jan, Y. N.; and Jan, L. Y. *Nature* 331: 142–145 (1988). "Expression of functional potassium channels from Shaker cDNA in *Xenopus* öocytes."

1305. Cook, N. S. *Trends in Pharmacol. Sci.* 9: 21–28 (1988). "The pharmacology of potassium channels and their therapeutic potential."

1306. Takumi, T.; Ohkubo, H.; and Nakanishi, S. *Science* 242: 1042–1045 (1988). "Cloning of a membrane protein that induces a slow voltage-gated potassium current."

1307. Norman, R. I.; Burgess, A. J.; Allen, E.; and Harrison, T. M. *FEBS Lett.* 212: 127–132 (1987). "Monoclonal antibodies identifies the 1, 4-dihydropyridine receptor associated with voltage-sensitive Ca^{++} channels detect similar polypeptides from a variety of tissues and species."

1308. Morton, M. E., and Froehner, S. C. *J. Biol. Chem.* 262: 11904–11907 (1987). "Monoclonal antibody identifies a 200-kDa subunit of the dihydropyridine-sensitive calcium channel."

1309. Miller, R. J. *Science* 235: 46–52 (1987). "Multiple calcium channels and neuronal function."

1310. Petersen, S. E.; Fox, P. T.; Posner, M. I.; Mintun, M.; and Raichle, M. E. *Nature* 331: 585–589 (1988). "Positron emission tomographic studies of the cortical anatomy of single word processing." Posner, M. I.; Petersen, S. E.; Fox, P. T.; and Raichle, M. E. *Science* 240: 1627–1631 (1988). "Visualization of cognitive operations in the human brain." Kosslyn, S. M. *Science* 240: 1621–1626 (1988). "Aspects of a cognitive science of mental imagery."

1311. Marshall, J. C. *Nature* 331: 560–561 (1988). "The lifeblood of language."

1312. Dudai, Y. *Trends in Neurosci.* 8: 18–21 (1985). "Genes, enzymes and learning in Drosophila."

1313. Dudai, Y. *Adv. in Cyclic Nucleotide & Protein Phosphorylation Research* 20: 343–363 (1986). "Cyclic AMP and learning in *Drosophila*."

1314. Tully, T. *Trends in Neurosci.* 10: 330–335 (1987). "*Drosophila* learning and memory revisited."

1315. Mishkin, M., and Appenzeller, T. *Sci. Amer.* 256(6): 62–71 (1987). "The anatomy of memory."

1316. John, E. R.; Tang, Y.; Brill, A. B.; Young, R.; and Ono, K. *Science* 233: 1167–1175 (1986). "Double-labeled metabolic maps of memory."

1317. Shashoua, V. E. *Advances in Biosci.* 61: 245–254 (1986). "The role of neurosecretory cells in learning and memory."

1318. Schmidt, R. "Biochemical participation of glycoproteins in memory consolidation after two different training paradigms in goldfish.'' In *Learning and Memory*, ed. H. Matthies, pp. 213–222. Oxford: Pergamon, 1986.

1319. Changeux, J.-P.; Klarsfeld, A.; and Heidmann, T. ''The acetylcholine receptor and molecular models for short- and long-term memory.'' Pages 31–84 in ref. 44.

1320. Rose, S.P.R. ''Obstacles and progress in studying the cell biology of learning and memory.'' In *Learning and Memory*, ed. H. Matthies, pp. 165–172. Oxford: Pergamon, 1986.

1321. Kandel, E. R. *Cellular Basis of Behavior*. San Francisco: W. H. Freeman, 1976. 727 pp.

1322. Kandel, E. R. ''Cellular mechanisms of learning and the biological basis of individuality.'' Chapter 62, pages 816–833 in ref. 17.

1323. Kandel, E. R. *Sci. Amer.* 223(1): 57–67 (1970). ''Nerve cells and behavior.''

1324. Howard, J., and Hudspeth, A. J. *Biophys. J.* 53: 411a (1988). ''Mechanical gating of transduction channels in otic hair cells.''

1325. Sachs, F. *Biophys. J.* 53: 410a (1988). ''Mechanical transduction and stretch-activated ion channels.''

1326. De Camili, P., and Greengard, P. *Biochem. Pharmacol.* 35: 4349–4357 (1986). ''Synapsin I: A synaptic vesicle-associated neuronal phosphoprotein.''

1327. Spira, M. E.; Blumenfeld, H.; Schacher, S.; and Sielgelbaum, S. A. *Neurosci. Abstr.* 13: 1238 (1987). ''Measurement of free intracellular calcium in *Aplysia* sensory neurons using Fura-2.''

1328. Bailey, C. H., and Chen, M. *Neurosci. Abstr.* 13: 617 (1987). ''The course of structural changes at identified sensory synapses during long-term sensitization.''

1329. Glanzman, D. L.; Kandel, E. R.; and Schacher, S. *Neurosci. Abstr.* 13: 617 (1987). ''Individual *Aplysia* sensory neurons visualized over time in cell culture: Attempts to correlate long-term synaptic and morphological changes.''

1330. Schacher, S.; Montarolo, P.; Castellucci, V.; Kandel, E. R.; and Goelet, P. *Biochem. Soc. Trans.* 15: 125–126 (1987). ''Molecular biological approaches to the relationship between short-term and long-term memory in *Aplysia*.'' Schacher, S.; Castellucci, V.; and Kandel, E. R. *Science* 240: 1667–1669 (1988). ''cAMP evokes long term facilitation in *Aplysia* sensory neurons that requires new protein synthesis.''

1331. Laufer, R., and Changeux, J.-P. *EMBO J.* 6: 901–906 (1987). ''Calcitonin-gene related peptide elevates cyclic AMP levels in chick skeletal muscle: possible neurotrophic role for a coexisting neuronal messenger.''

1332. Yovell, Y.; Kandel, E. R.; Dudai, Y.; and Abrams, T. W. *Proc. Nat'l. Acad. Sci. USA* 84: 9285–9289 (1987). ''Biochemical correlates of short-term sensitization in *Aplysia*: Temporal analysis of adenylate cyclase stimulation in a perfused-membrane preparation.''

1333. Chin, G. J.; Vogel, S. S.; Mumby, S. M.; and Schwartz, J. H. *Neurosci. Abstr.* 13: 315 (1987). ''Characterization and localization of G_s and G_o protein molecules in *Aplysia* neurons.''

1334. Kandel, E. R., and Schwartz, J. H. *Science* 218: 433–443 (1982). ''Molecular biology of learning: Modulation of transmitter release.''

1335. Wu, J.-Y.; Zecevic, D. P.; London, J. A.; Rioult, M.; and Cohen, L. B. *Neurosci. Abstr.* 13: 817 (1987). ''Optical measurement of neuron activity during the gill withdrawal reflex in *Aplysia*.''

1336. Mackey, S. L.; Glanzman, D. L.; Small, S. A.; Dyke, A. M.; Kandel, E. R.;

and Hawkins, R. D. *Proc. Nat'l. Acad. Sci. USA* 84: 8730–8734 (1987). "Tail shock produces inhibition as well as sensitization of the siphon-withdrawal reflex of *Aplysia*: Possible behavioral role for presynaptic inhibition mediated by the peptide phe-met-arg-phe-NH_2."

1337. Grinvald, A. *Ann. Rev. Neurosci.* 8: 263–305 (1985). "Real-time optical mapping of neuronal activity: From single growth cones to the intact mammalian brain."

1338. Shrager, P.; Chiu, S. Y.; Ritchie, J. M.; Zecevic, D.; Cohen, L. B., *Biophys. J.* 51: 351–355 (1987). "Optical recording of action potential propagation in demyelinated frog nerve."

1339. Cohen, L. *Nature* 331: 112–113 (1988). "More light on brains."

1340. Kauer, J. S. *Nature* 331: 166–168 (1988). "Real time imaging of evoked activity in local circuits of the salamander olfactory bulb."

1341. Alkon, D. I. *Sci. Amer.* 249(1): 62–74 (1983). "Learning in a marine snail."

1342. Grover, L. M., and Farley, J. *Behavior. Neurosci.* 101: 658 (1987). "Temporal order sensitivity of associative neural and behavioral changes in *Hermissenda*."

1343. Farley, J., and Auerbach, S. *Nature* 319: 220–233 (1986); 324: 702 (1986). "Protein kinase C activation induces conductance changes in *Hermissenda* photoreceptors like those seen in associative learning."

1344. Farley, J.; Resnick, D.; and Auerbach, S. *Neurosci. Abstr.* 13: 389 (1987). "PKC closes single voltage dependent K channels in *Hermissenda* type B photoreceptors."

1345. Schuman, E. M., and Farley, J. *Neurosci. Abstr.* 13: 389 (1987). "A PKC-inhibitor prevents in vitro conditioning produced changes in *Hermissenda* type B photoreceptors."

1346. Crow, T., and Forrester, J. *Proc. Nat'l. Acad. Sci. USA* 83: 7975–7978 (1986). "Light paired with serotonin mimics the effects of conditioning on phototactic behavior of *Hermissenda*."

1347. Crow, T., and Forrester, J. *Neurosci. Abstr.* 13: 389 (1987). "Inhibition of protein synthesis blocks long term plasticity in identified B-photoreceptors in *Hermissenda*."

1348. Alkon, D. L.; Acosta-Urquidi, J.; Olds, J.; Kuzma, G.; and Neary, J. T. *Science* 219: 303–306 (1987). "Protein kinase injection reduces voltage-dependent potassium currents."

1349. Greenberg, M. E.; Ziff, E. B.; and Greene, L. A. *Science* 234: 80–83 (1986). "Stimulation of neuronal acetylcholine receptors induces rapid gene transcription."

1350. Black, I. B.; Adler, J. E.; Dreyfus, C. F.; Friedman, W. F.; LaGamma, E. F.; and Roach, A. H. *Science* 236: 1263–1268 (1987). "Biochemistry of information storage in the nervous system."

1351. Blakely, R. D.; Ory-Lavollee, L.; Thompson, R. C.; and Coyle, J. T. *J. Neurochem.* 47: 1013–1019 (1986). "Synaptosomal transport of radiolabel from N-acetyl-aspartyl-[^{3}H]glutamate suggests a mechanism of inactivation of an excitatory neuropeptide." Blakely, R. D., and Coyle, J. T. *Int'l. Rev. Neurobiol.* 30: 39–46 (1988). "The neurobiology of N-acetylaspartylglutamate."

1352. Robinson, M. B.; Stauch, B. L.; and Coyle, J. T. *Neurosci. Abstr.* 13: 765 (1987). "Partial purification of N-acetylated-alpha-linked acidic dipeptidase (NAALADase): A quisqualate-sensitive peptidase that cleaves N-acetyl-aspartyl-glutamate to N-acetylaspartate and glutamate in vitro."

1353. Lauridsen, J.; Honore, T.; and Krøgsgaard-Larsen, P. *J. Med. Chem.* 28: 668–

672 (1985). ''Ibotenic acid analogues. Synthesis, molecular flexibility and *in vitro* activity of agonists and antagonists at central glutamic acid receptors.''

1354. Goldberg, O., and Teichberg, V. I. *J. Med. Chem.* 28: 1957–1958 (1985). ''A dipeptide derived from kainic and L-glutamic acids: A selective antagonist of amino acid induced neuroexcitation with anticonvulsant properties.''
1355. Noszek, J. C., and Siman, R. *Neurosci. Abstr.* 13: 1684 (1987). ''An excitatory amino acid activates calpain I and structural protein breakdown in vivo.''
1356. Glaser, T., and Kosower, N. S. *FEBS Lett.* 206: 115–120 (1986). ''Fusibility of erythrocytes is determined by a protease-protease inhibitor (calpain-calpastatin) balance.''
1357. Levis, R. A. *Biophys. J.* 53: 226a (1988). ''Single Na channel currents from squid giant axon following removal of fast inactivation by pronase.''
1358. Llinas, R., and Sugimori, M. *J. Physiol. (Lond.)* 305: 171 (1980). ''Non-inactivating Na channels of Purkinje cells?''
1359. Steinberg, I. Z. *Neurosci. Abstr.* 13: 178 (1987). ''Computer simulation of the effect of non-inactivating sodium channels on excitability.''
1360. Wilson, D. *Nature* 320: 313–314 (1986). ''Scotophobin resurrected as a neuropeptide.''
1361. Shyng, S.-L., and Salpeter, M. M. *Neurosci. Abstr.* 13: 150 (1987). ''Degradation of 'new' receptors inserted into denervated vertebrate neuromuscular junctions and their response to reinnervation.''
1362. Margiotta, J. F.; Berg, D. K.; and Dionne, V. E. *Proc. Nat'l. Acad. Sci. USA* 84: 8155–8159 (1987). ''Cyclic AMP regulates the proportion of functional acetylcholine receptors on chicken ciliary ganglion neurons.''
1363. Thompson, R. F. *Science* 233: 941–947 (1986). ''The neurobiology of learning and memory.''
1364. Livingstone, M. S. *Sci. Amer.* 258(1): 68–75 (1988). ''Art, illusion and the visual system.''
1365. Hubel, D. H., and Wiesel, T. N. *Nature* 221: 747–750 (1969). ''Anatomical demonstration of columns in the monkey striate cortex.''
1366. Blasdel, G. G., and Salama, G. *Nature* 321: 579–585 (1986). ''Voltage-sensitive dyes reveal a modular organization in monkey striate cortex.''
1367. Grinvald, A.; Lieke, E.; Frostig, R. D.; Gilbert, C. D.; Wiesel, T. N. *Nature* 324: 361–364 (1986). ''Functional architecture of cortex revealed by optical imaging of intrinsic signals.''
1368. Mountcastle, V. B. ''An organizing principle for cerebral function: The unit module and the distributed system.'' Pages 21–42 in ref. 25(d).
1369. Wurtz, R. H.; Goldberg, M. E.; and Robinson, D. L. *Progr. Neurosci.*, pp. 82–90. New York: W. H. Freeman, 1986.
1370. Waitzman, D. M.; Ma, T. P.; Optican, L. M.; and Wurtz, R. H. *Neurosci. Abstr.* 13: 394 (1987). ''Memory contingent responses of *superior collicus* movement cells.''
1371. Miashita, Y., and Chang, H. S. *Nature* 331: 68–70 (1988). ''Neuronal correlate of pictorial short-term memory in the primate temporal cortex.'' Miyashita, Y. *Nature* 335: 817–820 (1988). ''Neuronal correlate of visual associative memory in the primate temporal complex.''
1372. Fig. 22–6, opposite p. 269 in ref. 17.
1373. Alkon, D. L., and Rasmussen, H. *Science* 239: 998–1005 (1988). ''A spatial-

temporal model of cell activation.'' Gruener, R., and Hoeger, G. *The Physiologist, Suppl.* 31: S48–S49 (1988). ''Does vector-free gravity simulate microgravity? Functional and morphological attributes of clinorotated nerve and muscle grown in cell culture.''

1374. Squire, L. N. In *Memory and Brain*, pp. 17–18. New York: Oxford University Press, 1987. 315 pp.

1375. Crick, F. *Trends in Neurosci.* 5: 44–46 (1982). ''Do dendritic spines twitch?''

1376. Lisman, J. E. *Proc. Nat'l. Acad. Sci. USA* 82: 3055–3057 (1985). ''A mechanism for memory storage insensitive to molecular turnover: A bistable autophosphorylating kinase.''

1377. Proust, M. *Remembrance of Things Past: Swann's Way*, trans. C.K.S. Moncrieff, pp. 34–36. New York: Random House, 1928. Reprint, New York: Vintage, 1970.

1378. Müller, S. C.; Plesser, T.; and Hess, B. *Science* 230, 661–663 (1985). ''The structure of the core of the spiral wave in the Belousov-Zhabotinskii reaction.''

1379. Müller, S. C.; Plesser, T.; and Hess, B. *La Rabida Conference*, 1986. NATO Advanced Study Institute Series, 1987. ''Chemical waves and natural convection.''

1380. Epstein, I. R. *Chem. Engr. News* 65(13): 24–36 (1987). ''Patterns in time and space generated by chemistry.''

1381. Müller, S. C., and Plesser, T. *Lecture Notes in Biomathematics* 55: 246–253 (1984). ''Spatial pattern formation in thin layers of NADH-solutions,'' In ''Modelling of Patterns in Space and Time,'' workshop, Heidelberg, Germany, July 4–8, 1983.

1382. Müller, S. C.; Plesser, T.; and Hess, B. Proc. *Int'l. Conf. on Synergetics and Temporal Order*, Bremen, Germany, Sept. 1984. Berlin: Springer, 1986. ''Coupling of glycolytic oscillations and convective patterns.''

1383. Luther, R. *Zeit. für Elektrochem.* 12: 596 (1906). ''Propagation of chemical reactions in space.'' Translation: Arnold, R.; Showalter, K.; and Tyson, J. J. *J. Chem. Ed.* 64: 740–742 (1987). Discussion: Arnold, R., and Tyson, J. J. *J. Chem. Ed.* 64: 742–744 (1987). ''Luther's 1906 discovery and analysis of chemical waves.''

1384. Sutula, T.; Xiao-Xian, H.; Cavazos, J.; and Scott, G. *Science* 239: 1147–1151 (1988). ''Synaptic reorganization in the hippocampus induced by abnormal functional activity.''

1385. Anderson, J. A. *Nature* 322: 406–407 (1986). ''Neural models: Networks for fun and profit.'' Crick, F. *Nature* 337: 129–132 (1989). ''The recent excitement about neural networks.''

1386. Pellionisz, A., and Llinas, R. *Neuroscience* 4: 323–348 (1979). ''Brain modeling by tensor network theory and computer simulation. The cerebellum: Distributed processor for predictive coordination.''

1387. Hopfield, J. J. *Science* 233: 625–633 (1986). ''Computing with neural circuits: A model.'' See comments, *Science* 235: 1226–1229 (1987).

1388. Tank, D. W., and Hopfield, J. J. *Proc. Nat'l. Acad. Sci. USA* 84: 1896–1900 (1987). ''Neural computation by concentrating information in time.''

1389. Hopfield, J. J. *Proc. Nat'l. Acad. Sci. USA* 84: 8429–8433 (1987). ''Learning algorithms and probability distributions in feed-forward and feed-back networks.''

1390. Nadal, J. P.; Toulouse, G.; Changeux, J.-P.; Dehaene, S. *Europhys. Lett.* 1: 535–542 (1986). ''Networks of formal neurons and memory palimpsests.''

1391. Bachmann, C. M.; Cooper, L. N.; Dembo, L. N.; Zeitouni, O. *Proc. Nat'l. Acad. Sci. USA* 84: 7529–7531 (1987). "A relaxation model for memory with high storage density."

1392. Hartline, D. K. *Neurosci. Abstr.* 13: 1542 (1987). "Synetsim 3.0: A model for simulating restricted neural networks."

1393. Carson, J. C. *Photonics Spectra* 21: 158–161 (1987). "Smart sensors and dynamic stare."

1394. Sherratt, A., ed. *The Cambridge Encyclopedia of Archeology*, Cambridge, England: Cambridge University Press, 1980. 495 pp.

Author Index

Note: Numbers refer to Reference list.

Abadamian, D., 493
Abarbanel, R. M., 868, 929
Abdel-Latif, A. A., 1276
Abdulaev, N. G., 670, 724
Abelson, J., 866
Abood, L. G., 1033
Abrams, T. W., 1332
Achenbach-Richter, L., 66
Acosta-Urquidi, J., 1348
Acree, T. E., 428
Adams, A. J., 689
Adams, D. J., 1232
Adelman, G., 25(b), 25(c), 25(d)
Adler, J., 68, 73, 81, 82, 83, 84, 85, 88, 102, 103, 106, 107, 113, 137, 161, 175, 193, 213, 214, 237, 251
Adler, J. E., 1350
Adrian, E. D., 517
Agard, D. A., 817
Agnew, W. S., 1111, 1138, 1302
Agosta, W. C., 346
Ahlgren, J. A., 144
Ahmed, F. R., 886
Aizawa, S.-I., 223
Akabas, M. H., 414
Akiba, I., 556, 1254, 1256
Akiyama, Y., 402
Aksamit, R., 90
Al-Awqati, Q., 414
Albeck, A., 727
Albert, P., 1247
Albone, E. S., 507
Albuquerque, E. X., 965, 966, 1030, 1031, 1137, 1215
Alger, B. E., 1284
Alger, J. R., 177
Alkon, D. I., 1282, 1341, 1348, 1373
Alkondon, M., 965, 966
Allen, E., 1307
Allen, R. D., 28
Alliod, C., 909, 1069
Almassy, R. J., 1151
Almers, W., 1188
Altman, J., 48, 605, 609, 1280
Altner, H., 292
Alvarez, R. A., 795
Amoore, J. E., 440, 441, 442, 478, 489
Anan, M., 1219
Andersen, K. K., 496
Anderson, J. A., 1385
Anderson, P.A.V., 514
Anderson, P. S., 888
Anderson, R. A., 190, 221, 591
Andose, J. D., 883
Andrea, T. A., 889
Andreasen, G. L., 997
Angelides, K. J., 1196
Anholt, R., 1066
Anholt, R.R.H., 537, 546
Anon., 345
Aoqvist, J., 303
Aoshima, H., 1219
Appenzeller, T., 1315
Applebaum, S. L., 484
Applebury, M. L., 559, 655, 656, 693, 750
Aprison, M. H., 1234
Aracava, Y., 965, 966, 1215
Arendt, A., 749
Argos, P., 267, 562, 563, 928, 933
Armitage, J., 117
Armstrong, C. M., 1129, 1161
Arnaboldi, M., 683, 759
Arnold, R., 1383
Aronstam, R. S., 1052
Arora, K., 433
Artamonov, I. D., 670, 724
Arvidson, K., 352, 385
Asai, M., 857, 859
Asakura, S., 98, 219
Asato, A. E., 728, 729, 733
Ashkenazi, A., 1253, 1259, 1262
Ashworth, D. J., 1032
Aswad, D. W., 155
Atassi, M. Z., 1061
Atkinson, P. H., 563
Aton, B., 661
Attwell, D., 585
Attygalle, A. B., 286
Auerbach, S., 1343, 1344

Aujla, R., 531
Auld, D. S., 1123
Auld, V., 1107
Auyeung, A., 40
Aveldaño, M. I., 592, 593, 594, 595
Avenet, P., 417
Axel, R., 1241
Axelrod, D., 264
Azuma, K., 781
Azuma, M., 781

Baasov, T., 711
Bachmann, C. M., 1391
Bada, J. L., 376
Baehr, W., 693, 750
Baeyer, A., 802
Bagley, K. A., 664
Bahl, J., 1255
Bailey, C. H., 1328
Bailey, G. S., 1072
Baker, H., 464
Baker, T. C., 288
Bakry, N. M., 1030
Balaban, M., 188
Balch, W. E., 65
Ball, C. B., 103
Balla, T., 1281
Ballantine, M., 1224
Ballivet, M., 861, 908, 909, 913, 1069
Balogh-Nair, V., 664, 683, 714, 759, 782
Bamberger, M. J., 1000
Bandini, G., 828
Banerjee, D. K., 1179
Banks, B.E.C., 1072, 1073
Bannister, L. H., 457
Baraban, J. M., 1284
Barash, S., 1283
Barboy, N., 731
Barchan, D., 1070, 1071
Barchi, R. L., 1124, 1126, 1136, 1201
Barclay, P. L., 692
Bard, J., 328, 329, 333, 341
Barkas, T., 1060, 1069
Barkdoll, A. E., III, 739
Barlow, D. J., 1121
Barlow, H. B., 580
Barnard, E. A., 811, 855, 943, 1217, 1218
Barnes, W. H., 1180
Barrantes, F. J., 820, 823, 824, 825, 836, 1012
Barreau, C., 279
Barry, B., 786
Barry, B. A., 663
Bar-Sagi, D., 1139, 1198
Barton, R. J., 1181
Bartoshuk, L. M., 347, 364, 398, 430
Baryshev, V. A., 147, 186
Bash, P. A., 939
Bate-Smith, E. C., 470, 471, 472
Baudry, M., 50
Baukal, A. J., 1281
Baumann, H., 990
Bavoso, A., 1123
Baylor, D. A., 611
Beach, J. M., 659
Beam, K. G., 425, 1299
Bear, M. F., 2
Beauchamp, G. K., 328, 329, 330, 333, 341, 486
Becker, R. S., 757, 758, 790, 791
Beckner, S. K., 635
Bedale, W. A., 145, 211
Behling, R. W., 894
Beiderman, H., 1235
Beidler, L. M., 355, 362, 400, 407, 435
Bel, W., 386
Belardetti, F., 1295
Bell, T. W., 319, 320, 321
Beman, J., 199
Benedetti, E., 809, 1123
Benedetti, E. L., 819
Bennett, C. D., 553
Bennett, E. L., 43
Bennett, N., 680
Benolken, R. M., 591
Benovic, J. L., 553, 641, 1244, 1248, 1249
Benson, T. E., 523
Bentaboulet, M., 858
Benz, W., 462
Berens, P. H., 896
Berg, D. K., 1214, 1362
Berg, H. C., 72, 94, 95, 105, 189, 191, 212, 215, 221, 222, 224, 225, 226, 230, 240, 243, 245, 246, 249
Berg, P., 862
Berge, C. T., 761
Berglund, B., 481
Berglund, U., 481
Bergmann, H., 1169
Berkowitz, B. A., 1189
Berman, H. A., 997
Bernstein, D. T., 496
Berridge, M. J., 538, 1277
Berstein, G., 1257

Besharse, J. C., 586
Bespalov, I. A., 724
Bestiani, M. J., 552
Bestmann, H. J., 436(b)
Betz, H., 1018, 1208, 1216, 1233
Beyreuther, K., 1208
Bezanilla, F., 1129, 1130, 1131, 1161
Bialek, W., 1089
Bibikov, S. I., 147
Bignetti, E., 299
Birch, M. C., 297, 305
Birdsall, N., 1260
Birdsall, N.J.M., 807, 1255, 1261
Birge, R. R., 654, 668, 761, 768
Birnbaumer, L., 1266, 1267
Bisson, R., 1193
Bitensky, M. W., 620, 622, 628, 744
Blacher, R., 462
Black, I. B., 1350
Black, M. M., 30
Blair, L.A.C., 1217
Blake, A. D., 1245
Blake, J., 1024
Blakely, R. D., 1351
Blakemore, R., 65
Blanchard, S. G., 944
Blanco-Labra, A., 389
Blank, V., 104
Blasdel, G. G., 1366
Blatchly, R. A., 782
Blatt, Y., 1017, 1117
Blinks, J. R., 602
Bloch, R. J., 980
Block, S. M., 212, 215, 245
Blomquist, G. J., 298, 306
Bloom, F. E., 26
Blumenfeld, A., 1283
Blumenfeld, H., 1327
Blundell, T. L., 1097
Bodzansky, M., 952
Boehm, M. F., 376
Bogachuk, A. P., 949
Bogachuk, A. S., 670
Boger, M. B., 884
Bogomolni, R. A., 188
Boineau, J., 1255
Bol, J. F., 390
Bolanowski, M. A., 553, 734
Bolger, M. B., 996
Bollinger, J., 210, 272
Bolton, J. R., 754
Bon, F., 841
Bond, J. R., 4
Bond, M. W., 139
Bonen, L., 65
Bonen, R., 65
Bonner, J. T., 77
Bonner, T. I., 557, 1258
Bonora, G. M., 1123
Bonting, S. L., 792
Boos, W., 91
Boppré, M., 314, 316, 319
Borczuk, A., 126, 203
Borden, J. H., 291
Borges, S., 585
Borkovich, K. A., 191
Bormann, J., 1231, 1268
Boshart, G. L., 509
Boudreau, J. C., 365
Boulter, J., 908, 911, 913, 915, 1211
Bouma, C. L., 168
Bourne, H. R., 558, 636, 1235
Bourret, R. M., 191
Boustead, C. M., 978
Bouvier, M., 1244
Bovee-Geurts, P.H.M., 727
Boyd, A., 119, 271
Boyd, N. D., 810, 1048, 1049, 1050
Boyse, E. A., 328, 329, 330, 333, 341
Brabson, J. S., 1111
Brade, G., 104
Bradford, H. F., 1207
Bradley, D. F., 434
Brann, M. R., 557, 588, 1258
Brass, J. M., 93
Braswell, L. M., 1011
Bredberg, L., 794
Breitwieser, G. E., 1269
Brennen, C., 190
Brenner, H. R., 1037
Breton, J., 723
Brett, M., 692, 698
Bridges, C. D., 795
Brill, A. B., 1316
Brinbaumer, L., 744
Brisson, A., 826, 924
Briston, J. A., 624
Brock, C. J., 1226
Brodin, P., 1006
Brodwick, M. S., 1163
Broek, A., 663
Brokaw, C., 190
Brooks, B. R., 937
Brouwer, J. N., 384, 406

Brower, K. R., 482, 483, 499, 500
Brown, A. M., 1266, 1267
Brown, D. A., 94
Brown, G. B., 1148
Brown, I. G., 187
Brown, J. E., 602
Brown, J. H., 1255
Brown, K. S., 491, 494
Brown, L. D., 1153
Brown, M. F., 659
Brown, P. K., 646
Brown, R. D., 995, 997, 1034
Brown, R. E., 336
Brown, R. L., 287
Bruderlein, F. T., 885
Buckley, N. J., 557, 1258
Bugg, C. E., 1134, 1151
Bujo, H., 1004, 1005, 1254
Bullock, T. H., 22
Bunt-Milam, A. H., 742
Bunzow, J. R., 1247
Buonanno, A., 1023
Burchenal, J., 126
Burden, S. J., 832
Burgen, A.S.V., 807
Burgess, A. J., 1307
Burgess, G. M., 1277
Burks, C., 217
Burley, S. K., 394
Bursztain, S., 973
Burt, D. R., 855
Busch, C., 1005
Busch, G. E., 655
Buss, J. E., 1113
Butenandt, A., 284
Butler, A., 1300
Byck, R., 1178
Byers, G. W., 766, 767
Bystrova, M. F., 468

Cagan, R. H., 383, 436(a), 578
Cahalan, M. D., 1188
Cain, M. H., 624
Caine, D. B., 560
Cairns, J., 340
Calhoon, R. D., 738
Calladine, C. R., 227
Callahan, A. M., 196
Callender, R., 789
Callender, R. H., 661, 762
Canale-Parola, E., 151
Candelore, M. R., 1245, 1246
Capasso, S., 950, 951
Capon, D. J., 1253, 1259, 1262
Carlin, B., 844
Caron, M. G., 553, 554, 630, 641, 734, 1239, 1240, 1243, 1244, 1248, 1249
Carriker, J. D., 782
Carson, J. C., 1393
Cartaud, J., 809, 819, 841, 842
Carterette, E. C., 347, 348, 512, 515
Cascio, M., 932
Casida, J., 1220
Casida, J. E., 1228
Cassel, D., 1283
Castellucci, V., 1330
Castiglione-Morelli, M. A., 396
Catt, K. J., 1281
Catterall, W. A., 1096, 1103, 1106, 1107, 1125, 1128, 1135, 1140, 1182, 1194, 1195, 1197, 1199, 1200
Cavaggioni, A., 299
Cavazos, J., 1384
Chabala, L. D., 1043
Chabré, M., 583, 590, 597, 619, 643, 723
Chakravanti, D., 401
Chan, S. I., 1229, 1230
Chang, E. L., 953
Chang, F.-H., 636
Chang, H. S., 1371
Chang, H. W., 1015
Chang, J.-Y., 988
Chang, S. S., 423
Changeux, J.-P., 1, 44, 52, 53, 54, 56, 796, 808, 809, 813, 819, 831, 837, 841, 842, 858, 970, 988, 989, 1023, 1051, 1193, 1319, 1331, 1390
Chelsky, D., 116, 133, 157
Chem. & Engr. News, 366
Chemouilli, P., 831
Chen, E., 1285, 1286
Chen, J. G., 785
Chen, K. N., 65
Chen, M., 1328
Chen, T., 207
Chen, Z., 465, 466, 469, 528
Cheng, M. T., 1032
Cheung, A. T., 1079
Chin, G. J., 1333
Chinn, K., 1152
Chiou, F., 126
Chiu, S. Y., 1338
Choe, S., 938
Chorev, M., 1271, 1273

Chothia, C., 930, 956
Chou, P. Y., 925
Christie, M., 1247
Chrivia, J., 996
Chung, F.-Z., 1247
Churchland, P. S., 32
Churg, A. K., 708
Ciobotariu, A., 255, 256, 528
Cioe, L., 1224
Civelli, O., 1247
Clancy, A. N., 346
Clancy, M., 179
Clapham, D. E., 1291
Clark, B.F.C., 631, 632
Clark, L., 503
Clark, S. P., 690
Clarke, E., 16
Clarke, J. H., 1003
Clarke, P.B.S., 1076
Clarke, S., 89, 125, 131(b), 153, 155
Clarke, S. A., 152
Claudio, T., 861, 1091, 1092
Clauss, K., 375
Cobbs, W. H., 600, 612, 739
Coddington, J., 378
Codina, J., 641, 744, 1266, 1267
Cohen, F. E., 868, 929
Cohen, J. B., 810, 819, 830, 1047, 1048, 1049, 1050
Cohen, L., 1339
Cohen, L. B., 1335, 1338
Cohen, L. V., 588
Cohen, M. A., 142
Collins, S., 1239
Colombetti, G., 74(c)
Colquhoun, D., 1042
Compadre, C. M., 370, 373
Cone, R. A., 645, 649
Conley, M. P., 191, 222
Conner, W. E., 318
Connolly, J., 913, 1211
Conrad, M. P., 718
Conti-Tronconi, B., 1090
Conti-Tronconi, B. M., 853, 918, 1074, 1171
Cook, N. J., 614
Cook, N. S., 1305
Cookingham, R., 763
Cookingham, R. E., 764
Cooper, A., 669, 793
Cooper, L. N., 2, 1391
Cooper, T. M., 668
Copie, V., 717
Corbett, A. M., 1145
Corley, D. E., 1002
Cornelissen, B.J.C., 390
Costa, M.R.C., 1194
Cotman, C. W., 45
Coulton, J. W., 216
Courtin, J., 717
Courtin, J.M.L., 712, 716
Courtney, K. R., 1185, 1186
Coussens, L., 1285, 1286
Covarrubias, M., 1086
Cowman, A. F., 696, 697
Cox, C. C., 1292
Cox, D. F., 164
Cox, R., 922
Coyle, J. T., 1351, 1352
Crabtree, E. V., 509
Crammer, B., 369(a), 369(b)
Craven, R. C., 143
Creighton, T. E., 358, 879, 931
Cresson, E. L., 888
Creveling, C. R., 1183, 1184
Criado, M., 1025, 1062, 1065
Crick, F., 1375, 1385
Criswell, D. W., 499, 500
Critchfield, J. M., 621
Croteau, A. A., 664
Crouch, R., 762
Crouch, R. K., 658, 783
Crow, T., 1346, 1347
Cruz, C.H.B., 678
Culp, W., 1026
Culvenor, C.C.J., 317
Curcio, C. A., 587
Curry, B., 662, 663
Curtis, C.A.M., 1261
Curtis, D. R., 562, 691
Curtis, P. J., 1224
Cushman, D. W., 890, 891

Daemen, F.J.M., 792
Daeniker, H. U., 437
Dahl, M. K., 262
Dahl, M. M., 84
Dahlquist, F. W., 64, 116, 118, 121, 123, 124, 133, 136, 138, 139
Dale, H. H., 801, 806
Dale, N., 1295
Dale, R. E., 651
Daly, J. W., 965, 966, 967, 1137, 1183, 1184
Dalziel, A. W., 953

Damle, V. N., 1054
Daniel, K. W., 1239
Darlison, M. G., 855, 1218
D'Arrigo, J. S., 1172
Das, M., 1088
Das, P. K., 757, 758
Dascal, N., 1287
Dasgupta, P., 461
Davey, M., 457
Davidson, N., 912, 1019, 1107, 1110, 1287, 1288
Davies, J. T., 449
d'Avignon, D. A., 1189
Davis, C. G., 834
Davis, M., 11
Davis, R. W., 675
Davison, M., 692
Dean, G. E., 217, 218
Debnath, N. B., 401
De Camili, P., 1326
de Duve, C., 866
DeFranco, A. L., 100, 115
DeGrado, W. F., 898
De Grip, W. J., 716, 727
Dehaene, S., 1390
Dehn, R., 509
DeKozak, Y., 597
Delaleu, J. C., 498, 504
Delcour, A. H., 960
Demant, P., 326
Dembo, L. N., 1391
Demuth, D. R., 1224
Deneris, E., 1211, 1212
Deng, F. H., 789
Dennis, M., 988, 989
Denny, M., 733
DePalma, R. L., 832
DePamphilis, M. L., 213, 214
Deretic, D., 749
Dergachev, A. E., 724
Derguini, F., 686, 755, 785
Derguini, K., 704
DeRosa, M., 158
Deshpande, S. S., 965, 966, 1215
DeSimone, J. A., 516, 549
Desmeules, P. J., 958
Des Rosiers, M. H., 519
Devary, O., 1283
Devillers-Thierry, A., 831, 837, 858
De Vos, A. M., 388, 634
Diamond, I., 833, 834
DiBlasio, B., 1123
Dickinson, K., 531
Diehl, E. W., 319
Diehl, R. E., 553
Dietrich, P. S., 371
Dietrich, S. W., 884, 889
Dimroth, K., 773
Dingledine, R., 1294
Dinur, B., 683
Dionne, V., 996
Dionne, V. E., 425, 522, 1217, 1362
DiPaola, M., 994
Dixon, R.A.F., 553, 734, 1245, 1246
Dixon, S. F., 793
Doak, T. G., 123, 124, 136
Dobzhansky, T., 310
Dodd, G., 505, 534
Dodd, G. H., 457, 502, 531, 532
Dodd, J., 414
Doe, C. Q., 552
Dohlman, H. G., 553, 554, 734, 1243, 1244
Dollinger, G., 664
Dolly, J. O., 811, 815, 943
Dommes, V., 828
Donaldson, P., 1092
Donellan, J. F., 804
Doniach, S., 1016
Doolittle, R. F., 869
Doty, R. L., 484, 506
Doukas, A., 762
Douville, K., 1026
Downey, W., 1107
Dowsett, A., 1107
Drachman, D., 1056
Drachman, D. B., 1058, 1059
Dratz, E. A., 599, 682
Dravnieks, A., 480
Dressler, R. L., 322
Dreyfus, C. F., 1350
DuBois, G. E., 371, 372
Duch, D. S., 1192
Dudai, Y., 47, 1312, 1313, 1332
Dudley, P. A., 591
du Lac, S., 552
Dunbar, B. I., 421
Dunis, D. A., 586
Dunlap, K., 1265
Dunn, D., 686
Dunn, R., 1107
Dunn, S.M.J., 853, 944, 945, 1090
Duskin, D., 974
du Villard, X. D., 384
Dwork, A. J., 875

Dworkin, M., 78
Dwyer, T. M., 1232
Dyckes, D. F., 1095
Dyer, T. A., 65
Dyke, A. M., 1336

Ealick, S. E., 1134
Earnest, J. P., 938, 1013, 1014
Earnest, T., 954
Eaton, D. C., 920, 1163
Eberlein, W., 1255
Ebner, F. F., 2
Ebrey, T. G., 664, 725, 785
Eccles, J. C., 23
Eckstein, F., 528, 621, 745
Eddy, R. L., 675
Edgar, J. A., 315, 317
Edgeworth, P., 399
Edmundson, A. B., 874
Edwardson, J., 387
Edwardson, J. A., 1072
Egorov, I. K., 329
Eguchi, G., 98
Eigen, M., 311
Einterz, C. M., 727
Eisenbach, M., 181, 188, 192, 197, 198, 236, 247, 254, 255, 256, 257
Eisenberg, D., 870, 871, 872, 926
Eisenstein, L., 664, 686
Eisner, T., 318
Eldefrawi, A. T., 1030
Eldefrawi, M. E., 1030
Elliott, J. R., 1191
Elmer, L. W., 1196
Ely, B., 148
Endo, T., 1078, 1081
Engel, W., 1255
Engelhard, M., 1171
Engelman, D. M., 702, 1120
Engström, P., 120, 140, 206
Entine, G., 648
Epstein, I. R., 1380
Epstein, W., 161
Erickson, R. P., 363
Erspamer, V., 1270
Eshhar, Z., 1168
Eusebi, F., 969
Evans, K., 908, 911, 913, 915
Eveleigh, P., 1261
Evers, A. S., 1189
Eyring, G., 662, 663
Eyring, H., 1172
Fabsitz, R. R., 494
Fain, G. L., 606
Fairclough, R. H., 821, 958, 1016
Falke, J. J., 135, 1229, 1230
Fallon, J. R., 985, 986
Fang, J.-M., 755
Farley, J., 1342, 1343, 1344, 1345
Fasman, G. D., 925
Faure, J. P., 597
Faurion, A., 349
Favach, M. C., 812
Feigina, M. Yu, 670
Fein, A., 579
Feinlib, M., 494
Feinstein, P. G., 460
Feix, J. B., 658
Feldberg, W., 801
Feldmann, R. J., 563, 937, 1204
Feller, D. J., 1106
Fels, G., 1037, 1084, 1085, 1086, 1171
Ferrara, G. B., 551
Ferris, A. M., 508
Fersht, A., 936
Fesenko, E. E., 468
Festenstein, H., 326
Festinese, R., 158
Feucht, B. U., 163
Field, K. G., 67
Fieles, W. E., 1126, 1201
Filbin, M. T., 804
Fillingame, R. H., 172
Findlay, J.B.C., 692, 698
Fine, R., 706
Finer-Moore, J., 904, 939, 958
Finkelstein, A., 1202
Firestein, S., 572
Firestone, L. L., 1011
Fischbach, G. D., 973
Flanagan, S. D., 835
Fletterick, R. J., 868, 929
Flewelling, R. F., 710, 1001
Flockerzi, V., 1102, 1299
Fong, S.-L., 562, 691
Fong, T. M., 849, 955, 1008, 1014
Fontaine, B., 53, 54, 56
Fontecilla-Camps, J. C., 1151
Forrester, J., 1346, 1347
Forsen, S., 1006
Foster, K. W., 62, 755
Fox, G. E., 65
Fox, J., 1065
Fox, J. L., 1025

Fox, P. T., 1310
Fox, T. D., 1091, 1092
Fracek, S. P., Jr., 500, 501
Francke, U., 734, 1240, 1286
Franeschini, N., 788
Frank, J., 820, 823, 824
Frank, M. E., 426
Franke, R. R., 626
Franks, N. P., 1190
Fransen, R., 662
Fraser, C. M., 1247
Fraser, P. J., 1224
Fraser, S. E., 843
Freedman, K., 790, 791
Freese, M., 574
Frei, B., 567
Frelin, C., 1156, 1187
French, R. J., 1132, 1158
Friberg, U., 352
Fridkin, M., 1063, 1071
Friedman, L., 447
Friedman, M. P., 347, 348, 512, 515
Friedman, W. F., 1350
Friedman, Z. Y., 1271
Frielle, T., 553, 734, 1239
Fritz, L. C., 1107
Froehner, S., 1022
Froehner, S. C., 976, 977, 979, 981, 982, 983, 984, 1026, 1027, 1028, 1308
Frostig, R. D., 1367
Fuchs, S., 1063, 1070, 1071
Fudenberg, H. H., 551
Fujita, N., 855, 1091, 1092
Fujita, Y., 942, 1004
Fukuda, K., 555, 942, 1004, 1005, 1020, 1256
Fukuda, M. N., 687
Fuller, W. D., 377, 378, 379
Funakoshi, M., 416, 432
Fung, B.K.-K., 598, 615, 618, 636, 743
Furman, R. E., 612
Furrer, A., 567
Furukawa, S., 1078
Furutani, Y., 857, 860, 863, 864, 906, 907, 910

Gabai, V. L., 180
Gabe, E. J., 1180
Gallacher, D. V., 1278
Galletti, P., 154, 158
Gallucci, K. W., 80
Galperin, M. Y., 187
Gambacorta, A., 158
Gambiner, B., 456
Garber, S. S., 1146
Gargus, J. J., 237
Gaston, I. K., 288
Gawinowicz, M. A., 683
Gent, J. F., 364
Geoffroy, B., 1023
Gergel, M. G., 548
Gershoni, J. M., 814, 1063, 1070
Gesteland, R. C., 497, 515
Getchell, M. L., 450, 497, 516
Getchell, T. L., 450, 453
Getchell, T. V., 516, 549, 572
Ghiselin, M. T., 67
Giachetti, A., 1255
Gibbons, B., 445
Gibbs, R. G., 45
Giberson, R., 484
Gibson, J., 65
Giersch, W., 565, 566, 567, 568, 569
Gierschik, P., 741
Giersig, M., 828, 829
Giffen, K., 1300
Gilardi, R. D., 765
Gilbert, C. D., 1367
Gilbert, W., 865
Gilman, A. G., 637, 1236
Gilon, C., 1271, 1273
Gilson, M. K., 706
Giovannoni, S. J., 67
Giraldo, E., 1255
Girard, P. R., 546
Giraudat, J., 837, 858, 988, 989, 1193
Glagolev, A. N., 147, 174, 180, 186, 187, 235
Glanzman, D. L., 1329, 1336
Glaser, D., 406
Glaser, L., 851
Glaser, T., 1356
Glass, D. B., 155
Glemcorse, T. A., 855
Godfrey, P. P., 1277
Goedde, H. W., 493
Goelet, P., 1330
Goh, J. W., 40
Gold, G. H., 415, 536
Goldberg, M. E., 1369
Goldberg, O., 1354
Golden, J. W., 282
Goldin, A. L., 1107
Goldman, A., 1120

Goldman, D., 908, 911, 913, 915, 1075, 1211, 1212
Goldman, D. J., 109, 110, 111
Goldstein, D., 1255
Goldstein, R. F., 1089
Gomel, J., 841
Gomes, S. L., 148
Gonzalez-Oliva, C., 623
Gonzalez-Ros, J. M., 812
Goodman, C. S., 552
Goodman, M., 377, 378, 379, 380, 381, 424
Goodman, R., 1300
Görbitz, C. H., 395
Gordon, A. S., 833, 834
Gordon, D., 1195
Gordon, R. D., 1126, 1201
Goren, E., 1169
Gott, S., 780
Gottesman, G. S., 832
Gottlieb, D. I., 526
Götz, M., 885
Goulbourne, E. A., Jr., 234, 258, 259, 260
Gould, R., 1282
Goy, M. F., 102, 107, 137
Graedel, T. E., 476
Grandy, D. K., 1247
Grant, A. J., 289
Grant, A. O., 1203
Graves, B., 393
Gray, T. M., 719
Graziadei, P.P.C., 451
Green, J. M., 509
Greenberg, E. P., 149, 151, 234, 258, 259, 260
Greenberg, M. E., 1349
Greenblatt, R. E., 1117
Greene, L. A., 1349
Greengard, P., 547, 846, 847, 960, 972, 1326
Greer, C. A., 520, 523
Grenet, D., 1266
Grenier, F. C., 159, 169
Grenningloh, G., 1208
Griffin, R. G., 712, 713, 717
Griffiths, N. M., 471, 472
Grillo, M., 461
Grinnell, A., 22
Grinvald, A., 1337, 1367
Grisolia, S., 27
Grob, D., 1057
Groth, E. J., 5
Grover, L. M., 1342
Gruener, R., 1373
Grundström, T., 1006
Grunhagen, H. H., 1168
Grunwald, E., 770
Grunwald, G. B., 741
Gubler, B. A., 437
Gubler, U., 461
Guenet, J.-L., 1023
Guerrero, A., 318
Guerri, C., 27
Guillemette, G., 1281
Gund, P., 883, 892
Gund, T. M., 893
Gundelfinger, E. D., 1018, 1208
Gupta, R., 65
Guy, H. R., 873, 927, 1118, 1119
Gysin, R., 835

Hadcock, J. R., 629
Haga, K., 555, 556, 1252, 1257
Haga, T., 555, 556, 638, 1252, 1257
Hagins, W. A., 601
Hagler, A. T., 896
Hainfield, J., 923
Haken, H., 9
Hall, B. G., 342
Hall, D. O., 754
Hall, P. F., 338
Hall, S. W., 746, 748
Hall, Z., 1022
Hall, Z. W., 1024, 1064
Halliday, K. R., 628
Halpern, B., 402
Halpern, B. P., 357
Halvorsen, S. W., 1214
Hamada, M., 1149
Hamill, O. P., 1231
Hamilton, K. A., 514
Hamilton, S. I., 838
Hamilton, S. L., 920
Hamm, H. E., 749
Hammer, R., 1255
Han, D. P., 244
Hanei, M., 267
Hänicke, W., 823, 824
Hanke, W., 614
Hanley, M., 1114
Hanley, M. R., 1272, 1274, 1279
Hansen, K., 319, 320
Hanski, E., 530
Hara, T. J., 574, 576, 577
Harada, Y., 1238
Harayama, S., 74(b), 272

Harbison, G. S., 712, 713
Hardie, R. C., 787
Hargrave, P., 559
Hargrave, P. A., 562, 563, 627, 671, 676, 682, 687, 691, 749, 1250
Harkness, J., 623
Harold, F. M., 226
Harper, R., 470, 471, 472
Harris, W. A., 722
Harrison, T. M., 1307
Hart, Y., 1273
Hartig, P. R., 1288
Hartline, D. K., 1392
Hartshorne, R., 1104
Hartshorne, R. P., 1103
Harvey, P. H., 311
Harwood, C. S., 277
Hatada, M., 388, 392, 393
Hattori, S., 635
Haumont, P.-Y., 989
Havoundjian, H., 1221
Hawkins, R. D., 1336
Hawrot, E., 814, 1068
Hayaishi, O., 545
Hayashi, K., 1078
Hayashida, H., 1098, 1100
Haydon, D. A., 1191
Haynes, K. F., 288
Haynes, L., 740
Hazelbauer, G. L., 74(b), 83, 84, 88, 120, 140, 150, 191, 209, 210, 272, 273
Heath, R. R., 339
Heck, G. L., 516, 549
Hecker, E., 284
Hecker, R. V., 109
Hedlund, B., 513, 550
Heeremans, C., 713
Heichal, O., 1283
Heidman, W., 1235
Heidmann, O., 1023
Heidmann, T., 813, 988, 989, 1051, 1319
Heinemann, S., 861, 908, 911, 913, 915, 1075, 1211, 1212
Heinemann, S. F., 1210
Heldman, E., 1282
Heldman, J., 528
Helfand, S. L., 552
Hellekant, G., 406
Hellmann, S., 1041
Hemmings, H. C., Jr., 547
Henderson, R., 699, 700
Hendrickson, A. E., 587
Hendrix, D. E., 446, 573
Hendry, B. M., 1191
Henkin, R. I., 431, 434, 541
Hermans-Borgmeyer, I., 1018
Hermodson, M. A., 266, 267
Hermsmeier, M., 893
Hertling-Jaweed, S., 828
Herz, J. M., 995, 997, 998
Herzfeld, J., 712, 713
Hescheler, J., 1237
Hesketh, T. R., 1175
Hespell, R. B., 65
Hess, B., 1378, 1379, 1382
Hess, G. P., 960, 962, 963, 964, 1091, 1092
Hess, J. F., 191
Heyn, M. P., 701, 703
Higashi, T., 432
Higgins, C. F., 104
Hildebrand, J. G., 524, 525
Hildebrandt, J. D., 744
Hille, B., 897, 1142, 1173, 1177, 1232, 1263
Hills, J. I., 448
Hinds, J. W., 452
Hinds, P. L., 452
Hirasawa, K., 1275
Hirata, M., 1093, 1094
Hirono, C., 1293, 1298
Hirono, H., 1108
Hirose, T., 555, 556, 857, 859, 860, 864, 906, 907, 942, 1098, 1099, 1100, 1299
Hirota, N., 127, 239
Hirsh, J., 693
Hirshfield, J., 888
Hirth, L., 493
Hisanaga, Y., 1108
Ho, C.-T., 423
Ho, S.-P., 898
Ho, Y.-K., 743
Hobson, A. C., 106
Hochschwender, S., 1022, 1025
Hodgkin, A. L., 608, 1127
Hodgson, K. O., 1016
Hoeger, G., 1373
Hoffman, B. J., 1288
Hofmann, F., 417
Hofmann, K. P., 749
Hofstadter, L., 26
Hogg, R. W., 237, 266
Hogness, D. S., 672, 673, 674, 675
Hogue-Angeletti, R., 1126
Hökfelt, T., 53
Holley, A., 444, 498, 504, 571

Holloway, R. A., 668
Holtzman, E., 921, 922, 923
Holz, G. G., IV, 1265
Hong, M. K., 664
Honig, B., 661, 683, 685, 704, 706, 715, 759, 785
Honig, B. H., 710, 1001
Honore, T., 1353
Hood, L., 335
Hood, L. E., 854
Hooft van Huijsduijnen, R.A.M., 390
Hoogsteen, K., 888
Hopfield, J. F., 960
Hopfield, J. J., 36, 37, 38, 1387, 1388, 1389
Hopfinger, A. J., 421
Hopkins, R., 1287
Horn, R., 1132
Hornung, D. E., 518
Horowitz, R. M., 397
Horridge, G. A., 63
Horsley, S. B., 319
Hotani, H., 99
Hough, C., 387
Hough, L., 368
Houts, S. E., 183
Howard, J., 1324
Hoy, D. J., 509
Huang, T.-C., 423
Hubbard, R., 777
Hubbell, W. L., 598, 710, 1001
Hubel, D. H., 1365
Hubert, H. B., 494
Hucho, F., 827, 828, 829, 990, 991, 992, 993
Hudspeth, A. J., 1324
Huganir, R., 847, 960, 961, 1251
Hughes, J. R., 446, 573
Hulme, E., 1260
Hulme, E. C., 807, 1255, 1261
Humber, L. G., 885, 886, 887
Hummel, H. E., 284
Hunkapiller, M. W., 139, 854
Hunt, C. A., 888
Hunter, D. D., 1263, 1264
Huppert, D., 776
Hurley, J. B., 618, 621, 637, 640, 742
Hutchinson, J. A., 791
Huxley, A. F., 1127
Hwang, P. M., 301
Hyrnchuk, R. J., 1181

Iberall, A., 7
Ichiyama, A., 555, 556, 638, 1252, 1257
Ifune, C., 1287, 1300
Iino, T., 98
Iio, H., 1219
Ikan, R., 369(a), 369(b)
Ikeda, T., 910, 1100, 1101, 1109
Imae, Y., 86, 87, 184, 185, 233, 239
Imoto, K., 942, 1004, 1005, 1021
Imoto, T., 1093, 1094
Imoto, Y., 1267
Inayama, S., 857, 859, 860, 864, 906, 907, 942, 1098, 1099, 1100
Ingebritsen, T. S., 627
Ingraham, J., 70
Irvine, R. F., 538, 1277, 1278
Irving, C. S., 766, 767
Isenberg, K., 844
Ishihara, A., 191, 245
Ishii, H., 1219
Ito, I., 1293, 1298
Ivanyi, P., 337
Iversen, L., 1296

Jabloner, H., 421
Jackson, M. B., 1045
Jackson, T., 1272, 1279
Jacobs, G. H., 720, 753
Jaenicke, R., 877
Jaffé, H. H., 756
Jakinovich, W., Jr., 367, 404, 405
Jan, L. Y., 1301, 1303, 1304
Jan, Y. N., 1301, 1303, 1304
Jancarik, J., 393, 634
Janson, C. A., 152
Jaramillo, F., 968
Jastreboff, P. J., 523
Jegou, E., 473
Jelinski, L. W., 894
Jenkins, R. E., 1227
Jensen, H., 375
Jessell, T. M., 1241
Job, K., 473
Johansson, C., 1006
John, E. R., 1316
Johnson, B. A., 155
Johnson, D., 998
Johnson, D. A., 995, 996, 997, 1034, 1079
Johnson, G. L., 629
Johnson, M. S., 205
Johnston, J. W., Jr., 573
Jones, K., 694, 695
Jones, O. T., 1009, 1014

Jones, T. A., 303
Jorgensen, E. C., 884, 889
Jorgenson, J. W., 331
Joshi, S., 433
Joubert, F. J., 1078
Juillerat, M., 1060
Julius, D., 1241
Jurnak, F., 632
Jürss, R., 1037
Juszczak, E., 562, 691

Kaila, K., 1222
Kaiser, D., 78
Kaissling, K.-E., 295, 297
Kakitani, H., 685
Kakitani, T., 665, 685
Kalcheim, C., 974
Kaldany, R.-R., 875, 922
Kalina, R. E., 587
Kalmus, H., 490
Kamath, S. K., 370
Kaminski, L.-A., 291
Kanaoka, Y., 1100
Kandel, E. R., 17, 1295, 1321, 1322, 1323, 1329, 1330, 1332, 1334, 1336
Kanehisa, H., 409, 410, 411, 412
Kang, S., 1080
Kangawa, K., 555, 556, 638, 1100, 1299
Kanne, D. B., 1032, 1033
Kao, P. N., 875, 946, 994, 1159
Kaplan, N., 191
Kare, M. R., 436(a)
Karle, I. L., 765
Karle, J., 765
Karlin, A., 838, 839, 840, 875, 919, 921, 922, 923, 946, 994, 1054
Karlson, P., 285
Karpen, J. W., 621
Karplus, M., 760
Kathariou, S., 149
Kato, R., 1149
Kato, S., 219
Kauer, J. S., 520, 523, 1340
Kaufman, L., 582
Kaufman, R. J., 625
Kaupp, U. B., 614
Kawakami, K., 1098, 1099
Kawamura, M., 1098, 1099
Kawamura, S., 665, 666
Kawamura, Y., 422
Kayano, T., 864, 906, 910, 1100, 1101, 1102
Keane, M. G., 202
Kebabian, J. W., 560, 561
Kehry, M., 118
Kehry, M. R., 121, 123, 124, 136, 139
Keightley, C. A., 1175
Kelling, S. T., 357
Kellis, J. T., Jr., 936
Kelly, P. T., 972
Kemp, C. M., 596
Kempf, C., 1226
Kendall, K., 271
Kennedy, C., 519
Kennedy, C.E.J., 297, 305
Kennedy, L. M., 403
Kentgens, A. P., 716
Kerr, D. E., 876
Kersulis, G., 195
Khan, S., 100, 231, 232, 249
Khen, M., 528
Khew-Goodall, Y. S., 461
Khodorov, B. I., 1147
Khorana, H. G., 625, 626, 639
Kihara, M., 86, 176
Kikyotani, S., 859, 860, 863, 864
Kim, H. Y., 1095
Kim, S., 154, 156
Kim, S.-H., 388, 391, 392, 393, 634
Kimura, M., 312
King, G.G.S., 291
Kinghorn, A. D., 370, 373
Kinnamon, S. C., 425
Kirschfeld, K., 788
Kistler, J., 821
Kito, Y., 781
Kjaer, A., 495
Kjeldgaard, M., 631
Klare, M., 422
Klarsfeld, A., 52, 53, 54, 56, 1319
Kleckner, N. W., 1294
Kleene, S. J., 106, 113
Kliger, D. S., 727
Kline, T., 62
Klink, S., 1026
Klock, I. B., 742
Klymkowsky, M. W., 817, 818, 821
Knutsson, H., 613
Kobata, A., 688
Kobayashi, S., 1219
Kobayashi, T., 657, 667, 725
Kobilka, B. K., 553, 734, 1239, 1240, 1245
Kobilka, T. S., 1240
Kochhar, A., 1212
Kodama, H., 402

Koehn, R. K., 312, 313, 342
Koenig, B., 749
Koestler, A., 13
Kojima, M., 555, 1299
Kollman, P., 889
Kollmann, P. A., 884
Komoriya, A., 1115
Kondoh, H., 103
Konishi, M., 44
Konno, T., 1004, 1005
Koono, K., 785
Kopito, R. R., 1225
Koshland, D. E., Jr., 69, 74(a), 75, 89, 90, 96, 100, 108, 112, 114, 115, 122, 128, 131(a), 131(b), 132, 134, 135, 157, 170, 178, 182, 194, 195, 201, 208, 252, 253, 269, 270
Koskinen, A.M.P., 1053
Kosower, E. M., 275, 564, 705, 736, 737, 771, 772, 775, 776, 778, 856, 902, 903, 905, 940, 941, 1083, 1115, 1116, 1141, 1205
Kosower, N. S., 1356
Kosslyn, S. M., 1310
Kossmann, M., 263
Kostina, B., 670
Koyama, N., 529
Krabbendam, H., 388
Krafte, D., 1110
Kraoz, Z., 1116
Kristofferson, D., 958
Krodel, E. K., 810
Krøgsgaard-Larsen, P., 1353
Kromer, W., 1255
Kropf, A., 777, 780
Kropf, R., 467
Krueger, B. K., 1145, 1158
Kubalek, E., 826, 920
Kubo, T., 555, 556, 910, 1254, 1256
Kudelin, A. B., 670
Kuffler, S. W., 21
Kühn, H., 597, 617, 641, 680, 746, 748
Kunath, W., 828, 829
Kuno, M., 910, 942, 957, 1109, 1238
Kuntz, I. D., 868, 929
Kuo, J. F., 546
Kuo, S. C., 208
Kurasaki, M., 942, 1004, 1020, 1101
Kurihara, K., 400, 407, 529
Kurihara, Y., 402, 408
Kuroki, R., 1093
Kurosaki, T., 957
Kuwada, J. Y., 552
Kuzma, G., 1348
Kyte, J., 869

Labarca, P., 1066
Labows, J. N., 479
LaCour, T.F.M., 631, 632
Ladinsky, H., 1255
LaGamma, E. F., 1350
LaLancette, R., 821
Lamb, T. D., 606
Lamola, A., 655
Lancet, D., 418, 419, 443, 465, 466, 467, 469, 520, 527, 528, 530, 544
Land, E. G., 470, 471, 472
Lande, R., 311
Landis, D. J., 591
Lane, D. J., 67
Langley, J. N., 797
Langosch, H., 1216
Langridge, R., 939
Langs, D. A., 900
Lanyi, J. K., 704
Lapidus, I. R., 247
LaPolla, R. J., 912
LaRochelle, W. J., 979, 981, 982, 1027, 1028
Larsen, S. H., 237
Larson, J., 41
Lasek, R. J., 30
Lauder, B. A., 487
Laufer, R., 56, 1271, 1273, 1331
Läuger, P., 241, 242
Lauridsen, J., 1353
Laverty, T., 694
Law, J. H., 280
Lawrence, L. J., 1228
Lazard, D., 467
Lazdunski, M., 1156, 1187
Lazerson, A., 26
Lear, J. D., 898
LeBrun, E., 841
Lederer, E., 473
Lederer, F., 989
Lee, B., 937
Lee, C. A., 159
Lee, C.-H., 398
Lee, J. F., 371
Lee, K.-H., 302
Lee, L., 185
Lee, L.W.Y., 324
Lee, R. B., 15

Leermakers, P. A., 766, 767
Lefkowitz, R., 641
Lefkowitz, R. J., 553, 554, 630, 734, 1239, 1240, 1243, 1244, 1248, 1249
Legler, D. C., 1011
Leibel, W. S., 1011
Leibowitz, M. D., 1142
Lelj, F., 396
Lemley, A. T., 764
Lenci, F., 74(c)
Lengeler, J. W., 160, 166, 171
Le Nguyen, D., 1065, 1087
Lennon, V. A., 816, 1061
Lentz, T. L., 814, 1068
Leonard, J., 1110
Lerea, C. L., 742
Lester, H. A., 1019, 1035, 1043, 1107, 1110, 1287, 1288
Levine, A. J., 1288
Levine, M. A., 542, 543, 1179
Levinson, S. R., 1111, 1192
Levis, R. A., 1357
Levitan, E. S., 1217, 1289
Levitzki, A., 1242
Lev-Ram, V., 1168
Levy, W. B., 42
Lewandowski, G. A., 1183, 1184
Lewin, R., 281, 343, 609
Lewis, A., 763, 764
Lewis, A. J., 20
Lewis, B. J., 65
Lewis, J. W., 727
Li, Y., 1201
Liang, C.-J., 688
Liang, R. F., 893
Lieb, W. R., 1190
Liebman, P. A., 599, 623, 648
Lieke, E., 1367
Lightman, S., 1279
Lindemann, B., 417
Lindstrom, J., 845, 848, 920, 957, 1017, 1022, 1025, 1029, 1062, 1065, 1087, 1088
Lindstrom, J. M., 1092
Lindvall, T., 481
Lingle, C. J., 1040
Linse, S., 1006
Linzmeier, R., 200
Lipkin, V. M., 949
Lipkowitz, K. B., 1234
Lippmann, W., 886
Lisman, J. E., 1376
Liu, R.S.H., 728, 729, 733
Livingstone, M. S., 1364
Llinas, R., 1358, 1386
Lo, M-v. C., 1165, 1166
Lobel, P., 922, 923
Lochrie, M. A., 640
Lodish, H. F., 1225
Loewi, O., 799, 800
London, J. A., 1335
Longstaff, C., 738
Loppnow, G. R., 663
Loring, R. H., 987, 1213
Lottspeich, F., 990, 991, 993
Love, R. A., 958
Low, K. B., 70
Lowe, G., 224, 240
Lübbert, H., 1107, 1287, 1288, 1290
Luehrsen, K. R., 65
Lugtenburg, J., 662, 663, 712, 713, 716, 717, 732, 786
Lukas, R. J., 1036
Lund, A. E., 1154
Lunt, G., 1215
Lunt, G. G., 804
Lüscher, M., 285
Luther, M., 1088, 1383
Lutz, R. A., 1179
Luyten, W., 908, 1212
Luyten, W.H.M.L., 1210
Lyddiatt, A., 943
Lygonis, C. S., 488
Lynch, G., 41
Lyon, T. F., 888

Ma, T. P., 1370
MacDermott, A. B., 1241
Machida, C. A., 1247
MacIntosh, F. C., 803
Mackay, D.H.J., 896
Mackay, R. A., 509
Mackey, S. L., 1336
MacLean, C. M., 491
MacLeish, P. R., 613
MacLeod, P., 444
Macnab, R. M., 70, 76, 86, 96, 97, 100, 170, 176, 177, 217, 218, 223, 231, 232, 244
Macrides, F., 346
Maddox, J., 49
Maderis, A. M., 194
Madill, K. A., 179
Maeda, A., 555, 556, 666, 1254, 1256
Maeda, K., 184

Maelicke, A., 735, 916, 917, 1037, 1038, 1041, 1080, 1082, 1083, 1084, 1085, 1086, 1170, 1171
Magasanik, B., 70
Magee, T., 1114
Magrum, L. J., 65
Mahoney, W. C., 266, 267
Malbon, C. C., 629
Mandel, G., 1300
Maniloff, J., 65
Manna, C., 158
Manson, M. D., 93, 104, 222, 226, 230, 246, 262, 263
Marcus, B., 511
Marei, S., 1115
Margiotta, J. F., 1362
Margolin, Y., 181, 256, 257
Margolis, F. L., 450, 461, 462, 463, 464
Mark, G. P., 350
Marshall, J., 1107, 1217
Marshall, J. C., 1311
Martin, A. R., 21
Martin, G., 908, 911, 915
Martin, J. M., 1263, 1264
Martin, R. L., 693, 750
Martinez-Carrion, M., 812, 1003
Martynov, V. I., 670
Marx, J. L., 633
Mas, M. T., 691
Masai, J., 250
Masland, R. H., 584
Mason, J. R., 503
Mason, P., 908, 911
Masters, S. B., 1235, 1255
Masu, Y., 1238
Masukawa, L. M., 513, 550
Masutani, T., 1149
Mathies, R., 644, 662, 663, 769
Mathies, R. A., 663, 678, 679, 712, 713, 717, 732, 786
Matias, P. M., 634
Matsui, H., 1240
Matsumoto, H., 729
Matsumoto, S. G., 524
Matsumura, P., 197, 198, 199, 200, 217, 218
Matsuo, H., 555, 556, 638, 1100, 1299
Matsuura, S., 233, 239
Matsuzaki, O., 333
Matthew, J. B., 707
Matthews, B. W., 719
Matthews, G., 610
Matthews, H. R., 606
Mattia, C., 951
Matuoka, S., 677
Matzinger, P., 332
Maue, R. A., 522
Maurer, B., 565
Mauron, A., 909, 1069
Maxam, A. M., 865
May, R. M., 311
Mayne, K. M., 1019
Mayor, F., Jr., 1249
Mazzarella, L., 950, 951
McCain, D. A., 715
McCarren, M., 1284
McCarthy, M. P., 938, 1013
McCaskill, J., 311
McClure, F. L., 499
McCormick, D. J., 1061
McCormick, F., 631
McDowell, J. H., 562, 563, 627, 671, 691, 1250
McGarraugh, G. V., 371
McGeer, E. G., 23
McGeer, P. L., 23
McGuiness, T. L., 972
McHugh, E., 1140
McIntosh, K. R., 1059
McKinney, J. S., 1277
McKusick, V. A., 485
Mclaughlin, J. T., 1000
McLaughlin, M., 838
McMahan, U. J., 58, 985, 986
McNamee, M. G., 849, 955, 1002, 1009, 1010, 1014
McNeal, E. T., 1184
McNelly, N. A., 452
McWhirter, K. G., 492
Mead, D., 729
Meadow, N. D., 168
Meares, C. F., 642
Medynski, D. C., 636
Meinwald, J., 316, 318, 319, 320, 321
Meiri, H., 1115, 1116, 1168, 1169
Meiselman, H. L., 475, 508
Meister, M., 224, 240, 243
Melnechuk, T., 25(a), 25(b)
Menco, B.P.M., 457, 458
Menevse, A., 534
Merck Index, 413
Merlie, J., 845, 848
Merlie, J. P., 844, 851, 1023
Merrick, J. M., 142
Mesibov, R. E., 83

Messner, D. J., 1106, 1125
Metcalf, E. R., 323, 324, 325
Metcalf, R. A., 420
Metcalf, R. L., 323, 324, 325, 420
Metcalfe, J. C., 1175
Methfessel, C., 1004, 1020, 1021, 1086, 1133, 1208
Meunier, J-C., 819
Meyers, H-W., 1037
Miake-Lye, R. C., 1016
Miashita, Y., 1371
Micheletti, R., 1255
Michell, B., 539
Michell, R. H., 540
Michel-Villaz, M., 643, 680
Middlemas, D. S., 850
Middleton, P., 968
Mielke, D. L., 932
Mierke, D. F., 378, 379
Mieskes, G., 836
Mihm, G., 1255
Mikami, A., 555, 556, 1299
Milburn, M. V., 634
Miles, K., 847
Milfay, D., 833
Miller, C., 1146
Miller, D. M., III, 92
Miller, I., 353, 354
Miller, J. B., 178, 182, 252
Miller, J. G., 8, 447
Miller, K. W., 1011
Miller, R. J., 1309
Miller, R. T., 1235
Miller, S., 340
Miller, T. A., 284
Milligan, D. L., 134
Minamino, N., 638, 1100
Minke, B., 788, 1283
Minor, A. V., 535
Mintun, M., 1310
Miroshnikov, A. I., 670
Mishina, M., 555, 910, 942, 957, 1004, 1005, 1020, 1021, 1254, 1256
Mishkin, M., 34, 1315
Mistretta, C. M., 572
Mistrot-Pope, M., 288
Mitchell, P., 229
Mitchell, W. C., 323, 324, 325
Mitra, A., 1122
Mittmann, U., 1255
Miura, K., 634
Mixter-Mayne, K., 912
Miyata, T., 857, 859, 860, 1098, 1100
Mizukoshi, T., 432
Mizuno, T., 86, 184, 185
Mochizuki, N., 239
Moczydlowski, E. G., 164
Mohana Rao, J. K., 562, 563, 928, 933
Molday, R. S., 625, 690
Molinaro, M., 969
Mollevanger, C.P.J., 716
Mollon, J. D., 580, 721
Momoi, M. Y., 816
Moncrieff, R. W., 374
Monferini, E., 1255
Montagna, E., 1255
Montal, M., 1017, 1066, 1067, 1104, 1117
Montal, M. S., 1017
Montarolo, P., 1330
Montecucco, C., 1193
Montell, C., 694, 695
Montie, T. C., 141, 143
Monti Graziadei, G. A., 451
Mora, W. K., 163
Morgan, E. D., 286
Mori, Y., 1004, 1005
Morimoto, Y., 942, 957
Morris, A. P., 1278
Morris, R. W., 383
Morton, M. E., 1308
Morton, T. H., 503
Moses, A. M., 542, 543
Moskal, J. R., 39
Moskowitz, H. R., 439
Motto, M. G., 683
Mottonen, J., 207
Moulton, R. C., 141
Mountcastle, V. B., 1368
Mowbray, S. L., 108, 265
Mozell, M. M., 518
Muellenberg, C. G., 688
Mueller, P., 612
Muhn, P., 990
Müller, S. C., 1378, 1379, 1381, 1382
Müller-Fahrnow, A., 828
Mumby, S. M., 546, 1333
Mumford, R. A., 553
Muradin-Szweykowska, M., 713
Murnane, A. A., 981, 982
Murphy, R.L.W., 606
Murray, E. D., Jr., 155
Murray, L. P., 668
Murray, R.G.E., 216
Murray, W. J., 889

Mutoh, N., 239
Mutter, L. C., 1032
Myers, A. B., 768
Myers, C. W., 1137

Nadal, J. P., 1390
Nagakura, S., 657
Nagano, K., 1098, 1099
Naim, N., 418, 419
Nairn, A. C., 547
Nakai, J., 1005, 1254
Nakamura, T., 536, 785
Nakanishi, K., 62, 664, 683, 684, 686, 704, 714, 715, 755, 759, 762, 782, 785
Nakanishi, M., 1078
Nakanishi, S., 1238, 1306
Nakatani, K., 603, 604, 607
Nakayama, H., 1100, 1112
Nakayama, K., 1238
Nakayama, T., 1108
Namir, M., 1115
Nanayakkara, N.P.D., 373
Narahashi, T., 1144, 1152, 1153, 1154, 1155, 1160, 1167
Nathans, J., 672, 673, 674, 675
Nathanson, N. M., 1263, 1264
Navon, G., 894
Navratil, E., 800
Neary, J. T., 1282, 1348
Neely, A., 1040
Neer, E. J., 744, 1291
Nef, P., 909
Nei, M., 312, 313, 342
Neidhardt, F. C., 70
Neitz, J., 720, 753
Nelson, N., 1091, 1092
Nemethy, G., 975
Nerbonne, J. M., 1043
Nestler, E. J., 846
Netter, F. H., 19
Nettleton, D. O., 110, 145, 150, 211
Neubig, R., 830
Neubig, R. R., 810, 1047, 1050
Neugebauer, D.-C., 820, 823, 824
Neuman, R. C., Jr., 761
Neumann, D., 1063, 1070, 1071
Neumann, E., 1015
Neve, K. A., 1247
Newcomer, M. E., 303
Newman, M. J., 162
Nicholls, J. G., 21
Nicole-Moayeri, N., 986
Nieto-Sampiedro, M., 45
Ninomiya, Y., 432
Nirenberg, M., 741
Nishikawa, Y., 638
Nishimura, S., 634
Nishiyama, T., 1257
Nishizuka, Y., 1275
Nitkin, R. M., 985, 986
Niwano, M., 167(a)
Noble, L. L., 761
Noda, M., 638, 857, 859, 860, 863, 864, 906, 907, 910, 957, 1021, 1098, 1099, 1100, 1101, 1102, 1109, 1133
Noe, L. J., 791
Noguchi, S., 634, 1098, 1099
Nojima, H., 1098
Norman, R. I., 1307
Norskov-Lauritsen, L., 631
Norton, S., 27
Noszek, J. C., 1355
Notake, M., 906, 1100
Novoselov, V. I., 468
Novotny, M., 331
Nowlin, D. M., 150, 210
Nukada, T., 638
Numa, S., 555, 556, 638, 857, 859, 860, 863, 864, 906, 907, 910, 914, 942, 957, 1004, 1005, 1020, 1021, 1098, 1099, 1100, 1101, 1102, 1109, 1133, 1254, 1256, 1299
Nunn, B. J., 608
Nutley, M. A., 793
Nutter, T. J., 1196
Nyberg, K., 936
Nyborg, J., 631, 632
Nyfeler, R., 380

Oberthür, W., 990, 991, 993
O'Brien, B. J., 1196
O'Connell, R. J., 289, 290
O'Connor, C. M., 153
Odashima, K., 686, 785
Ogata, C. M., 391, 392
Ogawa, S., 427
Ohara, R., 1108
Ohkubo, H., 1306
Ohloff, G., 473, 474, 477, 565, 566, 567, 568, 569, 570
Ohta, T., 1098, 1099
Ohtani, H., 657, 725
Ohtsuka, E., 634
Okabe, K., 704

Okabe, M., 62, 714, 715, 755
Okai, H., 409, 410, 412
Okamoto, M., 219
Okayama, H., 862
Olds, J., 1348
Olive, J., 589
Olsen, G. J., 67
Olsen, J. W., 3
Olsen, S. J., 3
Olson, E. N., 844, 851
Olson, J. F., 157
Olson, J. S., 92
O'Malley, C. D., 16
Omirbekova, N. G., 180
Ondetti, M. A., 890, 891
Ono, K., 1316
Ookubo, K., 402
Oosawa, F., 250
Oosawa, K., 86, 87
Ophir, D., 469, 528
Oprian, D. D., 625, 626
Optican, L. M., 1370
Orchin, M., 756
Ordal, G. W., 71, 109, 110, 111, 144, 145, 150, 211, 248
Orkand, R., 22
Ornston, L. N., 277, 278
Ortiz de Montellano, P. R., 876
Ory-Lavollee, L., 1351
Oswald, R. E., 1000
Otagiri, K., 409, 412
O'Tousa, J. E., 693
Ottolenghi, M., 653, 727
Ottoson, D., 455
Ovchinnikov, Y. A., 948, 949
Ovchinnikov, Yu. A., 670, 724
Overbaugh, J., 340
Oxford, G. S., 1160, 1163, 1164

Pace, N. R., 67
Pace, U., 418, 419, 466, 527, 528, 530, 544
Packer, B. M., 61
Packer, O., 587
Padgett, W., 967
Paerl, H. W., 80
Paik, W. K., 154, 156
Pak, W. L., 693
Palczewski, K., 627, 1250
Palings, I., 662, 717, 732
Palti, Y., 1115, 1116, 1169
Palumbo, A., 1224
Pamingle, H., 568
Panasenko, S., 131(b)
Panasenko, S. M., 74(c)
Pande, C., 789
Paoni, N. F., 114
Papazian, D. M., 1301, 1303, 1304
Papermaster, D. F., 687
Pappin, D.J.C., 692, 698
Paraschos, A., 812
Pardoen, J. A., 712, 713, 716, 717, 732, 786
Park, C., 209, 272, 273
Parke, D., 278
Parker, K. R., 599
Parker, P. J., 1285, 1286
Parkes, J. H., 623
Parkinson, J. S., 129, 130(a), 130(b), 183, 196
Parnes, J. R., 91
Partridge, L., 311
Pasquale, E. B., 964
Pastore, A., 396
Patel, N., 62
Pates, R. D., 659
Patlak, C. S., 519
Patrick, J., 861, 908, 911, 913, 915, 1075, 1211, 1212
Patterson, P. M., 487
Paul, S. M., 1221
Pauron, D., 1187
Pavone, V., 1123
Payne, T. L., 297, 305
Pecher, A., 166
Pedder, E. K., 1261
Pedersen, P. E., 523
Pedone, C., 1123
Peebles, P.J.E., 4, 5
Peeredeman, A. F., 388
Pellionisz, A., 1386
Pelosi, P., 299
Peralta, E. G., 1253, 1259, 1262
Perriard, J.-C., 837
Persaud, K., 505
Persaud, K. C., 299, 516
Peskin, C. S., 1202
Peters, K., 656
Petersen, O. H., 1278
Petersen, S. E., 1310
Peterson, G. L., 1253, 1262
Petsko, G. A., 265, 394, 876, 905
Pettigrew, K. D., 519
Petty, R. L., 316
Pevsner, J., 300, 301, 459, 460
Pezzuto, J. M., 370, 373

Pfaff, D. W., 356
Pfaffinger, P. J., 1263
Pfaffman, C., 348, 361
Pfister, C., 597
Philipp, A. H., 885, 886, 887
Piantanida, T. P., 675
Pickenhagen, W., 567
Pilosof, D., 1095
Pink, J.R.L., 551
Piomelli, D., 1295
Plesser, T., 1378, 1379, 1381, 1382
Plonsey, R., 261
Plouet, J., 597
Plümer-Wilk, R., 1085, 1086, 1171
Pober, J. S., 622
Polak, E., 502
Pollard, H. B., 1179
Pollard, W. T., 678
Polonsky, J., 473
Ponder, M., 769
Poo, M.-M., 649, 843
Popot, J.-L., 702, 808, 841, 842
Porter, S., 983
Porzig, H., 1299
Posner, M. I., 1310
Postma, P. W., 160
Postma, S. W., 1182
Powell, K. A., 1299
Poynder, T. M., 534
Poyner, D., 1261
Poziomek, E. J., 509
Pratt, D. R., 920
Presta, L. G., 878
Prestwich, G. D., 283, 293, 294, 298, 304, 306, 307, 308
Price, S., 549
Prillinger, L., 292
Prinz, H., 1037, 1038, 1039
Prives, J., 1139, 1198
Prives, J. M., 971
Proust, M., 1377
Pryor, G. T., 509
Przybylska, M., 886
Pugh, E. N., 605
Pugh, E. N., Jr., 600, 739
Pugsley, T., 886
Puliti, R., 951
Putney, J. W., Jr., 1277

Quarton, G. C., 25(a), 25(b)
Quast, U., 1187
Quiocho, F. A., 92, 268

Raff, E. C., 67
Raff, R. A., 67
Raftery, M. A., 850, 852, 853, 854, 918, 944, 945, 1007, 1074, 1090, 1100, 1111, 1112
Raichle, M. E., 1310
Raleigh, D. P., 717
Ralston, E., 1024
Ralston, S., 920
Ramachandran, J., 855, 1024, 1253, 1259, 1262
Ramon y Cajal, 27
Randall, W. C., 888
Rando, R. R., 738
Rando, T. A., 1150
Rands, E., 553, 1245, 1246
Rane, S. G, 1265
Rao, V. J., 686, 755
Rapoport, H., 1053, 1215
Rapuano, M., 971
Rasenick, M. M., 620, 628
Rashin, M., 706
Rasmussen, H., 1373
Rath, P., 789
Ratnam, M., 1029, 1065, 1087
Ravid, S., 181, 192, 197, 236
Ravikumar, K., 901
Rawlins, J.N.P., 35
Raz, T., 255
Reale, V., 855
Recio-Pinto, E., 1192
Reed, R. R., 302, 460
Reese, T. S., 29
Reeves, J. P., 1267
Regan, J. W., 1240
Register, R. B., 1245, 1246
Regnier, F. E., 280
Reichardt, C., 773, 774
Reinoso-Suarez, F., 27
Reist, N. E., 985, 986
Remsen, J. V., Jr., 344
Remy, D. C., 778, 888
Renaud, J. F., 1187
Renk, G. E., 658
Renner, I., 166
Rentzepis, P. M., 655, 656
Repaske, D. R., 175
Rephaeli, A. W., 162
Resneck, W. G., 980
Resnick, D., 1344
Reuter, H., 1299
Revello, P. T., 130(a)

Revich, M., 519
Rhee, L., 1285, 1286
Rhee, L. M., 855
Rhein, L. D., 578
Rhodes, A. M., 420
Rhodes, J. B., 883
Richards, C. D., 1175
Richardson, D. C., 878
Richardson, J. S., 878
Richardson, M., 389, 1220
Riddiford, L. M., 293, 294, 296, 305
Riddle, D. L., 282
Rienitz, A., 1208
Ringe, D., 876
Rioult, M., 1335
Risley, E. A., 888
Ritchie, J. M., 1338
Ritchie, R. J., 228
Rittle, K. E., 888
Rivelli, M., 277, 278
Rivier, J., 1065, 1087
Rivlin, R. S., 475, 508
Roach, A. H., 1350
Robb, J. L., 793
Robert, B., 1023
Roberts, R. H., 1124
Robertson, B. E., 1181
Robertson, M., 334
Robinette, R. R., 491
Robinson, C. J., 531, 532
Robinson, D. L., 1369
Robinson, M. B., 1352
Robishaw, J. D., 637
Roche, C., 643
Rodbard, D., 1179
Rodman, H., 685, 715
Rodrigues, V., 433
Rodriguez, H., 855
Roelofs, W. L., 287
Roeske, W. R., 1255
Rogers, K. E., 461
Rojas, E., 1161, 1162
Rollins, C., 138
Romey, G., 1187
Ronen, D., 466
Rooman, M. J., 935
Roper, S. D., 425
Rose, G. D., 878
Rose, S.P.R., 1320
Roseman, S., 168
Rosenberg, E., 79
Rosenberg, L., 484
Rosenberg, R. L., 1138
Rosenthal, V., 1116
Rosenthal, W., 744, 1237
Rosenthal, Y., 1169
Rosenzweig, M. R., 43
Roser, B., 336
Ross, A. F., 971
Ross, M. J., 817
Rossie, S., 1195
Roth, B., 1069
Roth, S. H., 1176
Rothenberg, S., 889
Rothschild, K., 954
Rotmans, J. P., 792
Rotstein, N. P., 595
Rottenberg, H., 256
Rovera, G., 1224
Rowan, R., III, 760
Rubin, B., 890
Rubin, G. M., 694, 695, 696, 697
Rubinstein, C. T., 1283
Rudy, B., 1162, 1163
Rushton, W.A.H., 752
Russell, G. F., 448
Russell, S. T., 708, 709
Russo, A. F., 270
Rydel, J. J., 200

Saari, J. C., 794
Sachs, F., 1325
Sack-Kongehl, H., 828, 829
Sackmann, E., 647
Saederup, E., 1220
Safran, A., 1070
Saibil, H., 590
Saier, M. H., Jr., 159, 162, 163, 164, 165, 169
Saitoh, T., 970
Sakakibara, M., 1282
Sakakibara, S., 379
Sakina, N. L., 535
Sakmann, B., 1004, 1005, 1020, 1021, 1042, 1133, 1231
Sakmar, T. P., 626
Sakurada, O., 519
Salama, G., 1366
Salemme, F. R., 899
Sali, D., 936
Salkoff, L., 1300
Salmon, A., 659
Salomon, Y., 530
Salon, J., 1247

Salpeter, M. M., 987, 1361
Samson, F., 27
Sandblom, P., 303
Sarai, A., 665
Saranak, J., 62, 755
Sargent, P., 1065
Sargent, P. B., 1087
Sarin, V., 1025, 1065
Sarvey, J. M., 39
Sastry, B. R., 40
Sastry, L., 686
Sato, T., 351, 355
Satou, M., 575
Sawruk, E., 1018
Scarpellino, R., 398
Scattergood, W., 1246
Scavarda, N., 1300
Schacher, S., 1327, 1329, 1330
Schaechter, L. E., 727
Schaechter, M., 70
Schaefer, R., 482
Schafer, R., 483, 499, 500, 501
Scheraga, H. A., 975
Scheuer, T., 1106, 1128
Schiavi, G. B., 1255
Schick, G. A., 668
Schierberl, M. J., 508
Schiffer, M., 874
Schiffman, S. S., 360
Schimerlik, M. I., 1253, 1262
Schimz, A., 146, 276
Schirmer, R. H., 867
Schlegel, W., 437
Schleppnik, A. A., 510
Schlessinger, J., 1168
Schlitzer, J. L., 508
Schmidt, J., 1197, 1199
Schmidt, J. H., 971
Schmidt, R., 1318
Schmidt, R. F., 581
Schmied, B., 380
Schmitt, B., 1208
Schmitt, F. O., 25(a), 25(b), 25(c), 25(d)
Schnapp, B. J., 29
Schneider, D., 316, 319, 320, 474
Schoenborn, B. P., 840
Schoepfer, R., 1088
Schofield, P. R., 855
Schopf, J. W., 61
Schotland, D. L., 1126, 1201
Schreiber, M., 1085
Schroer, T. A., 29
Schudel, P., 437
Schuessler, R., 1255
Schuetze, S. M., 968, 1064
Schulman, L. H., 866
Schulman, L. S., 6
Schulte-Elte, K. H., 568
Schultz, G., 1237
Schulz, G. E., 867
Schuman, E. M., 1345
Schuster, P., 311
Schwartz, A., 1115
Schwartz, J. H., 17, 1295, 1333, 1334
Schwarz, T. L., 1301, 1303, 1304
Schwemer, J., 789
Schwende, F. J., 331
Schwob, J. E., 526
Scott, G., 1384
Scott, K. R., 574
Scott, T. R., 350
Sealock, R., 976, 1028
Seamon, K. B., 967
Sebbane, R., 845
Seeburg, P., 1218
Seeburg, P. H., 636, 855
Seedburgh, D., 490
Seetharamulu, P., 1118
Sefecka, R., 403
Sefton, B. M., 1113
Segall, J. E., 191, 212, 245, 246
Seiden, P. E., 6
Seiff, F., 701, 703
Sejnowski, T. J., 32
Selander, R. K., 313
Seldner, M., 5
Selinger, Z., 1271, 1273, 1283
Seyama, I., 1149
Shafir, I., 528
Shallenberger, R. S., 428, 429, 431
Shaman, P., 484
Shanbhag, S., 433
Shank, C. V., 678
Shapiro, L., 148
Shashoua, V. E., 1317
Shaw, P., 148
Sheetz, M. P., 29
Shemyakin, V. V., 949
Shepherd, G. M., 24, 454, 513, 520, 521, 523, 533, 550, 572, 1297
Sheridan, R. E., 1035, 1043
Sherman, M. Y., 174, 204
Sherman, M. Yu., 180
Sherratt, A., 1394

Sheves, M., 684, 711, 727
Shiba, T., 379
Shichi, H., 688, 689
Shichida, Y., 657, 677
Shigenaga, T., 409, 412
Shih, T. Y., 635
Shimada, K., 225
Shimizu, N., 715, 755
Shimizu, S., 906, 910, 1100
Shimokawa, K., 779
Shimuzu, S., 864
Shin, W.-C., 392
Shinohara, M., 519
Shioi, J.-I., 233
Shipolini, R. A., 1072, 1073
Shirley, S., 502
Shirley, S. G., 531, 532
Shotton, D. M., 31
Showe, L. C., 1224
Shows, T. B., 675
Shrager, P., 1165, 1166, 1338
Shushan, B., 479
Shuster, M. J., 1013
Shuvaeva, T. M., 949
Shyng, S.-L., 1361
Sibanda, B. L., 1097
Sibley, D. R., 1248
Sicard, G., 571
Siddiqui, O., 433
Siegel, S. W., 109
Siegelbaum, S. A., 1295, 1327
Siemen, D., 1041
Sigal, I. S., 553, 734, 1245, 1246
Sigworth, F. J., 1046, 1157
Siksorski, L., 484
Silk, J., 4, 5
Silver, M. L., 875
Silverman, M., 101
Siman, R., 1355
Simmons, D., 1075
Simmons, P. A., 453
Simms, S. A., 202, 274
Simon, J. R., 1234
Simon, M., 101, 271
Simon, M. I., 119, 191, 637, 640, 1287
Simons, K., 456
Sine, S. M., 1044
Singer, A. G., 346
Singer, W., 51
Singh, P. B., 336
Sitaramayya, A., 623, 747
Sklar, P. B., 300, 301, 537
Skolnick, P., 1221
Skulachev, V. P., 186, 187, 235
Slater, E. E., 553
Slessor, K. N., 291
Sloan, K. J., Jr., 587
Slocum, M. K., 130(b)
Slonczewski, J. L., 177
Small, S. A., 1336
Smith, B. H., 25(d)
Smith, C. D., 1292
Smith, D., 636
Smith, D. H., 1253, 1259
Smith, D. P., 691
Smith, G. M., 883
Smith, J. M., 309, 311, 312
Smith, S. O., 712, 713, 717
Snutch, T., 1107, 1110
Snutch, T. P., 1287, 1288
Snyder, M. A., 253
Snyder, S. H., 300, 301, 459, 460, 537, 546, 1284
Snyderman, R., 1292
Sobel, A., 809
Sokoloff, L., 519
Somers, D. E., 742
Soneira, R. M., 5
Soodak, H., 7
Sopata, C. S., 145
Sorbi, R. T., 299
Spalding, B. C., 1157
Sparrow, K., 131(b)
Spencer, M., 220
Spencer, S. R., 1171
Spiegel, A., 741
Spiegel, A. M., 542, 543
Spira, G., 1115, 1116
Spira, M. E., 1327
Spivak, C. E., 893, 1031
Sprecher, H., 593
Springer, J. P., 888
Springer, M. S., 102, 107, 137
Springer, W. R., 131(a)
Spudich, J. L., 188, 715
Squire, L. N., 1374
Squire, L. R., 33, 34
Squires, R., 1220
Stabel, S., 1285
Stackebrandt, E., 65
Stader, J., 217, 218
Stahl, G. L., 952
Stahl, T., 65
Stalder, R., 909

Staniszewski, C., 1249
Stanton, P. K., 39
Stark, W. S., 722
Starmer, C. F., 1203
Starzinski-Powitz, A., 1055
Staub, A., 203
Stauch, B. L., 1352
Stein, P. J., 620, 628
Stein, S., 875
Steinbach, J. H., 1044
Steinberg, I. Z., 1359
Steinmetz, M., 335
Steitz, T. A., 1120
Stengelin, S., 908, 911
Stephenson, F. A., 855
Stephenson, R. A., 371, 372
Stern, J. H., 613
Sternberg, M.J.E., 1097
Stetter, K. O., 66
Stevens, C. F., 1209
Steward, O., 42
Stewart, I., 12
Stewart, R. C., 64
Stewart, W. B., 523
Stimers, J. R., 1130
Stinson, M. W., 142
Stock, A., 201, 207
Stock, A. M., 274
Stock, J., 126, 201, 202, 203, 207
Stock, J. B., 132, 194, 195, 253, 274
Stockton, J. W., 1255
Stoffers, D. A., 546
Stolberg, J., 843
Stone, G., 509
Stoof, J. C., 561
Stover, E. W., 168
Stowe, M. K., 339
Strader, C. D., 553, 852, 854, 1245, 1246
Strader, D. J., 553
Strauss, H. C., 1203
Strauss, H. L., 718
Strichartz, G., 1174
Strichartz, G. R., 1150
Striem, B. J., 418, 419
Strittmatter, S. M., 459
Stroud, R., 1122
Stroud, R. M., 817, 818, 821, 822, 904, 938, 939, 958, 959, 1013, 1016, 1024
Stryer, L., 615, 616, 618, 619, 621, 642, 644, 681, 745
Stühmer, W., 1133
Suami, A., 427
Suddath, F. L., 1151
Suga, A., 427
Sugimori, M., 1358
Sugimoto, K., 556, 638
Sugiyama, H., 1077, 1108, 1293, 1298
Sullivan, K., 636
Sullivan, K. A., 1235
Sun, S., 462
Suss, E., 1283
Sutro, J. B., 1142, 1143
Sutula, T., 1384
Suzuki, H., 638, 1101, 1109
Swanson, K. L., 1215
Swanson, L. W., 1075
Swanson, S. M., 373
Sweeney, K., 148
Sweet, M. T., 1092
Sydor, W., 462
Sykes, B. D., 760
Szabo, G., 1269
Szalay, A. S., 5
Szmelcman, S., 251
Szupica, C. J., 193
Szuts, E. Z., 579

Tabushi, I., 779
Takagi, S. F., 512
Takahashi, H., 555, 556, 638, 857, 859, 860, 863, 864, 906, 907, 910, 942, 1098, 1099, 1100, 1101, 1102, 1109, 1299
Takahashi, T., 910, 942, 957, 1020, 1021
Takai, T., 906, 907, 910, 1020, 1021, 1100
Takashima, H., 859, 860
Takata, M., 427
Takei, K., 427
Takeshima, H., 1101, 1109, 1299
Takezoe, H., 650
Takumi, T., 1306
Talvenheimo, J. A., 1105
Tamaki, H., 1238
Tamiya, N., 1078, 1081
Tamkun, M., 1104
Tanabe, M., 509
Tanabe, T., 638, 857, 859, 860, 863, 864, 906, 907, 1100, 1299
Tanaka, J. C., 612
Tanaka, K., 906
Tanaka, K.-I., 907, 942
Tanaka, M., 689
Tancredi, T., 396, 950
Tang, Y., 1316
Tanguy, J., 1144

Tank, D. W., 37, 38, 960, 1388
Tanner, M.J.A., 1226, 1227
Tanner, R. S., 65
Tapia, O., 303
Tasaki, H., 402
Taylor, B. L., 74(c), 167(a), 167(b), 178, 205
Taylor, P., 995, 996, 997, 998, 999, 1034, 1079
Taylor, R. E., 1130
Tedesco, P., 226, 230
Teeter, J., 415
Teichberg, V. I., 1354
Teitelbaum, Z., 462
Tejedor, F. J., 1140, 1301
Tempel, B. L., 1303, 1304
Temussi, P. A., 396, 424
Teplow, D. B., 637
Terao, M., 957
Termini, J., 686
Terwilliger, T. C., 122, 128, 870, 871, 926
Theerasilp, S., 408
Theimer, E. T., 438, 441
Thipayathaana, P., 238
Thirup, S., 632
Thoelke, M. S., 145
Thoelke, M. W., 211
Thom, R., 10
Thomas, D., 674
Thomas, D. D., 642
Thomas, J. B., 552
Thomas, L., 327, 328, 329, 330, 333, 341, 1216
Thomas, R., 1223
Thommen, W., 566
Thompson, R. C., 1351
Thompson, R. F., 1363
Thompson, W. J., 57
Thomson, P., 692
Thornton, J. M., 934, 947, 1097, 1121
Thulin, E., 1006
Timpe, L. C., 1304
Tirindelli, R., 299
Tobimatsu, T., 942, 957
Toews, M. L., 113
Tokunaga, F., 665
Tolbert, L. P., 525
Tomiko, S. A., 1138
Tomlinson, G., 392
Tomosaki, K., 362
Tong, L., 634
Toniolo, C., 1123
Tonosaki, K., 416
Tosteson, D. C., 1123
Tosteson, M. T., 1123
Totaro, J. A., 888
Totty, N., 1285
Toulouse, G., 1390
Toyosato, M., 857, 859, 860, 863, 864
Toyoshima, C., 826
Träuble, H., 647
Trautwein, W., 1237
Treco, D., 913, 915
Trewhalla, J., 702
Tribhuwan, R. C., 205
Trifiletti, R. R., 459
Triggle, C. R., 805
Triggle, D. J., 805
Trummlitz, G., 1255
Tsai, M-C., 1030
Tsang, J. W., 380
Tsuda, M., 724, 725
Tsujimoto, K., 759
Tully, R. B., 5
Tully, T., 1314
Tumlinson, J. H., 339
Tuyen, V. V., 597
Tyson, J. J., 1383
Tzartos, S., 845
Tzartos, S. J., 1055, 1060, 1069

Uchiyama, H., 1257
Udgaonkar, J. B., 963, 964
Ueda, K., 545
Ueda, T., 1094
Uematsu, Y., 427
Ullman, S., 46
Ullrich, A., 1285, 1286
Umbarger, H. E., 70
Unwin, P.N.T., 699, 700, 826, 920
Urban, B. W., 1192
Urry, D. W., 895

Vacante, D., 200, 218
Valdes-Rodriguez, S., 389
Vale, R. D., 29
Valentine, R. C., 238
Vallejo, M., 1279
van den Berg, E., 732
van den Bergh, S., 4
van der Drift, C., 226
Vander Meer, R. K., 318
van der Wel, H., 279, 382, 384, 385, 386, 388, 406
Van der Werf, P., 112
Van Dop, C., 636

Van Dyke, C., 1178
Vandyke, T., 1288
van Gunsteren, W. F., 303
van Meer, G., 456
Van Rapenbusch, R., 841
Van Tol, H.H.M., 1247
Vassilev, P. M., 1128
Veeman, W. S., 716
Veltel, D., 1171
Venable, J. C., 301
Venkataraghavan, B., 1204
Venter, J. C., 1247
Verdoorn, T. A., 1294
Verlander, M. S., 377
Vernin, G., 359
Vestal, M. L., 1095
Vial, C., 473
Vicini, S., 1064
Vickroy, T., 1255
Vigne, P., 1156
Vittitow, J., 664
Vogel, S. S., 1333
Vogel, Z., 974
Vogler, A. P., 171
Vogt, M., 801
Vogt, R. G., 293, 294, 296, 298, 305, 308
Voipio, J., 1222
Voith, K., 885, 886, 887
Vollrath, D., 675
Volterra, A., 1295
Volz, K., 199
Vorobyeva, N. V., 180
Vostrowsky, O., 436(b)
Vuong, T. M., 619
Vyas, M. N., 268
Vyas, N. K., 268

Wada, K., 1211
Waechter, C. J., 1197
Waggoner, Y., 780
Waitzman, D. M., 1370
Wakabayashi, S., 666
Wald, G., 652, 660, 784
Walker, J. A., 722
Walker, J. H., 978
Wall, J., 839, 921, 923
Wallace, B., 932
Wallace, B. A., 901
Wallace, B. G., 59, 985, 986
Wallat, I., 701, 703
Wallimann, T., 836
Wang, A.-C., 551
Wang, C.-D., 1247
Wang, E. A., 269
Wang, J. K., 562, 671, 691
Wang, J. Y., 128
Warnick, J. E., 1137
Warren, C. B., 439
Warrick, H. M., 178
Warshel, A., 708, 709, 730, 731, 760
Warwick, R., 18
Wasserman, Z. R., 898
Watanabe, A. M., 1255
Waterfield, M. D., 1285, 1286
Waters, J., 1031
Waters, J. A., 893
Watson, M., 1255
Watt, D. D., 1151
Watters, D., 1084
Watzke, H., 1037
Waygood, E. B., 159, 169
Wei, A., 1300
Weinstock, R. S., 542, 543
Weishaar, R. E., 624
Weiss, E. R., 629
Weiss, R. M., 870, 871, 926
Welch, M., 247
Wells, R. G., 302
Werblin, F., 572
Werman, R., 1206
Westerhausen, J., 701, 703
Wetzel, N., 573
Weyand, I., 641
Wheatley, M., 1261
Wheeler, G. L., 620, 628
Wheeler, M. A., 628
Wheeler, T. G., 591
Whissell-Buechy, D., 489
White, G., 42
White, M. M., 1019
Whiting, P., 1088
Whittam, T. S., 313
Whitten, W. M., 322
Whittenberger, A., 780
Wideman, J., 875
Wiedmann, T. S., 659
Wiesel, T. N., 1365, 1367
Wikström, M., 173
Wilcox, W., 871
Wilden, U., 746, 748
Wilke, M. E., 1122
Willhalm, B., 566
Williams, M., 888
Williams, N. H., 322

Williams, P. L., 18
Williams, R.J.P., 880, 881, 882
Williams, R. W., 953
Wilson, D., 14, 1360
Wilson, K. R., 896
Wilson, M., 585
Wilson, P. T., 1068
Winkel, C., 732
Winslow, J. W., 1253, 1259, 1262
Winstein, S., 770
Winston, M. L., 291
Winter, B., 565, 568
Wise, D., 921
Wise, D. S., 839, 840
Withy, R. M., 1112
Witkop, B., 798, 1031, 1052
Wittmann-Liebold, B., 990, 993
Witzemann, V., 978
Wodak, S. J., 935
Woese, C. R., 65, 66
Wolf, H. R., 473
Wolfe, A. J., 191
Wolfe, R. S., 65
Wolff, E. K., 1086
Wolosin, J. M., 943
Wong, R. Y., 397
Wonnacott, S., 1215
Wood, C., 750
Wood, J. M., 179
Woodbury, J. W., 1172
Woodcock, A., 11
Worcester, D., 590
Worden, F. G., 25(c), 25(d)
Worley, J. F., III, 1158
Worley, P. F., 1284
Wormser, U., 1271, 1273
Worobec, S. W., 109
Wray, B. E., 976, 1028
Wright, H. N., 542, 543
Wright, R. H., 446
Wu, C. H., 1160
Wu, J.-Y., 1335
Wu, S. M., 585
Wu, T.Y.T., 190
Wunderer, Hj., 320
Wurtz, R. H., 1369, 1370
Wysocki, C. J., 341, 486

Xiao-Xian, H., 1384

Yadav, J. S., 893
Yager, P., 953
Yagi, A., 402
Yamada, H., 1093, 1094
Yamada, K., 1149
Yamagishi, S., 1108
Yamaguchi, M., 328
Yamamoto, D., 1167
Yamamoto, K., 185
Yamamoto, T., 957, 1122
Yamamura, H. I., 1255
Yamanaka, G., 745
Yamane, T., 894
Yamashita, K., 688, 1077
Yamazaki, A., 628, 744
Yamazaki, G., 620
Yamazaki, K., 328, 329, 330, 333, 341
Yang-Feng, T. L., 734, 1240, 1286
Yasuda, K., 427
Yatani, A., 1266, 1267
Yatsunami, K., 639
Yau, K.-W., 603, 604, 607, 609, 740
Yee, A. S., 1002
Yee, G. H., 1251
Yeh, J. Z., 1144, 1164, 1167
Yodh, N., 923
Yogeeswaran, G., 848
Yoshikami, S., 601
Yoshizawa, T., 657, 660, 665, 666, 677, 726
Yost, B., 835
Young, A. C., 557, 1258
Young, E. F., 938, 1024
Young, R., 1316
Young, S., 1285
Young, S. H., 843
Yovell, Y., 1332
Yu, H., 650

Zablen, L. B., 65
Zabrecky, J. R., 1007
Zaccaï, G., 702
Zamoyska, R., 332
Zani, B. M., 969
Zappia, V., 158
Zarilli, G., 62
Zarilli, G. R., 755
Zecevic, D., 1338
Zecevic, D. P., 1335
Zehavi, U., 418, 419
Zeitoun, I., 1168, 1169
Zeitouni, O., 1391
Zel'dovich, Y. B., 5
Zell, A., 1134
Zensen, M., 1208
Ziff, E. B., 1349

Zigmond, R. E., 1213
Zimanyi, L., 704
Zimmerman, A. L., 611
Zingsheim, H. P., 820, 823, 824
Zinkand, W. C., 1145
Zola-Morgan, S., 34
Zolotarev, A. S., 670, 724
Zopf, D., 1018
Zuker, C. S., 694, 695, 696, 697
Zweig, J., 319

Index

Numbers in italics refer to illustrations

absorption maxima: adjacent charge effects, 169
acetylcholine (ACh), 188, 271; binding, 188; conformation by NMR, 232; conformations, 188; historical aspects, 173; model combining site, 189, 194; as neurotransmitter, 173
acetylcholine receptor: α-subunit, mRNA increase, 18; primary structure, 178; purification, *175*; secondary structure, 180; structure, 175–183; turnover, 309
acetylcholine receptors: distribution, 173
acetylcholinesterase (AChE), 215
N-acetylglucosamine, 37
ACh binding site, 199
AChR: amphiphilic regions, 191; exobilayer portion, models, 205; hydrophobic transmembrane segments, *197*; ion channel elements, 197; pentagonal symmetry, 195; subunit composition, 176; subunit order, 195; two binding sites, 199
AChR dimers, 176
AChR model: bilayer portion, 198; exobilayer portion, 199; principles, 192
active amino acids: homologies, 226, 228
adenosine, 292
3′,5′-cyclic adenosine monophosphate (cAMP), 60, 66
adenosine triphosphate: and bacterial motor, 48
S-adenosylmethionine: glutamyl transferase, 55
S-adenosylmethionine (SAM), 32
adenylate cyclase, 117, 304
ADP-ribosylation, 289
adrenaline, 271
adrenergic receptor: hydrogen-bonding in binding, *286*
β_2-adrenergic receptor, 281
α_2-adrenergic receptor, 281
aequorin, 138
agonists, 173; for nAChR, 229
agrin, 18, 215, 319
Alcmaeon, 7
allomones, 59
amacrine cells: in eye, 133
Amanita muscaria, 287
ambergris, 106
Ambystoma tigrinum (tiger salamander), 114
amino acid residues: channel-active, 191
γ-aminobutyric acid, 271, 289
gamma-aminobutyric acid receptor, 273; states, 275
α-aminoisobutyrate, 27, 34
cAMP, 117, 304
cyclicAMP, 66
amphiphilic α-helix, 192
amphiphilic character, 198
amphiphilic ion channel, 191; element, *201*
amphiphilic peptide: assembly, 190
amphiphilic segments, 198
amygdala, 309
amyl acetate, 116
anatoxin-a, 229, 272
anesthetics, 259; general, 263; local, 259; natural, 263
anion exchange protein (AEP): and GABA receptor, 277
anosmias, 110–112; androst-16-en-3-one, 111; butanethiol, 111; explanation from model, 128; hydrogen cyanide, 111
antagonists, 173; for nAChR, 229
Antheraea polyphemus (wild silk moth), 68
anti-C_1^+ antibody, 259
antibodies: and AChR, 233; and sodium channels, 259
anxiolytics, 274; binding sites, 276
Aplysia californica (sea hare), 294, 299, 300, 301
aqueous humor, 132
arachidonate, 294
arginine bonding agent, 278
Aristotle, 7, 73
arrestin, 137, 140, 141
artichokes: and sweet taste, 90
Ascalaphus macaronius (owl fly), 170

asparagine, 123
Aspartame, 80, 85, 90; crystal structure, 90; diastereomer, 86
aspartate, 32, 41
aspartate chemoreceptor, 55
aspartate receptors: dimers, 33
L-aspartyl-L-phenylalanyl methyl ester (see Aspartame), 80
ATP, 141
atropine, 287, 288
attention, 316
attractant: 1-(4-hydroxyphenyl)-3-butanone, 71; N-acetyl-D-glucosamine, 26; L-aspartate, 26; D-galactose, 26; L-glutamate, 26; D-mannitol, 26; methyl eugenol, 71; L-serine, 26
auditory cortex, 299
autoimmune disease, 233
axon, 13, *14*
axon hillock, 16
axon terminals, 13, *14*
azidoquinacrine, 219

Bacillus subtilis, 32, 33, 34, 49
Bacillus thuringiensis, 241
bacteria: motility mechanism, 30; motor machinery, 47; smooth swimming, 27; tumbles, 27
bacterial flagellae, 48
bacterial motor: molecular mechanism, *50*; proton motive force, 48; rotor and stator, *49*; switching, molecular mechanism, *51*
bacteriorhodopsin: femtosecond experiments, 150
band 3 protein, 278
barbiturates, 274; bonding sites, 276
bathorhodopsin, 146, 147, 148
batrachotoxin (BTX), 255
BCNI, 220
belief systems, 318
Belousov-Zhabotinsky (B-Z): reaction, 314; wave, 315
benzodiazepines, 274; binding sites, 276
bilayer helices: α-subunit (nAChR), *200*; arrangement (in nAChR), 221; β-subunit (nAChR), *202*; δ-subunit, *204*; γ-subunit, *203*; geometric relationships, *221*; in sodium channel, 247
bipolar cells: in eye, 133
bitter taste, 77, 93; allyl thiocarbamide, 96; brucine, strychnine, solanidine, 93; picric acid, 93; 6-*n*-propylthiouracil, 97; salts, 78; thiourea, 93
blob cells, 310
blobs, 309
bolas spider: and female sex attractants, 72
bombykol, 60, 67, 68; as attractant, 63
Bombyx mori L., 60
Bos taurus (calf), 193, 227
Bowman's gland, *100*
Bradyrhizobium japonicum, 56
brain: dendritic arborization, 11; evolution, 19; major components, 11; number of cells, 10; principal sections, *12*
brain death, 7
bromoacetylcholine, 231
α-bungarotoxin (α-BTX), 234
bush baby, 132
1-butanethiol, 121
t-butylbicyclophosphorothionate, 277

Caenorhabditis elegans, 59
calcitonin gene-related peptide (CGRP), 18, 304
calcium channel, 307
calcium ion, 138
calpain I, 308
calpastatin, 308
camphor, 106
camphor odorants, 103
Canis familiaris, 5
carbodiimide, 258
carcinogenicity, 85
carveol, 106
carvone, 106
catecholamines: binding site groups, 284
caudal hair cells, 305, 306
Caulobacter crescentus, 34
Centuroides sculpturatus, 255
cephalad hair cells, 306
cerebellum, 9, 11, *12*, 309; cells, 13
cerebral cortex, 309
cerebrum, 9, 11, *12*
channel activators, 255
channel elements, 198, 216
channel inactivation blockers, 255
channel-active (CA) amino acids, 192; in sodium channel, 246
Chateau d'Yquem, 91
chemical communication, 60
chemical stimuli, 58
chemiosmotic hypothesis, 48
chemosensors, 25

chemotactic responses, 27; in higher vertebrates, 72
chemotactic stimulants, *29*
chemotaxis, 23–56; adaptation, 30, 34; adaptation, via methylation, 31; attractants, 25; blue light assay, 36; detection, 24; EII-mediated, 36; general scheme, 56; general scheme for MCP type, 39; in higher organisms, 59; methyl esterase, 33, 34; methyltransferase, 33, 34; in nematodes, 59; pH change, 40; pH increase and CCW motor response, 36; phosphotransferase systems (PTS), 35; repellents, 25; role in organisms, 23; signal transfer protein, 43
Chlamydomonas (single-cell algae), 21, 162
chloride ion channel, 275
chlorisondamine, 230
cholinergic synapses, 173; classes, 174; structure, 174
Chrysomelidae, 93
ciguatoxins, 255
cilia: on olfactory neurons, 99; in rod cells, 134
1,8-cineole, 116
citronellal, 65
citronellol, 108, *109*
clathrin, 15
clockwise rotation, 40
cocaine, 260, 261, 268
coelenterates, 305
collagen-like packing, 215
color blindness, 161
concanavalin A, 138
conditioned stimulus, 305
cone cells, 133, 148, 309; morphology: relation to rod cell, 134; number, 134; in retina, 133; structure, *135*; types, 136; wavelength sensitivity, 136
conformational changes: after rhodopsin excitation, 146; single group rotation theory, 183
μ-conotoxin, 255
convection cells, 315
conveyor belt, 319
counterclockwise rotation, 40
Creatonotos (moths), 70
cross-linking, 255, 280
cross-linking proximate groups, 235
cross-linking agent, 33
crystallins, 131
cucurbitacins, 93
cultural evolution, 319
cultural learning and memory, 320
cultural representations, 319
cup: nAChR model, 217
curly form: of flagellae, 31
cyclic disulfide, 201
1,2-cyclohexanedione, 278
cyclooxygenase, 294
Cynara scolymus, 90
cys-cys disulfide, 201

Dacus cucurbitae (melon fly), 71
Dacus dorsalis (Oriental fruit fly), 71
Danainae (butterflies), 70
decamethonium, 230
demethylase, 42
demethylation, 55
Democritus, 73
Denatonium benzoate, 93
dendrites, 14
Dendroaspis viridis, 234
α-dendrotoxin, 234
denervation, 272, 308
2-deoxyglucose, 116
dephosphorylation, 312
depolarization, 252
desacetylanisomycin, 307
Descartes, 299
desensitization, 231; model for nAChR, 233; of nAChR, 231
desensitized receptor, 192
1,2-diacylglycerol (DAG), 292, 293
Dictyostelium discoideum (slime mold), 66
Didinium (protozoan), 21
dihydropyridine, 295
diisocyanatostilbenedisulfonate (DIDS), 277
dimethyl disulfide, 113
Dioscoreophyllum cumminsii, 86
dishabituation, 300
disk membranes, 136
disk structures: formation, 134
disks: in rod cells, 134
disulfide agents, 280
disulfide bond: unusual, 200
disulfide links, 215; in AChR, 205
disulfides, 182, 192, 193, 205; in AChR, 205, 207
dithiothreitol, 230; reduction with, 182
dopamine, 271
dopamine β-hydroxylase, 309
Drosophila, 151, 299
Drosophila melanogaster, 297

E. coli, 33; swimming after change, 31
edge enhancement, 318
eel ion channel elements: arrangement, 249
elaidic acid, 39
electromagnetic waves, 58
electron microscopic images: AChR, 175
electronic transitions: in polyenes and polyenals, 162
electroolfactogram: (EOG), 101, 112, 114, 115
Electrophorus electricus (electric eel), 174, 238
emotion, 316
enantiomeric odorants, 108, 109
enantiomeric pairs: odor, 106
environmental niche, 70
Enzyme II receptors, 35
epileptic attacks, 17
epinephrine, 271
epitopes, 259; of sodium channel C_1^+, *260*
ethidium, 220
ethyl bromoacetate, 112
N-ethylmaleimide, 112, 231
Euglena, 21
evolution, 69; channels, 278; pheromonal changes, 71. *See also* brain; neurons
evolutionary persistence: of active amino acids, 226
exobilayer arrangement: nAChR α-subunit, *206*; nAChR β-subunit, *208*; nAChR δ-subunit, *210*; nAChR γ-subunit, *209*
exobilayer strand arrangement: nAChR α-subunit, *206*
eye: iris, 130; lens, 131, *132*; pupil, 130; dark adaptation time, 134; range in sensitivity, 136; retina, 131; structure, *132*; wavelength sensitivity, 136
eye motion, 311

facilitation: in *Aplysia*, 303
fetal γ-subunit, 227
flagella, 30; clockwise (CW) rotation, 31; counterclockwise (CCW) rotation, 30
flagellar motor mechanism: driving force, 48; molecular model, 49
flagellar motor structure, 47
flatworms (*Platyhelminthes*), 19
flower model: for nAChR, 233
flutterdance: of male moth, 60
folding rules: for proteins, 198, 240
food: and change in pheromones, 70; conversion to pheromones, 70
fovea, 133
frontalin, 66
functionally plausible substitutions, 226, 273; in sodium channel, 241
Fura-2, 302
fusiform cells, *12*

G-protein receptors: general model, 285; homologies, 281; sequence comparison, *283*; transmembrane helices, 286
G-protein superfamily, 284
G-proteins, 117, 120, 304; in frog epithelium, 120; inhibitory (Gi), 142; stimulatory (Gs), 142, 281, 292, 304; subunit composition, 281
$GABA_A$ receptor: activation mechanism, 276; model, 273
galactitol, 37
galactose binding protein, 27, 54; receptor, 55
Gallus domesticus (chicken), 193, 227
ganglion cells: in eye, 133
gating current, 253
GDP, 43, 120, 121
genetic code, 181
genetic homology, 192; in sodium channel, 239, 241, 243
genetic influence: on mating behavior, 71
genetics: of color blindness, 161; of olfactory response, 110; of taste, 94, 96
geraniol, 65
glucose chemotaxis receptor, 34
glutamate receptor, 308
L-glutamate, 294
glutamic acid, 271; taste, 95
γ-glutamyl carboxylic acid group: methylation site, 32
glutathione (GSH), 131
glycine, 271
glycine receptor, 280
cGMP, 118, 139, 142; hydrolysis per photon, 140
cyclicGMP, 43
goldfish, 299
gramicidin A, 190
granule cells, 11, 13
granule layer, 100
gravity fields: and synapse formation, 312
grayanotoxin (GTX), 255
GTP, 43, 120, 121, 141; analogue, 289
guanosine diphosphate (GDP), 281
guanosine monophosphate (5′-GMP), 95, 141
guanosine triphosphate (GTP), 281

3′,5′-cyclic guanosine monophosphate (cGMP), 118, 137, 138, 141
gustatory (taste) system, 73
gustatory cell, 74
Gymnema sylvestre, 91
Gymnemic acid, 91

habituation, 300, 302, 313; long-term, 302, 303; molecular processes, 303
Halobacterium halobium, 34, 39, 56
helical hydrophobic moment, 182
helical wheel, 182
helix: 3_{10}-, 182, *185*, 243, 266; 3_{10}- in sodium channel, 240; α-, *182*, *185*; collagen, 182; π-, *185*
helix representation, *181*
Hermissenda crassicornis, 292, 299, 305; learning, molecular basis of, 307; neural system, 306
hernandulcin, 83
12-HETE, 294
trans-10-*cis*-12-hexadecadienol-1. *See* bombykol
hexamethonium, 230
hierarchies, 3–7; in large systems, 4
hierarchy: level, 5; taxonomy of dogs, 5
Hildebrand, J. S., 4
hippocampus, 9, 10, 16, 17, 309, 313, 316
histamine, 271
holistic approach: to model construction, 199
Homo sapiens (human), 193, 227
honey, 83
horizontal cells: in eye, 133
hormone, 141
hot spots, 314, 319
Hovenia dulcis, 91
12-HPETE, 294, 295
human eye, 129
human pheromone, 72
hydra (*Coelenterates*), 19, 305
hydrogen cyanide (HCN), 123; anosmia, 111
hydrophobic helices: rhodopsin, 153; in sodium channel, 239, 243, *245*
hydroxyeicosatetraenoic acid, 294
hypnotics, 274; binding sites, 276
hypogeusia, 97
hypsorhodopsin, 147

Imipramine, 307
inner control element (ICE), 251
inositol 1,4,5-trisphosphate (IP_3), 287, 289, 292
intrabilayer polar interactions: in nAChR, 223
iodocyanopindolol, 284
iodopsins, 135, 136, 148; amino acid sequences, *153*; critical groups, 155; human, 149
ion channel: molecular model, *191*
ion channel element (nAChR), 198; α-subunit, *200*; β-subunit, *202*; δ-subunit, *204*; γ-subunit, *203*
ion channel elements, *200*; in sodium channel, 241
ion channels: molecular model, 189
ion passage: AChR channel, 216
ion-carrying segments, 192

Jacobson's organ, 101

kainate, 293, 294, 308
kairomones, 59, 93
kangaroos: and salt, 78
kinesin, 15
Koestler, A., 7
!Kung San, 7

lateral geniculate body, 131, 309, 310; (LGS), 129
lateral inhibition: in retina, 133
learning: biological basis, 309; generalized view, 311; model systems, 299
learning and memory, 3, 299–320; molecular scheme, 319
lectin-binding glycoproteins, 102
Leeuwenhoek, 23
Leiurus quinquestriatus, 238, 255
Lepidoptera, 70
levels: biological world, 8; inner program, 6; nervous system, 9; outer program, 6; physical/inorganic world, 6; relationships, 6; visual system, 130
lidocaine, 260, 269
ligand-activated G-protein receptors, 280; adrenergic, 281; muscarinic acetylcholine, 281; serotonin, 281; substance K, 281
ligand-activated phospholipase C-related receptors, 292
ligand-gated ion channel receptors, 271
ligand-receptor combination: classes, 186
light amplification: biochemical cycles, *141*; and hormone amplification, 141
light sensitivity: and dark adaptation, *135*; retinal cells, *135*
light-detecting cells: in retina, 133
limonene, 65
Limulus (horseshoe crab), 138

linalool, 108, *109*
lipids: unsaturated, in disk membranes, 137
12-lipoxygenase, 294
5-lipoxygenase, 294
Lippia dulcis, 83
local anesthetics, 261, 263, 268; and ICE, 251; mechanism, 261; and sodium channel, 262
locus coeruleus, 308
long-term memory, 302, 318, 320; and hippocampus, 17
long-term potentiation (LTP), 17
Lucretius, 98

M2 helices: labeling in nAChR, 217
Macaca fascicularis (macaque monkeys), 17, 299
magnocellular cells, 310
major histocompatibility complex: and pheromones, 71
malodor counteractant, 114
mannitol, 37
MCP proteins: methylation level, 32; numbers, 32
MCP receptors: scheme, *44*; structure and function, 53
mechanoreceptors, 300, 301, 303, 304
melatonin, 134
membrane anion exchange proteins (AEP): homologies, *279*
membrane crossing segments, 192
membrane potential change, 53
memory, 3, 19, 20, 316; consolidation, 18; generalized view, 313; long term, 3, 17; longevity, 4; molecular scheme, 318; short term, 3, 17; transient, 3. *See also* short-term memory
metarhodopsin II, 141, 149
methionine deprivation, 32
2-methoxy-3-isobutylpyrazine, 102, 116
methyl esterase, 46
N-methyl-D-aspartate (NMDA), 293, 294
N-methyl-D-aspartate (NMDA) receptors, 308
meythyl-accepting chemotaxis protein (MCP), 25; I, II and III, 32
methylation: carboxylate groups, 258; degree of, 33
methylesterase, 43. *See* chemotaxis
miraculin, 86, 91; amino acid sequence, 93
mitochondria: in rod cells, 134
molecular graphics, 261; model: eel sodium channel, 266
monellin, 80, 86, 87; backbone structure, 89; crystal structure, 88
monkeys: sweetness response, 86
moths: antennae, 63; cabbage looper, 61; pink bollworm, 61; sensory hairs, 63; silkworm, 63
mouse: urine odor and genetics, 71
mucopolysaccharide, 99, 101
mucus layer, *101*
Mus musculus (mouse), 193, 227
muscarine, 287
muscarinic acetylcholine receptor (mAChR1), 281; overview, 289
musk, 111
mustard gas, 112
mutagenicity, 85
myasthenia gravis, 233, 234
myo-inositol, 83
myxobacteria, 23

nAChR: extension above bilayer, 204; extra-junctional, 272; general structure of α-subunit, 225; phosphorylated sites, 207; resting state, 192
nAChR exobilayer strands: spatial organization, 205
nAChR superfamily, 193, 280
NADH, 315
Necturus maculosus (mud puppy), 145
nematodes, 59
neohesperidin dihydrochalcone, 90
nereistoxin, 229
nervous system, 7; structural elements, 7
neural cells, 10
neural circuit, *303*; in *Aplysia*, 302; for visual system, 310
neural network models, 317
neuroglia, 10
neurokinin A, 290, 291
neuron structure, *14*
neurons, 11; evolution, 19
neuropeptides, 296
neurotransmitter-receptor complexes, 172; binding sites, 186
neurotransmitters, 271; criteria for, 270
nicotine, 287
nicotinic acetylcholine receptor (AChR), 172–236; dynamic model, 217; electron micrograph, *177*; electron micrographic image analyses, 179; flower model, 233; ion channel, 191; ion channel model, 218; model,

192; number of amino acids, 193; polypeptide subunits, 192
nicotinic acetylcholine receptor family, 272
nicotinic AChR: subunit sequence homology, *196*
nigericin, 38
Noctiluca (dinoflagellate), 21
noncompetitive antagonists (NCA): binding site, 216
noradrenaline, 271, 281, 286, 289
norepinephrine, 271, 281, 286

octopus rhodopsin, 156
odor: alkyl isocyanides, 113; limburger cheese, 113
odorant classes: ethereal, floral, fruity, camphoraceous, minty, musky, pungent, almond, aromatic, and anise, and putrid, 103, *104*; EOG, 117; names, 103
odor components: number, 109
odor intensities, 107
odorants, 98; cyclooctane, 103; 1,4-dichlorobenzene, 103; enantiomeric, 105; hexachloroethane, 103; hexamethylethane, 103; methyl isobutyl ketone, 103; molecular shape, 103, *107*; quantitation, 107; reactive, 112
odors: threshold values, 109
olfactant, 98
olfaction, 128; infrared detection mechanism, 98
olfaction (smell), 98
olfactory axons: number, 99
olfactory binding proteins (OBP), 68, 102, 116, 118
olfactory bulb, *12*, 99; signal processing, 99; structure, 100
olfactory cilia, 101; frog, 102; number, 101
olfactory code, 121
olfactory cortex, 100
olfactory detector organ: nerve cells, 99; secretory cells, 99; size, 99
olfactory epithelium, *100*, 101; adenylate cyclase, 117
olfactory G-proteins, 119
olfactory neural signals, 114
olfactory neuron: structure, *101*
olfactory neurons, 99; characteristic frequency, 121; lifetime, 122; loss, 99; replacement, 99; small size, 99; specificity, 116, 122; structure, 99
olfactory receptor: and G-protein, 117; kinetic factors, 120; model, 124, *125*, 127; structural and dynamic model, 123
olfactory receptor molecules, 102
olfactory receptor specificity, 123, 124
olfactory response: molecular scheme, 119
olfactory rod, 99
olfactory stimuli, 113; classification, 103
olfactory system, 98; electrical signals, *115*; elements, *100*
Oncorhynchus nerka (himé salmon), 127
opsin: membrane helices, 283
opsin shift, 148, 149, 164; origin, 154; responsible groups, 149
optic fibers: number, 133
orchid flowers, 71
outer control element (OCE), 251
oxygen chemotaxis, 36
oxygen triads, *256*

parallel: agonist orientation, 232
parallel fibers, 13
Paroctopus defleini (octopus), 156
parvocellular cells, 310
D-penicillamine, 97
Peranema, 21
perception, 316; of color, depth, and movement, 309
Periplaneta americana (American cockroach), 64
perpendicular: agonist orientation, 232
persistent activators, 255
PET scan, 311
pH taxis, 37
phenylketonuria, 85
N-phenylmaleimide, 222
phenylthiocarbamide, 96
pheromonal binding proteins (PBP), 102
pheromonal communication, 71
pheromone: benzaldehyde, 64; binding protein, 68; detector sensitivity, 63; detector structure, 63; 2,3-dihydro-7-methyl-1H-pyrrolizidin-1-one, 64; (Z)-7-dodecenol, 62; (Z)-5-dodecenyl acetate, 62; 11-dodecenyl acetate, 62; mixture, 62; periplanone B, 64; receptor structure, 68; (Z)-7-tetradecenyl acetate, 62; (Z)-9-tetradecenyl acetate, 62
pheromone binding protein, 69
pheromones, 58, 59, 60–73; aggregating, 66; alarm, 65; molecule lengths, 62; recruiting, 65; sex, 64
phorbol-12,13-diacetate, 292

phosphatidylcholine, 137
phosphatidylethanolamine, 137
phosphatidylinositol, 137
phosphatidylinositol 4,5-bisphosphate ($PtdIP_2$), 292, 293
phosphatidylserine, 137
phosphodiesterase (PDE): in rod cell, 139
phosphoenolpyruvate, 35, 37
phospholipase C, 294
phosphorylation, 312; of AChR, 214; of sodium channel, 264
phosphotransferase (PTS) systems, 37. *See also* chemotaxis
photorhodopsin, 147
phototaxis, 305
phylogeny: prokaryotes, 22, *24*
picrotoxinin, 277
pirenzepine, 287, 288
pleated sheet: β-, *182*; antiparallel, *185*; parallel, *185*
point-charge model, 149, 150
polyenals: absorption maxima, 163
polyproline II structures, 215
positron emission tomography, 299
potassium channels, 297
proline taxis, 38
prolines: in C-terminal sequences (nAChR), 215; in cytoplasmic section β-subunit (nAChR), 212; in cytoplasmic section δ-subunit (nAChR), 214; in cytoplasmic section γ-subunit (nAChR), 213; in sodium channel, 264
pronase, 258
protein kinase C (PKC), 307
proton electrochemical potential, 48
Proust, 314
Pseudomonas aeruginosa, 34
Pseudomonas putida, 56
punctuated equilibrium, 69
Purkinje cells, 11, *13*
pyramidal cells, *12*
N-(1-pyrenyl)maleimide, 223
pyrethroids, 257, 278
pyrrolizidine alkaloids, 70

quinine, 93
quisqualate (QA), 293, 294, 308
QX-314, 260, 262

Ramon y Cajal, 11
Rana pipiens (frog), 145
Rattus rattus (rat), 193, 227
rebaudioside A, 80, *84*
receptive fields, 129
receptor cell, 20
receptors: aspartate, 25; dissociation constants, galactose, ribose, 27; galactose, 25; maltose, 27; removable, 25; ribose, 25; serine, 25
reflex pathway, 302
reiteration, 17
repellent odors: uses, 113
repellents, 56
representations, 3, 20; in culture, 318, 320; higher-order, 314; transient, 16
reproductive isolate, 70
resensitization, 233
retina, *132*, 133; blood supply, 134; structure, 133
11-*cis*-retinal, 163
11-*trans*-retinal, 144
retinol binding protein (RBP), 68, 102
retinol isomerase, 171
retinylidene group: conformations, 164
retinylidene imines, 164
retinylidene iminium system, 155
retinylidene moiety: in rhodopsin, 153
retinylidene N-butylimines: absorption maxima, 163
11-*cis*-retinylidene-ϵ-lysine, 161
11-*cis*-retinylidene-N-butylimine, 163
11-*cis*-retinylidene-N-butyliminium chloride, 163
retinylidene-ϵ-lysine: electronic transitions, 162
retro-inverso peptides, 85
reverberation, 313
Rhizobium trifolii, 56
rhodopsin, 125, 135, 136, 141, 143, 281; absorption maximum, 148; amino acid sequences, 148; bleaching by light, 144; in *Chlamydomonas*, 21; chromophore, 143, 161; cis-trans isomerization, 149; dichroic absorption, 144; enthalpy change to bathorhodopsin, 148; geometric change on photoisomerization, 157; glycosylation, 151; intermediates after excitation, 146; intermediates after light excitation, *147*; lateral diffusion rate, 145; lateral mobility, 137; location, 135; ovine, 151; rotational relaxation rate, 144; ungulate, 151
rhodopsin (RX), 138, 141
rhodopsin and iodopsin: structural and dynamic model, 148
rhodopsin kinase, 159

rhodopsins: amino acid sequences, *153*; critical groups, 155; human and bovine, 149
ribose: receptors, 27
rod cells, 133; light sensitivity, 134; morphology, 134; number, 133, 134; in retina, 133; structure, *135*; wavelength sensitivity, 136
rod outer segments (ROS), 134, *135*; and cGMP, 139
rose oxide, 108
Rousseau, 83

S. typhimurium, 33
saccadic motion, 311, 318
saccharin, 80, 84; bitter aftertaste, 93
Salmo gairdneri (rainbow trout), 127, 128
Salmonella typhimurium, 43, 55
salt taste, 77, 78
saxitoxin (STX), 255
Schiff bases, 161
Schwann cells, 15
α-scorpion toxin, 255
β-scorpion toxin, 255
scotophobin, 309
sea slugs, 300
second messenger mechanism, 287
semiochemicals, 59
sensitization, 302, 304; molecular processes, 304
sensory deprivation, 17
sensory hair: fine structure, 67
serine chemoreceptor, 55
L-serine, 39; adaptation, 34
serotonin (5-HT), 271, 292, 294, 303, 304, 307
serotonin (5-HT) receptor, 281, 294
SGR, 191; of lysine, 189
SGR theory, 186; background, 184
short-term memory, 302, 304, 318
shrimp taste, 78
signal molecules (see pheromone)
signal processing: in retina, 133
signal transfer protein. *See* chemotaxis
Silkworm moth, *66*
single group rotation (SGR) theory, 183
single group rotations (SGR), 154, 186, *187*, 194, 247, 248
sliding helix model, 265
smooth swimming, 40
snake neurotoxins, 234
snow crab flavor, 77
sodium binding site, 247, 268
sodium channel, 237–269; conductance, 251; control elements, *252*; dynamics, *253*; eel, 238; in information transfer, 238; ion permeabilities, 257; and lidocaine, 269; mechanism, 238; molecular arrangement, *248*; organization, 250; primary structure, 238; states, 254; stick model, 267, 268
sodium channels: rat brain, 238
sodium ion channel elements: in eel, *242*
solvent effect: on absorption maxima, 165
solvent polarity: definition, 166
solvent polarity parameters, 165; ET(30)-values, 166; Y-values, 165; Z-value basis, *167*; Z-values, 165, 166
sour taste, 77, 95
sphingomyelin, 137
Spirochaeta aurantia, 34
sponges (*Porifera*), 19
ST-protein, 43, *45*
statocyst, 305, 306
stellate cells, *12*
steviol: mutagenic action, 83
stevioside, 80, 83, *84*
stimulus: amplitude, 58; topology, 58
stimulus-receptor combination, 173
stimulus-response scheme, 20; multicellular organism, *20*; single cell, 22, *23*
Streptococcus strain, 49
structural models: as guide, 193
strychnine, 280; binding, 280
suberoyl bis-choline, 192, 230
substance K receptor, 281, 290
substance P, 290, 308
substantia nigra, 308
substantia nigra para reticulata, 311
succinoyl bis-choline, 192, 230
sucrose, 79, 80, 82
superfamilies, 270–298
superfamily: ligand-gated ion channel receptors, 271
superior collicus, 311
sweet proteins, 86
sweet substances: taste intensity, 80
sweet taste, 77, 79; amino acids, 79; cyclic sulfonamides, 79; dipeptides, 79; polyols, 79; sugars, 79; ureas, 79
sweet-tasting compounds: classes, 79, 80
sweeteners, 79; synthetic, 84
sweetness: dependence on ring size, 86; and hydrophilic-hydrophobic balance, 84
synapse, 13; level analysis, 174; neuron-neuron, 14; postsynapton, 13; presynapton, 13
synapses: in eye, 133

synapsin I, 15, 302, 303
Synsepalum dulcificum, 91

T.GDP, 139
T.GTP, 139
tachykinins, 290
tactile stimulus, 300
tapetum lucidum, 132
taste, 58, 73–97; mechanism, 95; nerve, 74; single receptor model, 78
taste bud, 73, *74*, *75*
taste modifiers, 90; chlorogenic acid, 90; cynarin, 90; and primate species, 91; ziziphin, 91
taste receptor cell, 74
taste receptors, 73; potential change, 76
taste signal: tastin, 76
taste stimuli, 73; classification, 77; response, 75
taxis, 23
telenzepine, 287
teratogenicity, 85
terbium(III) distribution, 224
tethered cells, 48
tethering: of bacteria, 31
tetrodotoxin (TTX), 255
thaumatin, 56, 80, 86, 87; backbone structure, 89
thaumatin I: crystal structure, *88*
Thaumatococcus daniellii, 86
thiol groups, 263
thiol protease, 308
thiol-disulfide status, 22
thiol-reactive agonists, 231
thiophenol, 121
tongue: taste receptors, 73
Torpedo californica (electric fish), 174, 193, 227; as AChR source, 174
Torpedo marmorata (electric fish), 174, 193, 227
Torpedo marmorata AChR: labeling, 216
Tortricidae, 61
toxin classes, 255
toxins, 256. *See also* oxygen triads
trace, 3, 16, 20
transducin, 140, 281; GTPase activity, 140; increase at night, 136; subunits, 139
transducin [T], 139
transducin.guanosine diphosphate (T.GDP), 141
trehalose, 26, 29
trigger region, 16
N-(4-trimethylammoniomethylphenyl)-maleimide, 231
triphenylmethylphosphonium ion, 48
d-tubocurarine, 189; binding site, 190
tumbling, 40
tyrosine hydroxylase, 307, 309

umami response: L-aspartic acid, 95; glutamic acid, 95; L-ibotenic acid, 95; monosodium glutamate, 95
umami taste, 77, 95

valinomycin, 38
vanillin, 110
veratridine, 308
visual cortex, *12*, 129, 131, 299
visual field, 129
visual pigments: absorption maxima, 171
visual system, 129–171, *131*
vitreous humor, 131
vomeronasal receptors, 101
Vorticella, 21

wavelength sensitivity: of eye, 136

Xenopus laevis öocytes, 205, 253, 272, 273, 288, 290, 293, 295

Z-value, 165